History OF THE Theory OF Numbers

Volume III

Quadratic and Higher Forms

Leonard Eugene Dickson

With a Chapter on the Class Number
BY G. H. CRESSE

Dover Publications, Inc.
Mineola, New York

Bibliographical Note

This Dover edition, first published in 2005, is an unabridged republication of the work originally published as Publication Number 256 in Washington, D.C. by the Carnegie Institute of Washington in 1919.

International Standard Book Number: 0-486-44234-9

Manufactured in the United States of America
Dover Publications, Inc., 31 East 2nd Street, Mineola, N.Y. 11501

PREFACE.

The favorable reception accorded to the first two volumes of this history has encouraged the author to complete promptly the present third volume which is doubtless the most important one of the series.

By a " form " we mean a homogeneous polynomial such as $f=ax^2+bxy+cy^2$, all of whose terms are of the same total degree in x and y. The arithmetical theory of forms has an important application to the problem to find all ways of expressing a given number m in a given form f, i. e., to find all sets of integral solutions x, y of $ax^2+bxy+cy^2=m$. For this application we do not consider merely the given form f, but also the infinitude of so-called equivalent forms g which can be derived from f by applying linear substitutions with integral coefficients of determinant unity. It is by the consideration of all these forms g that we are able to solve completely the proposed equation $f=m$. The theory needed for this purpose is called the arithmetical theory of forms, which is the subject of the present volume. This theory is applicable to most of the problems discussed in Volume II. The present methods have the decided advantage over the special methods described in Volume II in that they give at once also the solution of each of the infinitely many equations $g=m$. By thus treating together whole classes of equivalent equations, the present methods effect maximum economy of effort.

Enough has now been said to indicate that we are concerned in Volume III mainly with general theories rather than with special problems and special theorems. The investigations in question are largely those of leading experts and deal with the most advanced parts of the theory of numbers. Such a large number of the important papers are so recent that all previous reports and treatises (necessarily all very incomplete) are entirely out of date.

As in the case of the first two volumes, no effort has been spared to make the list of references wholly complete. There was now the additional burden of examining many titles which turned out to relate to the algebraic theory of forms, rather than the arithmetical theory.

The prefaces to the first two volumes gave a clear account of the leading results on the subjects treated. This was possible partly on account of the elementary nature of most of those subjects, and partly because the gist of the investigations could be embodied in definite theorems expressed without the use of technical terms. But in the case of Volume III, it is a question not primarily of explicit results, but chiefly of general methods of attacking whole classes of problems, the methods being often quite intricate and involving extensive technical terminology. Accordingly it is not possible to present in this preface a simple summary of the contents of the volume; perhaps the opening sentences of the preface have provided sufficient orientation in the subject. However, to each of the longer chapters is prefixed an appropriate introduction and summary.

The congruencial theory of forms, presented in the final chapter, is not only of special intrinsic interest, but has important applications in the theory of congruence groups and in the classic theory of genera of ordinary arithmetical forms as developed especially by Poincaré and Minkowski (Ch. XI).

The proof-sheets of each chapter were read critically by at least one specialist in the subject of the chapter. Mrs. Mayme I. Logsdon of the University of Chicago read the short Ch. XVII. Miss Olive C. Hazlett of Mount Holyoke College read Chs. XVIII and XIX. All of the earlier sixteen chapters were read by E. T. Bell of the University of Washington, by L. J. Mordell of the University of Manchester, England, and by A. Speiser of the University of Zürich, each a well known expert in the arithmetical theory of forms. Both Bell and Mordell compared many of the reports with the original papers. Such a comparison was made independently by the author.

G. H. Cresse devoted five years to the preparation of the report in Ch. VI on the difficult subject of the number of classes of binary quadratic forms, which involves many branches of pure mathematics. His historical and critical report was written as a thesis for the doctorate at Chicago, and will prove indispensable to future investigators in this difficult subject. Mordell read critically the manuscript of this long Ch. VI, compared the reports with the original papers, read also the proof-sheets, and on each occasion made numerous important suggestions. Bell, who read a portion of the manuscript and most of the proof-sheets, made valuable suggestions resulting in improvement in the presentation of the reports on papers which follow the methods of Hermite and Humbert. A. Ziwet of the University of Michigan translated for Cresse important portions of papers in Bohemian.

The volume has been much improved as to clearness and accuracy by the generous aid from Bell and Mordell, whose many investigations in this field made their cooperation most valuable. In particular, Mordell's criticisms proved his mastery of practically every subject in this extensive and intricate branch of mathematics.

Readers are requested to send notice to the author of errata or omissions, which will be published later as a supplement.

LEONARD EUGENE DICKSON
Professor of Mathematics
University of Chicago

TABLE OF CONTENTS.

CHAPTER I.

REDUCTION AND EQUIVALENCE OF BINARY QUADRATIC FORMS, REPRESENTATION OF INTEGERS.

Introduction.

There will be given in Chapters II–VIII reports of the literature on the following topics of the arithmetical theory of binary quadratic forms: Explicit values of x, y in $x^2+\Delta y^2=g$, composition, orders and genera, irregular determinants, number of classes of forms with integral coefficients, forms whose coefficients are complex integers or integers of a field, and their class number. The present chapter deals with the remaining, miscellaneous topics, the nature of which will be clear from the following brief summary.

Euler was the first to publish (in 1761, 1763) proofs of the facts that every prime of the form $6n+1$ can be represented by x^2+3y^2, and every prime $8n+1$ by x^2+2y^2. These and a few similar theorems had been merely stated by Fermat in 1654. In 1773, Lagrange proved many such facts by means of his general theory of reduction and equivalence of binary quadratic forms. In 1798, Legendre simplified and extended Lagrange's methods and tables, being aided largely by the reciprocity law for quadratic residues (although his proof of it was not quite complete).

In 1801, Gauss introduced many new concepts and extended the theory in various directions. His work has continued to occupy the central position in the literature, although many of his methods have since been materially simplified by Dirichlet, and to a less extent by Arndt and Mertens.

In 1851, Hermite developed his fundamental method of continual reduction. Closely related to it is the geometrical theory introduced by Smith in 1876 and applied by him to elliptic modular functions, later simplified by Hurwitz in 1894, by Klein in 1890, 1896, and by Humbert in 1916, 1917. A like goal was reached by Dedekind in 1877 and Hurwitz in 1881, both by means of the equivalence of complex numbers.

Selling gave in 1874 important methods of reducing both definite and indefinite forms. In 1880, Poincaré gave extensions of the methods of representing numbers $a+b\sqrt{D}$ by points, and forms by lattices. In 1881 and 1905, he constructed transcendental arithmetic invariants. Kronecker in 1883 and Stouff in 1889 studied reduction and equivalence under special types of substitutions. The investigation by Markoff in 1879 of the upper limits of the minima of forms was resumed by Schur and Frobenius in 1913 and Humbert in 1916. There are many further investigations of a special or miscellaneous character.

DEFINITIONS AND NOTATIONS.

Lagrange considered the general binary quadratic form

$$ax^2+bxy+cy^2 \tag{1}$$

with integral coefficients. Gauss restricted attention to

$$ax^2+2bxy+cy^2, \tag{2}$$

in which the middle coefficient is even, and called b^2-ac the *determinant* of (2). The form (1) is not only more general than (2), but has the special advantage of being more suitable for Dedekind's important correspondence between classes of forms (1) and certain sets of algebraic numbers determined by a root of $a\xi^2+b\xi+c=0$. The discriminant $d=b^2-4ac$ of this equation is called the *discriminant*[1] of (1), and plays an important rôle in the correspondence mentioned. Except when the contrary is expressly stated, the notation (a, b, c) will be used for (2), and not for (1).

If there exist integers x, y for which

$$ax^2+bxy+cy^2=m, \tag{3}$$

the form (1) is said to *represent* m. According as x and y are relatively prime or not, the *representation* (x, y) is said to be *proper* or *improper*.

If to the form f defined by (1) we apply a substitution

$$x=\alpha x'+\beta y', \qquad y=\gamma x'+\delta y', \tag{4}$$

with integral coefficients of determinant $\Delta=\alpha\delta-\beta\gamma\neq 0$, we obtain

$$f'=a'x'^2+b'x'y'+c'y'^2, \tag{5}$$

where

$$a'=a\alpha^2+b\alpha\gamma+c\gamma^2, \qquad c'=a\beta^2+b\beta\delta+c\delta^2, \tag{6}$$
$$b'=2a\alpha\beta+b(\alpha\delta+\beta\gamma)+2c\gamma\delta.$$

The discriminant d' of f' is equal to $\Delta^2 d$. Thus $d'=d$ if and only if $\Delta=\pm 1$. The latter is a necessary and sufficient condition that the inverse of the substitution (4) shall have integral coefficients. When this is the case, the forms f and f' evidently represent the same numbers and are said to be *properly* or *improperly equivalent*, according as $\Delta=+1$ or $\Delta=-1$.

Fermat[2] stated that he had a solution of a problem he had proposed to Frenicle de Bessy: To find in how many ways a given number is the sum of the two legs of a right triangle. In his reply, Frenicle[3] stated that every prime of the form $8n\pm 1$ is the sum of the two legs of a right triangle, and every number which is the sum of the legs of a primitive right triangle is of the form $8n\pm 1$. Every product of primes of the forms $8n\pm 1$ is the difference between the legs of an infinitude of primitive

---

[1] This is in accord with Dedekind, Kronecker, Weber, and others. But in the Encyclopédie des sc. math., t. I, v. 3, p. 101, p. 132, $4ac-b^2$ is called the discriminant of (1), and $ac-b^2$ that of (2).

[2] Oeuvres, 2, 1894, 221, 226 (§ 11); letters to Mersenne and Frenicle, June 15, 1641.

[3] Oeuvres de Fermat, 2, 1894, 231, 235; letters to Fermat, Aug. 2, Sept. 6, 1641.

right triangles. [The sum of the legs x^2-y^2, $2xy$ is $(x+y)^2-2y^2$ and their difference is $(x-y)^2-2y^2$ or $2y^2-(x-y)^2$.]

Fermat[4] stated that every prime $8n+1$ or $8n+3$ is of the form x^2+2y^2, that every prime $3n+1$ is of the form x^2+3y^2, and conjectured that the product of two primes ending in 3 or 7, and each having the form $4n+3$, is of the form x^2+5y^2. Fermat[5] stated that no prime $3n-1$ is of the form x^2+3y^2.

L. Euler[6] stated that, if x and y are relatively prime, $2x^2+y^2$ has no prime divisor other than 2 and primes of the forms $8n+1$ and $8n+3$; if $8n+1$ or $8n+3$ is a prime it is expressible in the form $2x^2+y^2$ in one and but one way. Every odd prime divisor of $2x^2-y^2$ is of the form $8n\pm1$.

Euler[7] stated many special empirical theorems on the representability of primes by $x^2\pm Ny^2$, where x and y are relatively prime (in connection with empirical theorems on the linear forms of the prime divisors of $x^2\pm Ny^2$, to be quoted under the quadratic reciprocity law in Vol. IV). Every prime $8n\pm1$ can be expressed in the form x^2-2y^2 in infinitely many ways. Every prime $12n\pm1$ can be expressed in either of the forms x^2-3y^2, $3x^2-y^2$ in infinitely many ways. Every prime $20n+1$ or $20n+9$, and the double of any prime $20n+3$ or $20n+7$ is of the form x^2+5y^2. Any prime $14n+k$ ($k=1$, 9 or 11) is of the form x^2+7y^2. Similar statements are made for x^2+Ny^2, $N=11$, 13, 17, 19. Any prime $24n+1$ or $24n+7$ is of the form x^2+6y^2; any prime $24n+5$ or $24n+11$ is of the form $3x^2+2y^2$; analogous results were stated for $N=2\cdot5$, $2\cdot7$, $3\cdot5$, $3\cdot7$, $5\cdot7$.

Euler[8] stated that every prime $8n\pm1$ is represented by $2t^2-u^2$ and by u^2-2t^2. A like result holds for 2 and any square, and hence for any product of primes $8n\pm1$, 2 and a square, since

$$(2\alpha^2-\beta^2)(2\gamma^2-\delta^2)=(2\alpha\gamma+\beta\delta)^2-2(\beta\gamma+\alpha\delta)^2, \qquad x^2-2y^2=2(x+y)^2-(x+2y)^2.$$

Chr. Goldbach[9] stated erroneously that any prime $4n+1$ is of the form dx^2+y^2 if d is any divisor of n. Euler[10] remarked that this is probably true if x, y are permitted to be rational, but not always for integers. Thus $89\neq 11x^2+y^2$ in integers, but

$$89=11\left(\frac{5}{2}\right)^2+\left(\frac{9}{2}\right)^2=11\left(\frac{4}{3}\right)^2+\left(\frac{25}{3}\right)^2.$$

At least for $d=1$, 2, 3, a representation by fractions implies one by integers. Euler[11] and Goldbach[11] discussed special methods of expressing a prime $4dk+1$ in the form dx^2+y^2, where x, y are rational.

---

[4] Oeuvres, 2, 1894, 313, 403–5; letters to Pascal, Sept., 1654, and to K. Digby, June, 1658 (French transl. of the second letter in Oeuvres, 3, 1896, 315–7). Wallis, Opera, 2, 1693, 858.

[5] Oeuvres, 2, 1894, 431; letter to Carcavi, Aug., 1659.

[6] Correspondance Mathématique et Physique (ed., P. H. Fuss), St. Pétersbourg, 1, 1843, 146, 149; letter to Goldbach, Aug. 28, 1742. Also, Euler[7], Theorems 4, 5, 42. Cf. this History, Vol. II, p. 260, Euler.[12]

[7] Comm. Acad. Petrop., 14, (1744–6) 1751, 151–181; Comm. Arith. Coll., St. Pétersbourg, 1, 1849, 35–49; Opera Omnia, (1), II, 194–222.

[8] Novi Comm. Acad. Petrop., 2, (1749) 1751, 49; Comm. Arith. Coll., 1, 1849, 69–70; Opera Omnia, (1), II, 236–7.

[9] Corresp. Math. Phys. (ed., Fuss), 1, 1843, 602, March 12, 1753.

[10] *Ibid.*, 604, April 3, 1753.

[11] *Ibid.*, 610, 615, 616, 619, 621, 625.

Euler[12] proved that every prime divisor of $2x^2+y^2$, where x and y are relatively prime, is itself of that form. First, if $N=2a^2+b^2$ is divisible by a prime $P=2p^2+q^2$, the quotient is of that form. For, P divides $a^2P-p^2N=a^2q^2-b^2p^2$ and hence divides one of its factors. Let $aq\pm bp=mP$. Then $a=mq+(2mp\mp b)p/q$. Hence $(2mp\mp b)/q$ is an integer, say $\mp n$. Then

$$b=nq\pm 2mp, \qquad a=mq\mp np, \qquad N=P(2m^2+n^2).$$

The last identity shows also that the product of several numbers of the form $2a^2+b^2$ is itself of that form. It now follows that if $2a^2+b^2$ is divisible by a number not of that form, the quotient is not a prime $2x^2+y^2$ nor the product of primes of that form.

To prove the main theorem, suppose that $2x^2+y^2$ (x and y relatively prime) has a prime divisor A' not of that form. We may set

$$x=mA'\pm a, \qquad y=nA'\pm b, \qquad 0\leqq a\leqq \tfrac{1}{2}A', \qquad 0\leqq b\leqq \tfrac{1}{2}A'.$$

Hence A' divides $A=2a^2+b^2<\tfrac{3}{4}A'^2$. The quotient and hence A has a prime factor B' not of the form $2r^2+s^2$. Also $B'<\tfrac{3}{4}A'$. In case $A>\tfrac{3}{4}B'^2$, we obtain as before a number $B=2c^2+d^2$ divisible by B', where c and d are relatively prime, and $B<\tfrac{3}{4}B'^2$; in the contrary case, we take $B=A$. Similarly, B has a prime divisor $C'<\tfrac{3}{4}B'$, where C' is not of the form $2x^2+y^2$ but is a divisor of a number $C=2e^2+f^2<\tfrac{3}{4}C'^2$, e and f relatively prime. Continuing thus we get smaller and smaller numbers $2z^2+w^2$ (z prime to w), divisible by numbers not of that form. But the small numbers $2z^2+w^2$ have all their divisors of that form.

Euler noted (§ 46) that the method is not applicable to mx^2+y^2 if $\tfrac{1}{4}(m+1)>1$, while for $m=3$, 2 divides $3x^2+y^2$ for $x=y=1$ but is not of that form, so that the method does not apply immediately to $3x^2+y^2$.

Euler[13] later noted the modification of the preceding method which proves that every prime divisor $\neq 2$ of $3x^2+y^2$ (x, y relatively prime) is itself of that form. We now have a prime divisor $A'\neq 2$ of $A=3a^2+b^2<A'^2$. If A is odd, the quotient A/A' has as before a prime factor $B'\neq 2$ not of the form $3r^2+s^2$. If A is even it is divisible by 4 and it is easily seen that $\tfrac{1}{4}(3a^2+b^2)$ is of the form $3c^2+d^2$. After thus removing all the factors 2, we are led to the first case.

Euler then concluded as follows that every prime $p=6n+1$ is of the form x^2+3y^2. Let a and b be any integers not divisible by p. Then p divides

$$a^{6n}-b^{6n}=(a^{2n}-b^{2n})(a^{4n}+a^{2n}b^{2n}+b^{4n}).$$

The second factor is of the form f^2+fg+g^2, which is equal to

$$(f+\tfrac{1}{2}g)^2+3(\tfrac{1}{2}g)^2, \text{ if } g \text{ is even}; \quad \left(\frac{f-g}{2}\right)^2+3\left(\frac{f+g}{2}\right)^2, \text{ if } f \text{ and } g \text{ are odd.}$$

But if p divides x^2+3y^2 it is of that form. The first factor $a^{2n}-b^{2n}$ is not always divisible by p; take $b=1$; the differences of order $2n$ of

$$2^{2n}-1, \quad 3^{2n}-1, \ldots, \quad (6n)^{2n}-1$$

are equal to $(2n)!$, so that these binomials are not all divisible by p.

[12] Novi Comm. Acad. Petrop., 6, (1756–7) 1761, 185–230; Comm. Arith. Coll., I, 1849, 174–192; II, 573–5 (with the first step of the proof modified); Opera Omnia, (1), II, 459–492.

[13] Novi Comm. Acad. Petrop., 8 (1760–1) 1763, 105–128; Comm. Arith. Coll., I, 287–296; Opera Omnia, (1), II, 556–575.

Euler stated that a similar method does not lead to a proof that every prime $8n+3$ is of the form x^2+2y^2, but applies to $8n+1$. In § 56 of his earlier paper, Euler[12] stated that he was not able to treat either case. Euler[14] elsewhere noted that a prime $8n+1$ divides

$$a^{4n}+b^{4n}=(a^{2n}-b^{2n})^2+2a^{2n}b^{2n}$$

for some values of a, b, not divisible by $8n+1$, and hence is of the form x^2+2y^2.

Euler[15] later gave a more elegant proof (like that for x^2+y^2 in this History, Vol. II, p. 231) that any divisor (except 2 when $m=3$) of x^2+my^2, $m=2$ or 3, x and y relatively prime, is of that form.

Jean Bernoulli[16], III, tabulated the primes a^2-10b^2 up to 3000.

E. Waring[17] proved that, if $N=a^2-rb^2$, then N^{2m+1} and N^{2m} can be expressed in [at least] $m+1$ different ways in the form p^2+rq^2. For,

$$N^{2m}=\frac{(s+t)^2}{4}(a^2+rb^2)^{2m-2l}-\frac{(s-t)^2}{4}(a^2+rb^2)^{2m-2l} \qquad (l=0, 1, \ldots, m),$$

where

$$s=(a+b\sqrt{-r})^{2l}, \qquad t=(a-b\sqrt{-r})^{2l}.$$

Let the expansion of s be $e+f\sqrt{-r}$. Then

$$p=e(a^2+rb^2)^{m-l}, \quad q=f(a^2+rb^2)^{m-l}.$$

To derive the expression for N^{2m+1} we have only to replace the exponents $2l$ in s, t by $2l+1$. Hence if $M, N, \ldots$ can be expressed in the form a^2+rb^2 in $m, n, \ldots$ ways, then $M^{\alpha}N^{\beta}\ldots$ can be expressed in the form p^2+rq^2 in

$$\tfrac{1}{2}(\alpha+1)\cdot\tfrac{1}{2}(\beta+1)\ldots mn\ldots$$

ways if $\alpha, \beta, \ldots$ are odd, while if any factor, as $\alpha+1$, is odd it must be replaced by $\alpha+2$. If $P=mp^2+q^2$ is a prime factor of $N=ma^2+b^2$, the quotient N/P has that form. From p^2N-a^2P, we see that $aq\pm bp$ is a multiple rP of P. Thus $a=rq+tp/q$, $t=mrp\mp b$. Hence $t=\mp sq$, $N=P(mr^2+s^2)$.

J. L. Lagrange[18] made the first general investigation of binary quadratic forms. He introduced no special terms such as are given in the Introduction above. We shall employ those terms in this report for the sake of brevity and for ease in comparison with later writings. In particular, Lagrange made no distinction between proper and improper equivalence.

Theorem I. If m is a divisor of $Pt^2+Qtu+Ru^2=mq$, where t and u are relatively prime, m can be represented properly by a form of discriminant $d=Q^2-4PR$.

---

[14] Corresp. Math. Phys. (ed., Fuss), 1, 1843, 628; letter to Goldbach, Aug. 23, 1755. Euler stated (*ibid.*, p. 622) that he had recently found a proof that any primes $8n+1$, $12n+1$ are of the respective forms x^2+2y^2, x^2+3y^2, but could not prove that primes $20n+1$ or $20n+9$ are of the form x^2+5y^2.

[15] Nova Acta Eruditorum, 1773, 193–211; Acta Acad. Petrop., 1777: II, 1780, 48–69; Comm. Arith. Coll., I, 541–4; Opera Omnia, (1), III, 224–9.

[16] Nouv. Mém. Ac. Berlin, année 1771 (1773), 323.

[17] Meditationes algebraicae, Cambridge, 1770, 205–6; ed. 3, 1782, 350–3.

[18] Recherches d'arithmétique, Nouv. Mém. Acad. Berlin, année 1773, 265–312; Oeuvres, III, 693–758.

Let g denote the g.c.d. of q and u, and write $q=gk$, $u=gx$. Then Pt^2, and hence P, is divisible by g. Write $P=Hg$. Thus

$$mk=Ht^2+Qtx+Rgx^2.$$

Since k and x are relatively prime, there exist integers θ and y for which $t=\theta x+yk$. Inserting this value of t, we get

$$mk=lx^2+bkxy+cky^2, \quad l=H\theta^2+Q\theta+Rg, \quad b=2H\theta+Q, \quad c=Hk.$$

Thus l is divisible by k. Write $l=ak$. Hence (3) follows. The discriminant of (3) is equal to d.

Theorem II. A form (1), whose discriminant is not a perfect square,[19] and in which b is numerically greater than either a or c, is properly equivalent to a form $a_1x_1^2+b_1x_1y_1+c_1y_1^2$ in which $|b_1|<|b|$.

For, if $|b|>|a|$, write $x=x_1+\mu y_1$, $y=y_1$. Then $b_1=b+2a\mu$ will be numerically $<b$ for a suitably chosen integer μ.

If either a_1 or c_1 is numerically $<b_1$, we pass to a third properly equivalent form $a_2x_2^2+b_2x_2y_2+c_2y_2^2$ in which $|b_2|<|b_1|$; etc. Since the integers $b, b_1, b_2, \ldots$ decrease numerically, the process must terminate. Hence we have the following conclusion:

Theorem III. Any form is properly equivalent to a form

$$(7) \qquad Ax^2+Bxy+Cy^2, \quad |B|\leqq|A|, \quad |B|\leqq|C|.$$

The given discriminant d of (7) may be negative or positive.

(i) Let $d=-K$, $K>0$. Then $4AC=B^2+K>0$. By (7), $|AC|\geqq B^2$. Hence $K\geqq 3B^2$. According as K is even or odd, B may take the even or odd integral values numerically $\leqq \sqrt{K/3}$. For each B, A and C are the factors, neither numerically $<|B|$, of $\frac{1}{4}(K+B^2)$.

(ii) Let $d>0$. Then $|AC|\geqq B^2$ requires that AC be negative. Thus $d\geqq 5B^2$. According as d is even or odd, B may take the even or odd integral values numerically $\leqq\sqrt{d/5}$. For each B, A and C are factors, neither numerically $<|B|$, of $\frac{1}{4}(B^2-d)$.

Hence for a given discriminant not a perfect square, there is a finite number of forms (7). In view also of Theorem I, the latter are spoken of as the forms of the divisors of $f=Pt^2+Qtu+Ru^2$. These forms of the divisors depend therefore only upon the discriminant d of f. This follows also from $4Pf=z^2-du^2$, where $z=2Pt+Qu$, whence the divisors of f are divisors of z^2-du^2.

For t^2+au^2, $a>0$, we have $P=1$, $Q=0$, $R=a$, $d=-K=-4a$. Hence B is even and will be replaced by $2B$. Since AC is positive, the form represents positive numbers only when A and B are positive. If $C<A$, we replace x by y and y by $-x$. Hence we may always take $C\geqq A$. We obtain the *reduced* form

$$(8) \qquad Ax^2+2Bxy+Cy^2, \quad C\geqq A\geqq 2\,|B|,$$

of given negative determinant $B^2-AC=-a$. Here $|B|\leqq\sqrt{a/3}$. Hence every positive divisor of t^2+au^2, with t and u relatively prime, can be properly represented

[19] The omission of this requirement led Lagrange to errors noted below. His proof fails if $a=0$. Similarly, given $|b|>|c|$, we write $x=x_1$, $y=y_1+\mu x_1$, and obtain $b_1=b+2c\mu$, which can be made numerically $<b$ if $c\neq 0$.

by one of the forms (8). For $a=1, 2, 3, 5$, these are x^2+y^2, x^2+2y^2, x^2+3y^2 and $2x^2\pm 2xy+2y^2$, x^2+5y^2 and $2x^2\pm 2xy+3y^2$, the double of the last being $(2x\pm y)^2+5y^2$. Hence for $a=2$ or 3 we have the results proved by Euler[12, 13] by a method not applicable when $a>3$.

Lagrange investigated (§ 22) the equivalence of (8) with another reduced form $A'x'^2+\dots$ of the same determinant $-a$ under the substitution

$$x=Mx'+Ny',\qquad y=mx'+ny'.$$

As shown in the Introduction above, $Mn-Nm=\pm 1$. Let $|M|>|N|$, since the argument with M and N interchanged applies when $|N|>|M|$. We can determine an integer $\mu \gtreqless 2$ such that $|M'|<|N|$ in $M=\mu N+M'$. Determine m' by $m=\mu n+m'$. For these values of M and m, we get $B'=\mu C'+B''$, where

$$B''=AM'N+B(M'n+Nm')+Cm'n.$$

By hypothesis, $C'\gtreqless 2|B'|$, whence $|B'|<C'$. Since $\mu\gtreqless 2$, B'' must be of sign opposite to $\mu C'$, and $|B''|>C'$. In terms of $y''=\mu x'+y'$, our substitution becomes $x=M'x'+Ny''$, $y=m'x'+ny''$, which has the determinant ± 1 and replaces (8) by $A''x'^2+2B''x'y''+C'y''^2$. Thus $A''C'-B''^2=a>0$, whence $A''>|B''|$, since $|B''|>C'$. Since $|N|>|M'|$, we may write $N=\mu' M'+N'$, where μ' is an integer $\gtreqless 2$, and $|N'|<|M'|$. Writing also $n=\mu' m'+n'$, $x''=\mu' y''+x'$, we see that our substitution becomes $x=M'x''+N'y''$, $y=m'x''+n'y''$, of determinant ± 1. As before it replaces (8) by a form $A''x''^2+2B'''x''y''+C''y''^2$ with $B''=\mu' A''+B'''$, whence $|B'''|>|A''|$, $|B'''|<|C''|$. The series of decreasing integers $|M|$, $|N|$, $|M'|$, $|N'|$, ... terminates with zero. Assuming that $N'=0$ for example, we readily conclude from the above relations that the two reduced forms are identical. But Lagrange failed to treat the case in which M and N are numerically equal (and hence equal to ± 1); in this case the series $M, N, M', \dots$ does not terminate with zero, but contains only terms ± 1. Contrary to his conclusion, $x=x'-y'$, $y=y'$ transforms $Ax^2+Axy+Cy^2$ into $Ax^2-Axy+Cy^2$. The only other exceptional case is that in which $C=A$ (Gauss,[35] Art. 172).

Similarly, every divisor of t^2-au^2 ($a>0$, t, u relatively prime) can be represented properly by one of the forms (some of which may be equivalent)

$$\pm Ax^2+2Bxy\mp Cy^2,\qquad C\gtreqless A\gtreqless 2|B|, \tag{9}$$

of determinant $B^2+AC=a$. Thus $|B|\leqq\sqrt{a/5}$. For $a=2$, there are two such forms x^2-2y^2 and $-x^2+2y^2$, and the former is transformed into the latter by $x=x'+2y'$, $y=x'+y'$. Hence every divisor of t^2-2u^2 is of the form x^2-2y^2.

Lagrange (§§ 23, 24) gave the following method of testing the equivalence of forms (9) having a given positive determinant a. For this purpose he used the notation

$$r'y^2+2qyy'-ry'^2,\qquad 2|q|\leqq r>0,\qquad 2|q|\leqq r'>0,\qquad rr'+q^2=a. \tag{10}$$

This form is transformed by $y=m'y'+y''$ into

$$f'=r'y''^2+2q'y''y'-r''y'^2,\qquad q'\equiv q+r'm',\qquad r'r''+q'^2=a.$$

If possible, select an integer $m'>0$ such that $2|q'|$ exceeds neither r' nor r''; then

f' is of type (10). If such a choice is impossible, take as m' the largest integer $<(\sqrt{a}-q)/r'$. Then $y'=m''y''+y'''$ transforms f' into

$$f''=r'''y''^2+2q''y''y'''-r''y'''^2, \qquad q''\equiv q'-r''m'', \qquad r''r'''+q''^2=a.$$

If possible, select an integer $m''>0$ such that $2|q''|$ exceeds neither r'' nor r'''; then f'' is of type (10). If such a choice is impossible, take as m'' the largest integer $<(\sqrt{a}+q')/r''$ and replace y'' by $m'''y'''+y''''$, etc. Note the alternations of signs, so that, for the next step, m''' is the largest integer $<(\sqrt{a}-q'')/r'''$. We find in this manner the finite period of forms (10) equivalent to one of them.

For example, $y^2-7y'^2$ is of type (10). Thus $q'=m'>0$, $r''=7-q'^2$. Since $2|q'|$ here exceeds $r'=1$, we take $m'=2$, the largest integer $<\sqrt{7}$. Then $q'=2$, $r''=3$. Next, $q''=2-3m''$, $r'''=(7-q''^2)/3$. Thus $2|q''|$ will not exceed r'' if and only if $m''=1$. We obtain $f''=2y''^2-2y''y'''-3y'''^2$, of type (10). The next transformed $2y''''^2+2y''''y'''-3y'''^2$ is of type (10). After another step, we obtain the initial form. Hence y^2-7z^2, $2y^2\pm 2yz-3z^2$ are equivalent to each other, but not equivalent to the negative of one of the three. The six exhaust the forms (9).

For $a=1$, x^2-y^2 is the only form (9). It is stated (§ 20) that every divisor of t^2-u^2, with t, u relatively prime, is of the form x^2-y^2. But the discriminant 4 is a perfect square and the general theory is not applicable (see Theorem II). Nor is the present conclusion correct. For $t=5$, $u=1$, $t^2-u^2=24$ has the divisor 6; if $6=x^2-y^2$, then $x\equiv y \pmod 2$, $x^2\equiv y^2 \pmod 4$.

The paper closes with a table* of the forms $py^2\pm 2qyz+rz^2(pr-q^2=a)$ of the odd divisors of t^2+au^2, and a table of the forms $py^2\pm 2qyz-rz^2(pr+q^2=a)$ of the odd divisors of t^2-au^2 for $a=1, \ldots, 31$, with a not divisible by a square.

Lagrange[20] noted that we can readily determine the integers b not exceeding $2a$ numerically such that all integers represented by $py^2+2qyz\pm rz^2$ are of the form $4an+b$, where $a=pr\mp q^2$. For example, if $q=0$, we may take $y=2mr\pm\rho$, $z=2m'p\pm\omega$, where ρ is chosen from $1, \ldots, r$, and ω from $1, \ldots, p$; then $py^2\pm rz^2=4an'+b'$, $b'=p\rho^2\pm r\omega^2$, and we may write $b'=4an''+b$, where $|b|\leqq 2|a|$.

It now follows from his former paper that every divisor of $t^2\pm au^2$ is of one of the forms $4an+b$. For the positive integers $a\leqq 30$ not divisible by a square, he listed in Table III, IV the values of b giving the forms $4an+b$ of the odd divisors prime to a of $t^2\pm au^2$, and in Tables V, VI the remaining values of b numerically less than and prime to $2a$. These tables are applied to the factoring of a number N. By expressing N or a multiple of N in the form $t^2\pm au^2$, where $0<a<31$, we have the possible forms $4an+b$ of a divisor of N.

He readily proved (p. 780) that if a prime $4n-1$ is [or is not] a divisor of a number of the form t^2-au^2, it is not [or is] a divisor of a number of the form t^2+au^2. For example, a prime $p=8n+3$ is not a divisor of a number of the form t^2-2u^2, by Table VI, and hence is a divisor of a number t^2+2u^2. Then, by Table I, p is representable by y^2+2x^2 [Fermat[4]]. There are proved similarly many theorems on the representation of primes $4n-1$ by $y^2\pm az^2$, $a\leqq 30$.

---

* Errata for $a=29$, 30 were noted in the next paper, p. 328, § 33, but were corrected in Oeuvres without comment.

[20] Nouv. Mém. Acad. Berlin, année 1775 (1777), 323–356; Oeuvres, III, 759–795.

A further device (p. 789) is necessary for the much more difficult case of primes $4k+1$. Let $p=4na+1$ be a prime. Then $x^{2na}+1$, being a factor of $x^{p-1}-1$, is divisible by p for a suitably chosen integer x. Write $x^n=r$, $r^2+1=s$. Then

$$R\equiv r^{2a}+1=s^a-as^{a-2}r^2+\tfrac{1}{2}a(a-3)s^{a-4}r^4-\ldots.$$

Hence r can be found such that R is divisible by p. For $a=2$, then $R=s^2-2r^2$, so that every prime $8n+1$ is a divisor of a number of the form t^2-2u^2 and hence of a number of the form t^2+2u^2. Thus, by results in the first paper, every prime $8n+1$ is representable by each of the forms y^2+2z^2, y^2-2z^2, $2z^2-y^2$. In this way it is shown that the primes $4an+b$ belonging to the divisors of $t^2\pm au^2$ are actually divisors of numbers of these forms when $b=9$, $a=5$, 10, and $b=1$, $a=1$, 2, 3, 5, 6, 7, 10, 14, 15, 21, 30. Hence for these cases there are obtained theorems on the representability of primes $4an+b$ by $t^2\pm au^2$.

P. S. Laplace[21] stated that no one had proved that every prime $p=8n+3$ is of the form x^2+2y^2 (Euler[13] having sought a proof in vain), and gave the following proof. By Fermat's theorem, p divides ab, where $a=2^{4n+1}-1$, $b=2^{4n+1}+1$. If p divides $a=2(2^{2n})^2-1^2$, it has the form $p=2u^2-s^2$. Since s is odd, $s^2=8l+1$. If $u=2i$, $p+1=8(i^2-l)$, whereas $p+1=4(2n+1)$. If $u=2i+1$, $p=8(i^2+i-l)+1$. Hence p divides $b=2(2^{2n})^2+1$ and thus is of the form x^2+2y^2.

It is proved similarly that every prime $8n+1$ is of the form $2x^2-y^2$, since a divisor of $2(2^{2n-1})^2+1$ is of the form x^2+2u^2.

L. Euler[22] observed that it is very remarkable that every prime $8n+1$ or $8n+3$ is expressible in one and but one way in the form a^2+2b^2. For proof, a prime $8n+1$ is a factor of $c^{8n}-1$ if not of c. Take c so that $8n+1$ is not a factor of $c^{4n}-1$. Hence it is a factor of

$$c^{4n}+1=(c^{2n}-1)^2+2(c^n)^2$$

and hence is expressible in the form a^2+2b^2. For a prime $8n+3$, use the product of $2^{4n+1}+1$ by $2^{4n+1}-1$ and note that a factor of the second is of the form $8k\pm1$.

Euler[23] repeated this proof for $8n+1$. If p is a prime $8n+3$, it divides one of $a\cdot a^{4n}\pm b\cdot b^{4n}$. Take $a=c^2$, $b=2d^2$. Thus p divides $A^2\pm2B^2$. But A^2-2B^2 is known to have only divisors of the form $8n\pm1$. Hence A^2+2B^2 is divisible by p.

A. M. Legendre[24] employed Lagrange's[18] method of reduction to a reduced form, but gave details only when the middle coefficient is even. The fact that no two reduced forms of the same negative determinant $-A$ are equivalent follows from the theorem that, if p, q, r are positive integers such that $2q$ exceeds neither p nor r, while $pr-q^2=A>0$, the least number represented by $py^2+2qyz+rz^2$ is the smaller of p and r. To prove the last theorem, note that this form exceeds $P=py^2-2qyz+rz^2$ when y, z are positive integers. Let $y\gtreqless z$. Replacing y by $y-1$, we see that P becomes

$$P'=P-2q(y-z)-y(p-2q)-p(y-1)<P,$$

21 De la Place, Théorie abrégée des nombres premiers, 1776, 29 pp. He apparently had not seen Lagrange's second paper.

22 Posth. paper, Comm. Arith. Coll., II, 606–7.

23 Opera postuma, 1, 1862, 158–9 (about 1783).

24 Théorie des nombres, 1798, pp. 69–76; ed. 2. 1808, pp. 61–67; ed. 3, 1830, I, pp. 72–80 (German transl. by H. Maser, I, 1893, pp. 73–81).

and P' is not negative since the factors of P are imaginary. By thus diminishing the larger of y, z, we finally obtain $y=z=1$, for which the value $p+r-2q$ of P is greater than or equal to the larger of p, r. Cf. Hermite.[53]

Legendre[25] investigated the equivalence of reduced forms $ay^2+2byz-cz^2$ of positive determinant $b^2+ac=A$, where a, c are positive and not less than $2b$. Use is made of the development into a continued fraction of a root of $ax^2+2bx-c=0$. In this way he computed his three-page Table I of reduced forms of positive determinant $A\leqq 136$, and his Table II of reduced forms $Ly^2+Myz+Nz^2$, where M is odd and M^2-4LN is positive and $\leqq 305$. Table III lists not merely the reduced forms $py^2+2qyz+ry^2$ whose determinant $a=q^2-pr$ is positive and $\leqq 79$ (called *quadratic divisors* of t^2-au^2), but also the linear forms $4ax+b$ of the odd divisors of t^2-au^2. Table IV gives the analogous material for t^2+au^2, $0<a=4n+1\leqq 105$. Table V lists the reduced forms $Ly^2+Myz+Nz^2$ with $4LN-M^2=a$ when M is odd, but with $LN-\frac{1}{4}M^2=a$ when M is even, for $0<a=4n+3\leqq 103$, as well as the linear forms $2ax+b$ of the odd divisors of t^2+au^2. Tables VI and VII list the reduced forms $py^2+2qyz+ry^2$ with $0<pr-q^2=2a\leqq 106$ and the linear forms $8ax+b$ of the odd divisors of t^2+2au^2. Errata have been noted by D. N. Lehmer[26] and A. Cunningham.[27]

A. Cayley[28] stated that Legendre's Tables I–VII are of comparatively little value since his classification of forms takes no account of the distinction between proper and improper equivalence, nor of their orders and genera.

Legendre[29] noted that if p is any divisor of t^2+au^2, and P is the quotient, where t, u are relatively prime, and hence also p, u, we can choose integers y, q such that $t=py+qu$. Hence

$$P=py^2+2qyu+ru^2, \qquad r=\frac{q^2+a}{p}.$$

Since P, as well as p, is a divisor of t^2+au^2, it follows that any divisor of t^2+au^2 is not only a divisor of x^2+a, but also of $py^2+\ldots$, and hence of a reduced form. A simple examination of the reduced forms shows that any divisor of one of the forms t^2+u^2, t^2+2u^2, t^2-2u^2 is of that form, respectively. It is then shown that the primes $8n+1$, $8n+3$ (and no others) are of the form y^2+2z^2, while the primes $8n\pm 1$ (and no others) are of the form y^2-2z^2.

Legendre[30] used the reduced forms $P=py^2+2qyz\pm rz^2$ of the divisors of $t^2\pm cu^2$ to find by trial the linear forms $4cx+a$ of the divisors [Lagrange[20]]. For, we need not try integers y, z exceeding $2c$, since if we replace y by $2c+y$ and z by $2c+z$, P becomes $P+4cM$. The reciprocity law also limits the form of the linear divisors. This work may, however, be abbreviated by use of the generalized reciprocity law.[31]

---

[25] Théorie des nombres, 1798, pp. 123–132; ed. 2, 1808, pp. 111–120; ed. 3, 1830, I, pp. 130–140 (Maser, I, pp. 131–142).
[26] Bull. Amer. Math. Soc., 8, 1902, 401.
[27] Messenger Math., 46, 1916–7, 51–52.
[28] British Assoc. Report for 1875, 326.
[29] Théorie des nombres, 1798, pp. 187–196; ed. 2, 1808, pp. 172–181; ed. 3, 1830, I, pp. 200–9 (Maser, I, pp. 200–210).
[30] *Ibid.*, pp. 243–277 (espec. p. 254); ed. 2, 1808, pp. 223–254 (232); ed. 3, 1830, I, 261–298 (272) (Maser, I, pp. 258–294). Mém. Acad. Roy. Sc. Paris, 1785, 524–559.
[31] Dirichlet, Zahlentheorie, § 52, or Matthews, Theory of Numbers, 1892, 50–53.

Legendre[32] proved that if $l=4cx+a$ is one of the possible linear forms of divisors of $t^2 \pm cu^2$, every prime A of the form l is a divisor of $t^2 \pm cu^2$ and therefore also of one the quadratic forms $P=py^2+2qyz \pm rz^2$ corresponding to l. By hypothesis, $P=l$, where the determinant $q^2 \mp pr$ is equal to the determinant $\mp c$ of $t^2 \pm cu^2$. Thus $P=4cX+A$. Multiplication by p gives $(py+qz)^2 \pm cz^2 = 4pcX+pA$. Hence * $(pA/\theta)=1$ if θ is any prime factor of c. If c is a product of two primes α, β, it is shown by means of the reciprocity law that $(\pm \alpha\beta/A)=+1$ for $A \equiv \pm 1 \pmod 4$, whence A is a divisor of $t^2+\alpha\beta u^2$. It is stated that this conclusion doubtless holds also when c is a product of more than two primes.

There are given various theorems on the number of ways in which a prime or its double can be represented by a quadratic form.

Legendre[33] spoke of $f=py^2+2qyz+rz^2$, of determinant $-c=q^2-pr$, not merely as a quadratic divisor of t^2+cu^2, but also as a *trinary* (quadratic) divisor of t^2+cu^2 in case f is expressible as a sum of three squares of linear functions of y, z with integral coefficients. A necessary condition is that c be a sum of three squares. Conversely, given such a c, we can find a trinary quadratic divisor of t^2+cu^2. If[34] c is a prime or the double of a prime, two distinct representations of c as a sum of three squares cannot correspond to the same trinary divisor of t^2+cu^2. A quadratic divisor f of t^2+cu^2 is called *reciprocal* if, for every integer N represented by f, c is a divisor of t^2+Nu^2. Two proofs are given of the principal theorem that every reciprocal divisor of t^2+Nu^2 is expressible as a trinary form in 2^{i-1} ways, where i is the number of distinct odd prime factors of N. Table VIII shows the trinary quadratic divisors of t^2+cu^2 for each $c \leqq 251$ which is expressible as a sum of three squares.

C. F. Gauss[35] (Arts. 147–9) applied the reciprocity law to find the linear forms of the divisors of x^2-A, where A may be taken free from square factors without loss of generality. If $A \equiv 1 \pmod 4$, the numbers of which A is a quadratic residue from $\frac{1}{2}\phi(A)$ arithmetical progressions $Ag+r_i$ $(i=1, \ldots, \frac{1}{2}\phi(A))$; those of which A is a non-residue form $\frac{1}{2}\phi(A)$ progressions $Az+n_i$. Here the r_i and n_i together give the numbers $<A$ and prime to A; r_i is a non-residue of an even number of the prime factors of A; each n_i is a non-residue of an odd number of the factors. Like results hold for $A=-Q$ or $A=\pm 2Q$, where $Q \equiv 1 \pmod 4$, the common difference for the progressions being $4Q$ or $8Q$, respectively.

Gauss (Art. 153) restricted his important investigation of binary quadratic forms to the case $F=ax^2+2bxy+cy^2$ whose middle coefficient is even, and designated it by (a, b, c). It is said (Art. 154) to have the *determinant* $D=b^2-ac$, assumed $\neq 0$, and to *represent* M if there exist integers $x=m$, $y=n$, for which

$$am^2+2bmn+cn^2=M.$$

* The Legendre symbol (k/θ) denotes $+1$ or -1 according as k is a quadratic residue or non-residue of θ, *i.e.*, $x^2 \equiv k \pmod \theta$ is or is not solvable.

[32] Théorie des nombres, 1798, pp. 278–303, 441–450; ed. 2, 1808, 255–278, 380–5; ed. 3, 1830, I, 229–325, II, 50–56 (Maser, I, pp. 294–321; II, pp. 50–57).

[33] Théories des nombres, 1798, pp. 321–400; ed. 2, 1808, pp. 293–339; ed. 3, 1830, I, pp. 342–396 (Maser, I, pp. 337–389). Cf. this History, Vol. II, p. 261.

[34] An omitted case in the proof was treated by T. Pepin, Jour. de Math., (3), 5, 1879, 21–30.

[35] Disquisitiones Arithmeticae, 1801; Werke, 1, 1863; German transl. by H. Maser, 1889; French transl. by A. C. M. Poullet-Delisle, 1807.

If m, n are relatively prime, then D is a quadratic residue of M. For, if μ, ν are integers such that $\mu m+\nu n=1$, then

$$M(a\nu^2-2b\mu\nu+c\mu^2)\equiv v^2-D(\mu m+\nu n)^2=v^2-D,$$

where $v=\mu(mb+nc)-\nu(ma+nb)$. Hence $D\equiv v^2 \pmod{M}$.

If (Art. 155) we employ a second set of integers μ', ν' such that $\mu' m+\nu' n=1$, and denote by v' the expression corresponding to v, we have $v'-v=(\mu'\nu-\mu\nu')M$. Moreover, we can choose μ', ν' so that v' will equal any assigned integer which is congruent to v modulo M. The representation of M by F with $x=m$, $y=n$, is said to *belong* to the particular root v of $v^2\equiv D \pmod{M}$.

Let (Art. 156) m_1, n_1 be another pair of relatively prime integers giving a representation of M by F. Write

$$\mu_1 m_1+\nu_1 n_1=1,\qquad v_1=\mu_1(m_1 b+n_1 c)-\nu_1(m_1 a+n_1 b).$$

If $v\equiv v_1 \pmod{M}$, the two representations of M are said to belong to the same root of $v^2\equiv D \pmod{M}$. They are said to belong to different roots if there do not exist integers μ, ν, μ_1, ν_1 for which $v\equiv v_1 \pmod{M}$. If $v\equiv -v_1$, the representations are said to belong to opposite roots.

If (Art. 157) the form $F=ax^2+2bxy+cy^2$ is transformed into $F'=a'x'^2+\ldots$ by a substitution with integral coefficients,

$$(11)\qquad x=\alpha x'+\beta y',\qquad y=\gamma x'+\delta y',\qquad \Delta=\alpha\delta-\beta\gamma\neq 0,$$

F is said to *contain* (*enthält*) F', while F' is contained in F. Then

$$D'\equiv b'^2-a'c'=D\Delta^2.$$

If also F' contains F, D and D' divide each other and have the same sign, so that $D=D'$, $\Delta=\pm 1$. Then F and F' are called *equivalent*.

According as Δ is positive or negative, the substitution is called *proper* (*eigentlich*) or *improper*, respectively. According as F is transformed into F' by a proper or improper substitution, F' is said to be *contained properly* or *improperly* in F.

If (Art. 158) F and F' have equal determinants and F' is contained in F, then F is contained in F' properly or improperly, according as F' is contained in F properly or improperly. If F and F' are contained properly (improperly) in each other, they are called *properly* (*improperly*) *equivalent*.

If (Art. 159) F contains F', and F' contains F'', then F contains F''. Let (11) replace F by F', and let

$$(12)\qquad x'=\alpha' x''+\beta' y'',\qquad y'=\gamma' x''+\delta' y'',\qquad \Delta'=\alpha'\delta'-\beta'\gamma'\neq 0,$$

replace F' by F''. By eliminating x', y', we obtain a substitution of determinant $\Delta\Delta'$ which replaces F by F''. Hence F contains F'' properly or improperly, according as Δ, Δ' are of like or opposite signs. The form (a, b, c) is improperly equivalent to its *opposite* form $(a, -b, c)$ and to (c, b, a), and properly equivalent to $(c, -b, a)$.

If (Art. 160) $b+b'\equiv 0 \pmod{c}$, (a, b, c) has the *right neighboring* form (c, b', c') provided their determinants are equal. The latter form has the former as a *left neighboring* form. The two are always properly equivalent, since the first is transformed into the second by

$$x=-y',\qquad y=x'+\frac{b+b'}{c}y',\qquad \Delta=+1.$$

If (Art. 161) the form (a, b, c) contains (a', b', c'), the g.c.d. of a, b, c is a divisor of a', b', c', and that of $a, 2b, c$ is a divisor of $a', 2b', c'$.

Given (Art. 162) two substitutions (11) and (12) both of which replace $F=(A, B, C)$ by the same form $f=(a, b, c)$, and such that Δ, Δ' have the same sign, we see from Art. 157 that $D\Delta^2=D\Delta'^2$, whence $\Delta=\Delta'$. Employ the abbreviations

$$(13)\quad a'=A\alpha\alpha'+B(\alpha\gamma'+\gamma\alpha')+C\gamma\gamma',\qquad c'=A\beta\beta'+B(\beta\delta'+\delta\beta')+C\delta\delta',$$
$$2b'=A(\alpha\beta'+\beta\alpha')+B(\alpha\delta'+\beta\gamma'+\gamma\beta'+\delta\alpha')+C(\gamma\delta'+\delta\gamma').$$

If m is the g.c.d. of $a, 2b, c$, we can choose integers $\mathfrak{A}, \mathfrak{B}, \mathfrak{C}$ such that $\mathfrak{A}a+2\mathfrak{B}b+\mathfrak{C}c=m$. Write

$$T=\mathfrak{A}a'+2\mathfrak{B}b'+\mathfrak{C}c',\qquad U=\mathfrak{A}(\alpha\gamma'-\gamma\alpha')+\mathfrak{B}(\alpha\delta'+\beta\gamma'-\gamma\beta'-\delta\alpha')+\mathfrak{C}(\beta\delta'-\delta\beta').$$

An extended computation shows that

$$(14)\qquad T^2-DU^2=m^2.$$

Given a pair of integral solutions of (14) and given the first substitution (11), it is shown by computation that the coefficients of the second substitution (12), having $\Delta'=\Delta$, are such that

$$(15)\qquad \begin{aligned} m\alpha'&=\alpha T-(\alpha B+\gamma C)U, & m\beta'&=\beta T-(\beta B+\delta C)U,\\ m\gamma'&=\gamma T+(\alpha A+\gamma B)U, & m\delta'&=\delta T+(\beta A+\delta B)U.\end{aligned}$$

The resulting substitution (12) actually transforms F into f, and has $\Delta'=\Delta$. In case the determinants D and d of F and f are not equal, the coefficients $\alpha', \ldots, \delta'$ need not be integers (cf. Art. 214). But the latter are integers for all integral solutions T, U when $D=d$. [While Lebesgue[50] obtained (15) by a simpler calculation, the main difficulty was avoided by Dirichlet[57] by finding all transformations of F into itself. Cf. Gravé.[162]]

An *ambiguous* form (a, b, c) is defined (Art. 163) to be one for which $2b$ is divisible by a. It has (c, b, a) as a left-neighboring form and hence is properly equivalent to the latter. But, by means of $x=y'$, $y=x'$, (c, b, a) is improperly equivalent to (a, b, c), which is therefore improperly equivalent to itself. Hence a form F will contain another form F' both properly and improperly if there exists an ambiguous form which contains F' and is contained in F.

The converse (Art. 164) of this theorem is true; the proof is long. In particular (Art. 165), if F and F' are both properly and improperly equivalent, there exists an ambiguous form equivalent to each (proved also in Art. 194). A simpler proof was given by Dirichlet.[60]

If (Art. 166) F contains F', every number representable by F' can be represented by F. For, if (11) transforms F into F', and if $F'=M$ for particular values of x', y', then $F=M$ for the corresponding values of x, y given by (11). Different pairs of values of x', y' correspond to different pairs x, y. In particular, if F, F'

are equivalent, a number can be represented in as many ways by one form as by the other; then if x', y' are relatively prime, also x, y are relatively prime.

If (Art. 167) F, F' are equivalent and have the determinant D, and if F' is transformed into F by

$$x'=\alpha x+\beta y, \qquad y'=\gamma x+\delta y, \qquad \alpha\delta-\beta\gamma=\pm 1,$$

and if $F=M$ for particular relatively prime numbers x, y, and hence $F'=M$ for the preceding values x', y', also relatively prime, then both representations of M belong to the same root or to opposite roots of $v^2\equiv D \pmod{M}$, according as the above substitution is a proper or improper one, i. e., $\alpha\delta-\beta\gamma=+1$ or -1.

If (Art. 168) $M\neq 0$ is represented by $ax^2+2bxy+cy^2$ with relatively prime values m, n, of x, y, and this representation belongs (Art. 155) to the root $v=N$, then (a, b, c) is properly equivalent to $G=(M, N, (N^2-D)/M)$.

If (Art. 169) m', n' are relatively prime numbers giving a second representation of M by (a, b, c) belonging to the same root N, then (a, b, c) is transformed into G by the further proper substitution

$$x=m'x'+y'(m'N-m'b-n'c)/M, \qquad y=n'x'+y'(n'N+m'a+n'b)/M.$$

Conversely, given any proper transformation (11) of F into G, we see that M is represented by F for $x=\alpha$, $y=\gamma$, and, since $\alpha\delta-\beta\gamma=1$, the root to which the representation belongs is equal to N. Hence given all proper transformations of F into G, we obtain all representations of M by F belonging to the same root N. Thus, by Art. 162, given one representation $x=\alpha$, $y=\gamma$, of M by F, all representations belonging to the same root N are furnished by

$$x=\frac{\alpha T-(\alpha b+\gamma c)U}{m}, \qquad y=\frac{\gamma T+(\alpha a+\gamma b)U}{m},$$

where m is the g.c.d. of a, $2b$, c, while T, U range over all integral solutions of $T^2-DU^2=m^2$.

Forms of negative determinant $-D$, where D is positive, are treated in Arts. 171-181 [cf. Lagrange[18]]. Any form of determinant $-D$ is properly equivalent to a *reduced* form (A, B, C) in which A is neither greater than either $\sqrt{\frac{4}{3}D}$ or C, nor smaller than $2|B|$. Two distinct reduced forms (a, b, c) and (a', b', c') of the same determinant $-D$ are properly equivalent (Art. 172) if and only if they are opposite forms and at the same time either ambiguous (with $2b=\pm a$) or $a=c=a'=c'$.

The number (Art. 174) of reduced forms of determinant $-D$ is finite; two simple methods of finding them are given.

If (Art. 175) we omit from the list of reduced forms of a given determinant $-D$ one of each pair of distinct properly equivalent forms, we obtain a representative of each *class* of forms, such that two forms of any class are properly equivalent, while forms in different classes are not. Forms whose outer coefficients a, c are both negative constitute as many classes as do the forms with a, c both positive, and the types of classes have no form in common; the former classes may be omitted.

Given (Art. 178) two properly equivalent forms F_0, f_0 of the same negative determinant, we can find a proper transformation of F_0 into f_0 by employing a series of right neighboring forms which starts with F_0 and ends with a reduced form F_m, and a similar series $f_0, \ldots, f_n$. Then there are two cases. First, F_m and f_n are identical or both opposite and ambiguous. Then F_{m-1} is a left neighboring form to $\bar{f}_{n-1}=(c, -b, a)$ if $f_{n-1}=(a, b, c)$. Then $F_0, F_1, \ldots, F_{m-1}, \bar{f}_{n-1}, \bar{f}_{n-2}, \ldots, \bar{f}_0, f_0$ constitute a series of neighboring forms, from which we can find a proper transformation of F_0 into f_0 by an algorithm (Art. 177) employed for continued fractions. Second, if F_m and f_n are opposite and their four outer coefficients are equal, $F_0, \ldots, F_m, \bar{f}_{n-1}, \ldots, \bar{f}_0, f_0$ constitute a series of neighboring forms.

If F, f are improperly equivalent, the form f' opposite to f is properly equivalent to F. From a proper transformation of F into f', we obtain an improper transformation of F into f by changing the signs of the coefficients of the second variable.

If (Art. 179) F and f are equivalent, we can find all transformations of F into f. We employ Art. 178 to obtain one such transformation, or both a proper and an improper one if F, f are both properly and improperly equivalent. Then all transformations of F into f follow by Art. 162 from the integral solutions of $t^2+Du^2=m^2$, where m is the g.c.d. of the coefficients $A, 2B, C$ of F. Since $B^2-AC=-D$ is negative, the solutions are

$$\begin{array}{lll} u=0,\ t=\pm m, & & \text{if } 4D>4m^2; \\ u=0,\ t=\pm m; & u=\pm 1,\ t=0, & \text{if } 4D=4m^2; \\ u=0,\ t=\pm m; & u=1,\ t=\pm\tfrac{1}{2}m;\ u=-1,\ t=\pm\tfrac{1}{2}m, & \text{if } 4D=3m^2; \end{array}$$

while $4D=2m^2$ or $4D=m^2$ are impossible.

To find (Art. 180) all representations of a given integer M by $F=ax^2+2bxy+cy^2$ of determinant $-D$, with x, y relatively prime, we seek the incongruent roots $\pm N$, $\pm N', \ldots$ of $N^2\equiv -D \pmod{M}$ and treat in turn each root as follows: If F is not properly equivalent to $f=(M, N, (D+N^2)/M)$, there exists no representation of M belonging to the root N (Art. 168). But if they are properly equivalent, we seek (Art. 179) the proper transformations (11) of F into f. Then $x=\alpha$, $y=\gamma$ give the representations of M by F belonging to the root N.

To obtain (Art. 181) the representations of M by F in which the g.c.d. of $x=\mu e$, $y=\mu f$ is $\mu>1$, note that $x=e$, $y=f$ give a relatively prime representation of M/μ^2 by F, and hence are found by Art. 180.

Application of this theory is made (Art. 182) to prove the theorems stated by Fermat,[4] and proved by Euler[13, 22] and Lagrange,[18] on the representation of primes by $x^2+ky^2 (k=1, 2, 3)$.

Forms whose determinant D is a positive integer, not a square, are treated in Arts. 183–212. Every such form is properly equivalent to a *reduced* form (A, B, C) in which $|A|$ lies between $\sqrt{D}+B$ and $\sqrt{D}-B$, while $0<B<\sqrt{D}$. Now (Art. 184) many reduced forms are equivalent. If (a, b, c) is a reduced form, a and c have unlike signs, and $|c|$, as well as $|a|$, lies between $\sqrt{D}+b$ and $\sqrt{D}-b$. Hence also (c, b, a) is reduced. Further, b lies between $\sqrt{D}$ and $\sqrt{D}-|a|$. Every reduced form has a single right (or left) neighboring reduced form.

The number (Art. 185) of all reduced forms of a given positive determinant D is finite and they are readily found by either of two methods.

If (Art. 186) F is a reduced form, F' its unique reduced right neighboring form, F'' that of F', etc., this series contains a form $F^{(n)}$ identical with F. If n is the least such value >1, $F, F', \ldots, F^{(n-1)}$ are distinct and are said to constitute the *period* of F.

The number n (Art. 187) is always even. All reduced forms of determinant D may be separated into periods, no two of which have a form in common. Two forms, as $(a, b, -a')$ and $(-a', b, a)$, are called *associated* forms if their coefficients are the same but taken in reverse order. They determine two associated periods. If the form associated with F occurs in the period determined by F, the period is said to be associated with itself; a necessary and sufficient condition is that the period contain exactly two ambiguous forms.

We may determine (Arts. 188–9) a proper substitution which replaces a form by any form of its period by means of an algorithm used in the theory of continued fractions.

If (Art. 191) the reduced form $(a, b, -a')$ of determinant D is transformed into the reduced form $(A, B, -A')$ of determinant D by the substitution (11) with the coefficients $\alpha, \beta, \gamma, \delta$, then $(\pm\sqrt{D}-b)/a$ lies between α/γ and β/δ if $\gamma\delta\neq 0$, where the upper sign is to be taken when either both limits have the same sign as a, or one has the same sign as a and the other is zero, but the lower sign is to be taken when neither of the limits has the same sign as a. Likewise, $(\pm\sqrt{D}+b)/a'$ lies between γ/α and δ/β, with a similar determination of the sign.

In Art. 193 the preceding theorem is applied at length to prove the fundamental result that two properly equivalent reduced forms belong to the same period (simpler proofs by Dirichlet[57] and Mertens[105]).

Given (Art. 195) any two forms Φ and ϕ with the same determinant, we can decide whether or not they are equivalent by employing respective reduced forms F and f and observing whether or not F or its associated form occurs in the period of f, the equivalence of Φ and ϕ being proper or improper in the respective final cases. All forms of a given determinant constitute as many classes as there are periods, those which are properly equivalent constituting a class.

Given (Art. 196) two properly equivalent forms, we can find by the method of Art. 183 a proper substitution which transforms one of them into the other. From one such substitution we can find (Arts. 197, 203–4) all by Art. 162; we need the least positive solution of $t^2-Du^2=m^2$. This can be found (Arts. 198–9) if we are given any form (M, N, P) of determinant D such that the g.c.d. of $M, 2N, P$ is m. We pass to its reduced form $f=(a, b, -a')$ and determine its period $f, f', \ldots, f^{(n-1)}$, whence $f^{(n)}\equiv f$. By Arts. 188–9 we find a proper substitution replacing f by $f^{(n)}$. The same is true of the identity substitution $x=x'$, $y=y'$. From these two substitutions we obtain a solution t, u by Art. 162. The positive values of t, u are proved to give the least positive solution T, U. From the latter (Art. 200), we obtain all positive solutions t_e, u_e from

$$t_e\pm\sqrt{D}u_e=m\left(\frac{T}{m}\pm\sqrt{D}\frac{U}{m}\right)^e \qquad (e=0, 1, 2, \ldots).$$

We may now (Art. 205) find all representations of a given number by a given form by the method of Arts. 180–1.

Forms whose determinant is a square h^2 are treated in Arts. 206–212. Such a form is properly equivalent to one of the $2h$ reduced forms $(A, h, 0)$, $0\leqq A\leqq 2h-1$, no two of which are properly equivalent. If F, F' are properly equivalent forms of determinant h^2, we readily find one, and then every, proper substitution which transforms F into F'. Conditions for improper equivalence are obtained in Art. 210. All representations of a given number by a given form are readily determined (Art. 212).

Given (Art. 213) a form f of determinant D and a form F of determinant De^2, where e is an integer >1, to decide whether or not f contains F properly, i. e., can be transformed into F by a proper substitution (11) with $\alpha\delta-\beta\gamma=e>0$, we have only to decide whether or not F is properly equivalent to one of the $m_1+m_2+\dots$ forms derived from f by the substitutions

$$x=m_ix'+k_iy', \quad y=n_iy' \qquad (k_i=0, 1, \dots, m_i-1),$$

where m_1, m_2, ... denote the positive divisors (including 1 and e) of e, while $e=m_1n_1=m_2n_2=\dots$. If (Art. 214) f contains F properly, we can readily find all proper substitutions of f into F. A more practical solution was given by Arndt.[48] Cf. Pepin.[156]

A form (Art. 215) of determinant zero is expressible as $m(gx+hy)^2$, where g, h are relatively prime integers, so that the theory reduces essentially to that of a linear form.

Legendre[36] proved that every reduced form of determinant $-N$ represents at least one integer $<N$ and prime to N or $\frac{1}{2}N$.

G. L. Dirichlet[37] discussed in an elementary manner, in connection with the biquadratic character of 2 modulo p,

$$p=8n+1=t^2+2u^2=\phi^2+\psi^2,$$

where p is a prime and ψ may be taken to be a multiple of 4. If m is the g.c.d. of $\phi=m\phi'$ and $u=mu'$, it is shown that

$$t+\psi=EK, \quad t-\psi=FL, \quad m^2=EF, \quad \phi'^2-2u'^2=KL,$$

where E and F are relatively prime, whence E is a square and $E\equiv 1 \pmod 8$. Since K is an odd divisor of $\phi'^2-2u'^2$, $K\equiv 1$ or 7, $t+\psi\equiv 1$ or 7. Thus if $\psi\equiv 0$, $t\equiv 1$ or 7; but, if $\psi\equiv 4$, $t\equiv 3$ or 5 (mod 8). Cf. Jacobi.[45]

C. F. Gauss[38] noted that the positive definite * form $ax^2+2bxy+cy^2$ represents the square of the distance between any two points in a plane whose coordinates, with respect to two axes making an angle whose cosine is $b/\sqrt{ac}$, differ by $x\sqrt{a}$, $y\sqrt{c}$. Let x, y take only integral values. Then the form relates to a parallelogrammatic system of points which lie at the intersections of two sets of parallel lines. The lines of each set are at equal distances apart, one interval measured parallel to the lines of the second set being $\sqrt{a}$, and the other interval measured parallel to the lines of the first set being $\sqrt{c}$. Hence the plane is divided into equal parallelograms of area

* With a, c and $\delta=ac-b^2$ all positive, whence $af=(ax+by)^2+\delta y^2$, so that the form f represents only positive integers.

36 Theorie des nombres, ed. 2, 1808, pp. 407–411; ed. 3, 1830, II, pp. 80–85 (German transl. by Maser, II, pp. 79–84).

37 Jour. für Math., 3, 1828, 40; Werke, I, 69, 70.

38 Göttingische gelehrte Anzeigen, 1831, 1074; reprinted, Jour. für Math., 20, 1840, 318; Werke, II, 1863, 194. Cf. Klein.[149]

$\sqrt{ac-b^2}$, whose corner points constitute the set of points. Any such set of points can be arranged parallelogrammatically in infinitely many ways, the corresponding quadratic forms being all equivalent. If one form can be transformed into another, but is not equivalent to it, the second relates to a set of points forming only a part of the set to which the first form relates [compare their determinants]. For two improperly equivalent forms the parallelograms are equal but arranged in reverse order (as if the plane were folded over).

F. Minding[39] proved that a reduced form (a, b, c) of negative determinant, where $a, 2b, c$ have no common divisor, represents a given prime number in a single way, apart from changes of sign of x or y and, if $a=c$, also their interchange, unless $a=2b$, when from one representation we obtain a second by replacing x by $x+y$, y by $-y$. The proof is long and employs Legendre's[24] result that the least number represented by the reduced form is the smaller of a, c. He discussed forms of positive determinant (pp. 148–171) by the continued fraction for a root of a quadratic, somewhat as had Legendre,[25] but developed the theory more fully.

Dirichlet[40] recalled that Lagrange[20] was able to prove only in special cases his conjectured (converse) theorem that every prime $4n+1$ which is of one of the possible linear forms of the divisors of $f=t^2+cu^2$ is actually a divisor of f, and that Legendre[32] showed that this theorem depends upon the reciprocity law. For simplicity, Dirichlet restricts c to be $\pm p$, where p is a prime. Then the linear forms of the divisors of f constitute 1 or 2 groups according as $\pm p \equiv -1$ or $+1 \pmod 4$. The characteristic properties of the individual quadratic forms belonging to a group cannot be expressed by the linear forms of the primes represented by the quadratic form, but depend upon an element not previously introduced into the theory. Let $a=8k+1$ be a prime which is a quadratic residue of two primes p, q of the form $4n+1$. By a *quadratic divisor* $4n+1$ is meant one which represents no odd number not of the form $4n+1$. Hence p and q are each represented by one and but one quadratic divisor $4n+1$ of $F=t^2+au^2$. Assume that both p and q are represented by the same quadratic divisor. Then their product pq is known to be of the form F. By use of the reciprocity law, it is proved that

$$\left(\frac{t}{a}\right)=\pm\left(\frac{t}{p}\right)\left(\frac{t}{q}\right),\qquad \left(\frac{u}{p}\right)\left(\frac{u}{q}\right)=\pm 1,$$

where both signs are $+$ if p, q are both of the form $8n+1$ or both of the form $8n+5$, while the signs are unlike if one of p, q is of the form $8n+1$ and the other of the form $8n+5$. Two primes are said to be in ***biquadratic reciprocity*** if each is either a biquadratic residue of the other or each a biquadratic non-residue; but in ***biquadratic non-reciprocity*** if one prime is a biquadratic residue of the other, and the other a biquadratic non-residue of the first. Then the above result is shown to imply that p, q are either both in biquadratic reciprocity with a or both in biquadratic non-reciprocity with a. The complete theorem is the following: If a is a prime $8n+1$, the primes represented by the same quadratic divisor $4n+1$ of t^2+au^2 are all in biquadratic reciprocity with a or all in biquadratic non-reciprocity with a. In the respective cases a quadratic divisor $4n+1$ is said to be of the first or second type

[39] Anfangsgründe der Höheren Arithmetik, Berlin, 1832, 105–110.
[40] Abh. Akad. Wiss. Berlin, 1833, 101–121; Werke, I, 195–218.

(class). The divisor t^2+au^2 itself is proved to be of the first type. With Legendre, $(2a, \beta, \gamma)$ and $(a, \beta, 2\gamma)$ are called *conjugate divisors*. If the prime $a=8n+1$ is expressed in the form $\phi^2+\psi^2$, any two conjugate divisors $4n+1$ of t^2+au^2 are of the same type or different types according as $\phi+\psi\equiv\pm1$ or $\pm5 \pmod 8$. A self-conjugate divisor is of the form $4n+1$ or $4n+3$ according as $\phi+\psi\equiv\pm1$ or $\pm5 \pmod 8$.

A. Göpel[41] proved that if A is a prime $4n+3$ or its double, we can obtain the representation of A by $\phi^2\pm2\psi^2$ from the development of $\sqrt{A}$ into a continued fraction. To go more into details, let first A be a prime $8n+3$ or its double; then in the continued fraction for $\sqrt{A}$ occur always three successive complete quotients

$$(16)\qquad \frac{\sqrt{A}+I_0}{D_0},\qquad \frac{\sqrt{A}+I}{D},\qquad \frac{\sqrt{A}+I'}{D'},$$

in which D is $\frac{1}{2}D_0$, $\frac{1}{2}D'$ or $\frac{1}{2}(D_0+D')$, and $A=I^2+2D^2$ in the first two cases, while in the third case

$$A=\tfrac{1}{4}(I-I')^2+2D^2=\tfrac{1}{16}(D_0-D')^2+2D^2,$$

where $I-I'$ is even and D_0-D' is divisible by 4. Next, if A is a prime $8n+7$ or its double, the continued fraction for $\sqrt{A}$ has two successive complete quotients given by the first two numbers (16); then $D+D_0=2I$, $A=2I^2-\frac{1}{4}(D-D_0)^2$. This method is strictly analogous to that used by Legendre (this History, Vol. II, p. 233) to obtain $4n+1=D^2+I^2$. The method was further generalized by Stern.[61] Cf. Hermite,[53] Smith,[79, 108] Cantor,[84] and Roberts.[100]

G. L. Dirichlet[42] noted that the number of representations of a positive odd integer n by x^2+2y^2 is double the excess of the number of divisors $\equiv1$ or $3 \pmod 8$ of n over the number of divisors $\equiv-1$ or $-3 \pmod 8$. Also, the number of representations with $0\leqq 3y<2x$ of a positive odd number n by x^2-2y^2 is the excess of the number of divisors $\equiv\pm1 \pmod 8$ of n over the number of divisors $\equiv\pm3 \pmod 8$. These results were obtained as special cases of general theorems on representation by any quadratic form. He[43] later deduced them from the following: If n is an odd number prime to D and if $\sigma=1$ or 2, the number of all representations of σn by all the forms of a complete system of representative primitive forms of determinant D (properly or improperly primitive according as $\sigma=1$ or 2) is $\kappa\Sigma(D/\delta)$, summed for all the divisors δ of n, where (D/δ) is the Legendre-Jacobi symbol ±1, while $\kappa=1$ if $D>0$, $\kappa=4$ if $D=-1$, $\kappa=6$ if $D=-3$ and $\sigma=2$, $\kappa=2$ in all remaining cases.

---

41 De aequationibus secundi gradus indeterminatis, Diss., Berlin, 1835; reprinted, Jour. für Math., 45, 1853, 1–14. Report by Jacobi, *ibid.*, 35, 1847, 313–5; Werke, II, 1882, 145–152, who separated the $A<1000$ into three lists corresponding to the cases $D=\frac{1}{2}D_0$, $\frac{1}{2}D'$, $\frac{1}{2}(D_0+D')$; French transl., Jour. de Math., 15, 1850, 357–362; Nouv. Ann. Math., 12, 1853, 136–8 (170–1, where Lebesgue remarked that the solvability of $A=\phi^2\pm2\psi^2$ was proved otherwise by Legendre, Théorie des nombres, ed. 3, 1830, I, pp. 305–6.

42 Jour. für Math., 21, 1840, 3, 6; Werke, I, 463, 466. The same results were deduced similarly by H. Suhle, De quorundam theoriae numerorum, Diss. Berlin, 1853; and by L. Goldschmidt, Beiträge zur Theorie der quadratischen Formen, Diss. Göttingen, Sondershausen, 1881, who deduced expressions for the number of lattice points in an ellipse or hyperbola.

43 Zahlentheorie, § 91, 1863; ed. 2, 1871, p. 226; ed. 3, 1879, 228; ed. 4, 1894, 229.

A. L. Cauchy[44] derived from the theory of elliptic functions the identity

$$(1+2t+2t^4+2t^9+\dots)(1+2t^3+2t^{12}+2t^{27}+\dots)$$
$$\equiv 1+2\left(\frac{1-t}{1-t^3}t+\frac{1+t^2}{1+t^6}t^2+\frac{1-t^3}{1-t^9}t^3+\frac{1+t^4}{1+t^{12}}t^4+\dots\right)$$

and stated that it implies that the number of sets of positive, negative, or zero integral solutions x, y of $x^2+3y^2=n$ is

$$N=(-1)^{n+1}\sum(-1)^{a+n/a}\frac{\sin 2\pi a/3}{\sin 2\pi/3},$$

summed for all divisors a of n. If n is not divisible by 3 and if n has an odd number of divisors, then N is odd. If n is an odd prime, the formula gives $\frac{1}{2}N=1\pm 1$, where the sign is the same as in $n\equiv\pm 1 \pmod 3$. Cf. Genocchi.[56]

C. G. J. Jacobi[45] noted that Dirichlet's[37] theorem is equivalent to the following: If a prime $8h+1$ is represented by the two forms

$$f=(4m+1)^2+16n^2, \qquad g=(4m'+1)^2+8n'^2,$$

then $m+n$ and n' are both even or both odd. From infinite series arising from elliptic functions, this theorem appears in the following more general form. For every number P, the excess of the number of solutions of $f=P$ in which $m+n$ is even over the number in which $m+n$ is odd is equal to the excess of the number of solutions of $g=P$ in which n' is even over the number in which n' is odd. Other theorems are said to follow for pairs of forms x^2+ky^2, but are not explicitly stated. However, at the end of the paper is a collection of formulas expressing equality of two infinite series in q whose exponents are two different ones of the forms $x^2+ky^2\,(k=1, 2, 3, 6)$, or $x^2+ky^2\,(k=1, 2)$ with $2x^2+3y^2$, etc.

L. Wantzel[46] established a g.c.d. process for numbers $x+y\sqrt{c}$, where x, y, c are integers, $c=-3, -2, -1, 2, 3, 5$, and thence proved that a prime p which divides x^2-cy^2 is of that form, except when $p=2$, $c=-3$, $c=5$ [and $c=3$]. Likewise for x^2+7y^2, x^2-13y^2, x^2-17y^2.

P. L. Tchebychef[47] made a complete determination of the linear forms of the divisors of $x^2\pm ay^2$. These linear forms are tabulated for $a=1, \dots, 101$, but with practically the same errata (corrected in the 1901 edition) as in Legendre's[25] table.

F. Arndt[48] gave a new treatment of the problem of Gauss (Arts. 213–4, with f, F interchanged), since his solution involved impracticable computations. Let $F=(A, B, C)$, of determinant D, be transformed into $f=(a, b, c)$, of determinant De^2, by a substitution with the coefficients $\alpha, \beta, \gamma, \delta$, where $\alpha\delta-\beta\gamma=e>1$. Let n be the positive g.c.d. of $\alpha=n\alpha_0$ and $\gamma=n\gamma_0$. Then $e=nm$. Choose integers β_0, δ_0 so that $\alpha_0\delta_0-\beta_0\gamma_0=1$. Then $\beta=\beta_0 m+k\alpha_0$, $\delta=\delta_0 m+k\gamma_0$. We may assume that β_0, δ_0 were

[44] Comptes Rendus Paris, 19, 1844, 1385 (17, 1843, 580, with the four plus signs in the second and fourth fractions changed to minus); Oeuvres, (1), VIII, 384 (64).

[45] Jour. für Math., 37, 1848, 61–94, 221–254; Werke, II, 1882, 217–288. Cf. H. J. S. Smith, Report Brit. Assoc. for 1865, 322, seq., Arts. 128-9; Coll. Math. Papers, I, 311–321.

[46] Soc. Philomatique de Paris, Extraits des Procès-Verbaux des Séances, 1848, 19–22.

[47] Theorie der Congruenzen, in Russian, 1848, 1901; in German, 1889, 209–237, 255–272.

[48] Archiv Math. Phys., 13, 1849, 105–112.

chosen so that $0 \leqq k \leqq m-1$. From the expressions for a, b, c in terms of $A, B, C, \alpha, \ldots, \delta$, we get

$$A\alpha_0^2+2B\alpha_0\gamma_0+C\gamma_0^2=\frac{a}{n^2}\equiv A', \quad A\alpha_0\beta_0+B(\alpha_0\delta_0+\beta_0\gamma_0)+C\gamma_0\delta_0=\frac{bn-ak}{mn^2}\equiv B',$$
$$A\beta_0^2+2B\beta_0\delta_0+C\delta_0^2=\frac{cn^2-2bkn+ak^2}{m^2n^2}\equiv C'.$$

Thus F is transformed into $F'=(A', B', C')$ by the substitution with the coefficients $\alpha_0, \beta_0, \gamma_0, \delta_0$ of determinant unity. By the values of A', B', n divides $b=nb'$.

Hence if we choose any divisor n of $e=nm$ whose square divides $a=n^2A'$ and hence also $b=nb'$ (in view of $b^2-ac=De^2$), and if we find that $(b'-A'k)/m$ and $(c-2b'k+A'k^2)/m^2$ are not both integers (B' and C') for some value of k between 0 and $m-1$, inclusive, we conclude that f is not contained in F. But if there exist values of n, k for which the preceding conditions of divisibility are all satisfied, and if one of the forms $F'=(A', B', C')$ is properly equivalent to F, then and only then is f contained in F properly. Furthermore, if we obtain all proper transformations $\alpha_0, \beta_0, \gamma_0, \delta_0$ of F into F', all proper transformations of F into f have the coefficients $n\alpha_0$, $m\beta_0+k\alpha_0$, $n\gamma_0$, $m\delta_0+k\gamma_0$, and all these transformations are distinct.

Ch. Hermite[49] gave an elementary proof by continued fractions that if p is any divisor of x^2+Ay^2, a suitably chosen power of p can be represented by x^2+Ay^2.

V. A. Lebesgue[50] obtained the formulas (15) of Gauss (Art. 162) by a simpler calculation, but under the assumption that the determinants Δ, Δ' of the two substitutions which replace F by f are both equal to ± 1. From the two sets of values of a, b, c in terms of A, B, C, we find that

$$AU=m(\delta\gamma'-\delta'\gamma), \qquad 2BU=m(\alpha\delta'-\alpha'\delta+\beta'\gamma-\beta\gamma'), \qquad CU=m(\beta\alpha'-\beta'\alpha),$$

for some integer U, since the quantities in parenthesis are proportional to A, $2B$, C, whose g.c.d. is denoted by m. Write $2T$ for $m(\alpha\delta'+\alpha'\delta-\beta'\gamma-\beta\gamma')$. Solving these equations, we obtain Gauss' formulas (15) with m replaced by $\pm m$ and

$$T+BU=m(\alpha\delta'-\beta\gamma'), \qquad T-BU=m(\alpha'\delta-\beta'\gamma), \qquad T^2-DU^2=m^2.$$

G. L. Dirichlet[51] employed Gauss'[38] geometrical representation of a positive binary quadratic form $lx^2+2mxy+ny^2$. A fundamental parallelogram is called *reduced* if no one of its sides exceeds either diagonal. Given any lattice whose points are the intersections of two systems of equidistant parallel lines, we can arrange the points parallelogrammatically so that the fundamental parallelogram $POQR$ is reduced. Without loss of generality we may evidently assume that angle POQ is not obtuse and that $OP \leqq OQ$. Write $OP=\sqrt{l}$, $OQ=\sqrt{n}$. Then the minimum distance of points of the lattice from O is $\sqrt{l}$. If P is one of the points at this distance, the (second) minimum distance from O of points not on OP is $\sqrt{n}$. The first minimum occurs only at P, and the second only at Q or at points P', Q' symmetrical with them

[49] Jour. de Math., 14, 1849, 451–2; Oeuvres, I, 274–5.
[50] Nouv. Ann. Math., 8, 1849, 83–86. We interchange a and A, b and B, c and C, to accord with Gauss' notations.
[51] Jour. für Math., 40, 1850, 213–220; Werke, II, 34–41.

with respect to O, except when a side is equal to a diagonal or another side. Write $\cos POQ=m/\sqrt{ln}$. Then

$$0 \leqq 2m \leqq l, \qquad 2m \leqq n,$$

which are Lagrange's[18] conditions for a reduced form. The points of the plane which are nearer to O than to any other lattice point are the points (and no others) which lie inside the hexagon whose sides are perpendicular bisectors of OP, OQ, OS, OP', OQ', OS', where OS is parallel and equal to PQ, and S' is symmetrical to S with respect to O.

P. L. Tchebychef[52] noted that Euler used forms of negative determinants to test for primes. In practice, it is simpler to use positive determinants D. Let x^2-Dy^2 be a form all of whose quadratic divisors are of the form $\lambda x^2-\mu y^2$. Let N be a number prime to D and having the form of a linear divisor contained in a single quadratic form $f=\pm(x^2-Dy^2)$. Then N is a prime if, within the limits

$$(17) \qquad 0 \leqq x \leqq \sqrt{\frac{(a\pm 1)N}{2}}, \qquad 0 \leqq y \leqq \sqrt{\frac{(a\mp 1)N}{2D}},$$

where a is the least $x>1$ satisfying $x^2-Dy^2=1$, there is a single representation of N by f, and if, in this representation, x and y have no common factor. In all other cases, N is composite. He tabulated, for each $D \leqq 33$ without a square factor, the limits (17) and the linear forms of N.

Ch. Hermite[53] proved that any form $f=ax^2+2bxy+cy^2$ with real coefficients of negative determinant $b^2-ac=-D$ is equivalent (under linear transformation with integral coefficients of determinant unity) to a reduced form $F=AX^2+2BXY+CY^2$ in which $2B$ is numerically less than A and C. From the forms obtained from f by all such transformations select the set in which the coefficient of X^2 is a minimum and from this set select the form F in which the coefficient of Y^2 is a minimum; then F is reduced. The proof is like that by Legendre[24] (for the case of integral coefficients). If $B>0$, the first, second and third minima of $AX^2-2BXY+CY^2$ for integral values of X, Y are A, C, $A-2B+C$.

Next, let $f=a(x+\alpha y)(x+\alpha' y)$ have a positive determinant, so that α, α' are real. With f associate the positive definite form

$$\phi=(x+\alpha y)^2+\lambda(x+\alpha' y)^2,$$

where λ is a positive real variable. The totality (f) of reduced forms is defined to be the set of forms obtained by applying to f all the substitutions with integral coefficients of determinant unity which replace ϕ by a reduced form when λ varies continuously from 0 to $+\infty$. To carry out this *continual reduction* of ϕ, start with a reduced form r of ϕ whose extreme coefficients (the first two minima of ϕ) are distinct. When λ, increasing continuously, reaches a value beyond which r ceases to reduced, on account of the interchange of the second and third minima, one of the substitutions P: $x=X+Y$, $y=Y$, or P^{-1} will reduce r. But if there was an interchange of the first and second minima, the substitution $Q: x=Y, y=-X$ will reduce r. Hence the forms in (f) are obtained from f by a succession of these substitutions.

---

[52] Jour. de Math., 16, 1851, 257–282; Oeuvres, I, 73–96. Exposition by Mathews.[137]
[53] Jour. für Math., 41, 1851, 193–5, 203–213 (§§ III, VII–XI); Oeuvres, I, 167–8, 178–189.

A reduced form in (f) is called *principal* or *intermediate*, according as the extreme coefficients of the corresponding ϕ are equal or distinct. The principal reduced forms correspond to those of Gauss.

Let the coefficients of f be integers. Since the coefficients of the forms in the set (f) are limited (Hermite[1] of Ch. XIV), there is only a finite number of reduced forms (f). Hence in the continual reduction of ϕ we ultimately reach a form already obtained and the set (f) is composed of a finite period of forms repeated an infinitude of times. Each form of the period is a right neighbor to its predecessor. While these periods are not exactly the same as the periods with Gauss (Art. 187), they may be computed similarly. The method of continual reduction leads also to all the transformations into itself of a reduced form in a more natural manner than by Gauss (Art. 162).

In § XI it is shown that D is representable by $2x^2+y^2$ if $a^2-Db^2=-2$ is solvable, and similar theorems (cf. Göpel[41]).

V. Bouniakowsky[54] took the residues modulo 4 of the terms of his relation (10), p. 284 of Vol. I of this History. Hence

$$\sigma(n)+\sigma(2)\sigma(n-2)+\sigma(3)\sigma(n-4)+\ldots+\sigma(k+1)\sigma(n-2k)+\ldots\equiv M \pmod 4,$$

where $\sigma(n)$ is the sum of the divisors of n. Let $n+2$ be the prime $P=16e+7$. Then $M\equiv 2 \pmod 4$. Then no two terms on the left are odd, whence one term, say $\sigma(k+1)\sigma(q)$, must be $\equiv 2 \pmod 4$, where $q=P-2-2k$. Thus one of the factors is odd. Either q is not a square, and $k+1=ra^2$, where $r=1$ or 2, whence $\sigma(q)\equiv 2 \pmod 4$, $q=Qc^2$, or *vice versa*, $q=b^2$, $k+1=rQc^2$. Here Q is a prime $4l+1$. Hence P is of one of the forms $2u^2+Qv^2$, u^2+Qv^2, u^2+2Qv^2, the last two being excluded by the reciprocity law.* Hence every prime $16e+7$ is of the form $2u^2+Qv^2$, where Q is a prime $8l+5$. This is said to establish a relation of equality between two primes, no such case being previously known, while the reciprocity law is merely a relation of congruence between primes. J. Liouville[55] noted that a similar method shows that the double of a prime $8\mu+3$ is always expressible as the sum of a square and the product of another square by a prime $8k+5$.

A. Genocchi[56] wrote N_2 for the number of sets of positive integral solutions of $x^2+3y^2=n$ and N_1 for the number of sets in which one unknown is zero. Then Cauchy's[44] identity gives

$$N_1+2N_2=(-1)^n(d_1-d_2-d_3-d_4),$$

where d_1 (or d_2) is the number of divisors d of n whose complementary divisors n/d are of the form $3m+1$ and differ from them by an odd (or even) number, while d_3 (or d_4) is the number of divisors whose complementary divisors are of the form $3m+2$ and differ from them by an odd (or even) number. Similarly,

$$(1+2t+2t^4+2t^9+\ldots)(1+2t^2+2t^{2.4}+2t^{2.9}+\ldots)$$
$$\equiv 1+2\left(\frac{t}{1-t}+\frac{t^3}{1-t^3}-\frac{t^5}{1-t^5}-\frac{t^7}{1-t^7}+\ldots\right)$$

* Or at once by taking residues modulo 8.

54 Mém. Acad. Sc. St. Pétersbourg (Sc. Math. Phys.), (6), 5, 1853, 319-320 (being Pt. I of tome VII of the full series VI of Sc. Math. Phys. Nat.).

55 Jour. de Math., (2), 2, 1857, 424; proof, (2), 3, 1858, 84-88 (249 for a proof of Bouniakowsky's theorem; cf. History, Vol. II, p. 331).

56 Nouv. Ann. Math., 13, 1854, 167-8.

implies for $x^2+2y^2=n$ that $N_1+2N_2=d_1+d_3-d_5-d_7$, where d_j is the number of divisors of the form $8n+j$ of n (cf. Stieltjes[112]).

G. L. Dirichlet[57] simplified the theory of forms $f=(a, b, c)$ whose determinant D is positive and not a square. Of the two roots $(-b\mp\sqrt{D})/c$ of $a+2b\omega+c\omega^2=0$, that with the upper sign is called the *first root* belonging to f, and that with the lower sign the *second root*. One root and D completely determine f. Let a substitution

$$(18)\qquad \begin{pmatrix}\alpha & \beta\\ \gamma & \delta\end{pmatrix}:\quad \begin{matrix}x=\alpha X+\beta Y,\\ y=\gamma X+\delta Y,\end{matrix}\qquad \alpha\delta-\beta\gamma=1,$$

replace f by the (properly) equivalent form $F=(A, B, C)$. Then if ω and Ω are both first roots or both second roots of f, F,

$$\omega=\frac{\gamma+\delta\Omega}{\alpha+\beta\Omega}.$$

Conversely, this relation and $\alpha\delta-\beta\gamma=1$ imply that the substitution (18) replaces f by F. The form f is called *reduced* if the absolute value of the first root exceeds unity and that of the second root is less than unity and if the roots have opposite signs. Then $0<b<\sqrt{D}$, ac is negative, and the first root is of the same sign as a.

A reduced form (a, b, a') has a unique reduced right neighboring (contiguous) form (a', b', a''). Here b' is the unique integer between $\sqrt{D}$ and $\sqrt{D}-|a'|$ for which $b'\equiv -b \pmod{a'}$. Writing $b'=-b-a'\delta$, we have $\omega=\delta-1/\omega'$, where ω, ω' are the first roots of the two forms. Hence any form determines a period of $2n$ forms (as in Gauss, Arts. 186–7). Choose the initial form ϕ_0 of the period so that its first coefficient is positive. Consider two consecutive forms $\phi_\nu, \phi_{\nu+1}$ of the period. Then the sign of the first root ω_ν of ϕ_ν is that of $(-1)^\nu$. Let k_ν denote the greatest integer $<|\omega_\nu|$ and write $\delta_\nu=(-1)^\nu k_\nu$. Then

$$\omega_\nu=\delta_\nu-\frac{1}{\omega_{\nu+1}},\quad (-1)^\nu\omega_\nu=(k_\nu, k_{\nu+1}, \ldots, k_{2n+\nu-1};\ k_\nu, \ldots)\equiv k_\nu+\frac{1}{k_{\nu+1}+\ldots},$$

which is a periodic continued fraction. If two reduced forms ϕ_0, Φ_0, whose first coefficients are positive, are equivalent under a substitution (18), their first roots ω_0, Ω_0 are such that

$$\omega_0=(k_0, k_1, \ldots),\qquad \Omega_0=(K_0, K_1, \ldots),\qquad \omega_0=\frac{\gamma+\delta\Omega_0}{\alpha+\beta\Omega_0},$$

where the k's and K's are positive. It is shown by the last relation that $\omega_0=(\lambda, m, \ldots, r, \sigma, \Omega_0)$, where $m, \ldots, r$ are positive and even in number. Inserting Ω_0, we can give to the continued fraction for ω_0 its normal form in which all elements are positive. If none of the elements following K_ν were disturbed by the normalization, the number of elements preceding K_ν was varied by an even number $2h$. Since the final result is identical with $(k_0, k_1, \ldots)$, we have $K_\nu=k_{2g+2h+\nu}$. It follows readily that $\Phi_0=\phi_{2m}$ for a certain minimum $2m$, so that Φ_0 is in the period of ϕ_0. This

[57] Abh. Akad. Wiss. Berlin, 1854, 99–115; French transl., Jour. de Math., (2), 2, 1857, 353 (with additions); Werke, II, 139–158, 159–181. Zahlentheorie, §§ 72–82, 1863; ed. 2, 1871, pp. 170–197; ed. 3, 1879, pp. 172–199; ed. 4, 1894, pp. 173–200.

furnishes a simple proof of the most difficult theorem of the theory: forms belonging to different periods are not equivalent (Gauss,[35] Art. 193).

Given one transformation of f into F, we may evidently reduce the problem to find all the transformations of f into F to the simpler problem of finding all transformations of f into itself [an important logical advance over Gauss,[35] Art. 162]. The details are quoted on p. 376 of Vol. II of this History.

The number of forms in the period of ϕ_0 is either equal to the number of k's in the least period in the continued fraction for ω_0 or is double that number, according as $(a, b, c)=\phi_0$ is not or is equivalent to $(-a, b, -c)$ [Zahlentheorie, § 83, long foot-note].

G. Oltramare[58] proved that every divisor of a^2+kb^2 of like form can be expressed in one and but one way in the form x^2+ky^2 where x, y constitute a solution of $ax+bky=z(x^2+ky^2)$, with x relatively prime to y and k. Every prime μ or one of its multiples $g\mu$ ($g<2\sqrt{\frac{1}{3}k}$ if $k>0$, $g<1-k$ if $k<0$) can be expressed in the form x^2+ky^2 if $-k$ is a quadratic residue of μ, but no multiple of μ can be so expressed if $-k$ is a quadratic non-residue [original erroneous, p. 160]. Various special cases are noted.

V. A. Lebesgue[59] noted that if $ax^2+2bxy+cy^2$ is known to take its minimum value a' for $x=m$, $y=n$, we can find at once the reduced form. Let $mn_0-m_0n=1$, $x=mx'+m_0y'$, $y=nx'+n_0y'$. We get $f'=a'x'^2+\dots$ with $a'\leqq c'$, by hypothesis. If $a'\gtreqless 2b'$, f' is reduced. In the contrary case, replace m_0, n_0 by $m_0+m\nu$, $n_0+n\nu$, and determine ν so that $(a', b'+a'\nu, c'')$ is reduced. But if a' is not given, use the method of reduction due to Gauss[35] (Arts. 171, 177).

G. L. Dirichlet[60] gave a very simple proof of the theorem of Gauss (Arts. 164–5): Given an improper substitution (of determinant -1) of a form into itself, we can always find an equivalent ambiguous form (i. e., one in the same class and having its middle coefficient $2b$ divisible by the first coefficient a).

A. Stern[61] noted that while Göpel[41] limited the statement of his theorem to the case in which A is a prime $8n+3$ or its double, his proof leads to the generalization that if 2 is the middle term of the continued fraction for $\sqrt{A}$ and if there is an even number of terms in the period of partial denominators, we may deduce the explicit values of x, y in one of the representations (the only one in Göpel's case) of A by x^2+2y^2. Proof is given (p. 78) by continued fractions of the known theorem that any prime expressible in the form x^2+ky^2, $k>0$, has only one such representation. Göpel distinguished three types of numbers represented by x^2+2y^2. Similarly (p. 82) there are three types for x^2-2y^2 and (p. 97) six types for x^2+3y^2.

R. Lipschitz[61a] proved that every primitive form of determinant Dd^2 is contained in a form of determinant D.

G. Mainardi[62] gave a more direct solution than had Gauss (Art. 162) of the problem to find all transformations of one form into another, given one such transformation.

[58] Jour. für Math., 49, 1855, 142–160.
[59] Jour. de Math., (2), 1, 1856, 403–5.
[60] Jour. de Math., (2), 2, 1857, 273–6; Werke, II, 209–214.
[61] Jour. für Math., 53, 1857, 54–102.
[61a] Jour. für Math., 53, 1857, 238.
[62] Atti Institute Lombardo Sc. Let. ed Arti. Milan, 1, 1858, 106–7.

V. A. Lebesgue[63] proved that a prime cannot be represented in two ways by x^2+ky^2. If $x^2+k=py$ ($k=1$, 2, or 3), where p is a prime [$p>2$ if $k=3$], he proved by developing x/p into a continued fraction that every divisor of x^2+k is of the form a^2+kb^2. [For $k=1$, Hermite, p. 237 of Vol. II of this History.]

A. M. Legendre[64] stated in effect, but attempted no proof, that a quadratic form which represents all the numbers represented by another form can be transformed into the latter. Cf. Schering,[65] Bauer.[146]

E. Schering[65] noted that the last theorem may fail unless the forms are properly primitive. Proof is made by use of the composition of forms and of Dirichlet's theorem on the infinitude of primes in an arithmetical progression. Call the g.c.d. O of A, B, C the *order* of (A, B, C). The latter is of the Eth *kind* if OE is the g.c.d. of $A, 2B, C$, whence $E=1$ or 2, according as $(A/O, B/O, C/O)$ is properly or improperly primitive.

Theorem I. If (a, b, c) is of order o, determinant d, and of the eth kind, and represents all the numbers which can be represented by (A, B, C), which is of order O, determinant D, and of the Eth kind, then EO is divisible by eo, and $E^2D/(e^2d)$ is the square of an integer.

Theorem II. In order that the form $f=(2a, 2b/e, 2c)$ of order $2/e$, determinant d, and of the eth kind, shall represent all numbers which can be represented by $F=(2A, 2B/E, 2C)$ of order $2/E$, determinant D, and of the Eth kind, it is necessary and sufficient that f contain F when $E \leqq e$; but if $e=1$, $E=2$, it is necessary that, when a, b are assumed odd (as may be assumed without loss of generality), $(2a, b, c/2)$ contain F, that the number of properly primitive classes of determinant $d/4$ shall not exceed the number of improperly primitive classes of the same determinant, and that $D \not\equiv 1 \pmod 8$.

Theorem III. A necessary and sufficient condition that the forms in Theorem I shall represent the same numbers is that at least one of the forms shall contain the other, that $oe=OE$, $e^2d=E^2D$, and that, if $e \neq E$, D/O^2 be of the form $8k+5$, and that for this number as determinant there are as many improperly primitive as properly primitive classes.

J. Liouville[66] stated that the double of any prime $24\mu+7$ can be expressed in an odd number of ways by $x^2+q^{4l+1}y^2$, where x, y are positive odd integers, and q is a variable prime $24\nu+3$ not dividing y; also similar theorems.

Liouville[67] stated that if m is a prime of either of the forms $16k+7$ or $16k+11$, there exists at least one pair of distinct primes p, q of the form $8\nu+3$ such that

$$2m=p^{4\alpha+1}x^2+q^{4\beta+1}y^2$$

has positive odd solutions prime to p, q. If there are several such pairs p, q, their number is odd when q, p is not distinguished from p, q. If[68] $m=4\mu+1$ and

[63] Exercises d'analyse numérique, 1859, 109–112.

[64] Théorie des nombres, ed. 3, 1830, I, p. 237–8. Granted that, for every pair of relatively prime integers y and z, we can find integers t and u for which $t^2+u^2=2fy^2+2gyz+fz^2$ [false if $f=y=3$, $z=1$] Legendre stated without proof that the equation becomes an identity in y, z for $t=Ay+Bz$, $u=My+Nz$, where A, B, M, N are integers.

[65] Jour. de Math., (2), 4, 1859, 253–270; Werke, I, 87–102.

[66] Jour. de Math., (2), 4, 1859, 399, 400; (2), 6, 1861, 28–30.

[67] *Ibid.*, (2), 5, 1860, 103–4.

[68] *Ibid.*, 119–120; (2), 7, 1862, 19–20.

$p=8n+1$ are primes, the number of sets of primes p and odd integers x, y not divisible by p of $2m=x^2+p^{4l+1}y^2$ is $\equiv \mu+\sigma+1 \pmod 2$, where σ denotes the number of primes $4s+1$ which divide ab, where $2m=a^2+b^2$; he stated also a similar theorem. He[69] stated many theorems of the following type: Any prime $8\mu+5$ can be expressed in an odd number of ways in the form $2x^2+p^{4l+1}y^2$, x, y odd, p a variable prime $8\nu+3$ not dividing y. He[70] stated several theorems of the type that the product of any prime $8k+3$ by any prime $8h+5$ can be expressed in an old number of ways in the form $x^2+q^{4l+1}y^2$, x, y odd, q a variable prime $8\mu+3$ not dividing y.

Liouville[71] stated that the quadruple of a prime $8\mu\pm 3$ can be expressed in an odd number of ways in the form $x^2+p^{4l+1}y^2$, $x\equiv \pm 1 \pmod 8$, y odd and positive, p a variable prime $8\nu+3$ not dividing y.

He[72] stated that, if N is the number of decompositions (disregarding the signs of x, y) of a prime m into $4x^2+q^{4l+1}y^2$, x, y odd, q a variable prime $8\nu+5$ not dividing y, then $N\equiv b \pmod 2$, where $m=a^2+8b^2$, uniquely; also many similar theorems.

He[73] considered the number N of decompositions of the product of a prime $m=8\mu+3$ by the square of a prime $a=8\nu+7$ into the form $x^2+2p^{4l+1}y^2$, where x is not divisible by a, and y not divisible by the variable prime p, and stated that N is odd or even according as a is a quadratic residue or non-residue of m.

He[74] stated that if m is a prime $8\mu+3$, m^4 can be expressed an odd number of ways in the form $16x^2+p^{4l+1}y^2$, x, y positive, p a variable prime not dividing y; likewise for $2m^2=x^2+p^{4l+1}y^2$, x, y odd.

Finally,[75] he stated that if m is any prime $20k+3$ or $20k+7$, $8m$ can be expressed in an odd number of ways in the form $5x^2+p^{4l+1}y^2$, x, y odd and positive, p a variable prime not dividing y; and likewise for $x^2+5p^{4l+1}y^2$.

H. J. S. Smith[76] gave a résumé of the work of Gauss. In § 89 he pointed out the basis of the theorem of Gauss (Art. 162), the principles underlying whose proof are concealed. Let $f=(a, b, c)$ be transformed into $F=(A, B, C)$ by two substitutions

$$X_0=\alpha_0x+\beta_0y,\ Y_0=\gamma_0x+\delta_0y; \qquad X_1=\alpha_1x+\beta_1y,\ Y_1=\gamma_1x+\delta_1y,$$

of equal determinants, so that $f(X_0, Y_0)=f(X_1, Y_1)=F(x, y)$. In view of the relation between the first roots ω, Ω of f and F (Dirichlet[57]), it follows that

$$\frac{\gamma_0+\delta_0\Omega}{\alpha_0+\beta_0\Omega}=\frac{\gamma_1+\delta_1\Omega}{\alpha_1+\beta_1\Omega}$$

is the same equation as $A+2B\Omega+C\Omega^2=0$. In other words, $X_0Y_1-X_1Y_0$ is the product of $F(x, y)$ by a constant. Another function with the same property is

$$G=aX_0X_1+b(X_0Y_1+X_1Y_0)+cY_0Y_1,$$

as follows from the formula of composition of $f(X_0, Y_0)$, $f(X_1, Y_1)$ in Gauss (Art.

---

69 Jour. de Math., (2), 5, 1860, 139–140, 300–2, 309–312, 387–392; (2), 6, 1861, 7–8; (2), 7, 1862, 17–18; (2), 8, 1863, 137–140.
70 *Ibid.*, (2), 5, 1860, 303–4; (2), 6, 1861, 185–206; (2), 7, 1862, 21–22.
71 *Ibid.*, (2), 6, 1861, 1–6, 93–96; (2), 8, 1863, 85–88, 102–4; (2), 9, 1864, 135–6.
72 *Ibid.*, (2), 6, 1861, 31–32, 55–56, 97–112, 147–152, 219–224.
73 *Ibid.*, 207–8.
74 *Ibid.*, (2), 7, 1862, 23–24, 136.
75 *Ibid.*, (2), 9, 1864, 137–144.
76 Report British Assoc. for 1861, 292–340; Coll. Math. Papers, 1, pp. 163–207.

229, report in Ch. IV). Let m be the positive g.c.d. of A, $2B$, C. Let U, T be the g.c.d.'s of the coefficients of x^2, xy, y^2 in $X_0Y_1-X_1Y_0$ and G respectively, the signs of U, T being such that

$$\frac{F(x,y)}{m}=\frac{X_0Y_1-X_1Y_0}{U}=\frac{G}{T}.$$

We obtain at once Gauss' formulas (14), (15), expressed in the present notations.

There is given (§ 93) a treatment of forms of positive determinant not materially different from that of Dirichlet.[57]

G. Skrivan[77] gave other proofs of the elementary results of Gauss (Art. 154, and final paragraph of Art. 166).

L. Kronecker[78] proved that if D, D', ... are the distinct numbers for which an odd prime p can be represented by x^2+Dy^2, $x^2+D'y^2$, ..., with y odd, and if (a_1, b_1, c_1), (a_2, b_2, c_2), ... give all the reduced properly primitive, positive forms of determinants $-D$, $-D'$, ..., then the congruences $a_iz^2+2b_iz+c_i\equiv 0 \pmod p$ have p distinct roots, each occurring twice, provided we count as two each root of those congruences in which the coefficient of z^2 does not have the same absolute value as one of the remaining two coefficients. If $p=4n+3$, there is no properly primitive ambiguous form (with $a=\pm 2b$ or $a=c$) having one of the determinants $-D$, $-D'$, ..., so that the roots form a complete set of residues modulo p.

If p is an odd prime and d is a positive integer $<\sqrt{p}$ and if (a_i, b_i, c_i) are the positive reduced forms of determinant d^2-p in which one of the outer coefficients is odd and the middle coefficient is $\gtreqless 0$, then the expressions $(\pm b\mp d)/a$ form a complete set of residues modulo p if we admit the four combinations of signs except for $a=2b$ (when we take only the minus sign before b) and for $a=c$ (when we take only the upper signs). This theorem is said to admit of generalization to a composite p.

A summary of the rest of the paper is given in Ch. VI.

H. J. S. Smith[79] gave an exposition of the results of Gauss[38] and Dirichlet[51] on the geometrical representation of quadratic forms of negative determinants. He applied continued fractions to the representation of numbers by quadratic forms by the method used for special cases by Göpel.[41]

V. Simerky[80] gave an exposition of the theory of binary quadratic forms.

J. Liouville[81] stated that, for $k=10$, 18, 22, 28 or 58, every prime of the form A^2+2kB^2, where B is odd, can be expressed in an odd number of ways in the form $kx^2+p^{4l+1}y^2$, where x, y are positive odd integers, and p is a prime not dividing y.

He[82] stated that every prime m of the form $4A^2+5B^2$, where A is odd, is expressible in the form $2(10x+k)^2+p^{4l+1}y^2$ an odd number of ways, where y is positive and odd, p is a prime not dividing y, and $k=3$ or 1 according as $m\equiv 1$ or 9 (mod 40).

C. Traub[83] investigated the primes represented properly by x^2-Dy^2, where

[77] Archiv Math. Phys., 38, 1862, 259.
[78] Monatsber. Akad. Berlin, 1862, 302–311. French transl., Ann. Sc. École Normale Supér., (3), 3, 1866, 287–294.
[79] Report Brit. Assoc. for 1863, 768–86; Coll. Math. Papers, I, 263–8, 283–8.
[80] Abh. K. Böhmischen Gesell. Wiss. Prag, 12, 1863, 193–259 (in Bohemian).
[81] Jour. de Math., (2), 10, 1865, 281–296.
[82] *Ibid.*, (2), 11, 1866, 41–48.
[83] Theorie der Sechs einfachsten Systeme complexer Zahlen, Progr. Lyceums in Mannheim, 1867, 1868, 20–26, 81–86.

$D=-1, \pm 2, \pm 3, 5$, after showing that there exists a g.c.d. process for algebraic numbers $a+b\sqrt{D}$.

G. Cantor[84] proved, without the use of continued fractions, the double theorem due under restrictions to Göpel.[41] If D is a prime $p=8n+3$ [or $8n+7$] or its double, and if $D=\phi^2+2\psi^2$ [or $\phi^2-2\psi^2$], where $\psi\equiv 1 \pmod 4$ when $D=p$, but $\psi\equiv 1$ or $3 \pmod 8$ when $D=2p$, then $(-2\psi, \phi, \psi)$ [or $(2\psi, \phi, \psi)$] is equivalent to $(1, 0, -D)$. For, -2 [or $+2$] is representable by s^2-Dt^2.

P. Bachmann[85] based a theory of quadratic forms on the representation of numbers, making no use of algebraic transformation. Two forms of the same determinant D are called equivalent if every number m which can be represented by one of them has also a representation by the other which belongs to the same root of $z^2\equiv D \pmod m$. Two forms equivalent by this definition are equivalent according to Gauss' definition. Pell's equation enters as by Smith.[76] This theory was amplified in his[86] text.

F. Vallès[87] stated that every prime $13n-1$, $13n-3$ or $13n-4$ can be expressed in one of the forms $\pm(x^2-13y^2)$; if the prime is also of the form $4N+1$, it can be expressed also in the form x^2+13y^2. He stated analogous theorems on primes $7n+k$. The theorem that a prime of the form $5n\pm 1$, also of the form $8N+1$ or $8N+5$, is expressible by x^2+5y^2 is equivalent to the known theorem that a prime $20k+1$ or $20k+9$ is expressible by x^2+5y^2.

L. Lorenz[88] employed the identities

$$\sum_{m,n=-\infty}^{+\infty} q^{m^2+2n^2}=1+2\sum_{m=0}^{\infty}\sum_{n=1}^{\infty}\{q^{(8m+1)n}+q^{(8m+3)n}-q^{(8m+5)n}-q^{(8m+7)n}\},$$

$$\Sigma q^{m^2+3n^2}=1+2\Sigma\Sigma\{q^{(3m+1)n}-q^{(3m+2)n}+2q^{(3m+1)4n}-2q^{(3m+2)4n}\}$$

to show that the number of solutions of $N=m^2+2n^2$ is double the excess of the number of divisors of the forms $8n+1$, $8n+3$ of N over the number of the forms $8n+5$, $8n+7$; and that the number of solutions of $N=m^2+3n^2$ is

$$2(N_{3m+1}-N_{3m+2})+4(N_{12m+4}-N_{12m+8}),$$

where N_k denotes the number of divisors of the form k of N.

J. Liouville[89] stated that, if (x, y) is unaltered by the change of sign of x or y,

$$\Sigma f(\alpha+3\beta, \alpha-\beta)=2\Sigma f(i, i'),$$

where α, β range over all sets of integral solutions of $\alpha^2+3\beta^2=m$, where m is a given odd integer, while i, i' range over all sets of positive odd integral solutions of $i^2+3i'^2=4m$.

B. Minnigerode[90] employed Dirichlet's[57] definition of the first and second roots of a form f of positive discriminant D, proved that any f is equivalent (by passing to

[84] Zeitschrift Math. Phys., 13, 1868, 259–61.
[85] Zeitschrift Math. Phys., 16, 1871, 181–9.
[86] Bachmann, Zahlentheorie, I, 1892, 165–213.
[87] Bull. Sc. Soc. Philomatiques de Paris, 1870–1, 191–3; L'Institut, Jour. Universal des Sc. et des Soc. Sav. en France et à l'Etranger, 40, 1872, 1957.
[88] Tidsskrift for Mathematik, (3), 1, 1871, 106–8.
[89] Jour. de Math., (2), 18, 1873, 142–4. Cf. Bell.[193]
[90] Göttinger Nachr., 1873, 619–652.

successive right neighboring forms) to one whose first root exceeds 2 and whose third coefficient is greater than or equal to the first coefficient (apart from signs), and called such a form (and those in the same period) *reduced* (although not in the sense of Gauss or Hermite[53]). Using continued fractions with negative quotients, he developed the first root of a reduced form F, the g.c.d. of whose coefficients is σ, to obtain all transformations of F into itself and hence to deduce all solutions of $t^2-Du^2=\sigma^2$ from the least positive solution.

W. Göring[91] proved that any prime $6m+1$ is represented by x^2+3y^2 in one and but one way.

E. Selling[92] considered a positive form (A, K, B) in which A, B and the invariant $I=AB-K^2$ are all positive, while A, K, B are real. The conditions $-A\leqq 2K\leqq A\leqq B$ for a reduced form adopted by Lagrange and Gauss are here replaced by the conditions that K is not positive and $-K$ is not greater than A or B. If (A, K, B) is one reduced form and if $A+H+K=0$, $B+K+G=0$, $C+G+H=0$, then (A, K, B), (B, G, C) and (C, H, A) are reduced and are permuted cyclically by the substitution $\binom{\ \ 0\ 1}{-1\ 1}$. These three forms are simultaneously reduced forms if no one of G, H, K is positive (or if A, B, C are positive and the sum of any two is not less than the third). There are only three reduced forms in a class; for, if (A, K, B) is reduced, A, B, C are the least numbers represented properly by forms of the class. Every class contains reduced forms.

Employing Gauss'[38] geometrical representation of positive forms, we see that $G\leqq 0$, $H\leqq 0$, $K\leqq 0$ imply that the exterior angles of the triangle formed by the lines of lengths $\sqrt{A}$, $\sqrt{B}$, $\sqrt{C}$ are obtuse, so that the triangle is acute. Let $\rho x+\sigma y$, $\rho' x+\sigma' y$ be the conjugate complex factors of $Ax^2+2Kxy+By^2$, and write $\rho=\xi+\xi_1 i$, $\sigma=\eta+\eta_1 i$. Then

$$(19)\qquad \xi^2+\xi_1^2=A,\qquad \xi\eta+\xi_1\eta_1=K,\qquad \eta^2+\eta_1^2=B.$$

Consider an indefinite form (a, k, b) with the invariant $I=k^2-ab>0$. Denoting its factors by $(\xi\pm\xi_1)x+(\eta\pm\eta_1)y$, we have

$$(20)\qquad \xi^2-\xi_1^2=a,\qquad \xi\eta-\xi_1\eta_1=k,\qquad \eta^2-\eta_1^2=b.$$

Take any set of real numbers ξ, ξ_1, η, η_1 satisfying (20) and insert them in (19); we obtain a positive form (A, K, B) corresponding to (a, k, b). Necessary and sufficient conditions for corresponding forms are

$$AB-K^2=k^2-ab,\qquad bA-2kK+aB=0,$$

i. e., their determinants are equal except as to sign, and their simultaneous invariant vanishes. If the same substitution $\binom{\alpha\ \beta}{\gamma\ \delta}$ be applied to an indefinite form and to the corresponding positive form, the new positive form corresponds to the new indefinite form. An indefinite form (a, k, b) with $a>0$ is called reduced if the corresponding positive form (A, K, B) is reduced whatever set of real solutions of (20) is employed, with the restriction that K is zero for one of these sets. The purpose of the last restriction and $a>0$ is to insure that only one of the forms (a, k, b), (b, g, c),

---

[91] Math. Annalen, 7, 1874, 382.
[92] Jour. für Math., 77, 1874, 143–164; revision (in French) in Jour. de Math., (3), 3, 1877, 21–42.

(c, h, a) shall be reduced, where h, g, c are defined by $a+h+k=0$, $b+k+g=0$, $c+g+h=0$. It is proved by use of Gauss' geometric interpretation that an indefinite form is reduced if and only if its first coefficient is positive and third coefficient is negative.

Applying to a reduced form (a, k, b) the substitutions $\binom{1\ 1}{0\ 1}$, $\binom{1\ 0}{1\ 1}$, we obtain

$$(a,\ a+k,\ a+2k+b), \qquad (a+2k+b,\ k+b,\ b),$$

the first or second of which is reduced, according as $a+2k+b$ is negative or positive. Any reduced form determines in this manner a finite period of reduced forms. A special reduced form is one for which not only a and b, but also $a-2k+b$ and $a+2k+b$, are of opposite signs; as noted by Lipschitz (Berichte Akad. Wiss. Berlin, 1865, 184), these conditions are equivalent to those of Gauss and Dirichlet[57] for a reduced form.

Selling compared his definition of reduced forms with that of Hermite[53], and considered at length the case in which $I=k^2-ab$ is a square. Cf. Voronoï[42] of Ch. XI.

E. Hübner[93] noted that, of the conditions that $ax^2+2bxx_1+a_1x_1^2$ become $cy^2+c_1y_1^2$ for $x=\alpha y+\beta y_1$, $x_1=\alpha_1 y+\beta_1 y_1$, two determine c and c_1, and the third is

$$a\alpha\beta+b(\alpha\beta_1+\alpha_1\beta)+a_1\alpha_1\beta_1=0.$$

If a, b, a_1 have no common factor, all solutions of the third condition in which α is prime to α_1, and β to β_1, are found by assigning relatively prime values to α, α_1 and taking β_1 and $-\beta$ as the quotients of $a\alpha+b\alpha_1$ and $b\alpha+a_1\alpha_1$ by their g.c.d.

If m^e and m^{e+k} are representable properly by a form of determinant $-D$, when m is prime to D, some power of m between m^e and m^{2e+k} is representable properly by $(1, 0, D)$. If $D=8n-1>0$, at least one power 2^k is representable properly by $(1, 0, D)$; the least k is $\leqq 2H-1$, where H is the number of classes of forms of determinant $-D$. If D is not divisible by the prime p, and if $-D$ is a quadratic residue of p, some power p^k is representable properly by $(1, 0, D)$; the least k is $\leqq 2H-1$.

R. Gent[94] proved that, if $n>1$ is an odd integer, the number of decompositions of $4n$ into x^2+3y^2, where x and y are odd and positive, is the excess of the number of divisors of the form $3h+1$ of n over the number of the form $3h+2$. He tabulated all decompositions $4p=x^2+3y^2$ for the possible $p<500$.

The difficulty with the determinant -7 is that there are two reduced forms $(1, 0, 7)$ and $(2, 1, 4)$. He conjectured that the number of solutions of $8n=x^2+7y^2$, $x>0$, $y>0$, is the excess of the number of divisors of n which are quadratic residues of 7 over the number which are quadratic non-residues of 7.

H. J. S. Smith[95] associated with the form (a, b, c) of determinant $N=b^2-ac>0$ the semi-circle

$$[a, b, c]:\ a+2bx+c(x^2+y^2)=0, \qquad y>0.$$

Two points ω and Ω in the upper half H of the complex plane are called *equivalent* if

$$\omega=\frac{\gamma+\delta\Omega}{\alpha+\beta\Omega}, \qquad \alpha\delta-\beta\gamma=1, \tag{21}$$

---

[93] Ueber die Transformation einer homog. binaeren quad. Form in ein Aggregat von 2 Quadraten, Progr., Memel, 1875.

[94] Zur Zerlegung der Zahlen in Quadrate, Progr., Liegnitz, 1877.

[95] Atti R. Accad. Lincei, Mem. fis. mat., (3), 1, 1876–7, 136–49; abstr. in Transunti, (3), I, 68–69; Coll. Math. Papers, II, 224–241.

where $\alpha, \ldots, \delta$ are integers such that $\beta \equiv \gamma \equiv 0 \pmod 2$, $\alpha \equiv \delta \equiv 1 \pmod 4$. Such a substitution is called *normal.* If two forms are equivalent under a normal substitution $\binom{\alpha\ \beta}{\gamma\ \delta}$, the corresponding circles are equivalent under it. Every point of H is equivalent to one and but one reduced point, i. e., in the reduced region [fundamental polygon] Σ lying above the circles

$$Q: x^2+y^2-x=0, \qquad Q^{-1}: x^2+y^2+x=0$$

and on or between the lines $P: x=1$ and $P^{-1}: x=-1$. Replace the points of a circle $[a, b, c]$ by the corresponding reduced points; the circle is thereby replaced by a series of disconnected circular arcs which are of six types, designated by (PP^{-1}), (QQ^{-1}), (PQ), (PQ^{-1}), $(P^{-1}Q)$, $(P^{-1}Q^{-1})$, each specifying the two boundaries of Σ which terminate the arc. The first two correspond to intermediate, and the last four to principal reduced forms of Hermite.[53] If we deform Σ into a closed surface such that P coincides with P^{-1}, and Q with Q^{-1}, the series of disconnected arcs will become a continuous curve, which represents the class of forms equivalent to (a, b, c). For Smith's application of this theory to elliptic modular functions and class number, see the latter topic in Ch. VI.

C. F. Gauss[96] employed the substitution (21) subject to the same congruencial conditions with $\omega=i/t'$, $\Omega=i/t$, and also a region obtained from Smith's Σ by rotation clockwise through 90°. Gauss considered also all substitutions (21) in connection with modular functions.

R. Dedekind[97] called ω and Ω equivalent numbers if they are connected by a relation (21) where $\alpha, \ldots, \delta$ are any integers. All complex numbers equivalent to a given one are equivalent to each other and are said to form a *class.* Every complex number $\omega=x+yi$, $y>0$, and every rational real number, is equivalent to one (and in general to only one) number $\omega_0=x_0+y_0i$ for which $x_0 \leqq \frac{1}{2}$, $x_0 \geqq -\frac{1}{2}$, $x_0^2+y_0^2 \geqq 1$, whence the point ω_0 lies in the region between the two lines parallel to the y-axis and at a distance 1/2 from it, and at the same time is above the circle of radius unity and center at the origin. The proof is similar to that of the existence of one (and in general only one) equivalent reduced binary quadratic form of negative determinant. Application is made to elliptic modular functions (cf. Dedekind[128] of Ch. VI).

E. de Jonquières[98] employed representations of N^2 by z^2+tv^2 and relations between them and representations of N by x^2+tu^2 to deduce the latter representations.

T. Pepin[99] proved that a primitive form $f=(a, b, c)$, whose determinant D has no square factor, represents an infinitude of integers p prime to D. Let α be the g.c.d. of $a=a'\alpha$, $D=D'\alpha$. If p is a quadratic residue of D, then $\alpha D'$, $\alpha a'$, $-a'D'$ are quadratic residues of a', D', α, respectively, and conversely. A necessary and sufficient condition that f shall represent squares is that it can represent positive integers which are prime to D and are quadratic residues of D.

S. Roberts[100] applied continued fractions to the representation of numbers by $x^2+\Delta y^2$, chiefly for $\Delta>0$. It is a sequel to the papers by Göpel[41] and Smith.[79]

---

96 Post. fragment, 1827; Werke, III, 1876, 477–8 (386); VIII, 1900, 105 (remarks by R. Fricke).
97 Jour. für Math., 83, 1877. 269–273. Cf. Hurwitz,[106] Mathews,[137] Weber.[145] Exposition by A. L. Baker, Amer. Math. Monthly, 8, 1901, 163–6.
98 Comptes Rendus Paris, 87, 1878, 399–402; Assoc. franç., 7, 1878, 40–49.
99 Atti Accad. Pont. Nuovi Lincei, 32, 1878–9, 81–87.
100 Proc. London Math. Soc., 10, 1878–9, 29–41. Partial report in Vol. II, p. 383.

A. Korkine and G. Zolotareff[101] stated that the precise limit of the minima of all binary quadratic forms of positive determinant D for integral values, not all zero, of x, y is $\sqrt{\frac{4}{5}D}$, which is the minimum of $f_0=\sqrt{\frac{4}{5}D}(x^2-xy-y^2)$ and forms equivalent to it; while the precise limit of the minima for all other forms is $\sqrt{\frac{1}{2}D}$. Proof by Humbert.[183]

A. Markoff[102] stated that Korkine had communicated to him the fact that $\sqrt{\frac{1}{2}D}$ is the minimum of forms equivalent to $f_1=\sqrt{\frac{1}{2}D}(x^2-2xy-y^2)$. Here Markoff proved that $\sqrt{100D/221}$ is the precise limit of the minima of all forms equivalent to neither f_0 nor f_1, and is the minimum of forms equivalent to

$$f_2=\sqrt{\frac{4D}{221}}\,(5x^2-11xy-5y^2),$$

and obtained a continuation of this series down to f_9. By the use of continued fractions, he proved that if l is a given number $>\frac{2}{3}$, there is only a finite number of classes of forms, of a given determinant D, whose values are not numerically $<l\sqrt{D}$. This number of classes increases indefinitely when l approaches $\frac{2}{3}$. For another statement of this result and a similar one, see Schur.[174] Another theorem by Markoff will be quoted under Frobenius.[175]

H. Poincaré[103] associated with a definite form $F=am^2+2bmn+cn^2$ a parallelogrammatic lattice R whose points have the coordinates

$$x=m\sqrt{a}+n\,\frac{b}{\sqrt{a}},\qquad y=n\sqrt{\frac{ac-b^2}{a}}.$$

To a form F' equivalent to F corresponds a lattice R' with the same points as R. We may convert R into R' by a rotation about the origin through a certain angle θ, called the angle of the transformation. Given θ, we can compute the coefficients of the transformation. An arithmetical covariant of a form is defined as a function of its coefficients which is equal to the product of the analogous function of the coefficients of any equivalent form by a function of the angle θ of the transformation. If we know a covariant, we may test the equivalence of two forms by finding θ and then the coefficients of the transformation. The covariants

$$\sum_{m,\,n=-\infty}^{+\infty}\left\{m\sqrt{a}+n\left(\frac{b+\sqrt{b^2-ac}}{\sqrt{a}}\right)\right\}^{-2k}$$

may be computed by means of infinite series or a definite integral.

H. Poincaré[104] showed how the usual representation of definite quadratic forms by lattices may be applied to indefinite forms. Let $A=\begin{bmatrix} a & b\\ c & d\end{bmatrix}$ denote the lattice composed of the points with the coordinates

(22) $\qquad x=am+bn,\quad y=cm+dn$ (m, n ranging over all integers).

[101] Math. Annalen, 6, 1873, 369–370; Korkine's Coll. Papers, 1, 1911, 296. Report in Ch. XI.[18]

[102] *Ibid.*, 15, 1879, 381–406; 17, 1880, 379–399.

[103] Comptes Rendus Paris, 89, 1879, 897–9. For details on related matters, see his papers.[104, 109, 157]

[104] Jour. école polyt., t. 28, cah. 47, 1880, 177–245. Report of Parts IV, V is made in Ch. III.[31]

Call $ad-bc$ the *norm* of the lattice. A lattice A is called a *multiple* of a lattice B if all points of A belong to B. Two lattices are *equivalent* if all points of each belong to the other. If $A'=AR$ under matrix multiplication, lattice R is said to be the *ratio* of A' to A. Proof is given of Eisenstein's theorem that every lattice with integral elements is equivalent to one with $d=0$. The system of points common to two lattices is called their least common multiple. The g.c.d. of A and A' is defined to be the system of points

$$\text{(23)} \qquad \begin{bmatrix} a & b & a' & b' \\ c & d & c' & d' \end{bmatrix}: \qquad \begin{aligned} x&=am+bn+a'm'+b'n', \\ y&=cm+dn+c'm'+d'n', \end{aligned}$$

where m, n, m', n' take all integral values. If $d=d'=0$, it is shown how to find the l.c.m. and g.c.d., they being of the form $\begin{bmatrix} \cdot & \cdot \\ \cdot & 0 \end{bmatrix}$. A *prime* lattice is one whose norm is a prime. There is a study of special lattices composed of the points whose coordinates are integers satisfying a congruence $ax+by\equiv 0 \pmod{c}$.

Part II (pp. 200–9) deals with the representation of numbers $a+b\sqrt{D}$ by points. If $D<0$ it is usually represented by the point m with the coordinates $a, b\sqrt{-D}$ in a plane P. Take a plane Q cutting P along the x-axis and making with it a dihedral angle equal to arc cos $1/\sqrt{-D}$. The coordinates of the projection of m on Q are a, b. Whether D is positive or negative, the point (x, y) is taken as the representative of $x+y\sqrt{D}$, whose modulus and argument are defined as

$$\sqrt{x^2-Dy^2}, \qquad \frac{1}{\sqrt{-D}} \text{ arc tan } \left(\frac{y}{x}\sqrt{-D}\right).$$

If real, the modulus is the ratio of the vector from the origin O to $C=(x, y)$ to the vector from O to the intersection of the former vector with the ellipse or hyperbola $\xi^2-D\eta^2=1$. But if $x^2-Dy^2<0$, the modulus is $\sqrt{-1}$ times the ratio of the similar vectors for $\xi^2-D\eta^2=-1$. The real component of the argument is double the area comprised between the vector OC, the x-axis, and $\xi^2-D\eta^2=1$. The modulus of a product is the product of the moduli of the factors. The argument of a product is the sum of the arguments (properly chosen) of the factors.

Part III (pp. 209–226) treats of the representation of forms by lattices. The lattice of points (22) is said to *represent* the form $(am+bn)^2-D(cm+dn)^2$. Then any form $aN^2+2bNM+cM^2$ is represented by the lattice

$$\begin{bmatrix} b/\sqrt{a} & \sqrt{a} \\ \sqrt{\dfrac{b^2-ac}{Da}} & 0 \end{bmatrix}.$$

In the usual representation of a definite form (with $D<0$), the corner element is $\sqrt{(ac-b^2)/a}$; it is here divided by $\sqrt{-D}$ to obtain the projection on the plane Q of the lattice placed in the plane P. For $D<0$ the present representation differs from the classic one only in multiplying the coordinates of points of the lattice by fixed factors. Henceforth let $D>0$. By a *fundamental triangle OAB* is meant (p. 222) one formed by the origin O and two points A and B of the lattice and containing in its interior no other point of the lattice. It is called an *ambiguous triangle* if the first "asymptote" $\sqrt{D}x=y$ is interior to the angle at O of the triangle, while the second asymptote $-\sqrt{D}x=y$ is exterior to this angle. If we complete the parallelogram $OABC$ whose half is a fundamental triangle OAB, the triangles OAC and OBC are aso fundamental and are called *derived triangles* of OAB. Also, OAB is said to be the *primitive* of OAC. Any lattice has ambiguous triangles. If a triangle is ambiguous, one and only one of its two derived triangles is ambiguous, and a single one of its two primitives is ambiguous. Hence there exist infinitely many ambiguous triangles forming a *period* such that each of them is the derived of the preceding triangle and the primitive of the following one. Each triangle of the period has a side in common with the following triangle. Thus in general several consecutive triangles of the period have a common side and are said to form a series, so that the period is divided into series. The last triangle of a series is the first one of the next series; such a triangle belonging to two series corresponds to a reduced form.

The lattice (23) is designated by $Am+Bn+A'm'+B'n'$, where

$$A=a+c\sqrt{D},\quad B=b+d\sqrt{D},\quad A'=a'+c'\sqrt{D},\quad B'=b'+d'\sqrt{D}.$$

In particular, the lattice (22) is $Am+Bn$. Employing also the conjugates

$$\bar{A}=a-c\sqrt{D},\quad \bar{B}=b-d\sqrt{D}$$

of A, B, we see by the first sentence of the report on Part III that the quadratic form (in the variables m, n)

$$(Am+Bn)(\bar{A}m+\bar{B}n)=(am+bn)^2-D(cm+dn)^2$$

is represented by either of the lattices $Am+Bn$, $\bar{A}m+\bar{B}n$.

F. Mertens[105] gave an elementary proof, without using continued fractions (as had Dirichlet[57]), of Gauss' theorem (Art. 193) that two properly equivalent reduced forms of positive determinant belong to the same period. Cf. Frobenius,[173] Mertens.[139, 190]

A. Hurwitz[106] developed the theory of Dedekind's[97] fundamental region and noted that it leads to a simple geometrical theory of the reduction of quadratic forms.

K. Küpper[107] investigated primes $p=m^2+kn^2$.

H. J. S. Smith[108] proved, by means of the functions $(q_1, \ldots, q_n)$ giving the numerators of the continued fraction with the quotients $q_1, q_2, \ldots$, that every prime $12n+7$ is of the form x^2+3y^2, and every prime $12n+11$ is of the form $3x^2-y^2$. By the same method he treated the representation of primes by $2x^2\pm y^2$. Cf. Göpel.[41]

H. Poincaré[109] obtained arithmetical invariants of a linear or quadratic form $F(x, y)$, i. e., functions of the coefficients of F which are unaltered under every linear substitution $x=\alpha x'+\beta y'$, $y=\gamma x'+\delta y'$, where α, β, γ, δ are integers such that $\alpha\delta-\beta\gamma=1$. Consider

$$\phi_k(q)\equiv\Sigma(qm+n)^{-2k},$$

---

[105] Jour. für Math., 89, 1880, 332–8.
[106] Math. Annalen, 18, 1881, 528–540.
[107] Casopis math. fys., Prag, 10, 1881, 10 (Bohemian).
[108] Coll. Math. in memoriam D. Chelini, Milan, 1881, 117–143; Coll. Math. Papers, II, 287–311.
[109] Assoc. franç. av. sc., 10 (Alger), 1881, 109–117. See Poincaré.[157]

summed for all pairs of integers m, n except 0, 0. It can be expressed as a definite integral and is holomorphic if q is not real. Further,

$$\Sigma(am+bn)^{-2k}=\frac{1}{b^{2k}}\phi_k\left(\frac{a}{b}\right)$$

is an arithmetical invariant of $ax+by$. We may employ * ϕ_1 to test the equivalence of $F=ax^2+2bxy+cy^2$ and $F'=a'x'^2+\dots$ of the same negative determinant $-D=b^2-ac$. Thus F and F' are the moduli of

$$l=x\sqrt{a}+y\frac{b+i\sqrt{D}}{\sqrt{a}},\quad l'=x'\sqrt{a'}+y'\frac{b'+i\sqrt{D}}{\sqrt{a'}}.$$

Let the above substitution transform F into F', and hence l into $\lambda l'$, where λ is a constant to be determined. Identify l with $bx+ay$ and employ its above invariant for $k=1$. Thus

$$\frac{1}{a}\phi_1\left(\frac{b+i\sqrt{D}}{a}\right)=\frac{1}{\lambda^2a'}\phi_1\left(\frac{b'+i\sqrt{D}}{a'}\right), \tag{24}$$

which determines λ. Write $\lambda=\mu+i\nu$. Since l is transformed into $\lambda l'$,

$$\begin{aligned} \alpha a+\gamma b&=\mu\sqrt{aa'}, & (\beta a+\delta b)\sqrt{a'}&=(\mu b'-\nu\sqrt{D})\sqrt{a},\\ \gamma\sqrt{D}&=\nu\sqrt{aa'}, & \delta\sqrt{Da'}&=(\nu b'+\mu\sqrt{D})\sqrt{a}. \end{aligned} \tag{25}$$

Equations (24), (25) give the coefficients $\alpha, \beta, \gamma, \delta$ of the substitution which transforms F into F', they being assumed equivalent.

To test their equivalence, compute the ϕ_1's in (24) with an approximation sufficient to insure solutions $\alpha, \dots, \delta$ of (25) with an error of less than 1/2. Then we have their exact values as integers. The forms are equivalent if and only if the resulting substitution transforms l into $\lambda l'$.

For a quadratic form F of determinant $b^2-ac=D>0$, let t, u be the least positive integers such that $t^2-Du^2=1$. Then F has the invariant

$$\Sigma(am^2+2bmn+cn^2)^{-k},$$

summed for all pairs of integers m, n for which $m>0$, $n \gtreqless 0$, $m/n<u/t$, while k is a given integer>1. This series may be expressed as a double integral. The same is true of a series defining a certain arithmetical invariant of a pair of linear forms.

J. Hermes[110] gave an algorithm for computing δ in passing from a form to a reduced right neighboring form (Dirichlet[57]).

Hermes[111] considered a chain $\phi_0, \phi_1, \dots, \phi_n$ of neighboring forms of positive determinant D, where ϕ_{i-1} is transformed into ϕ_i by $\begin{pmatrix} 0 & 1\\ -1 & \delta_i\end{pmatrix}$. Then $\begin{pmatrix} \alpha & \beta\\ \gamma & \delta\end{pmatrix}$ replaces ϕ_0 by ϕ_n if

$$\alpha=\pm[\kappa_2,\dots,\kappa_{n-1}],\quad \beta=\pm[\kappa_2,\dots,\kappa_n],\quad \gamma=\pm[\kappa_1,\dots,\kappa_{n-1}],\quad \delta=\pm[\kappa_1,\dots,\kappa_n],$$

* Since the series for ϕ_1 is only semiconvergent, further definition of ϕ_1 is needed. It is essentially log Δ, where Δ is the modular invariant of Klein-Fricke, Theorie der Elliptischen Modulfunctionen, 1890. The limit for $k=1$ of the series for ϕ_k is evaluated in Weber's Algebra, III, p. 560.

110 Archiv Math. Phys., 68, 1882, 432–9.

111 Jour. für Math., 95, 1883, 165–170.

in the notation of continued fractions (Gauss, Art. 27), where $\kappa_{2l+1}=\delta_{2l+1}$, $\kappa_{2l}=-\delta_{2l}$. Let Φ_0, of determinant D, be transformed into the same ϕ_n by using $\lambda_1, \ldots, \lambda_m$ instead of κ's. If ϕ_0 is transformed into Φ_0 by $\left(\begin{smallmatrix} A & B \\ \Gamma & \Delta \end{smallmatrix}\right)$, then

$$\Gamma/A=(\kappa_1, \ldots, \kappa_{n-1}, 0, \mp\lambda_{m-1}, \ldots, \mp\lambda_1),$$

where the upper or lower signs hold according as the total number of elements of this continued fraction is odd or even. Hence

$$A=\pm[\kappa_2, \ldots, \mp\lambda_1], \qquad B=\pm[\kappa_2, \ldots, \mp\lambda_1, h],$$
$$\Gamma=\pm[\kappa_1, \ldots, \mp\lambda_1], \qquad \Delta=\pm[\kappa_1, \ldots, \mp\lambda_1, h].$$

Two forms which lead to different periods are not equivalent.

T. J. Stieltjes[112] noted that if $F(n)$ is the number of representations of n by x^2+2y^2, and if d_i is the number of divisors of the form $8k+i$ of n, then (Genocchi[56])

$$F(n)=2(d_1+d_3-d_5-d_7),$$
$$F(1)+F(2)+\ldots+F(n)=2\left\{\left[\frac{n}{1}\right]+\left[\frac{n}{3}\right]-\left[\frac{n}{5}\right]-\left[\frac{n}{7}\right]+\ldots\right\}.$$

The sum in brackets is transformed into $S+S_1-\lambda\phi(\lambda)$, where S is a like sum ending with $\pm[n/(2\lambda-1)]$, while

$$S_1=\phi\left[\frac{n+1}{2}\right]+\ldots+\phi\left[\frac{n+\lambda}{2\lambda}\right], \qquad \lambda=[\tfrac{1}{4}(\sqrt{8n+1}+1)], \qquad \phi(x)=2\sin^2 \pi x/4.$$

L. Kronecker[113] noted that every substitution with integral coefficients of determinant unity is the product of one of the following six

$$I=\begin{pmatrix}1 & 0\\ 0 & 1\end{pmatrix}, \begin{pmatrix}0 & -1\\ 1 & 0\end{pmatrix}, \begin{pmatrix}1 & 1\\ -1 & 0\end{pmatrix}, \begin{pmatrix}-1 & 1\\ 0 & -1\end{pmatrix}, \begin{pmatrix}0 & -1\\ 1 & 1\end{pmatrix}, \begin{pmatrix}1 & 0\\ -1 & 1\end{pmatrix}$$

by a substitution congruent to the identity I modulo 2. By interchanging the two columns we get six substitutions of determinant -1 to one of which any substitution of determinant -1 is congruent modulo 2. Two forms are called *completely* (or *incompletely*) *equivalent* if they can (or cannot) be transformed into each other by a substitution $\equiv I \pmod 2$. Thus to any form of negative determinant belong five properly, but incompletely, equivalent forms and six improperly and incompletely equivalent forms derived by the above substitutions. Corresponding to a form of negative determinant, there may not exist one completely equivalent to it and satisfying the conditions $c^2 \gtreqqless a^2 \gtreqqless (2b)^2$ for a Lagrange reduced form; however, to every form corresponds a completely equivalent form whose coefficients $a_0, \ldots$ satisfy the conditions $a_0^2 \gtreqqless b_0^2$, $c_0^2 \gtreqqless b_0^2$, such a form being here called a *reduced* form. Two such reduced positive forms (i. e., with outer coefficients positive and not smaller than the absolute value of the middle coefficient) can be completely equivalent only if they both occur in the above set of 12 forms. All the forms of the same negative determinant which are completely equivalent are said to constitute a *class*. The number of classes equals the number of reduced forms (a_0, b_0, c_0) if we retain only one of two reduced forms $(a_0, \pm a_0, c_0)$ and only one of $(a_0, \pm c_0, c_0)$.

112 Comptes Rendus Paris, 97, 1883, 891.
113 Abh. Akad. Wiss. Berlin, 1883, II, No. 2; Werke, II, 433–444.

E. Cesàro[114] concluded that the number of positive integral solutions of $Ax^2+Bxy+Cy^2=n(A>0, C>0)$ is in mean $\pi/(2\delta)-B/\delta^2$, where $\delta^2=4AC-B^2$. According to Encyclopédie des sc. math., t. I, vol. 3, p. 350, there was an error in the limits of integration, and the correct value is

$$\frac{\pi}{2\delta}-\frac{1}{\delta}\operatorname{arc\,tan}\frac{B}{\delta}.$$

Note, however, that Gegenbauer[123] agrees with Cesàro.

T. Pepin[115] proved that every prime $m=8l+3$ is of the form x^2+2y^2 by use of the sum of the divisors of m and the sum of the odd divisors of each $m-n^2$.

Pepin[116] applied general theorems of Dirichlet[42] (§ VII) on the number of representations by a quadratic form of negative determinant D to show that the number of solutions of $2^\alpha m=x^2+2y^2$ is double the excess of the number of divisors of the forms $8l+1$ or $8l+3$ of m over the number of divisors of the form $8l+5$ or $8l+7$. If m is odd and not divisible by 3, the number of representations of $3^\beta m$ by x^2+3y^2 is $2\omega(m, 3)$, where $\omega(m, 3)=\Sigma(-3/i)$, in which i ranges over the divisors of m, and $(-3/i)$ is a Jacobi-Legendre symbol ± 1 of quadratic reciprocity; the number of representations of $2^\alpha m$ $(\alpha>0)$ by $2x^2+2xy+2y^2$ or by x^2+3y^2 is $6\omega(m, 3)$. The number of representations of $2^\alpha 3^\beta m$ by that one of the forms $(1, 0, 6)$, $(2, 0, 3)$, which is suitable to it, is the double of $\Sigma(-6/i)$.

L. Gegenbauer[117] proved that the number of ways any integer r can be represented by a system of binary quadratic forms of discriminant Δ is the product of the number of linear automorphs of a form of discriminant Δ by the sum of the numbers of solutions of $x^2\equiv\Delta \pmod{d_2}$, where d_2 ranges over those divisors of r whose complementary divisors are squares. The number of ways an odd integer r can be represented by x^2+2y^2 is double the number of decompositions into two relatively prime factors of those divisors of r which have only prime factors $8s+1$ or $8s+3$ and whose complementary factors are squares. Similarly for x^2+3y^2.

The number of representations, by quadratic forms of discriminant Δ, of those divisors of a number, whose complementary divisor is the product of an even number of primes, exceeds the number of representations of the remaining divisors by the product of the number of automorphs of a form of discriminant Δ by the excess of the number of those divisors d_2, with complementary square divisors, for which Jacobi-Legendre's symbol (Δ/d_2) is $+1$, over the number of divisors d_2 for which $(\Delta/d_2)=-1$. There are corollaries for x^2+2y^2 and x^2+3y^2 like those for x^2+y^2 quoted in this History, Vol. II, pp. 247–8.

A. E. Pellet[118] noted that if $A=-1, \pm 2, \pm 3, 5, -7, -11, 13$, so that factorization in the field defined by $\sqrt{A}$ is unique, every divisor of m^2-An^2 or of $m^2+mn+n^2(1-A)/4$, according as $A\equiv -1$ or $+1 \pmod 4$, can be given the same form apart from sign.

---

114 Mém. Soc. R. Sc. de Liège, (2), 10, 1883, No. 6, 197–9.
115 Atti Accad. Pont. Nuovi Lincei, 37, 1883–4, 42.
116 *Ibid.*, 38, 1884–5, 163–170. Cf. Pepin.[131]
117 Sitzungsber. Akad. Wiss. Wien (Math.), 90, II. 1884, 437–448.
118 Comptes Rendus Paris, 98, 1884, 1482.

F. Cajori[119] noted that, if $2b$ is divisible by neither a nor c, and if (a, b, c) is of negative determinant $-\Delta$ and is transformed into $(a', 0, c')$ by $\binom{\lambda\ \mu}{\nu\ \rho}$, $\lambda\rho-\mu\nu=\pm 1$, then $a\lambda\mu+b(\lambda\rho+\mu\nu)+c\nu\rho=0$. The quadratic in μ obtained by eliminating λ has real roots if $a^2 \gtreqless 4\nu^2\rho^2\Delta$. The quadratic in ρ obtained by eliminating ν has real roots if $c^2 \gtreqless 4\mu^2\lambda^2\Delta$. Each coefficient of the substitution is numerically >0.

J. C. Fields[120] considered the preceding question when $\lambda\rho-\mu\nu=+1$, and proved easily that there must be two numbers α and γ whose product is $-\Delta$ such that $a\rho^2-\gamma\nu^2=a$, $a\lambda+b\nu\equiv 0 \pmod{a}$ for integers ν, ρ. If integers ν, ρ can be found, the two forms are equivalent. He stated similar conditions for the equivalence of (a, b, c) to (a', b', a') or to $(a', \frac{1}{2}a', c')$.

L. Kronecker[121] proved that the number of representations of n by a quadratic form of negative determinant D is in mean $\pi/\sqrt{-D}$.

J. W. L. Glaisher[122] proved by means of products of infinite series that the excess of the number of representations of $24n+3$ by x^2+2y^2, in which x, y are both of the form $12m\pm 1$ or both of the form $12m\pm 5$, over the number of representations in which x is of one form and y of the other is equal to double the excess of the number of representations of $8n+1$ by x^2+2y^2, in which x is odd and y is a multiple of 4, over the number of representations in which x is odd and $y\equiv 2 \pmod{4}$; either excess may be positive or negative.

L. Gegenbauer[123] gave a simple proof of E. Cesàro's[114] theorem that, if $a>0$, $b \gtreqless 0$, $c>0$, $\Delta=4ac-b^2>0$, the mean number of those representations of an integer by $(a, -b, c)$, of negative discriminant $-\Delta$ and with like sign for the values of the two variables, is $\pi/\sqrt{\Delta}+2b/\Delta$. He found expressions for the mean number of divisors of x and of y in the various representations by ax^2+by^2 of an integer in a given interval or of an integer with s digits, also when the divisors are of specified character, such as divisible by λ, or divisible by no σth power.

Gegenbauer[124] found, for the various representations of an integer with s digits by ax^2+cy^2, where $a>0$, $c>0$, the mean number of divisors of x such that the divisors are $\equiv\alpha \pmod{\lambda}$ and have complementary divisors $\equiv\beta \pmod{\mu}$, and also for other conditions on the divisors.

J. Vivanti[125] proved that a form $(a, b, -c)$ of positive determinant D, not a square, is reduced if a and b are positive and $c=a+b$, and called it a *Null form*. A Null form occurs in the system of reduced forms of determinant D if and only if the exponents of the prime factors of the form $6n+5$ and 2 of D are all even. If $(a, b, -c)$ is a Null form, also $(b, a, -c)$ is a Null form; they are improperly equivalent by means of $x=x'$, $y=x'-y'$; they are properly equivalent if and only if there exist integral solutions of $x^2D/d^2-3=y^2$, where d is the g.c.d. of a, b. He[126] gave obvious theorems on these Null forms. The conditions for integral

---

[119] Johns Hopkins Univ. Circular, 4, 1885, 122 (in full).
[120] *Ibid.*, 5, 1885, 38.
[121] Sitzungsber. Akad. Berlin, 1885, 775. Cf. H. Brix, Monatshefte Math. Phys., 21, 1910, 309–325 (p. 325); Dedekind, Jour. für Math., 121, 1900, 115. Cf. Landau.[148]
[122] Quar. Jour. Math., 20, 1885, 96.
[123] Sitzungsber. Akad. Wiss. Wien (Math.), 92, II, 1885, 380–409.
[124] *Ibid.*, 93, II, 1886, 90–105.
[125] Zeitschrift Math. Phys., 31, 1886, 273–282.
[126] *Ibid.*, 32, 1887, 287–300.

solutions of $x^2D-3=y^2$ are not known. In practice, it is solved by means of Pell's equation and the theory of binary forms. For $1<D<1000$, there are 147 values of D for which a primitive Null form exists.

G. Wertheim[127] presented the theory mainly from the standpoint of Gauss, but followed Dirichlet's[57] proof that reduced forms of positive determinant of different periods are not equivalent.

X. Stouff[128] employed any fixed set of n integers (called *modules*) relatively prime in pairs. Let L denote their product, P_i any of the 2^n products of the modules in which each enters only once as a factor, and P_i' the product of the remaining modules, so that $P_iP_i'=L$. The substitutions

$$z'=\frac{\alpha P_i z+\beta L}{\gamma z+\delta P_i},\qquad \begin{vmatrix}\alpha P_i & \beta L\\ \gamma & \delta P_i\end{vmatrix}=P_i, \tag{26}$$

obtained from the various P_i and all sets of integers α, β, γ, δ for which the determinant is P_i, form a group H. In fact, the product of (26) by the analogous substitution with the parameters α', β', γ', δ', P_j is such that the four coefficients are divisible by the product P_h of the modules common to P_i and P_j, and the resulting substitution is of the form (26) with P_i replaced by $P_k=P_iP_j/P_h^2$. Such a group H has a finite number of generators.

Consider another such group G with Q_i in place of P_i. In order that G be a subgroup of H it is necessary and sufficient that each module of G be a product of modules of H (each entering only once as a factor) and that the product of all the modules of G be equal to that for H. Then G is an invariant subgroup of H. For example, the group defined by the modules 2, 3, 5 has as invariant subgroups those defined by the modules 2, 15; 6, 5; 3, 10; 1, 30.

If a, b, c are integers such that a, c, bP_i have no common factor, $ax^2+bP_ixy+cLy^2$ is called a form attached to the group G. This form is called equivalent to the form $a'x_1^2+b'P_ix_1y_1+c'Ly_1^2$ (in which a', b', c' are integers) obtained from it by applying the corresponding homogeneous substitution

$$x=\frac{\alpha Q_j x_1+\beta L y_1}{\sqrt{Q_j}},\qquad y=\frac{\gamma x_1+\delta Q_j y_1}{\sqrt{Q_j}},$$

of determinant unity. All forms equivalent to a given one are said to form a *class*. There is given a process to select one or more representatives of each class and to test their equivalence. Since that process is laborious, it is desirable to enumerate the classes without knowing their representatives. Under certain restrictions the class number is found by Dirichlet's classic method (Dirichlet[19] of Ch. VI). There is suggested a generalization to several sets of modules and also to substitutions with irrational coefficients.

A. Hurwitz[129] called a pair of numbers x, y *reduced* if the point (x, y) lies in the region R composed of two infinite strips, the first bounded by the lines $x=2$, $y=r$, $y=r-1$, and the second by $x=-2$, $y=-r$, $y=1-r$, where $r=\frac{1}{2}(3-\sqrt{5})$, the first strip extending to the right of $x=2$, and the second to the left of $x=-2$, the points

[127] Elemente der Zahlentheorie, 1887, 237–374.
[128] Annales Fac. Sc. Toulouse, 3, 1889, B. 1–28.
[129] Acta Math., 12, 1889, 397–401.

$(3-r, r)$ and $(-3+r, -r)$ of the boundary being the only ones counted as belonging to R. A form (a, b, c) of positive determinant, not a square, is called a *reduced* form if its roots form a reduced pair of irrational numbers. Every form is equivalent to a reduced form. Two reduced forms of determinant D are equivalent or not according as they belong to the same period or not. While these theorems had been proved by use of the ordinary development into continued fractions, they are here proved by use of a development of any real number x_0 into a continued fraction by means of $x_0=a^0-1/x_1$, $x_1=a_1-1/x_2$, $\ldots$, where a_n is an integer chosen so that x_n-a_n lies between $-1/2$ and $+1/2$.

*T. Pepin[130] discussed the number of representations by (1, 0, 8) and (1, 0, 16).

T. Pepin[131] derived his[116] former results from formulas of Liouville (this History, Vol. II, Ch. XI). He added a theorem on the number of representations by a form of determinant -12.

F. Klein[132] called a form $ax^2+2bxy+cy^2$, of negative determinant $D=b^2-ac$, *reduced* if its root $(-b+i\sqrt{-D})/a$ is a reduced number in the sense of Dedekind[97] (i. e., represents a point lying in the region R defined by him). A form of positive determinant D is called reduced if R is crossed by the semi-circle whose diameter joins the points of the x-axis representing the real roots (values of x/y) of the form. For the second case $D>0$, this representation of the form by a semi-circle goes back to Smith,[95] but with now a simpler definition of equivalent points and a simpler fundamental region.

L. K. Lachtine[133] called a form of positive determinant *reduced* if its positive root is developable into a pure continued fraction with only positive elements, and gave a method of finding all automorphs of a reduced form without use of a Pell equation.

P. V. Prebrazenskij[134] showed that Lachtine's method differs from Dirichlet's in that Lachtine does not exclude improper equivalence.

Prebrazenskij[135] treated together forms of positive and negative determinants and divided the classes of forms of any determinant D into two types. Those of the first type (*vollkommen*) are such that if a prime is representable by such a form, it can be represented by the form in four ways if $D<0$ and in four infinitudes of ways if $D>0$. For the second type a prime can be represented in only two ways or in two infinitudes of ways, respectively. For $D=-11$, there are two classes of the first type and one of the second. He[136] later extended this method to complex variables. For D negative, imaginary periods occur. The Pell equation can be replaced by an addition theorem.

G. B. Mathews[137] developed Dedekind's[97] theory of reduction of complex numbers and its application to the geometrical theory of quadratic forms. He discussed Poincaré's[104] use of lattices (nets) in the geometrical reduction of indefinite forms, employing "hyperbolic complex quantities" $x+yj$, where $j^2=1$, $1\cdot j=j\cdot 1=j$.

130 Memorie Pont. Accad. Nuovi Lincei, 5, 1889, 131-151.
131 Jour. de Math., (4), 6, 1890, 8-11.
132 Klein-Fricke, Elliptischen Modulfunctionen, I, 1890, 243-260; II, 1892, 160-9. Summary in Encyclopédie des sc. math., t. I, vol. 3, 116-9.
133 Math. Soc. Moscow, 14, 1890, 487-526; 15, 1891, 573 (Russian).
134 *Ibid.*, 15, 1891, 118-121.
135 Bull. Imper. Soc. Univ. Moscow, 65, 1890, 62-83 (Russian).
136 X. Vers. Russ. Nat., 331 (Jahrbuch Fortschritte der Math., 1898, 176).
137 Theory of Numbers, 1892, 103-131.

K. Th. Vahlen[138] proved that every prime $8n+1$ or $8n+3$ is of the form a^2+2b^2 and every prime $6n+1$ is of the form a^2+3b^2 by using the fact that the number of representations of an odd integer s as a sum of 4 squares is 8 times the sum of the divisors of s, account being taken of permutations and signs of the roots of the squares.

F. Mertens[139] gave a simple proof of the chief theorem on reduced forms of positive determinant: two reduced forms are equivalent if and only if they belong to the same period. Use is made of the fact that if $(a, b, -c)$ and $(a', b', -c')$ have positive first coefficients and negative third coefficients and the first form is transformed into the second by $\binom{\alpha\ \beta}{\gamma\ \delta}$, then $\alpha\delta>0$.

M. Lerch[140] found series which give arithmetical invariants of $\sigma m+\tau n$ and $F=am^2+2bmn+cn^2$, the latter being a positive form of negative determinant $b^2-ac=-\Delta$. In particular (cf. Hurwitz[143] of Ch. VI),

$$K(s)=\sum_{m,\,n}{}' F^{-s}$$

satisfies the reciprocity relation

$$\Gamma(s)\left(\frac{\sqrt{\Delta}}{\pi}\right)^s K(s)=\Gamma(1-s)\left(\frac{\sqrt{\Delta}}{\pi}\right)^{1-s} K(1-s).$$

A. Hurwitz[141] gave a preliminary account of his[142] theory.

Hurwitz[142] gave a geometrical theory of reduction of forms, not necessarily with integral coefficients, of positive or negative determinants. The new principle consists in first investigating the degenerate forms and then applying the conclusions to the general form. A form $f=(a, b, c)$ of vanishing determinant D corresponds to a point λ with the homogeneous coordinates $a:b:c=1:-\lambda:\lambda^2$ on the conic $D=0$, which by choice of the coordinate system may be taken as a circle K such that the points, 0, 1, ∞ appear at the vertices of an inscribed regular triangle. If r, s, u, v are integers for which $rv-su=\pm1$, the line joining the points r/u and s/v is called an *elementary chord* of K. A triangle inscribed in K is called elementary if its three sides are elementary chords. Consider all rational numbers r/u whose numerators and denominators taken positively do not exceed the positive integer n; the corresponding points r/u of K are the vertices of a convex polygon P_n called the nth *Farey polygon,* the parameters of whose successive vertices form the nth Farey series (this History, Vol. I, pp. 155–8). Thus each side of P_n is an elementary chord and conversely. Now each elementary chord is a side of just two elementary triangles which lie on opposite sides of the chord. The elementary triangles which can be formed from the vertices of P_n cover it fully without overlapping. As n increases indefinitely, P_n approaches K. Hence all the elementary triangles cover the interior of K without overlapping.

---

138 Jour. für Math., 112, 1893, 32–33.

139 Sitzungsber. Akad. Wiss. Wien (Math.), 103, IIa, 1894, 995–1004.

140 Rozpravy Akad. Fr. Josefa, Prag, 2, 1893, No. 4 (in Bohemian). Fortschritte Math., 1893–4, 790–1.

141 Math. Papers Chicago Congress of 1893, 1896, 125–132. French transl. in Nouv. Ann. Math., (3), 16, 1897, 491–501.

142 Math. Annalen, 45, 1894, 85–117. Exposition by Klein[149] (pp. 173–266).

Under a linear transformation with integral coefficients

$$S:\quad x=\alpha x'+\beta y',\qquad y=\gamma x'+\delta y',\qquad \alpha\delta-\beta\gamma=1,$$

the point λ is transformed into a point λ' such that

$$\lambda=\frac{\alpha\lambda'+\beta}{\gamma\lambda'+\delta},\qquad \lambda'=\frac{\delta\lambda-\beta}{-\gamma\lambda+\alpha}. \tag{27}$$

By computing the points into which are transformed the end points r/u and s/v of an elementary chord, we see that the latter is transformed into an elementary chord. If p, q, r are the vertices of any elementary triangle, then

$$p=\frac{\beta}{\delta},\qquad q=\frac{\beta+\alpha}{\gamma+\delta},\qquad r=\frac{\alpha}{\gamma},$$

being three successive terms of a Farey series. By (27), S replaces this triangle by the elementary triangle T with the vertices 0, 1, ∞, which has exactly three automorphs

$$\begin{pmatrix}1&0\\0&1\end{pmatrix},\qquad \begin{pmatrix}0&-1\\1&-1\end{pmatrix},\qquad \begin{pmatrix}1&-1\\1&0\end{pmatrix}. \tag{28}$$

Every quadratic form of negative determinant is represented by a point inside the circle K and will be called *reduced* if that point lies inside or on the boundary of T. The representative point certainly is inside or on some one elementary triangle; hence each of the three transformations which carry this triangle to T will replace the form by a reduced form. The reduced forms whose representative points lie inside (and not on the boundary of) T therefore occur in triples of equivalent forms, viz., by (28),

$$(a,b,c),\qquad (c,\,-b-c,\,a+2b+c),\qquad (a+2b+c,\,-a-b,\,a).$$

Two such reduced forms are equivalent only if they belong to the same triple. The reduced forms whose representative points lie on the sides of T fall into sets of six equivalent forms.

Next, every $f=(a, b, c)$ of positive determinant is represented by a point outside the circle K, or preferably by its polar $az-2by+cx=0$ with respect to K. The parameters of its intersections with K are the roots

$$\lambda_1=\frac{-b+\sqrt{D}}{a},\qquad \lambda_2=\frac{-b-\sqrt{D}}{a}\quad\text{of } a\lambda^2+2b\lambda+c=0.$$

Call a form *reduced* if its first root λ_1 is positive and its second root λ_2 is negative (and hence if $a>0$, $c<0$). Any elementary chord which intersects the line $\lambda_1\lambda_2$ can be transformed into 0∞ by two transformations, one of which evidently carries f into a reduced form. Hence every f is equivalent to a reduced form.

Let Δ be any elementary triangle crossed by the line $\lambda_1\lambda_2$, and σ, σ' the sides met by $\lambda_1\lambda_2$ such that a point travelling from λ_2 to λ_1 along $\lambda_1\lambda_2$ crosses σ on entering Δ and σ' on leaving it. Call σ' the right neighboring chord to σ. Hence, starting with any elementary chord σ_0 meeting $\lambda_1\lambda_2$, we can form a series of elementary chords $\ldots, \sigma_{-1}, \sigma_0, \sigma_1, \sigma_2, \ldots$, each meeting $\lambda_1\lambda_2$, each being the right neighboring chord to its predecessor, and including all elementary chords meeting $\lambda_1\lambda_2$. To each σ_i

corresponds a unique transformation S_i which replaces f by a reduced form ϕ_i. Hence we get a unique chain of reduced forms belonging to f and equivalent to f. It is proved that two forms of positive determinant are equivalent if and only if to each of them belongs the same chain of reduced forms.

It is stated that the method may be extended to forms with complex coefficients and to quadratic forms in n variables.

H. W. Lloyd Tanner[143] recalled that the usual distinction between proper and improper automorphs A is by the sign of the determinant of A. Dirichlet noted that A is proper or improper according as it changes a linear factor of the form into a multiple of itself or of the other factor. An equivalent criterion is that A is improper if its square or higher power is unity, otherwise proper. [But forms of discriminant $-4\sigma^2$ or $-3\sigma^2$ have proper automorphs whose fourth or sixth powers (and no lower powers) are unity. All that he proved was that, if an automorph is improper, its square is unity, and conversely except for $x'=\pm x$, $y'=\pm y$. This is a corollary to Dirichlet's remark.] If $\theta_i^2-2b\theta_i+c=0$, and if u, v are integers for which $u^2+2buv+cv^2=1$, the substitution obtained by equating the rational parts and also the irrational parts of

$$x+y\theta_1=(u+v\theta_1)(\zeta+\eta\theta_1)$$

is a proper automorph of $(1, b, c)$. An improper automorph follows from

$$x+y\theta_1=(u+v\theta_1)(\zeta+\eta\theta_2).$$

Ch. de la Vallée Poussin[144] called (a, b, c) a *reduced* form if its coefficients are positive and a root of $a\omega^2+2b\omega+c=0$ is developable in a simply periodic negative continued fraction (whose incomplete quotients after the first are all negative integers). Every form is equivalent to a reduced form, etc., as in Gauss' theory.

H. Weber[145] gave an exposition of Dedekind's[97] theory of reduced complex numbers and applied it to numbers $x+y\sqrt{d}$, where x, y, d are rational, using Dedekind's region for a reduced number when d is negative, but for d positive the region defined by

$$0<y\sqrt{d}-x<1<y\sqrt{d}+x.$$

His deduction of Gauss' theorem on the equivalence of indefinite quadratic forms holds only for forms with integral coefficients, since it depends upon the periodicity of the development into continued fractions.

M. Bauer[146] gave an elementary proof that if two properly primitive forms represent (properly or improperly) the same numbers they have equal determinants and, if their determinants are negative, they are equivalent. Cf. Schering.[65]

F. Mertens[147] proved that, if $f=ax^2+2bxy+cy^2$ is a positive form of negative determinant $-\Delta=b^2-ac$, then $\Sigma 1/f$, summed for all pairs of integers x, y not both

[143] Messenger Math., 24, 1895, 180–9.
[144] Annales Soc. Sc. Bruxelles, 19, I, 1895, 111–3.
[145] Lehrbuch der Algebra, I, 1895, 371–401; ed. 2, I, 1898, 414–445. Cf. Archiv Math. Phys., (3), 4, 1903, 193–212.
[146] Math. és Termés. Ertesítö, 13, 1895, 316–322. Extract in Math. u. Naturw. Berichte aus Ungarn, 13, 1897, 37–44.
[147] Sitzungsber. Akad. Wiss. Wien (Math.), 106, IIa, 1897, 411–421.

zero for which $f(x, y) \leqq n$, has for large values of n the asymptotic expression

$$\frac{\pi}{\sqrt{\Delta}}\left\{\log n+2C-\log 4\Delta-2\log\frac{\theta_1\left(\frac{2}{3}, \frac{a}{3}\right)\theta_1\left(\frac{2}{3}, \frac{\beta}{3}\right)}{3\sqrt{a}}+\epsilon\right\},$$

where ϵ is of the order of magnitude of $1/\sqrt{n}$, C is Euler's constant 0.57721..., a and $-\beta$ are the roots of $f(x, 1)=0$, and

$$\theta_1(x, \omega)=-i\sum_{m=-\infty}^{\infty}(-1)^m e^k, \quad k\equiv(2m+1)^2\pi\omega i+(2m+1)\pi x i.$$

E. Landau[148] gave other proofs of the last result on Kronecker's[121] limit formula with references.

F. Klein[149] chose rectangular axes such that three consecutive vertices of Gauss'[38] fundamental parallelogram are

$$(0, 0), \quad (\sqrt{a}, 0), \quad \left(\frac{b}{\sqrt{a}}, \frac{\sqrt{-D}}{\sqrt{a}}\right),$$

where D is the (negative) determinant of the positive definite form (a, b, c). To extend this representation to indefinite forms with $D>0$, $a>0$, we employ a parallelogram three of whose vertices have the preceding coordinates with $\sqrt{-D}$ replaced by $\sqrt{D}$. The form now represents the hyperbolic distance $\sqrt{(x-x_1)^2-(y-y_1)^2}$ between points (x, y) and (x_1, y_1) of the lattice, one being the origin. He also represented (p. 177) the form by the point[150] with the homogeneous coordinates a, b, c.

H. Minkowski[151] gave another geometrical theory of indefinite forms by use of a chain of parallelograms representing a chain of substitutions $\binom{p\ r}{q\ s}$.

C. Cellérier[152] noted that the problem to represent $p=5n\pm1$ by x^2-xy-y^2 reduces, by the substitution $t=2x-y$, $u=y$, to the solution of

$$t^2-5u^2=4p.$$

If one solution a, β of the latter is known, all solutions are found by

$$t+u\sqrt{5}=(a\pm\beta\sqrt{5})(f+g\sqrt{5})/2,$$

when f, g range over all solutions of $f^2-5g^2=4$. Hence from $f=3$, $g=-1$, we get a second solution having $u=(3\beta-a)/2$. By repetitions of the process, we must reach a solution with $a>5\beta$. Hence there exists a solution with

$$\beta<a/5<\sqrt{p/5},$$

and only one such solution. The work of finding it is abbreviated by noting that $\sqrt{4p}<a<\sqrt{5p}$, and $(a^2-4p)/5$ is a square (β^2).

---

148 Jour. für Math., 125, 1903, 165, seq. Cf. M. Lerch, Archiv Math. Phys., (3), 6, 1903, 85–94.
149 Ausgewählte Kapitel der Zahlentheorie, 1, 1896, 70. Summary in Math. Annalen, 48, 1897, 562–588.
150 Cf. R. Fricke and F. Klein, Automorphen Funktionen, 1, 1897, 491; Hurwitz.[142]
151 Geometrie der Zahlen, 1, 1896, 164 (196).
152 Mém. Soc. Physique Hist. Nat. de Genève, 32, 1894–7, No. 7 (end).

E. de Jonquières,[153] by a slight modification of Gauss (Art. 162), showed how to obtain solutions of $t^2-Du^2=-m^2$ if we are given a transformation of $F=(A, B, C)$ into (a, b, c) and a transformation, whose determinant has the same sign as the former, of F into $(-a, b, -c)$. Such a pair of transformations exist when D is a prime $4k+1$.

A. Thue[154] proved that every prime divisor >2 of x^2+2y^2 is of that form, and similarly for x^2+3y^2.

A. Cunningham[155] easily proved that $2^2\cdot 7\cdot 13\cdot 19\cdot 31$ is the least N for which $x^2+3y^2=N$ has 24 sets of positive integral solutions.

T. Pepin[156] supplemented the investigation by Gauss (Arts. 213–4) of a form $f=(a, b, c)$ of determinant D transformable into $F=(A, B, C)$ of determinant De^2 by a proper substitution of determinant e, by showing that every form of determinant De^2 is obtained by transformation from some form of determinant D.

H. Poincaré[157] called a uniform function $\phi(a, b)$ an arithmetical invariant of $ax+by$ if it is unaltered by every linear substitution with integral coefficients of determinant unity. An example is

$$\Sigma(am+bn)^{-2k} \qquad (k>1),$$

summed for all pairs of integers m, n except 0, 0. If, as in this example, $\phi(a, b)$ is homogeneous of order $-2k$ in a, b, and if k is an integer >1, then $\phi(a, 1)$ is a thetafuchsian function corresponding to the modular group. Let $H(x, y)$ be a rational function, homogeneous of order $-2k$ in x, y. Then, if the summations extend over all sets of integers $\alpha, \beta, \gamma, \delta$ for which $\alpha\delta-\beta\gamma=1$,

$$\Sigma H(\alpha a+\beta,\ \gamma a+\delta)=\Sigma(\gamma a+\delta)^{-2k}H\left(\frac{\alpha a+\beta}{\gamma a+\delta}, 1\right)$$

is a thetafuchsian series and $\Sigma H(\alpha a+\beta b, \gamma a+\delta b)$ is an arithmetical invariant, under specified conditions for convergence.

Consider a rational function $H(z, z')$ and the series

$$\Sigma H\left(\frac{\alpha z+\beta}{\gamma z+\delta}, \frac{\alpha z'+\beta}{\gamma z'+\delta}\right)(\gamma z+\delta)^{-2k}(\gamma z'+\delta)^{-2k},$$

summed for all sets of integers $\alpha, \beta, \gamma, \delta$ for which $\alpha\delta-\beta\gamma=1$, and under specified conditions for convergence (including the fact that k is an integer >1). Write

$$H(a, b, a', b')=b^{-2k}b'^{-2k}H\left(\frac{a}{b}, \frac{a'}{b'}\right).$$

Then

$$(29) \qquad \Sigma H(\alpha a+\beta b,\ \gamma a+\delta b,\ \alpha a'+\beta b',\ \gamma a'+\delta b')$$

is obviously an arithmetical invariant of the two forms $l=ax+by$, $l'=a'x+b'y$. It will be an invariant of the quadratic form ll' if $H(z, z')$ is symmetric in z, z'. As a

[153] Comptes Rendus Paris, 127, 1898, 596–601, 694–700.
[154] Forh. Vid. Selsk. Kristiania, 1902, No. 7, 21 pp. (Norwegian).
[155] Math. Quest. Educ. Times, 3, 1903, 28–29.
[156] Jour. de Math., (6), 1, 1905, 333–346.
[157] Jour. für Math., 129, 1905, 89–150. For corrections and additions, Annales fac. sc. Toulouse, (3), 3, 1911, 125–149. Cf. Poincaré.[103] His related papers in Acta Math., 1, 3, 5, 1882–4, and other journals, are reprinted in his Oeuvres, II.

generalization, let $H(a, b, a', b')$ be any rational function, homogeneous of degree $-4k$ in a, b, a', b'; then the series (29), when convergent, is an invariant of l, l'.

He amplified his[109] earlier investigation of invariants of an indefinite form $F=am^2+2bmn+cn^2$ of positive determinant and studied not only ΣF^{-s} but also the series $\Sigma \pm q^F x^m y^n$.

O. Spiess[158] proved that if (A, B, C), (A', B', C') have the determinants D, D' and

$$A+2B+C=A'+2B'+C',$$

then for all values of α_1, α_2 we can determine β_1, β_2 such that

$$A(t-\alpha_1)^2+2B(t-\alpha_1)(t-\alpha_2)+C(t-\alpha_2)^2= A'(t-\beta_1)^2+2B'(t-\beta_1)(t-\beta_2)+C'(t-\beta_2)^2,$$

identically in t. In order that there shall hold at the same time the similar identity with A and A', ..., C and C' interchanged, it is necessary and sufficient that $D=D'$ or $D=\pm D'$, according as $A+2B+C$ is or is not zero.

Th. Pepin[159] gave an exposition of the classic theory of binary quadratic forms.

J. Sommer[160] applied algebraic numbers to prove that every prime $\equiv 1$ or 3 (mod 8) is representable by x^2+2y^2, $x>0$, $y>0$, in one and but one way, likewise a prime $\equiv 1 \pmod 3$ by x^2+3y^2, while $x^2-2y^2=8n\pm 1=$prime has an infinitude of solutions.

A. Aubry[160a] gave a summary on quadratic forms without exact references.

H. Minkowski[160b] proved that if $f=ax^2+2bxy+cy^2$ is a positive form with $\delta=ac-b^2>0$, $a>0$, we can assign integral values not both zero to x and y such that $f\leqq 2\sqrt{\delta/3}$. The equality sign holds only when f is equivalent to $\sqrt{\delta}(x^2+xy+y^2)$. He gave a simple geometrical interpretation by means of the thickest packing of circles.

L. E. Dickson[161] obtained necessary and sufficient conditions that two pairs of binary quadratic forms with coefficients in any field (or domain of rationality) F shall be equivalent under linear transformation with coefficients in F.

D. A. Gravé[162] let $\binom{\alpha\ \beta}{\gamma\ \delta}$ and $\binom{\alpha'\ \beta'}{\gamma'\ \delta'}$ be substitutions of determinants e and e' which transform (A, B, C) into (a, b, c) and (a_1, b_1, c_1), and transform $(\bar{A}, \bar{B}, \bar{C})$ into $(\bar{a}, \bar{b}, \bar{c})$ and $(\bar{a}_1, \bar{b}_1, \bar{c}_1)$, respectively. Define $a', c', 2b'$ by (13) of Gauss. Let $\bar{a}', \bar{c}', 2\bar{b}'$ denote the analogous functions of $\bar{A}, \bar{B}, \bar{C}$ with γ and β, γ' and β', interchanged. Then

$$\Omega\equiv \bar{A}a+2\bar{B}b+\bar{C}c=A\bar{a}+2B\bar{b}+C\bar{c},$$

---

158 Archiv Math. Phys., (3), 9, 1905, 340–4.

159 Memorie Pontif. Accad. Romana Nuovi Lincei, 24, 1906, 243–288; 25, 1907, 83–107 (reduction for negative determinant, table of linear forms of divisors of x^2+A for $A=1, \ldots, 31$); 27, 1909, 309–351 (positive determinant, Pell equation); 28, 1910, 307–348 (periods of reduced forms, equivalence of primitive forms of positive determinant); 29, 1911, 319–339 (distribution of classes into genera; see report under quadratic reciprocity law).

160 Vorlesungen über Zahlentheorie, 1907, 125–7; French transl. by A. Lévy, Paris, 1911, 132–4.

160a L'enseignement math., 9, 1907, 289–294, 431–2, 436, 442–3.

160b Diophantische Approximationen, 1907, 55.

161 Amer. Jour. Math., 31, 1909, 103–8.

162 Comptes Rendus Paris, 149, 1909, 770–2. His $\alpha_1, \ldots, \delta_1$ are here replaced by $\alpha', \ldots, \delta'$ to fit Gauss' notations.

from which we obtain Ω' and Ω_1 by replacing $a, \ldots, \bar{c}$ by $a', \ldots, \bar{c}'$ and by $a_1, \ldots, c_1$, respectively. We have the identity

$$\Omega'^2 - \Omega\Omega_1 = D\bar{N}^2 + \bar{D}N^2 - D\bar{D}\Delta,$$

where $D = B^2 - AC$, $\bar{D} = \bar{B}^2 - \bar{A}\bar{C}$,

$$\Delta = 2ee_1 + (\alpha'\delta + \alpha\delta' - \beta\gamma' - \beta'\gamma)^2,$$
$$N = A(\alpha\beta' - \beta\alpha') + B(\beta'\gamma + \alpha\delta' - \alpha'\delta - \beta\gamma') + C(\gamma\delta' - \gamma'\delta),$$

while $\bar{N}$ is the same function of $\bar{A}, \bar{B}, \bar{C}$ with β and γ, β' and γ' interchanged. By means of this identity we readily obtain all the details of Gauss (Arts. 162–6). For, if $a_1 = a$, etc., we have $e_1^2 = e^2$. If $e = e_1$, then $N^2 - D\Delta = 0$ and $\Omega'^2 - \Omega\Omega_1 = D\bar{N}^2$. If $e = -e_1$, we conclude that ambiguous forms exist.

F. de Helguero[163] made an elementary study, without the theory of quadratic forms, of the numbers representable properly by $x^2 + xy + y^2 = \{x, y\}$. If $a = \{x_1, y_1\}$ and $b = \{x_2, y_2\}$, then $ab = \{\alpha, -\beta\} = \{\alpha, \gamma\} = \{\beta, -\gamma\}$, where

$$\alpha = x_1x_2 + y_1y_2 + x_1y_2, \quad \beta = x_1x_2 + y_1y_2 + x_2y_1, \quad \gamma = x_2y_1 - x_1y_2.$$

These and the three representations of ab derived by interchanging x_2 and y_2 are distinct in general. These six representations are proper if a is prime to b. If a is a prime, every proper representation of ab is obtained from one proper representation of a and all proper representations of b by means of these product formulas. A prime p is representable by $\{x, y\}$ if and only if $p = 3h + 1$ or $p = 3$. Every product of powers of n distinct primes $3h + 1$, or 3 times such a product, has exactly $3 \cdot 2^{n-1}$ different proper representations by $x^2 + xy + y^2$.

J. V. Uspenskij[164] applied an algorithm closely analogous to ordinary continued fractions to the reduction of indefinite binary quadratic forms and obtained the periods of reduced forms more rapidly than had Gauss.

G. Fontené[165] noted that every prime factor $\neq 3$ of a number of the form $x^2 + xy + y^2$ is of the form $3k + 1$.

H. C. Pocklington[166] gave an elementary discussion of the form of the prime divisors of $mx^2 + ny^2$ and of any divisor of $ax^2 + 2bxy + cy^2$.

L. Aubry[167] proved that every divisor of $x^2 + ky^2$ (x prime to y) which is not $< 2\sqrt{k/3}$ for $k > 0$ or not $< \sqrt{k}$ [meaning $\sqrt{|k|}$] for $k < 0$ is of the form $f = (v^2 + ku^2)/d$, where $d \leqq 2\sqrt{k/3}$ for $k > 0$ or $d \leqq \sqrt{k}$ for $k < 0$. He readily transformed f into a reduced form. Use is made of the lemma that

$$X^2 + kY^2 = DE, \quad Xq - uE = \pm pY, \quad Xp - vE = \mp kqY, \quad E = \frac{1}{d}(p^2 + kq^2)$$

imply

$$D = \frac{1}{d}(v^2 + ku^2), \quad Xu - qD = \mp vY, \quad Xv - pD = \pm kuY.$$

As corollaries, every divisor of $X^2 + 1$ is of the form $u^2 + v^2$, and every odd divisor of $X^2 + k$ is of the form $v^2 + ku^2$ for $k = 3$ or 7.

---

163 Giornale di Mat., 47, 1909, 345–364. Many misprints.
164 Applications of continuous parameters in the theory of numbers, St. Petersburg, 1910.
165 Nouv. Ann. Math., (4), 10, 1910, 217.
166 Proc. Cambridge Phil. Soc., 16, 1911, 13–16.
167 Assoc. franç. av. sc., 40, 1911, 55–60.

J. Schatunovsky[168] proved that, if $x^2+Dy^2\equiv 0 \pmod p$ has relatively prime solutions, also $u^2+D\equiv 0 \pmod p$ has a solution. All odd divisors of a^2+2b^2 are $\equiv 1$ or $3 \pmod 8$. If D is a prime $4n+3$ and p is an odd divisor of a^2+Db^2, then $(p/D)=+1$. If D is a prime $4n+1$ and p is a divisor of a^2+Db^2, then

$$\left(\frac{p}{D}\right)=(-1)^{\frac{p-1}{2}}.$$

He tabulated the linear forms of divisors of a^2+Db^2 for $D=1, 2, 3, 7, 11, 19, 27, 43, 67, 163$.

P. Bernays[169] gave an elementary exposition of Landau's[170] asymptotic enumeration of the primes $\leqq x$ representable by a primitive quadratic form. Passing from primes to arbitrary integers, he obtained an asymptotic expression for the number of positive integers $\leqq x$ representable properly by a class of forms, and similarly when the representation is either proper or improper. In each case the limit for $x=\infty$ of the ratio of the numbers of representations by two different classes is unity, so that, asymptotically, equally many numbers are representable by the various classes of discriminant D.

D. N. Lehmer[171] studied pencils $\alpha A+\beta B$ of forms, where the base forms A, B are binary quadratic forms in x, y with integral coefficients, while α and β each ranges over all integers. As base forms we may also take $\bar{A}=\alpha A+\beta B$, $\bar{B}=\gamma A+\delta B$, where $\alpha, \ldots, \delta$ are integers of determinant ± 1. Use is made also of linear substitutions on x, y with integral coefficients of determinant ± 1, and consequently of invariants and covariants of the pair of base forms.

A. Châtelet[172] presented Hermite's continual reduction and principal reduced forms from the standpoint of matrices.

G. Frobenius[173] considered forms $(a, b, c)=ax^2+bxy+cy^2$ of positive discriminant $b^2-4ac=D=R^2$, $R>0$, with any real coefficients. It is called *reduced* if $b<R$, $b>R-2|a|$, $b>R-2|c|$. Every reduced form $\phi_0=(a_0, b_0, -a_1)$ has a unique right-neighboring reduced form $\phi_1=(-a_1, b_1, a_2)$ and a unique left-neighboring reduced form ϕ_{-1}, and hence a chain $\ldots, \phi_{-1}, \phi_0, \phi_1, \phi_2, \ldots$. Any two forms are called *equivalent* if transformable into each other by linear substitutions with integral coefficients of determinant unity. By use of the continued fraction for a rational number, it is proved that two equivalent reduced forms belong to the same chain. It is next proved that every form is equivalent to a reduced form. Finally, if $\phi_\lambda=(a_\lambda, b_\lambda, a_{\lambda+1})$, where λ ranges from $-\infty$ to $+\infty$, constitute the chain of reduced forms determined by ϕ, the a_λ include all numbers which are representable by ϕ and are numerically $\leqq \frac{1}{2}R$ (Lagrange[18]). The present method of reduction is said to be essentially that by Mertens[105] and to furnish an introduction to the following paper by Schur.

---

[168] Der grösste gemein. Theiler von algebraischen Zahlen zweiter Ordnung, Diss. Strassburg, Leipzig, 1912, 51–58.
[169] Ueber die Darstellung . . . primitiven, binären Quad. Formen . . . , Diss., Göttingen, 1912.
[170] Math. Annalen, 63, 1907, 202.
[171] Amer. Jour. Math., 34, 1912, 21–30.
[172] Leçons sur la théorie des nombres, 1913, 95–102.
[173] Sitzungsber. Akad. Wiss. Berlin, 1913, 202–211.

I. Schur[174] wrote $K(\phi)$ for the class of all forms obtained from $\phi=ax^2+bxy+cy^2$ (with real coefficients) by linear substitutions with integral coefficients of determinant unity. If q is a constant $\neq 0$, the class $K(q\phi)$ is said to be *proportional* to the class $K(\phi)$. Consider only forms ϕ of positive discriminant $b^2-4ac=D$ and which vanish for no set of integers x, y except 0, 0. When ϕ ranges over all forms of its class, let A and B be the lower limits of the $|a|$ and $|b|$, respectively. Markoff[102] had proved that for all classes K of discriminant D, the least point (value) of condensation of $Q'=\sqrt{D}/A$ is 3, and there exist infinitely many non-proportional classes for which $Q'=3$. Here is proved the similar theorem that the least point of condensation of $Q''=\sqrt{D}/B$ is $2+\sqrt{5}$. Only for the classes proportional to that containing $(1, 1, -2-\sqrt{5})$ is $Q''=2+\sqrt{5}$. A formula is given for the Q'' less than $2+\sqrt{5}$. Call $\mu=(a, b, -c)$ a *minimal* form if $0\leqq b\leqq a\leqq c$, $a>0$, and if a is the least number represented by $|\mu|$. In every minimal form μ, other than $(a, a, -a)$, we have $c\gtreqless 2a+b$. For all minimal forms, the least point of condensation of $t=\sqrt{D}/b$ is $2+\sqrt{3}$, while t is less than the latter only for three forms:

$$(a, a, -a), \quad (a, a, -3a), \quad \left(a, a, -\frac{3+2\sqrt{3}}{2}a\right).$$

G. Frobenius[175] quoted a theorem by Markoff[102] in the following explicit form: Let $\psi=ax^2+bxy+cy^2$ have any real coefficients such that $D=b^2-4ac>0$. Let M be the least value of $|\psi|$ for integral values of x, y. Then for the totality of forms ψ, the least point of condensation of $\sqrt{D}/M$ is 3. If $\sqrt{D}<3M$, the product of ψ by a suitable factor is properly or improperly equivalent to a form

$$\phi=px^2+(3p-2q)xy+(r-3q)y^2,$$

where p, q, r are positive integers such that[176] $p^2+p_1^2+p_2^2=3pp_1p_2$ has integral solutions, while $\pm q$ is the absolutely least residue of p_1/p_2 modulo p, and r is given by $pr-q^2=1$. For ϕ, $M=p$, $\sqrt{D}<3M$. But if the ratios of the coefficients of ψ are not all rational, $\sqrt{D}\gtreqless 3M$. Frobenius called p a *Markoff number*, studied its properties, and gave explicit expressions for p, q, r in terms of the partial denominators of a continued fraction. But he did not treat the general Markoff theorem.

R. Bricard proved that every prime $8q\pm1$ is of the form x^2-2y^2 by a method described in this History, Vol. II, p. 255.

H. N. Wright[177] tabulated the reduced forms of negative determinant $-\Delta$ for $\Delta=1, \ldots, 150, 800, \ldots, 848$. The values of b, c occur at the intersection of the row giving Δ and the column giving a.

G. H. Hardy[178] wrote $r(n)$ for the number of representations of n by $\alpha x^2+\beta xy+\gamma y^2$, where $\alpha>0$, $\Delta^2=4\alpha\gamma-\beta^2>0$, and wrote

$$R(x)=\sum_{n\leqq x} r(n)=\frac{2\pi x}{\Delta}+P(x).$$

He proved there exists a positive constant K such that each of the inequalities

174 Sitzungsber. Akad. Wiss. Berlin, 1913, 212–231. *Cf.* Frobenius.[173]
175 *Ibid.*, 458–487.
176 On this and analogous Diophantine equations, see History, Vol. II, p. 697.
177 University of California Publications, Math., 1, 1914, No. 5, 97–114.
178 Quar. Jour. Math., 46, 1915, 282–3.

$P(x) > Kx^{1/4}$, $P(x) < -Kx^{1/4}$ is satisfied by values of x surpassing all limit. He gave an explicit analytic expression for $R(x)$ as an infinite series involving Bessel's function J_1.

G. Humbert[179] considered positive reduced forms $f=ax^2-2bxy+cy^2$, whence $a>0$, $c>0$, $ac-b^2>0$, $2|b|\leqq a \leqq c$. Without loss of generality we may take $b \gtreqless 0$. As known, the first, second, and third minima of f are a, c, $a-2b+c$. It is proved that the fourth minimum is $a+2b+c$, the fifth is $4a-4b+c$, the sixth is the least of $4a+4b+c$ and $a-4b+4c$, while the seventh is the greater of the latter two numbers. Another proof was given by G. Julia.[180]

J. G. Van der Corput[181] stated and W. Mantel[181] proved that if $P=AX^2+2BXY+CY^2$ is divisible by $p=ax^2+2bxy+cy^2$, and if $AC-B^2=ac-b^2$, there exist integers X_1, Y_1, x_1, y_1 such that

$$P(AX_1^2+2BX_1Y_1+CY_1^2)=p(ax_1^2+2bx_1y_1+cy_1^2).$$

J. G. Van der Corput[182] discussed the forms $A=\pm(a^2+mab+nb^2)$ for the cases $m^2-4n=-8, -7, -4, -3, 5, 8, 12, 13, 17$. By adding to a a multiple of b, we see that we may alter m by a preassigned even integer, and hence take $m=0$ or 1. Every divisor of A is representable in the same form A. Primes of certain linear forms are representable by A. If $-\omega_1$ and $-\omega_2$ are the values of a/b for which $A=0$, all the representations by A of $p_1^{s_1}\ldots p_\nu^{s_\nu}$, where the prime p_ν is equal to $\pm(a_\mu^2+ma_\mu\beta_\mu+n\beta_\mu^2)$, are obtained by multiplying

$$a+b\omega_1=(t+u\omega_1)\prod_{\mu=1}^{\nu}(a_\mu+\beta_\mu\omega_1)^{s_\mu-w_\mu}(a_\mu+\beta_\mu\omega_2)^{w_\mu},$$

by the similar relation with ω_1, ω_2 interchanged, in which t, u is the general set of solutions of $t^2+mtu+nu^2=1$, $s_\mu \gtreqless w_\mu \gtreqless 0$. By this theorem there is found the number of representations by a^2+b^2 and a^2+2b^2. Many references are given to sources of various cases of the theorems proved.

G. Humbert[183] proved the theorems stated by Korkine and Zolotareff.[101]

G. Humbert[184] proved that the number of Hurwitz[142] reduced indefinite forms which are equivalent to (a, b, c) is equal to the sum of the incomplete quotients of the minimum period of the ordinary continued fraction for the positive root ω of $a\omega^2+2b\omega+c=0$, or double that sum, according as ω is not or is modularly equivalent to $-\omega$, i. e., (a, b, c) is not or is equivalent to $(-a, b, -c)$. This is the analogue of the theorem (quoted at the end of the report on Dirichlet[57]) on Gauss reduced forms (here proved on p. 128).

The Hermite[53] reduced forms include principal (corresponding to Gauss reduced forms) and secondary, and are represented by circles which penetrate the classic fundamental domain D_0 of the modular group (Klein,[132] Smith[95]). Let (a, b, c) be an indefinite primitive form such that a root ω of $a\omega^2+2b\omega+c=0$ is positive. Humbert proved that, if $h_1, \ldots, h_k$ are the incomplete quotients of the minimum

---

179 Comptes Rendus Paris, 160, 1915, 647–650.
180 *Ibid.*, 162, 1916, 151–4.
181 Wiskundige Opgaven, 12, 1915–18, 166–8 (in Dutch).
182 Nieuw Archief voor Wiskunde, (2), 11, 1915, 45–75 (in French).
183 Jour. de Math., (7), 2, 1916, 164.
184 Jour. de Math., (7), 2, 1916, 104–154. Summary in Comptes Rendus Paris, 162, 1916, 23–26, 67–73.

period of the ordinary continued fraction for ω, the total number of Hermite reduced forms equivalent to (a, b, c) is $\Sigma(1+h_j)$, provided the two reduced forms (α, β, γ) and $(-\alpha, \beta, -\gamma)$ are not regarded as distinct, while a reduced form whose representative circle passes through one of the summits $\tau=\pm\frac{1}{2}+i\sqrt{3}/2$ of the domain D_0 is counted as two forms. He found the number of principal Hermite reduced forms which are equivalent to (a, b, c), viz., those whose representative circle cuts the circle, through the points τ, having the radius unity and center at the origin.

In the main paper only (pp. 124–130), he called an indefinite form a *Smith reduced form* if its representative circle cuts the two straight sides, $x=0$, $x=1$, of the initial triangular domain. The principal Smith reduced forms are in (1, 1) correspondence with the Gauss reduced forms.

G. Humbert[185] gave a theory of reduction much simpler than that of Smith[95] by employing equivalence with respect to the following different subgroup Γ of the classic modular group. A substitution

$$z'=\frac{\lambda z+\nu}{\mu z+\rho},\qquad \lambda\rho-\mu\nu=1,$$

is in Γ if $\lambda, \ldots, \rho$ are ordinary integers such that $\lambda+\rho$ and $\mu+\nu$ are even. The fundamental domain D_0 of Γ is the region in the half plane above the ξ-axis which is bounded by the lines $\xi=\pm1$ and the semi-circle of radius unity and center at the origin.

A positive form (a, b, c) of proper order (i. e., with a and c not both even) is called *reduced modulo* 2 if (i) a and c are odd and (ii) if its representative point is in D_0 or on the part of its boundary at the left of the η-axis, i. e., if $|b|\leqq a\leqq c$, and $b \gtreqless 0$ if either sign is equality. There is one and only one reduced form equivalent in the ordinary sense to a given form. But a positive form of improper order is called reduced modulo 2 if conditions (ii) hold, there being three reduced forms equivalent in the ordinary sense to a given form.

An indefinite form of proper (improper) order is called reduced modulo 2 if a and c are both odd (even) and if the semi-circle $a(\xi^2+\eta^2)+2b\xi+c=0$ representing (a, b, c) penetrates the domain D_0, i. e., if at least one of the numbers $a(a\pm2b+c)$ is negative.

Principal reduced forms are those whose representative circles cut the curved side of D_0, i. e., if $a\equiv c \pmod 2$ and $(a+c)^2-4b^2<0$. For a proper order they constitute one or two periods, according as u_0 is odd or even in the least positive solution of $t_0^2-Du_0^2=1$. For an improper order they constitute 1, 2 or 3 periods, depending on the residues of $b^2-ac \pmod 4$ and $u_0 \pmod 2$.

Humbert[186] considered indefinite principal reduced forms modulo 2 having $b>0$. Then $\beta=b-\frac{1}{2}|a+c|$ is positive. By equating coefficients of infinite series, it is shown that the sum of the values of 2β for all the principal reduced forms of proper order of given determinant N is

$$\Sigma 2(-1)^{\frac{m_1-1}{2}}(m_2-m_1) \text{ if } N\equiv 2 \text{ or } 3 \pmod 4,$$

$$\Sigma 2m(-1)^{\frac{m_1-1}{2}}+k \text{ if } N\equiv 0 \text{ or } 1 \pmod 4,$$

185 Comptes Rendus Paris, 165, 1917, 253–7 (157, 1913, 1358–62, for indefinite forms only).
186 *Ibid.*, 298–304.

summed for all the classes of positive forms of proper order of determinant $-N$, where m_1 and m_2 are the odd minima and m the even minimum of the class, while k denotes 0 or $-2N$, according as N is not or is a square. There are analogous results for improper orders. For applications, see Humbert[355] of Ch. VI.

U. Scarpis[187] proved that if (a, b, c) is a reduced form of negative determinant and (p, q, r) is any equivalent form, either (i) the two forms do not have extreme coefficients in common and then p and r exceed c and $|q|>b$, or (ii) they have in common one of the extreme coefficients and then the remaining coefficients of (p, q, r) are not less numerically than the corresponding coefficients of (a, b, c). Thus in a class H of equivalent forms, the reduced form has the minimum coefficients. Let Σ be a set of those forms of H which have the same first coefficient a, and S the system of those forms of Σ which are parallel forms (their middle coefficients b being congruent modulo a). It is shown that the forms of H can be separated into sets each composed of one or more systems such that the first coefficients which are constant in each set are arranged in order of magnitude, and a process is given for finding the $(n+1)$th set when the first n sets are known.

L. J. Mordell[187a] proved that ϕ is an ambiguous form if there exist integers x, y for which the partial derivatives ϕ_x and ϕ_y are both divisible by ϕ.

M. Amsler[188] applied continued fractions and Farey series (Hurwitz[142]) to obtain theorems on reduced forms.

J. A. Gmeiner[189] gave a single process of finding reduced forms whether the determinant is positive or negative.

F. Mertens[190] gave a simple proof independent of continued fractions of Gauss' theorem that equivalent reduced forms of positive determinant belong to the same period.

A. Cunningham[191] noted that, since every prime, $10\omega\pm1$ can be expressed in the form t^2-5u^2, it is evidently expressible in a single way [infinitely many ways] in the form $2C-C'$ and in infinitely many ways in the form $2C'-C$, where

$$C=\frac{x^3-y^3}{x-y},\quad C'=\frac{x^3+y^3}{x+y},\quad 2C-C',\quad 2C'-C=\left(x\pm\frac{3}{2}y\right)^2-5\left(\frac{y}{2}\right)^2,$$

Again, $6Q-5S=(2C-C')^2$, $6Q'-5S=(2C'-C)^2$, where

$$Q=\frac{x^5-y^5}{x-y},\quad Q'=\frac{x^5+y^5}{x+y},\quad S=\frac{x^6+y^6}{x^2+y^2}.$$

P. Epstein,[192] starting from *elementary* continued fractions whose partial denominators are all $+1$ and partial numerators are ±1, was led to the group B of linear fractional *even* substitutions generated by $U: x'=1+1/x$ and $V: x'=1-1/x$. Two irrational numbers are equivalent under B if and only if their developments into elementary continued fractions coincide after a certain place. A substitution

187 Periodico di Mat., 32, 1917, 150–8.
187a Messenger Math., 47, 1917–9, 71.
188 Bull. Soc. Math. France, 46, 1918, 10–34.
189 Sitzungsber. Akad. Wiss. Wien (Math.), 127, IIa, 1918, 653–698; 128, IIa, 1919, 957–1005.
190 *Ibid.*, 127, IIa, 1918, 1019–34. Cf. Mertens.[105]
191 Proc. London Math. Soc., (2), 18, 1920, XXV, XXVI.
192 Jour. für Math., 149, 1919, 57–88; 151, 1920, 32–62.

generated by U, V and $R: x'=1/x$ is even or odd according as the product contains an even or odd number of factors R. The even and odd substitutions together give all substitutions with integral coefficients of determinants 1 or -1. There is developed a theory of reduction and equivalence of indefinite binary quadratic forms under the group B of even substitutions.

E. T. Bell[193] noted that, if $f(u, v) \equiv f(-u, -v)$,

$$\Sigma tf(n_1+n_2,\ n_1-3n_2)=\Sigma tf(2n_2,\ 2n_1),\qquad t\equiv 1+(-1)^{n_1+n_2},$$

summed for all integral solutions n_1, n_2 of $n_1^2+3n_2^2=n$. Let $N(n)$ be the number of solutions of the latter equation. For $n=2r$, and for summations extending over all the solutions,

$$\Sigma n_1^2=3\Sigma n_2^2=rN(2r),\qquad 2\Sigma n_1^4=18\Sigma n_1^2n_2^2=18\Sigma n_2^4=3r^2N(2r).$$

As known, $N(2r)=6\Sigma(-3/d)$, summed for the odd divisors d of r.

NOTES FROM L'INTERMÉDIAIRE DES MATHEMATICIENS.

Rodallec, 20, 1913, 25–27, expressed a prime $A=8n+7$ in the form a^2-2b^2 and a prime $A=8n+3$ in the form $\alpha^2+2\beta^2$ by use of the continued fraction for $\sqrt{A}$.

"Quilibet," 22, 1915, 240, noted errata in Table IV of the linear forms of the divisors of $x^2 \pm \Delta y^2$ in Cahen's Théorie des Nombres, 1900.

E. Malo, 22, 1915, 248–251, discussed forms of negative determinant capable of representing the same number in several ways.

M. Rignaux, 24, 1917, 7, stated that all representations of a number by a form of positive determinant are given by one or more recurring series with the same law $T_{n+1}=qT_n-T_{n-1}$, where q is found by developing the first root of the form into a continued fraction.

G. Métrod, 24, 1917, 9–13, reported known results on the representation by ny^2-x^2. He proved that, whether Δ is positive or negative, any prime p is representable by $x^2-\Delta y^2$ if Δ has no square factor and is numerically $\leqq \frac{1}{2}(p-1)$, except for $p=2$, 3, 5.

A. Gérardin, 25, 1918, 59–61, expressed numbers c^4-1 by x^2-2y^2.

In Vol. I of this History references were given for the application of binary quadratic forms to the solution of binomial congruences by Gauss and Legendre, p. 207, by Smith, p. 210, and by Cunningham, p. 219. On the number of representations of an odd integer by x^2+2y^2, see Glaisher,[141] p. 318. For the applications of binary quadratic forms to factoring (with material on congruent forms and idoneal numbers), see pp. 361–5, 369 (Gauss), and 370 (Schatunovsky). For Lucas' results on the divisors of ax^2+by^2, see pp. 396–401. A binary quadratic form represents an infinitude of primes, pp. 417–8, p. 421 (Frobenius).

In Vol. II, Chapters XII, XIII, XX, and p. 546, occurs material on the representation of numbers by binary quadratic forms.

---

[193] Bull. Soc. Math. de Gréce, II, 2, 1921, 70–74. Cf. Liouville.[89]

CHAPTER II.

EXPLICIT VALUES OF x, y IN $x^2+\Delta y^2=g$.

In his Disquisitiones Arithmeticae, Art. 358 (Maser's transl., pp. 428-433), Gauss solved $4p=t^2+27u^2$ in terms of numbers arising in the theory of the three periods of pth roots of unity, p being a prime $3m+1$.

In his study of biquadratic residues, Gauss (this History, Vol. II, p. 234) obtained, by applying *cyclotomy* (the theory of roots of unity), the residues modulo p of x, y in $x^2+y^2=p$, when p is a prime $4k+1$.

In a posthumous MS. (Werke, X_1, 1917, 39), Gauss stated the first result by Jacobi,[1] the second by Stern,[5] and those by Stern.[12]

C. G. J. Jacobi,[1] as an application of cyclotomy, found that, if $p=3n+1$ is a prime, we have $4p=a^2+27b^2$, where a is the *absolutely least residue* (between $-\frac{1}{2}p$ and $+\frac{1}{2}p$) modulo p of $-(n+1)(n+2)\ldots(2n)/n!$, and that this residue is $\equiv 1 \pmod 3$. If $p=7n+1$ is a prime, then $p=L^2+7M^2$, where L is the absolutely least residue modulo p of $\frac{1}{2}(2n+1)(2n+2)\ldots(3n)/n!$, and this residue is $\equiv 1 \pmod 7$.

A. L. Cauchy[2] employed a prime p, a prime divisor n of $p-1$, the least integer $m\equiv \pm 2A_{(n+1)/4} \pmod n$, where A_j is the jth Bernoullian number, a primitive root s of n, the number n' of roots $<\epsilon=(n-1)/2$ of $x^\epsilon\equiv 1 \pmod n$, and the number $n''=n-n'-1$ of roots $<\epsilon$ of $x^\epsilon\equiv -1$. Set $\omega=(p-1)/n$, $(s)=s!$. Then, if $n=4k+3$, $x^2+ny^2=4p^m$ is solvable in integers and is verified by

$$x\equiv(\omega)(s^2\omega)(s^4\omega)\ldots(s^{n-3}\omega) \text{ or } (s\omega)(s^3\omega)\ldots(s^{n-2}\omega) \pmod p,$$

according as $n'<n''$ or $n'>n''$. In the respective cases, $m=(n-1-4n')/d$ or $(4n'-n+1)/d$, where $d=2$ or 6, according as $n\equiv 7$ or $3 \pmod 8$.

M. A. Stern[3] proposed the following problem: If p is a prime $6n+1=a^2+27b^2$, prove that

$$(n+1)(n+2)\ldots(2n)\equiv -(2n+1)(2n+2)\ldots(3n) \pmod p.$$

T. Clausen[4] proved this formula in which the sign is correctly minus if $p\equiv 7 \pmod{12}$, but should be plus if $p\equiv 1 \pmod{12}$.

M. A. Stern[5] employed a prime $p=6n+1$ for which $p=a^2+3b^2$, $4p=g^2+27h^2$, quoted Jacobi's[1] result (with n replaced by $2n$), and noted that

$$\pm g\equiv\frac{\{(2n+1)\ldots(3n)\}^2}{1\cdot 2\ldots(2n)},\qquad \pm 2a=\frac{(2n+1)\ldots(3n)}{n!} \pmod p,$$

the last sign being such that $\pm a\equiv 1 \pmod 3$.

---

[1] Jour. für Math., 2, 1827, 69; Werke, VI, 1891, 237. Second theorem stated also in a letter to Gauss, Feb. 8, 1827; Werke, VII, 1891, 395–8.
[2] Bull. des Sc. Math. Phys. Chim. (ed., Férussac), 15, 1831, 137–9.
[3] Jour. für Math., 7, 1831, 104.
[4] *Ibid.*, 8, 1832, 140.
[5] Jour. für Math., 9, 1832, 97.

C. G. J. Jacobi[6] proved by cyclotomy that, if $p=8n+1=c^2+2d^2$ is a prime, c is the absolutely least residue modulo p of *

$$-\tfrac{1}{2}\left(\frac{p+1}{2}\right)\left(\frac{p+3}{2}\right)\ldots\{\tfrac{5}{8}(p-1)\}/\left(\frac{p-1}{8}\right)!.$$

His theory implies similar results for $p=x^2+\Delta y^2$, first enunciated by Cauchy.[9] He tabulated the values of c, d for $p<6000$, and the values of A, B in $p=6n+1=A^2+3B^2$, p a prime <12000 (errata[30]).

V. A. Lebesgue[7] proved the earlier result by Jacobi[1] by a study of the solutions of $x^3-cy^3\equiv d \pmod{p=3n+1}$. He proved Gauss' theorem.

M. A. Stern[8] proposed the following problem: If p is a prime $8n+1$, so that $p=a^2+b^2$, show that (besides Gauss' result) we have

$$\pm 2a\equiv\frac{(3n)(3n-1)\ldots(2n+1)}{n!} \pmod{p},$$

the sign being + only when $a=4m+1$, $b=8r$, $n=2s$, or $b=8r+4$, $n=2s+1$; $a=4m+3$, $b=8r$, $n=2s+1$, or $b=8r+4$, $n=2s$.

A. L. Cauchy[9] proved that if Δ is an integer $\equiv 3 \pmod 4$ not divisible by a square, and if $p=\Delta n+1$ is an odd prime, and if a, b range over the integers less than Δ and prime to Δ for which, respectively,

$$\left(\frac{a}{\Delta}\right)=+1,\quad \left(\frac{b}{\Delta}\right)=-1,$$

in Jacobi's quadratic residue symbols, then[10]

$$4p^\mu=x^2+\Delta y^2,\qquad \pm x\equiv\frac{1}{\Pi_a\{(an)!\}} \pmod{p},\qquad \mu=\frac{\Sigma b-\Sigma a}{\Delta}.$$

The results of Gauss and Jacobi[1, 6] for $\Delta=1$, 2 3 are noted[11] as special cases. For another statement and generalization of Cauchy's results, see the report on Stickelberger.[25]

M. A. Stern[12] remarked that he had published[13] prior to Jacobi[6] the equivalent result $\pm 2c\equiv(4n)\ldots(3n+1)/n! \pmod p$, which he now reduced to the simpler form $\pm 2c\equiv(2n)\ldots(n+1)/n! \pmod p$, the upper sign holding if $c\equiv 1$, 3 (mod

* For $p=17$, the expression becomes $-\frac{1}{2}.45\equiv\frac{1}{2}.6 \pmod{17}$, and $c=3$.

[6] Monatsber. Akad. Berlin, 1837, 127–136; Jour. für Math., 30, 1846, 166–182; Werke, VI, 1891, 256, 268–271; Opuscula Math., I, 1846, 318–9, 329–332. French transl., Nouv. Ann. Math., 15, 1856, 337–351.

[7] Jour. de Math., 2, 1837, 279, 283.

[8] Jour. für Math., 18, 1838, 375–6.

[9] Mém. Institut de France, 17, 1840, 249–768; Oeuvres, (1), III, 1–450 (especially notes II, III, XIII). Comptes Rendus Paris, 9, 1839, 473, 519; 10, 1840, 51, 85, 181, 229; Oeuvres, (1), IV, 504–13; (1), V, 52–81, 85–111. Bull. sc. math., phys., chim. (ed., Férussac), 12, 1829, 205.

[10] As enunciated by H. J. S. Smith, Report Brit. Assoc. for 1863, 768–786; Coll. Math. Papers, I, 273. He gave the companion theorem $p^\mu=x^2+\Delta y^2$, the residue of x being half of that in the text, where $p=4\Delta n+1$, a and b being integers $<4\Delta$ and prime to 4Δ for which $(-\Delta/a)=+1$, $(-\Delta/b)=-1$. In his proofs, Smith followed the method barely indicated by Jacobi,[6] rather than Cauchy's.

[11] Mém. Institut de France, 17, 1840, 724–30; Oeuvres, (1), 3, 410–8.

[12] Jour. für Math., 32, 1846, 89, 90.

[13] Jahrbücher für Wiss. Critik, 1831, 679.

8), and the lower sign if $c \equiv 5, 7 \pmod 8$. Finally, if $p=8n+3=c^2+2d^2$ is a prime, and the sign is taken as before, then

$$\mp 2c \equiv (4n+1)\ldots(3n+2)/n! \pmod p.$$

These results were found by induction and not proved.

G. Eisenstein[14] supplemented the last result by remarking that $c=4\tau+1-2n$, where τ is the number of values of z in the series $1, 2, \ldots, \frac{1}{2}(p-1)$ for which $1+zi$ is the residue of an eighth power modulo p. If $q=7n+2=A^2+7B^2$ is an odd prime, or $q=7n+4=A^2+7B^2$, then

$$2A \equiv \frac{-(3n)(3n-1)\ldots(2n+1)}{n!} \text{ or } \frac{(3n+1)\ldots(2n+2)}{n!} \pmod q,$$

respectively. Jacobi[1] had treated primes $7n+1$.

V. A. Lebesgue[15] simplified proofs by Cauchy[9] (notes I, V) of theorems by Jacobi[6] on cyclotomy, in particular on $4p=a^2+27b^2$.

C. G. Reuschle[16] gave, in Part B, Table III, the values, for primes $p=6n+1$, of A, B in $p=A^2+3B^2$ for $p \leqq 13669$ and for those greater primes $p<50000$ for which 10 is a cubic residue, and the values of L, M in $4p=L^2+27M^2$ for $p \leqq 5743$. For primes $p=8n+1$, Table IV gives the values of c, d in $p=c^2+2d^2$ for $p \leqq 12377$ and for the primes $p<25000$ for which 10 is a biquadratic residue; also the values of a, b in $p=a^2+b^2$ for primes $p=4n+1 \leqq 12377$ and for the larger primes <25000 for which 10 is a quadratic residue. His Table A of factors a^n-1 and Table C of primitive roots and exponents are described on p. 383 and p. 190 of Vol. I of this History.

P. Bachmann[17] gave an exposition of many of the preceding results, including a proof of Cauchy's[9] theorem for the case in which Δ is a prime.

Th. Pepin[18] obtained $4p=x^2+3y^2$ by cyclotomy if p is a prime $3\omega+1$. Cauchy's[11] statement (p. 724, p. 412) that y is always divisible by 3 is proved by use of Cauchy's function $R_{11}=\Sigma\rho^k$, summed for $s=1, \ldots, p-2$, where $k=\text{ind. } s(s+1)$, and ρ is a primitive cube root of unity.

G. Oltramare[19] stated that every prime $p=8m+3$ can be given the form x^2+2y^2, where $x \equiv \pm 2^{2m}\psi(m) \pmod p$, $\psi(m)$ denoting $(m+1)\ldots(2m)/(m!)$. Every prime $p=6m+1$ can be given the form x^2+3y^2, $x \equiv \pm 2^{m-1}\psi(m)$. Every prime $p=14m+1$ can be given the form x^2+7y^2, where

$$x \equiv \pm 2^{6m-1} \frac{\psi(2m)\,\psi(3m)}{\psi(m)} \pmod p.$$

Four such theorems are stated for x^2+6y^2, three for x^2+5y^2, and one for x^2+15y^2. If $p=2am+1$ or $p=4am+1$ is a prime representable by x^2+ay^2, it is stated that

$$x \equiv \pm \tfrac{1}{2} A^m \psi(m)^{c_1} \psi(2m)^{c_2} \ldots \psi(am)^{c_a} \pmod p,$$

---

14 Jour. für Math., 37, 1848, 111, 126. Proof repeated by Smith, Coll. Math. Papers, I, 280–3.

15 Comptes Rendus Paris, 39, 1854, 593–5; Jour. de Math., 19, 1854, 289; (2), 2, 1857, 152. Cf. reports on same in this History, Vol. II, pp. 305–6.

16 Math. Abhandlung, enthaltend neue Zahlentheoretische Tabellen sammt einer dieselben betreffenden Correspondenz . . . Jacobi, Progr. Stuttgart, 1856, 61 pp. Described by Kummer in Jour. für Math., 53, 1857, 379, and in Report British Assoc. for 1875, p. 325. Errata.[30]

17 Die Lehre von der Kreistheilung, 1872, 122–150, 235–7, 279–294.

18 Comptes Rendus Paris, 79, 1874, 1403.

19 Comptes Rendus Paris, 87, 1878, 734–6.

where A is an algebraic function of m, while $c_1, \ldots, c_a$ are integers $<p-1$, and $a \leqq (p-1)/(4m)$.

T. Pepin[20] proved the following generalizations of Cauchy's[9] theorems: If a is any (unknown) odd divisor of t^2-D, every power of a, whose exponent is divisible by the double of the number of classes in the principal genus, can be properly represented by x^2-Dy^2; the same is true when the exponent is equal to or is a multiple of the number of properly primitive classes of determinant D. When a is a given divisor of t^2-D, the problem to find the values of m for which $a^m=x^2-Dy^2$ is solvable can be treated by the elements of the theory of quadratic forms. We find that, if $\mu, \mu', \ldots$ are the numbers of classes in the periods generated by the classes $A, A', \ldots$ of determinant D by which a given odd divisor a of t^2-D can be properly represented, then all powers of a, whose exponents are divisible by one of the $\mu, \mu', \ldots$, can be properly represented by x^2-Dy^2, and no other powers of a can be so represented.

L. Kronecker[21] proved that, if p is a prime $6n+1$, then in $4p=a^2+27b^2$ we have $a^2=(3c-p+2)^2$, where $(p-1)c$ is the number of distinct sets s_1, s_2, s_3 chosen from $1, 2, \ldots, p-1$ such that $s_1+s_2+s_3$ is divisible by p and also $9(s_1^3+s_2^3+s_3^3)$ is a cubic residue of p.

T. J. Stieltjes[22] deduced from Jacobi's[1] theorem the fact that, in $p=3n+1=c^2+3d^2$, c is the residue between $-\frac{1}{2}p$ and $\frac{1}{2}p$ when $2^{n-1}\binom{2n}{n}$ is divided by p, while $c-1$ is divisible by 3. Replacing n by $2m$, we have

$$c \equiv (-1)^{\frac{m^2+m}{2}} 2^{m-1}\binom{2m}{m} \quad (\text{mod } p=6m+1).$$

Except as to sign, this congruence was stated without proof by Oltramare.[19]

S. Réalis[23] recalled that a product $p=8q+1$ of primes of that form is expressible in the form $2x^2+y^2$. Then $3p=2(x+y)^2+(y-2x)^2$. Since x is even and y odd, $3p=2(2a+1)^2+(2b+1)^2$. All solutions of $2x^2+y^2=p$ are given by $x=\frac{2}{3}(a-b)$, $y=\frac{2}{3}(2a+b)+1$, where a, b take all values for which $p=\frac{8}{3}[a^2+a+(b^2+b)/2]+1$.

It is stated that if p is odd or double an odd integer, and if $p=2x^2+y^2$, then

$$x=p-\tfrac{1}{3}[m(m+1)+n(2n-1)], \qquad y=p-\tfrac{2}{3}\left[\frac{m(m+1)}{2}+n(n+1)\right].$$

Analogous theorems are given for $p=3x^2+y^2$.

F. Goldscheider[24] proved that if $p=8n+1=A^2+2B^2$ is a prime then $A\equiv\frac{1}{2}[5n]_n$ (mod p), where $[a]_n=a(a-1)\ldots(a-n+1)$. If $q=8h+3=a^2+2b^2$ is a prime, the least residue modulo q of $\frac{1}{2}[2h]_h$ is $(-1)^{h+1}b$, where b is taken $\equiv 1$ (mod 4); that of $2[4h]_h$ is $(-1)^{h+1}a$, where $a\equiv 1$ (mod 4).

L. Stickelberger[25] obtained a generalization of Cauchy's[9] theorem by applying the theory of ideals and a generalized cyclotomic resolvent. Let m be any positive integer divisible by no odd square >1 and such that either $m\equiv 3$ (mod 4) or $m\equiv 4$

---

[20] Atti Accad. Pont. Nuovi Lincei, 33, 1879–80, 50–59.
[21] Jour. für Math., 93, 1882, 364.
[22] Amsterdam Verslagen en Mededeel. K. Akad. Wetenschappen, (2), 19, 1884, 105–111; French transl., Annales Fac. Sc. Toulouse, 11, 1897, No. 4, 6 pp. In No. 3, p. 65, is a new proof of Jacobi's[1] theorem.
[23] Nouv. Ann. Math., (3), 5, 1886, 113–122.
[24] Das Reziprozitätsgesetz der Achten Potenzreste, Progr., Berlin, 1889, pp. 26, 27.
[25] Math. Annalen, 37, 1890, 358–361.

or 8 (mod 16) (so that $-m$ is a fundamental discriminant of a binary quadratic form). Proof is given of Cauchy's[26] result that, except for $m=3, 4, 8$, the sum of the $\frac{1}{2}\phi(m)$ positive integers $k(k<m)$, for which Jacobi's symbol $(-m/k)=+1$, is divisible by m. Write $\Sigma k=mK$, $\Sigma k'=mK'$, where $k'=m-k$ is not one of the k's. Thus $K+K'=\frac{1}{2}\phi(m)$. Let p be a prime congruent modulo m to one of the k's. It is proved that both $4p^{K'+K}$ and $4p^{K'-K}$ are representable by x^2+my^2, and that, for the second,

$$x\equiv\Pi P\left(\frac{pk'}{m}\right) \pmod{p}, \qquad P(t)=1\cdot 2\ldots[t],$$

$[t]$ being the largest integer $\leqq t$, where the product extends over all positive integers k' less than m and prime to m for which $(-m/k')=-1$. This proves for any prime p for which $(-m/p)=+1$ what Cauchy proved for the case $p\equiv 1 \pmod{m}$. In case also m is a prime,

$$(-p)^K x\equiv(-1)^{\frac{1}{2}\phi(m)} \pmod{m}.$$

H. Scheffler,[27] generalizing Eisenstein,[14] expressed in terms of binomial coefficients the residues of the numbers giving a representation of any prime $pn+m$ by a quadratic form.

H. W. Lloyd Tanner[28] tabulated the least solutions of $p=\frac{1}{4}(X^2-5Y^2)$ for each prime $p=10\mu+1<10{,}000$. The omission of $p=3371$, $X=127$, $Y=23$ was noted by Cunningham.[30]

H. Weber[29] proved, by use of the equation for the three periods of nth roots of unity and also by use of numbers determined by a cube root of unity, that if n is a prime $\equiv 1 \pmod 3$, then $n=a^2-ab+b^2$, $4n=A^2+27B^2$, whence $n=x^2+3y^2$.

A. Cunningham's[30] main table (pp. 1–240) gives for each prime $p<100{,}000$ the prime factors of $p-1$ and the values of $a, \ldots, M$ in $p=a^2+b^2$, c^2+2d^2, A^2+3B^2, $\frac{1}{4}(L^2+27M^2)$; also for $p=e^2-2f^2<25{,}000$, and

$$p=x^2-5y^2,\quad \tfrac{1}{4}(X^2-5Y^2),\quad t^2+7u^2,\quad \tfrac{1}{4}(\nu^2+11\omega\ ),\quad p<10{,}000.$$

Tables (pp. 241–256) give for $p<10{,}000$ the decompositions

$$p=x^2-Dy^2(D=3, 6, 7, 10, 11),\quad p=x^2+dy^2(d=5, 6, 10).$$

Errata are noted in the earlier shorter tables by Jacobi[6] and Reuschle.[16]

A. S. Werebrüsow[31] gave a table of the representations of numbers <1000 by x^2+xy+y^2.

E. Jacobsthal, in 1907, gave the actual values of a, b in $p=a^2+b^2$, quoted in this History, Vol. II, p. 253.

L. von Schrutka[32] proved that every prime $p=6n+1$ is of the form a^2+3b^2 and gave expressions for the absolutely least residues of a and b modulo p. The method is that of Jacobsthal, just cited.

For Kronecker's solution of $U^2+DV^2=4p$ by cyclotomy, see p. 140.

---

26 Mém. Institut de France, 17, 1840, 525–588 (Notes VII, VIII); Oeuvres, (1), III, 292.
27 Die quadratische Zerfällung der Primzahlen, Leipzig, 1892.
28 Proc. London Math. Soc., 24, 1893, 256–262.
29 Lehrbuch der Algebra, 1895, I, 579–82, 593; ed. 2, 1898, I, 628–632, 643.
30 Quadratic Partitions, London, 1904. Errata, Messenger Math., 34, 1904–5, 132–6; 46, 1916–7, 68–69.
31 Math. Soc. Moscow, 25, 1905, 417–437.
32 Jour. für Math., 140, 1911, 252–265.

CHAPTER III.

COMPOSITION OF BINARY QUADRATIC FORMS.

Diophantus made use of the formula

$$(x^2+y^2)(x'^2+y'^2)=X^2+Y^2, \qquad X=xx'\pm yy', \qquad Y=xy'\mp x'y,$$

in his study of right triangles having integral sides (this History, Vol. II, p. 225). We shall follow the history of the general composition of two binary quadratic forms $f(x, y)\cdot f'(x', y')=F(X, Y)$, where X, Y are bilinear functions of x, y and x', y'. Gauss was the first to treat the general problem; his proofs rest on long computations and devices whose origin was not divulged. Of first importance in the applications in Ch. IV is the fact that we are dealing with composition of classes of forms: if f_1 and f'_1 are equivalent to f and f', respectively, and if their compound is F_1, then F_1 and F are equivalent. Accordingly, Dirichlet[14] and Dedekind[27] simplified the algebraic work by replacing the forms to be compounded by equivalent forms so related that their compound is found at once. Another basis of classification of the papers of this subject is the extent to which the bilinear substitution is in the foreground; it plays the dominant rôle in the elegant theory by Speiser,[46] following Dedekind.[39] The papers best suited for a first introduction to the theory are probably those by Dedekind,[27] Speiser,[46] Mertens,[37] Pepin,[30] and Smith.[19]

The Hindu Brahmegupta[1] (born 598 A. D.) and L. Euler[1] employed the composition

$$(1) \qquad (x^2-ey^2)(x'^2-ey'^2)=X^2-eY^2, \qquad X=xx'\mp eyy', \qquad Y=xy'\mp x'y,$$

in their work on the solution of $e\xi^2+1=\eta^2$ in integers.

Euler[2] gave the formula, which reduces to (1) for $a=1$, $c=-e$,

$$(2) \qquad (ax^2+cy^2)(x'^2+acy'^2)=aX^2+cY^2, \qquad X=xx'-cyy', \qquad Y=axy'+x'y,$$

which follows from

$$(x\sqrt{a}\pm y\sqrt{-c})(x'\pm y'\sqrt{-ac})=X\sqrt{a}\pm Y\sqrt{-c}.$$

A. J. Lexell[3] noted that

$$(\alpha ap^2+\beta bq^2)(abr^2+\alpha\beta s^2)=ab(apr\pm\beta qs)^2+a\beta(aps\mp bqr)^2,$$
$$(\alpha a^2+2\beta ab+\gamma b^2)(ap^2+2\beta pq+\gamma q^2)=x^2+2\beta xy+\alpha\gamma y^2,$$
$$x=\alpha ap-\gamma bq, \qquad y=aq+bp+2bq\beta/\alpha.$$

---

[1] This History, Vol. II, 346, 355 (507, 554 for cases $e=-2$, $e=-3$).

[2] Algebra, St. Petersburg, 2, 1770, Ch. 11, §§ 173–180; French transl., Lyon, 2, 1774, pp. 208–218; Opera Omnia, (1), I, 420–5.

[3] Euler's Opera postuma, 1, 1862, 159–160 (about 1767).

A. M. Legendre[4] reduced to the case (1) the composition of

(3) $$\Delta = py^2 + 2qyz + rz^2, \qquad \Delta' = p'y'^2 + 2q'y'z' + r'z'^2,$$

when p and p' are relatively prime and $pr - q^2 = p'r' - q'^2 = a$. Write $py + qz = x$, $p'y' + q'z' = x'$. Then $p\Delta = x^2 + az^2$, $p'\Delta' = x'^2 + az'^2$, and

$$pp'\Delta\Delta' = (xx' \pm azz')^2 + a(xz' \mp x'z)^2.$$

But he desired a formula of the type

(4) $$\Delta\Delta' = pp'Y^2 + 2\phi YZ + \psi Z^2, \qquad pp'\psi - \phi^2 = a.$$

From this, $pp'\Delta\Delta' = (pp'Y + \phi Z)^2 + aZ^2$. Hence we have the conditions

$$pp'Y + \phi Z = xx' \pm azz', \qquad Z = xz' \mp x'z.$$

In the first condition, replace a by its value in (4), and x, x' by their expressions above; we get

$$Y = \left(y + \frac{q \pm \phi}{p} z\right)\left(y' + \frac{q' - \phi}{p'} z'\right) \pm \psi zz'.$$

The two fractions will be integers, $\pm n$ and $-n'$, if $\phi \equiv pn \mp q = p'n' + q'$. This equation has integral solutions n, n' since p, p' are relatively prime. Since $\phi = pn \mp q$ and $q^2 + a = pr$, $\phi^2 + a$ is divisible by p. Similarly, it is divisible by p'. Hence $\psi = (\phi^2 + a)/(pp')$ is an integer. Hence if p and p' are relatively prime integers, the forms (3) with the same determinant may be compounded into the form (4) with integral coefficients by the substitution

(5) $$Y = (y \pm nz)(y' - n'z') \pm \psi zz', \qquad Z = (py + qz)z' \mp (p'y' + q'z')z.$$

C. F. Gauss[5] discussed (Art. 235) without restrictions the problem of the transformation of a form

(6) $$F = AX^2 + 2BXY + CY^2$$

into the product of two forms

(7) $$f = ax^2 + 2bxy + cy^2, \qquad f' = a'x'^2 + 2b'x'y' + c'y'^2$$

by means of a (bilinear) substitution

(8) $$X = pxx' + p'xy' + p''yx' + p'''yy', \qquad Y = qxx' + q'xy' + q''yx' + q'''yy',$$

whose eight coefficients, as well as those of F, f, f', are all integers. In particular, if the six numbers

(9) $$\begin{matrix} P = pq' - qp', & Q = pq'' - qp'', & R = pq''' - qp''', \\ S = p'q'' - q'p'', & T = p'q''' - q'p''', & U = p''q''' - q''p''' \end{matrix}$$

have no common divisor, F is said to be *compounded* of f, f'.

Denote their determinants by D, d, d', respectively. Let M be the positive g.c.d. of A, $2B$, C; m that of a, $2b$, c; and m' that of a', $2b'$, c'.

[4] Theorie des nombres, 1798, 421–2; ed. 3, 1830, II, § 358, pp. 27–29 (German transl. by H. Maser, 2, 1893, pp. 28–29).

[5] Disquisitiones Arithmeticae, 1801; Werke, I, 1863, pp. 239–267, 272, 371; German transl. by H. Maser, 1889, pp. 229–255, 260, 356.

By means of the nine conditions on the coefficients which result from $F=ff'$, and after introducing auxiliary numbers, Gauss proved that $d=Dn^2$, $d'=Dn'^2$, where n and n' are rational numbers such that $m'n$ and mn' are integers. Hence (i) the determinants D, d, d' are proportional to the squares of integers, and (ii) D is an exact divisor of dm'^2 and $d'm^2$. Next (iii),

(10) $P=an', \quad R-S=2bn', \quad U=cn', \quad Q=a'n, \quad R+S=2b'n, \quad T=c'n.$

Hence the g.c.d. k of P, Q, R, S, T, U divides mn' and $m'n$, while any common factor of the last two divides k, so that k is the g.c.d. of mn' and $m'n$. This implies (iv) that Dk^2 is the g.c.d. of dm'^2, $d'm^2$.

Next, it is shown that mm' divides Mk^2 and that M divides mm'. Hence (v) if F is compounded of f and f', so that $k=1$, then $M=mm'$. Finally, if $k=1$, then (vi) F is derived from a *properly primitive form* (i. e., the g.c.d. of A, B, C is equal to the divisor M of F) if and only if both f and f' are derived from properly primitive forms.

During the proof of the fifth of these six conclusions, Gauss deduced

(11) $Ann'=q'q''-qq''', \quad 2Bnn'=pq'''+qp'''-p'q''-q'p'', \quad Cnn'=p'p''-pp'''.$

Conversely, he stated that if (10) and (11) hold, where n, n' have arbitrary values, then $F=ff'$ under the substitution (8).

Composition is commutative. For, if we interchange f with f', p' with p'', and q' with q'', and hence P with Q, T with U, n with n', and change the sign of S, we find that conditions (10) are merely permuted, while the right members of (11) are unaltered, so that the values of A, B, C are unaltered.

In Art. 236, Gauss showed how to find a substitution (8) and a form (6) which is compounded of two given forms (7) whose determinants d, d' are in the ratio of two squares. Let D, taken with the same sign as d, d', be the g.c.d. of dm'^2, $d'm^2$. Since their quotients by D are relatively prime integers whose product is a square, each quotient is a square, so that d/D and d'/D are rational squares n^2, n'^2. Evidently nm' and $n'm$ are relatively prime integers. Also, an', cn', $a'n$, $c'n$, $bn'\pm b'n$ are integers. Choose four integers $\mathfrak{Q}, \mathfrak{Q}', \mathfrak{Q}'', \mathfrak{Q}'''$ at random, but such that the left members of (12) are not all zero, and let μ be the g.c.d. of those members:

$$(12)\quad \begin{cases} \mathfrak{Q}'an' + \mathfrak{Q}''a'n + \mathfrak{Q}'''(bn'+b'n) = \mu q, \\ -\mathfrak{Q}an' + \mathfrak{Q}'''c'n - \mathfrak{Q}''(bn'-b'n) = \mu q', \\ \mathfrak{Q}'''cn' - \mathfrak{Q}a'n + \mathfrak{Q}'(bn'-b'n) = \mu q'', \\ -\mathfrak{Q}''cn' - \mathfrak{Q}'c'n - \mathfrak{Q}(bn'+b'n) = \mu q'''. \end{cases}$$

Since $q, \ldots, q'''$ have unity as their g.c.d., there exist integers $\mathfrak{P}, \ldots, \mathfrak{P}'''$ such that $\mathfrak{P}q+\ldots+\mathfrak{P}'''q'''=1$. Denote by $p, \ldots, p'''$ the values obtained from the left members of (12) by replacing the $\mathfrak{Q}$'s by $\mathfrak{P}$'s. Finally, define A, B, C by (11); they are shown to be integers. Since equations (10) are proved to hold, we have $F=ff'$ under substitution (8).

If (Art. 237) F is transformed into ff' by (8) and if f' contains the form f'', then F is transformable into ff''. Here f', given by (7), is said to contain f'' if it is transformed into f'' by a substitution

(13) $x'=\alpha x''+\beta y'', \qquad y'=\gamma x''+\delta y'',$

with integral coefficients. By eliminating x', y' between (8) and (13), we evidently obtain a substitution which transforms F into ff''. If d'' is the determinant of f'', $d''=d'e^2$, where $e=\alpha\delta-\beta\gamma$. Thus $n'''^2=d''/D=e^2n'^2$. The sign of n'' is stated to be the same or opposite to that of n' according as e is positive or negative, i. e., according as f' contains f'' properly or improperly.

Similarly (Art. 238), if F is contained in F' and if F is transformable into ff', then F' is transformable into ff'.

If (Art. 239) F is compounded of f, f', and if F' has the determinant D' and is transformed into ff' in such a manner that the square roots $\mathfrak{n}, \mathfrak{n}'$ of d/D' d'/D' have the signs of n, n', respectively, then F' contains F properly. Use is made of the following lemma (Art. 234). Given two matrices whose elements are integers,

$$M=\begin{pmatrix} a & a' & a'' \dots a^{(n)} \\ b & b' & b'' \dots b^{(n)} \end{pmatrix}, \qquad N=\begin{pmatrix} c & c' & c'' \dots c^{(n)} \\ d & d' & d'' \dots d^{(n)} \end{pmatrix},$$

such that every (two-rowed) determinant from N is equal to the product of the corresponding determinant from M by a constant integer k, and such that the determinants from M have no common factor >1, there exist four integers $\alpha, \beta, \gamma, \delta$ such that

$$\begin{pmatrix} \alpha & \beta \\ \gamma & \delta \end{pmatrix} M=N, \qquad \begin{vmatrix} \alpha & \beta \\ \gamma & \delta \end{vmatrix}=k.$$

Henceforth it is assumed that n, n' are taken positive. From the compositions $ff'=F$, $f''F=\mathfrak{F}$, $f''f=F'$, $F'f'=\mathfrak{F}'$, it follows (Arts. 240–1) that $\mathfrak{F}$ and $\mathfrak{F}'$ are properly equivalent. In other words, $f''(ff')$ and $(f''f)f'$ are properly equivalent, so composition is associative in the sense of equivalence.

In particular (Arts. 242–3), let the forms f and f' to be compounded have equal determinants $d=d'$, let m be prime to m', and let n and n' be positive. Then in Art. 236, $n=n'=1$, $D=d$, and we may choose $\mathfrak{Q}=-1$, $\mathfrak{Q}'=\mathfrak{Q}''=\mathfrak{Q}'''=0$. Thus μ is the g.c.d. of $a, a', b+b'$, which are assumed to be not all zero. Then

$$(14) \qquad A=\frac{aa'}{\mu^2}, \qquad B\equiv b\left(\text{mod } \frac{a}{\mu}\right), \qquad B\equiv b'\left(\text{mod } \frac{a'}{\mu}\right).$$

If a/μ and a'/μ are relatively prime, these congruences have a single root B between 0 and $|A|-1$. Then, for $C=(B^2-D)/A$, (A, B, C) is compounded of f, f'. If a is prime to a', then $\mu=1$, and we obtain (Art. 243, I) the following useful result: If (a, b, c) and (a', b', c') have the same determinant D, if a is prime to a', and if the g.c.d. of $a, 2b, c$ is prime to that of $a', 2b', c'$, the form compounded of them is (A, B, C), where

$$A=aa', \quad B\equiv b \pmod{a}, \quad B\equiv b' \pmod{a'} \quad C=(B^2-D)/A.$$

This case occurs if the first form is the principal form, whence $a=1$, $b=0$, $c=-D$; we may taken $B=b'$, whence $C=c'$. Hence if we compound the principal form with any form f' of the same determinant, we obtain f'.

The compound of two opposite properly primitive forms (a, b, c) and $(a, -b, c)$ is the principal form of the same determinant.

All forms (Art. 224) properly equivalent to the same form are said to constitute a *class*. To the forms (a, b, c) of one class K correspond the opposite forms

$(a, -b, c)$ which determine another class K', said to be *opposite to* K. Every form of K is improperly equivalent to every form of K'. If K and K' coincide, K is opposite to itself and (Art. 164) contains an ambiguous form $f=(a, b, c)$ with $2b$ divisible by a and hence is called an *ambiguous class*. Conversely, any ambiguous class is opposite to itself; for, f has (c, b, a) as a left-neighboring form and hence is properly equivalent to it, and yet is obviously improperly equivalent to it, so that f is improperly equivalent to itself.

If (Art. 249) f, g are forms of a class and if f', g' are forms of a class, the forms compounded of f, f' and of g, g' belong to the same class (as proved in Art. 239). Hence we may speak of *composition of classes*. All forms equivalent to $(1, 0, -D)$ constitute the *principal class* 1 of determinant D, which plays the rôle of unity in composition. An ambiguous class was just seen to be its own opposite class. Hence, by Art. 243, the principal class arises by *duplication* (composition with itself) of any properly primitive ambiguous class. Conversely, if the principal class arises by duplication of a properly primitive class K, then K is an ambiguous class.

If K, K' are opposite classes and L, L' are opposite classes, then the class compounded of K, L is opposite to that compounded of K', L'. Hence the compound of two ambiguous classes is ambiguous. While Gauss denoted the compound of two classes K and K' by $K+K'$, we shall follow later[6] custom and denote it by KK' or $K\times K'$. In particular, $K\times K$ will be written K^2.

If (Art. 306, II–IV) m is the least positive integer for which $C^m=1$, the classes $1, C, C^2, \ldots, C^{m-1}$ constitute the *period* of C. The classes C^j and C^{m-j} are opposite. If m is even, $C^{\frac{1}{2}m}$ is opposite to itself and hence is an ambiguous class. If μ is the g.c.d. of m and h, the period of C^h contains m/μ classes. The number of powers of C whose periods contain m/μ classes is the Euler function $\phi(m/\mu)$ which counts those positive integers $<m$ having with m the g.c.d. μ.

J. T. Graves[7] proved that we may express the product $f_1f_2f_3$ in the form f_4, where $f_i=ax_i^2+bx_iy_i+cy_i^2$. This was proved by J. R. Young[8] by means of

$$af_1f_2=f_0,\qquad x_0=ax_1x_2-cy_1y_2,\qquad y_0=ax_1y_2+ay_1x_2+by_1y_2.$$

Similarly, $af_0f_3=f_4$. Hence, by multiplication, $a^2f_1f_2f_3=f_4$. Write X_4 for x_4/a, Y_4 for y_4/a.

H. W. Erler[9] assumed that there are two periods, one of $m=p^\mu$ classes and the other of p^ν classes, where p is a prime and $\mu \gtreqless \nu$, having only the principal class 1 in common, and noted that we can derive $p^\mu\cdot p^\nu$ distinct classes whose mth power is 1. If for the same determinant there are two periods of m and n classes, where m and n are relatively prime, we obtain by composition a period of mn classes.

F. Arndt[10] proved that if $P, \ldots, U$ are six given integers satisfying

$$PU-QT+RS=0, \tag{15}$$

[6] A. C. M. Poullet-Delisle in his French transl. of Gauss' Disq. Arith., Paris, 1807, 274, 286.
[7] Phil. Magazine, London, (3), 26, 1845, 320.
[8] Trans. Royal Irish Acad., 21, II, 1848, 337.
[9] De periodis, quae compositione formarum quadraticarum ejusdem determinantis fiunt, Progr. Züllichau, 1847, 16 pp.
[10] Archiv Math. Phys., 13, 1849, 410–8. Of the simpler proofs by Speiser,[46] the second is the shorter.

we can find integral solutions $p, \ldots, q'''$ of the six equations (9). First, let P, Q, R be not all zero and denote their g.c.d. by e, and that of P, Q, R, S, T, U by k. Select any divisor μ of k and write λ for the integer e/μ. Select any two integers p, q whose g.c.d. is λ. Determine five integers $\phi_1, \psi_1, p_1, q_1, r_1$ such that

$$p\phi_1 - q\psi_1 = e, \qquad p_1P + q_1Q + r_1R = e,$$

and write

$$(16) \qquad \begin{cases} eq' = \phi_1 P + q_1 qS + r_1 qT, & ep' = \psi_1 P + q_1 pS + r_1 pT, \\ eq'' = \phi_1 Q + r_1 qU - p_1 qS, & ep'' = \psi_1 Q + r_1 pU - p_1 pS, \\ eq''' = \phi_1 R - p_1 qT - q_1 qU, & ep''' = \psi_1 R - p_1 pT - q_1 pU. \end{cases}$$

Then $p, \ldots, q'''$ are integers satisfying (9), and furnish all solutions when P, Q, R are not all zero. There is a similar, simpler discussion of the cases $P=Q=R=0$, with S, T not both zero, or both zero with $U \neq 0$.

Application is made to Gauss' problem (Art. 236) to find a substitution (8) and a form F compounded of given forms f, f'. No use is made of Gauss' $\mathfrak{Q}, \ldots, \mathfrak{Q}'''$, $\mathfrak{P}, \ldots, \mathfrak{P}'''$. We define $P, \ldots, U$ by (10) and see that their g.c.d. is 1. The problem is to solve (9) and (11) for integral values of $p, \ldots, q''', A, B, C$. When P, Q, R are not all zero, we apply the result quoted above with $k=1$. Thus $\mu=1$, $\lambda=e$. Condition (15) is satisfied identically by the present values (10). For the numbers $q', \ldots, p'''$ given by (16), the values of A, B, C in (11) are shown to be integers.

F. Arndt[11] noted that, to find all forms F' transformable into the product of two given forms f, f', given one form F compounded of f, f', we need only find all forms F' which contain F properly. Given one transformation of F' into ff', we obtain all such transformations as in Gauss, Arts. 237–9.

More interesting is his first theorem: If F, f, f' have the same determinant and if $F=ff'$ under the substitution $p, p', p'', p'''; q, q', q'', q'''$, defined by (8), then f' is transformed into the product of F and the form $(a, -b, c)$, opposite to $f=(a, b, c)$, by the substitution $-q', p', q''', -p'''; q, -p, -q'', p''$. For, we have (10) and (11) with $n=n'=1$, which may be rewritten to give the corresponding conditions for $f'=F\cdot(a, -b, c)$ under the second substitution.

This theorem leads to a solution of the problem to find all forms f' of the same determinant D as two given forms F, f such that F is compounded of f, f'. After finding one form $\mathfrak{F}$ compounded of F and $(a, -b, c)$, it remains to determine the forms f' which contain $\mathfrak{F}$ properly. The details of this determination occupy many pages (pp. 433–467). The solution is more general than that by Gauss (Art. 251) who restricted F, f to be forms in the same order.

A. Cayley[12] gave a formula of composition under a substitution which involves, in both z_1 and z_2, the misprints of x_1x_2 for x_1y_1, and y_1y_2 for x_2y_2. When his corrected formula is changed to Gauss' notations, it states that the form F in (6) is the product of the forms f, f' in (7) under the substitution (8) provided that, in the notations (9),

$$a=P, \quad 2b=R-S, \quad c=U, \quad a'=Q, \quad 2b'=R+S, \quad c'=T,$$
$$A=q'q''-qq''', \quad 2B=pq'''+p'''q-p'q''-p''q', \quad C=p'p''-pp'''.$$

[11] Archiv Math. Phys., 15, 1850, 429–480.
[12] Jour. für Math., 39, 1850, 14–15; Coll. Math. Papers, I, 532–3.

Since these are Gauss' relations (10) and (11) for the case $n=n'=1$, and since Gauss stated (Art. 235, end) that (10) and (11) imply that $F=ff'$ under the substitution (8), Cayley's result is implied by Gauss'. Cayley derived it by means of his hyperdeterminants.

M. Bazin[13] readily obtained a necessary condition that the form (6) be transformed into the product of the forms (7) by the substitution (8). Take $x'=1, y'=0$; then $X=px+p''y$, $Y=qx+q''y$ must transform F into $a'f$. Hence $DQ^2=da'^2$, $d/D=n^2$, and, similarly, $d'/D=n'^2$, where n and n' are rational. Thus $Q=a'n$, $T=c'n$, $P=an'$, $U=cn'$, as in (10). The nine conditions for $F=ff'$ are written briefly by employing the partial derivatives of $F(p, q)$, $F(p', q')$, $F(p'', q'')$ and $F(p''', q''')$, and are shown to imply $S=b'n-bn'$, $R=b'n+bn'$, as in (10). He gave essentially Art. 236 of Gauss in changed notations and proved the results in Arts. 239–241.

G. L. Dirichlet[14] gave a new and elementary exposition of composition. Call the roots $\zeta, \zeta', \ldots$ of $u^2 \equiv D$ modulis $m, m', \ldots$, respectively, *concordant* if there exists a root Z of $u^2 \equiv D \pmod{mm'\ldots}$ such that $\zeta \equiv Z \pmod{m}$, $\zeta' \equiv Z \pmod{m'}$, ... The root $Z=(mm'\ldots, Z)$ is said to be *composed* of the roots $\zeta=(m, \zeta)$, (m', ζ'),... and is denoted by $(m, \zeta)(m', \zeta')$.... Consider only forms $\phi \equiv ax^2+2bxy+cy^2$ of determinant D in which a, b, c have no common divisor and a, c are not both even (since a slight modification of the discussion applies to the contrary case). If $\phi=m$ for x, y relatively prime, and if integers ξ, η are chosen so that $x\eta-y\xi=1$, Gauss (Arts. 154–5, report in Ch. I) proved that $\zeta=(ax+by)\xi+(bx+cy)\eta$ is a root of $u^2\equiv D \pmod{m}$ and that we obtain the same root if we vary ξ, η; we say that the root (m, ζ) *belongs* to this representation of m by ϕ. Consider two such forms ϕ, ϕ' of the same determinant D and let m, m' be any two odd integers prime to D and representable by ϕ, ϕ', respectively, in such a manner that the roots (m, ζ), (m', ζ') to which these representations belong are concordant. The whole difficulty consists in proving that the representations of mm' which belong to the root $(m, \zeta)(m', \zeta')$ are made by the forms of a single class however m and m' are varied. This is the fundamental theorem on composition.

After transforming the forms into equivalent forms whose first coefficients are relatively prime, we have (a, b) concordant with (a', b'). The latter condition alone is sufficient for the sequel. Let (aa', B) be composed of (a, b) and (a', b'), so that $B\equiv b \pmod{a}$, $B\equiv b' \pmod{a'}$, $D=B^2-aa'C$, where C is an integer. Hence ϕ and ϕ' are equivalent to

$$ax^2+2Bxy+a'Cy^2=\phi, \qquad a'x'^2+2Bx'y'+aCy'^2=\phi'.$$

Hence $\phi\phi'=aa'X^2+2BXY+CY^2\equiv\psi$ for

$$X=xx'-Cyy', \qquad Y=axy'+a'x'y+2Byy'. \tag{17}$$

If $\phi=m$ and $\phi'=m'$ for x, y relatively prime integers and likewise for x', y', it is

---

13 Jour. de Math., 16, 1851, 161–170. His notations are here changed to accord with those of Gauss.

14 De formarum binariarum secundi gradus compositione, Berlin, 1851. Reprinted in Jour. für Math., 47, 1854, 155–160; Werke, II, 1897, 105–114. French transl., Jour. de Math., (2), 4, 1859, 389–398.

proved that (17) are relatively prime and hence give a proper representation of mm' by ψ, and that the root (mm', Z) to which this representation belongs is composed of the roots (m, ζ), (m', ζ') to which belong the representations of m, m'.

W. Šimerka[15] proceeded as had Legendre but without prefixing the factor 2 to the middle coefficients of the forms. He studied (p. 51) the period defined by the powers of a form. He applied (pp. 66–67) composition to the solution of $ax^2+bxy+cy^2=pz^m$.

F. Arndt[16] treated the problem of Gauss (Art. 236) to find a substitution and a form F which is transformed into the product of two given forms f, f' whose determinants d, d' are in the ratio of two rational squares. By the method of Bazin,[13] he proved that $DQ^2=a'^2d$, $DP^2=a^2d'$. Assuming that a/m, a'/m' are relatively prime to D, it follows that D is a common divisor of dm'^2, $d'm^2$. It is assumed that D is their g.c.d., whence F is said to be compounded of f, f'. It is shown that P, Q, R, S, T, U defined by (10) are integers without a common divisor. Let μ be the g.c.d. of P, Q, R. It is proved that there exists an integer B (determined up to a multiple of A) satisfying the three congruences[17]

$$\frac{P}{\mu}B\equiv\frac{ab'}{\mu},\quad \frac{Q}{\mu}B\equiv\frac{ba'}{\mu},\quad \frac{R}{\mu}B\equiv\frac{bb'+Dnn'}{\mu}\left(\text{mod } A=\frac{aa'}{\mu^2}\right), \tag{18}$$

in which the indicated fractions are all integers. Further, $C=(B^2-D)/A$ is an integer. Hence the coefficients of

$$\left\{\begin{aligned} X&=\mu xx'+\frac{\mu(b'-Bn')}{a'}xy'+\frac{\mu(b-Bn)}{a}yx' \\ &\qquad\qquad\qquad\qquad+\frac{\mu}{aa'}[bb'+Dnn'-B(b'n+bn')]yy', \\ Y&=\frac{an'}{\mu}xy'+\frac{a'n}{\mu}yx'+\frac{b'n+bn'}{\mu}yy' \end{aligned}\right. \tag{19}$$

are integers. Furthermore,

$$AX+BY+\sqrt{D}Y=\frac{1}{\mu}(ax+by+ny\sqrt{D})(a'x'+b'y'+n'y'\sqrt{D}). \tag{20}$$

Hence $F=ff'$ under substitution (19).

It is stated that we obtain all transformations of F into ff' if we multiply the second member of (20) by $(t+u\sqrt{\Delta})/\omega$, where Δ is the quotient of D by the square of the g.c.d. of A, B, C, while $\omega=1$ or 2 according as F is derived from a properly or improperly primitive form by multiplying its coefficients by the same integer.

By duplication of a properly primitive ambiguous class there arises the principal class. For, if f belongs to the former class, we may assume that $2b$ is divisible by a, whence $\mu=a$, $A=1$, $B=0$.

If N and N' are numbers representable by f and f', respectively, we can find a form compounded of them which represents NN'/μ^2.

[15] Sitzungsber. Akad. Wiss. Wien (Math.), 31, 1858, 33–67.

[16] Jour. für Math., 56, 1859, 64–71. Reproduced in G. B. Mathews' Theory of Numbers, 1892, 149–158.

[17] Applicable for composition in general and not merely to the special case considered by Gauss (Arts. 242–3).

L. Schläfli[18] noted that the equation used by Gauss (Art. 235) to prove that A, B, C are integers can be given a remarkable symbolic form involving partial derivatives of the second order. No application is made of the complicated formula.

H. J. S. Smith[19] employed the determinants (resultants)

$$\Delta=\begin{vmatrix}\frac{\partial X}{\partial x} & \frac{\partial X}{\partial y}\\ \frac{\partial Y}{\partial x} & \frac{\partial Y}{\partial y}\end{vmatrix}\equiv\begin{vmatrix}px'+p'y' & p''x'+p'''y'\\ qx'+q'y' & q''x'+q'''y'\end{vmatrix},\qquad \Delta'=\begin{vmatrix}\frac{\partial X}{\partial x'} & \frac{\partial X}{\partial y'}\\ \frac{\partial Y}{\partial x'} & \frac{\partial Y}{\partial y'}\end{vmatrix}$$

of the bilinear substitution (8), considered first as a linear substitution on x, y, and second as one on x', y' as variables. Since F becomes $f'f$ under the former substitution of determinant Δ, the determinant f'^2d of the transformed form $f'f$ is equal to the product of the determinant D of F, by Δ^2, if d is the determinant of f. Thus $f'^2d=\Delta^2D$. Thus $m'^2d=\delta^2D$, if δ is the g.c.d. of the coefficients of the quadratic form Δ. Similarly, if δ' is that of Δ', and we employ the second substitution above, we have $f^2d'=\Delta'^2D$, $m^2d'=\delta'^2D$. We obtain at once Gauss' first four conclusions. The idea underlying this proof is due to Bazin[13] and Arndt.[16] Cayley's[12] identity is given as a relation to be verified by the direct multiplication of Δ by Δ'. Its comparison with $\Delta\Delta'=nn'F$, which follows from the above equations involving Δ^2 and Δ'^2 yields Gauss' formulas (11). In § 108, Smith made clearer the nature of Gauss' discussion (Art. 236): Given integers P, Q, R, S, T, U without a common factor and satisfying $PU-QT+RS=0$, we seek eight integers $p, \ldots, q'''$ satisfying the six equations (9). To this end we seek two fundamental sets of solutions x_i of

$$x_1U-x_2T+x_3S=0,\qquad -x_0U+x_2R-x_3Q=0,$$
$$x_0T-x_1R+x_3P=0,\qquad -x_0S+x_1Q-x_2P=0.$$

The left members of Gauss' relations (12), with an' replaced by P, etc., from (10), give a set of solutions with $\mathfrak{Q}, \ldots, \mathfrak{Q}'''$ arbitrary (denoted by $\theta_0, \ldots, \theta_3$ by Smith).

Smith defined composition of n forms and proved that if F is compounded of $\phi, f_3, \ldots, f_n$, and if ϕ is compounded of f_1 and f_2, then F is compounded of $f_1, \ldots, f_n$. He deduced the congruences (18) of Arndt[16] from Gauss' general solution of the problem of composition. He gave a full report of the method of Dirichlet.[14]

Smith[20] solved the problem to find all matrices of n rows and $n+m$ columns with integral elements, given the integral values not all zero of all its n-rowed minors. Gauss (Arts. 236, 279) treated the cases $n=2$, $n+m=4$ or 3, without indication of the origin of his solution. The case $m=1$ had been solved by C. Hermite.[21] M. Bazin[22] had treated the main problem, but without specifying integral values for the elements and given minors.

Smith[23] stated that if (a, b, c) and (a', b', c') are primitive forms of determinants D and D', and divisors m and m', and vanishing joint invariant $ac'-2bb'+ca'$, then

---

18 Jour. für Math., 57, 1860, 170–4.
19 Report British Assoc. for 1862, 503–515; Coll. Math. Papers, I, 231–246. Exposition by G. B. Mathews, Theory of Numbers, 1892, 140–9.
20 Phil. Trans. London, 151, 1861, 302; Coll. Math. Papers, I, 377.
21 Jour. für Math., 40, 1850, 264.
22 Jour. de Math., 16, 1851, 145–160.
23 Report British Assoc. for 1863, p. 783; Coll. Papers, I, p. 284, § 123.

m^2D' and m'^2D can be represented primitively by the duplicates of (a, b, c) and (a', b', c'), respectively. An elementary proof was given by G. B. Mathews.[24]

E. Schering[25] proved that for every determinant D there exist fundamental classes by whose repeated composition with one another arise every class of determinant D and each in a single way if we permit no more compositions of a class than the number of classes in its period. [In modern[26] terminology, the abelian group whose elements are the classes of forms of determinant D has a set of independent generators.]

R. Dedekind[27] called two forms (a, b, c) and (a', b', c') of the same determinant D *concordant* (*einig*) if $a, a', b+b'$ have no common divisor. Under the latter condition and $b^2 \equiv D \pmod{a}$, $b'^2 \equiv D \pmod{a'}$, there exists an integer B satisfying the three congruences

$$B \equiv b \pmod{a}, \qquad B \equiv b' \pmod{a'}, \qquad B^2 \equiv D \pmod{aa'},$$

and two such solutions B are congruent modulo aa'. Moreover $a, a', 2B$ have no common divisor. Write C for $(B^2-D)/(aa')$. Then any one of the infinitude of (parallel) forms (aa', B, C) of determinant D is said to be *compounded* of (a, b, c) and (a', b', c').

Note that (a, b, c) and (a', b', c') are equivalent to $(a, B, a'C)$ and (a', B, aC) respectively, while the latter are concordant and are compounded into the same form (aa', B, C). In view of the substitution

$$X = xx' - Cyy', \qquad Y = (ax+By)y' + (a'x'+By')y,$$

we have

$$[ax+(B+\sqrt{D})y][a'x'+(B+\sqrt{D})y'] = aa'X + (B+\sqrt{D})Y.$$

Change the sign of $\sqrt{D}$ and multiply. Hence

$$(ax^2+2Bxy+a'Cy^2)(a'x'^2+2Bx'y'+aCy'^2) = aa'X^2+2BXY+CY^2.$$

The following generalization is called the fundamental theorem. If the concordant forms (a, b, c) and (a', b', c') are equivalent to the concordant forms (m, n, l) and (m', n', l') respectively, the form (aa', B, C) compounded of the first two is equivalent to the form (mm', N, L) compounded of the last two. If σ is the *divisor* of (a, b, c), i. e., the positive g.c.d. of $a, 2b, c$, then σ is relatively prime to the divisor σ' of the concordant form (a', b', c'), and $\sigma\sigma'$ is the divisor of the form (aa', B, C) compounded of them.

Conversely, if K, K' are two classes of forms of the same determinant D and with relatively prime divisors σ, σ', there exist two concordant forms belonging respectively to K, K', which are called *concordant classes*. In view of the fundamental theorem, we may speak of the class KK' compounded of the classes K and K'. Composition of classes obeys the associative law, so that $KK'K''$ has a single meaning.

---

24 Quar. Jour. Math., 27, 1895, 230.

25 Abh. Gesell. Wiss. Göttingen, 14, 1869, 3–16; Werke, 1, 1902, 135–148.

26 G. Frobenius and L. Stickelberger, Jour. für Math., 86, 1879, 217–262.

27 Dirichlet-Dedekind, Zahlentheorie, Suppl. X, §§ 145–9, 1871, 1879, 1894. Reproduced for forms $ax^2+bxy+cy^2$ by Weber, Göttingen Nachr., 1893, 55–57, with errata in (2) and (9). L. Bianchi, Atti R. Accad. Lincei, Rendiconti, (4), 5, I, 1889, 589–599, noted that this theory applies unchanged to forms with complex coefficients with a, a' relatively prime.

Let (a, b, c) be a properly primitive form of determinant D. By compounding it with the concordant form (c, b, a), we get $(ac, b, 1)$. But (c, b, a) is evidently equivalent to $(a, -b, c)$, while $(ac, b, 1)$ is equivalent to $(1, -b, ac)$ and hence to $(1, 0, -D)$. Hence the compound of two properly primitive opposite classes yields the principal class. Thus if H is a properly primitive class, $HK=HL$ implies $K=L$.

Since any two properly primitive classes are concordant, they may be compounded. Any set of properly primitive classes constitute a *group* if the compound of any two of them belongs to the set.

Dedekind[28] derived Gauss' general type of composition, but for generalized forms $ax^2+bxy+cy^2$, from a study of a *modul* $[\alpha, \beta]$, composed of all linear homogeneous functions with integral coefficients of two integral algebraic numbers α and β of a quadratic field.

Dedekind[29] established a correspondence between binary quadratic forms of discriminant d and ideals of a quadratic field of discriminant d such that composition of forms corresponds to multiplication of ideals. The correspondence between classes of forms representing positive numbers and narrow classes of ideals is (1, 1). The details are similar to those of Dedekind.[28]

T. Pepin[30] proved the six conclusions (i)–(vi) of Gauss (Art. 235) without his long computation. Assigning any particular integral values to x', y', we may write the substitution (8) in the form

$$X=\alpha'x+\beta'y, \qquad Y=\gamma'x+\delta'y \qquad (\alpha'=px'+p'y', \ldots). \tag{21}$$

Then

$$\alpha'\delta'-\beta'\gamma'=Qx'^2+(R+S)x'y'+Ty'^2. \tag{22}$$

Since (21), transforms F into $f'f=f'ax^2+\ldots$ of determinant f'^2d,

$$f'^2d=D(\alpha'\delta'-\beta'\gamma')^2. \tag{23}$$

Similarly, by assigning integral values to x, y, (8) becomes

$$X=\alpha x'+\beta y', \qquad Y=\gamma x'+\delta y' \qquad (\alpha=px+p''y, \ldots). \tag{24}$$

Then

$$\alpha\delta-\beta\gamma=Px^2+(R-S)xy+Uy^2, \tag{25}$$

$$f^2d'=D(\alpha\delta-\beta\gamma)^2. \tag{26}$$

By (23) and (26), (i) determinants D, d, d' of F, f, f' are in the ratios of three integral squares. Let M, m, m' denote the divisors of the three forms. Write g for f'/m', and h for the quotient of the second member of (22) by the g.c.d. K of its coefficients Q, $R+S$, T. By (23), $dm'^2g^2=DK^2h^2$. Since a'/m', $2b'/m'$, c'/m' have no common divisor, we can determine x', y' so that g is prime to any assigned integer,

---

[28] Dirichlet-Dedekind, Zahlentheorie, Suppl. XI, ed. 2, 1871, §§ 169–170, pp. 488–497. In ed. 3, 1879, § 181, p. 611; ed. 4, 1894, § 187, p. 640, he used a normalized modul $[m, m\omega]$ and obtained a specialized composition, essentially that of Arndt.[16] Cf. Mertens.[37]

[29] *Ibid.*, ed. 2, 1871, 468–9; ed. 3, 1879, 549, 626; ed. 4, 1894, 585, 655. Cf. H. Weber, Math. Annalen, 48, 1897, 459–462; Algebra, III, 1908, 330–7; R. König, Jahresber. d. Deutschen Math.-Vereinigung, 22, 1913, 239–254; J. Sommer, Vorlesungen über Zahlentheorie, 1907, 197-220 (French transl. by A. Lévy, Paris, 1911, 205–229); P. Bachmann, Grundlehren der Neueren Zahlentheorie, 1907, 248; ed. 2, 1921; R. Fricke, Elliptische Funktionen, 2, 1922, 148.

[30] Atti Accad. Pont. Nuovi Lincei, 33, 1879–80, 6–36.

say DK^2. Thus DK^2 must divide dm'^2. Similarly, we can determine x', y' so that h is prime to dm'^2, whence dm'^2 must divide DK^2. Hence we have (27_1). Similarly, if K' denotes the g.c.d. of P, $R-S$, U, we have (27_2):

$$(27) \qquad DK^2=dm'^2, \qquad DK'^2=d'm^2.$$

Hence (ii) D is a common divisor of dm'^2 and $d'm^2$. By (i), $d=Dn^2$, $d'=Dn'^2$, where n, n' are rational. By (22)–(26), the right members of (22) and (25) are the products of f' by n and f by n', respectively, identically in x', y' and x, y, respectively. This proves Gauss' relations (10). It follows readily that the g.c.d. k of $P, \ldots, U$ is that of K, K'. Gauss' conclusions (iv)–(vi) are now easily proved.

If $F=ff'$ by substitution (8) and if $F'=F$ by

$$(28) \qquad X'=\delta X-\beta Y, \qquad Y'=-\gamma X+\alpha Y, \qquad \alpha\delta-\beta\gamma=1,$$

then $F'=ff'$ by the substitution obtained from (28) by elimination of X, Y by means of (8). In Y' the coefficient $\alpha q-\gamma p$ of xx' is zero when $\alpha=p/\mu$, $\gamma=q/\mu$, where μ is the g.c.d. of p, q. Then we can find an infinitude of pairs of integers β, δ such that $\alpha\delta-\beta\gamma=1$. Hence if $F=ff'$, there exist infinitely many forms equivalent to F which are transformable into ff' by means of a substitution (8) having $q=0$. Without loss of generality, we take $q=0$ henceforth.

For $x'=1$, $y'=0$, $F=ff'$ becomes

$$A(px+p''y)^2+2B(px+p''y)q''y+Cq''^2y^2\equiv a'(ax^2+2bxy+cy^2).$$

Multiplying corresponding coefficients by nn' and replacing an', $a'n$, $2bn'$, cn' by their values from (10), we obtain (11) with $q=0$, the first of which is equivalent to $Ap^2=aa'$.

Thus (10), (11) and $n=\pm\sqrt{d/D}$, $n'=\pm\sqrt{d'/D}$ are necessary conditions for $F=ff'$ under substitution (8). They are verified to be sufficient conditions for $Ap^2F=aa'ff'$, or

$$p^2(AX+BY)^2-p^2Y^2D=[(ax+by)^2-Dn^2y^2][(a'x'+b'y')^2-Dn'^2y'^2].$$

This follows from Arndt's[16] relation (20) with μ replaced by p. The conditions that the latter be an identity in x, y, x', y', $\sqrt{D}$ are readily verified to follow from (10), (11).

To prove that there exist forms F compounded of two given forms f, f', it remains to verify when $k=1$ that conditions (10), (11), with $q=0$, are solvable for integers $p, \ldots, q'''$, A, B, C. Then the g.c.d. of q', q'', q''' is 1, whence p is that of

$$pq'=an', \qquad pq''=a'n, \qquad pq'''=bn'+b'n,$$

which therefore determine integers p, q', q'', q'''. Let L, L', L'' be any three integers for which, provisionally,

$$(29) \qquad Lq'+L'q''+L''q'''=1, \qquad Lp'+L'p''+L''p'''=0.$$

Then

$$p'=L'S+L''T, \qquad p''=L''U-LS, \qquad p'''=-LT-L'U$$

are known by (10). For these values we see that (29_2) is satisfied identically in L, L', L'', S, T, U. Hence (29_1) is the only condition on the L's and it can be satisfied by an infinitude of sets of integers L, L', L'' since the q's have no common divisor. The longer proof (pp. 21–24) that A, B, C are integers is simplified by replacing f, f' by equivalent forms in which $K'a/m, Ka'/m'$ are relatively prime.

The compound of f, f' is determined (p. 28) by congruences equivalent to those of Arndt.[16] Composition of classes is proved to be associative.

H. Poincaré[31] employed a second kind of multiplication $L_1 \otimes L_2$ of two lattices which is commutative and differs from the ordinary multiplication of lattices (or matrices):

$$(Am+Bn) \otimes (A'm'+B'n') = AA'\mu_1+AB'\mu_2+BA'\mu_3+BB'\mu_4,$$

where $\mu_1, \ldots, \mu_4$ range independently over all integers,[32] so that the second member, as well as each factor on the left, is a lattice (see the notations explained at the end of the report in Ch. I[104] on the earlier part of this memoir). As there explained, the lattice $AM+BN$ represents the quadratic form

$$F = (AM+BN)(\overline{A}M+\overline{B}N).$$

Suppose that F is transformable into the product of two other quadratic forms

$$f = (\alpha m+\beta n)(\overline{\alpha}m+\overline{\beta}n), \qquad f_1 = (\alpha_1\mu+\beta_1\nu)(\overline{\alpha}_1\mu+\overline{\beta}_1\nu)$$

by a bilinear substitution

$$M = pm\mu+p'm\nu+p''n\mu+p'''n\nu, \qquad N = qm\mu+q'm\nu+q''n\mu+q'''n\nu,$$

i. e., $F \equiv ff_1$ identically when M, N are replaced by the latter expressions. This identity is easily seen to imply that *

$$(\alpha m+\beta n)(\alpha_1\mu+\beta_1\nu) \equiv \lambda(AM+BN),$$

where λ is a number independent of m, n, μ, ν. Hence we obtain four relations $\alpha\alpha_1 = \lambda(Ap+Bq), \ldots$. Multiplying them by arbitrary numbers $M_1, \ldots, M_4$ and adding, we get

$$L \equiv \alpha\alpha_1 M_1+\alpha\beta_1 M_2+\beta\alpha_1 M_3+\beta\beta_1 M_4 = \lambda AR+\lambda BS,$$
$$R = pM_1+\ldots+p'''M_4, \qquad S = qM_1+\ldots+q'''M_4.$$

Thus the numbers of L all occur among the numbers of the lattice $(\lambda AM+\lambda BN)$, where M and N are arbitrary. The converse is true if the six determinants (9) have no common factor. For, we can then choose integers $M_1, \ldots, M_4$ such that $R=1$, $S=0$, whence $L=\lambda A$, and similarly a set of M's for which $L=\lambda B$, so that every number of the lattice $(\lambda AM+\lambda BN)$ occurs in L. Hence these two lattices are identical. This proves that if F is the compound of f and f_1, the lattice of F is the second product of the lattices of f and f_1, so that composition of forms reduces to multiplication of lattices. The theorems of Gauss are now readily proved.

---

* Or the relation obtained by replacing one or both of the factors on the left by their conjugates.

[31] Jour. école polyt., t. 28, cah. 47, 1880, 226–245.

[32] For the special values $\mu_1 = mm'$, $\mu_2 = mn'$, $\mu_3 = nm'$, $\mu_4 = nn'$, the second member represents the products of the numbers represented by the two factors on the left. But the second member usually represents also further numbers.

Finally there is deduced a correspondence (more precisely stated by Dedekind[29]) between ideals and quadratic forms of determinant D not divisible by a square. Represent $\lambda+\mu\sqrt{D}$ by the lattice

$$\begin{bmatrix}\lambda & \mu D\\ \mu & \lambda\end{bmatrix}.$$

This lattice is included among the lattices called *ideals*, viz., those with integral elements and having $\epsilon=1$, where $\epsilon=p/\mu$, p being the *norm* of the lattice (the determinant of its four elements) and μ the g.c.d. of the coefficients of the quadratic form represented by the lattice. The product of two ideals is the second product of their matrices. The initial theorems on ideals are established.[33]

S. Levänen[34] proceeded as had Dedekind,[27] but permitted also improper equivalence. He derived Legendre's[4] formula (5) and analogous ones.

F. Klein[35] recalled Gauss' representation of a class of positive binary quadratic forms by a point-lattice (Gauss[38] of Ch. I). To the h classes correspond h point-lattices with the origin as a common point. It is stated that, after a proper choice of the direction of the x-axis with respect to the lattices, the product of any complex numbers of any lattice L_1 by any one of a lattice L_2 is always a complex number of a definite lattice L_3. But the distances from the origin to the points of L_1 are the square roots of the values of the forms of a class. Hence we have a geometrical interpretation of composition of quadratic forms. Like results hold for indefinite forms if we take

$$\sqrt{(x-x')^2-(y-y')^2}, \qquad \text{arc cos}\,\frac{xx'-yy'}{\sqrt{x^2-y^2}\cdot\sqrt{x'^2-y'^2}}$$

as the "distance" between the points (x, y) and (x', y'), and the "angle" between the lines joining them to the origin. Klein[36] later gave details.

F. Mertens[37] gave a very clear exposition of the composition of any two quadratic forms of the same determinant and relatively prime divisors. The theory, which is essentially that by Dedekind,[28] depends on a theorem of Dedekind[38] which is here proved in detail. If a system S of linear forms $f_1, \ldots, f_m$ in $x_1, \ldots, x_n$ with integral coefficients is of rank n, there exist linear forms

$$\omega_i=c_{i1}x_1+\ldots+c_{ii}x_i \qquad (i=1, \ldots, n)$$

with integral coefficients such that $0\leqq c_{ik}<|c_{kk}|$ for each $i>k$, while the ω's are linear functions of the f's with integral coefficients and vice versa. The sign of each c_{kk} may be chosen at pleasure; in other respects the ω's are uniquely determined by S. The system $\omega_1, \ldots, \omega_n$ is called a *reduced system* of S.

33 Cf. Poincaré, Comptes Rendus Paris, 89, 1897, 344–6 (obscure preliminary paper).
34 Öfversigt af Finska Vetenskaps-Soc. Förhandlinger, Helsingfors, 35, 1892–3, 57–68.
35 Göttingen Nachr., 1893, 106–9. Lectures on Mathematics (Evanston Colloquium), New York, 1894, 58–66.
36 Ausgewählte Kapitel der Zahlentheorie, II, 1897, 94–221 (see 131–2).
37 Sitzungsber. Akad. Wiss. Wien (Math.), 104, IIa, 1895, 103–143.
38 Dirichlet's Vorlesungen über Zahlentheorie, ed. 3, 179, § 165, 486–493; ed. 4, 1894, § 172, 518–520. Dedekind employed the terminology of moduls. Cf. Dedekind, Bull. Sc. Math. Astr., (2), 1, 1877, 17–41.

Consider two forms of the same determinant D,

$$f=ax^2+2bxy+cy^2, \qquad f'=a'x'^2+2b'x'y'+c'y'^2,$$

and relatively prime divisors m, m'. Then $af, a'f'$ are the products of

$$f_0=ax+(b+\sqrt{D})y, \qquad f'_0=a'x'+(b'+\sqrt{D})y'$$

by the linear forms derived by changing the sign of $\sqrt{D}$. The coefficients of $f_0f'_0$ are

$$(30) \qquad aa', \qquad ab'+a\sqrt{D}, \qquad a'b+a'\sqrt{D}, \qquad D+bb'+(b+b')\sqrt{D}.$$

By the above theorem, this system of four forms in 1 and $\sqrt{D}$ has a reduced system c_{11}, $c_{21}+c_{22}\sqrt{D}$ in which c_{22} is positive and c_{11} has the same sign as aa'. Evidently c_{22} is the g.c.d. ν of $a, a', b+b'$. Thus $\nu=a\alpha+a'\beta+(b+b')\gamma$ for suitable integers α, β, γ. Write $A=aa'/\nu^2$. It is readily proved that $c_{11}=\nu A$, and $c_{21}=\nu B$, where the integer B is the least positive residue modulo A of

$$\{ab'\alpha+ba'\beta+(D+bb')\gamma\}/\nu.$$

The reduced system is therefore νA, $\nu(B+\sqrt{D})$. Hence the numbers (30) are equal to $p_i\nu A+q_i\nu(B+\sqrt{D})$ for $i=1, 2, 3, 4$. Write

$$X=p_1xx'+p_2xy'+p_3yx'+p_4yy', \qquad Y=q_1xx'+q_2xy'+q_3yx'+q_4yy'.$$

Thus $f_0f'\equiv\nu AX+\nu(B+\sqrt{D})Y$. Multiplying this by the identity obtained by changing the sign of $\sqrt{D}$, we see that $(B^2-D)/A$ is an integer C and hence that

$$ff'=AX^2+2BXY+CY^2\equiv F.$$

This compound F of f and f' is also the compound of f' and f. It is easily verified that F has the determinant D and divisor mm'. Given the forms $f_1, \ldots, f_\mu$ of the same determinant D and divisors $m_1, \ldots, m_\mu$ which are relatively prime in pairs, we evidently obtain a unique compound $f_1\ldots f_\mu$ which is independent of the sequence in which the f's are taken.

If f and f' are of the same determinant and have relatively prime divisors and if they are equivalent to g and g' respectively, then ff' is equivalent to gg'. The proof (pp. 118–121) is direct and employs the linear factors of the four forms.

Call a form (a, b, c) *restricted* (*schlicht*) if $0\leqq b<|a|$. Let

$$F=n\left(ax^2+\frac{2b}{\sigma}xy+cy^2\right), \qquad \sigma=1 \text{ or } 2,\ b \text{ odd if } \sigma=2,$$

be a restricted form of divisor n. We seek all primitive restricted forms $f=(A, B, C)$ of determinant D whose compound $f\omega$ with the *simplest form*

$$\omega=\left(n, \quad \frac{n(\sigma-1)}{\sigma}, \quad \frac{n^2(\sigma-1)^2-D\sigma^2}{n\sigma^2}\right)$$

of divisor n and determinant D is F. Necessary and sufficient conditions are

$$A=a\nu^2, \qquad B=nb/\sigma+sa\nu, \qquad C=(B^2-D)/A,$$

where ν is a divisor of n, and s is an integer. Conversely, these conditions imply that

$f\omega = F$. This f is primitive if n is chosen as a product $\nu\mu$, where μ is prime to a, and ν contains only prime factors of a, and if s is chosen so that $n^{-1}F(s, \mu)$ is prime to $a\nu^2$, which is a possible choice. Since the last fact is not altered when s is increased by a multiple of ν, we may assume that $0 \leqq B < |A|$, so that f is restricted and primitive. Denote the resulting solution f by f_0. If f_0' is its opposite form and if f is any solution, then $f_0'f = \phi$ satisfies $\phi\omega = \omega$ and is restricted and primitive. Since the compound of $f_0\phi$ with ω is F, the problem has been reduced to the case $F = \omega$, and is readily treated. The application to class-number is quoted under Mertens[237] of Ch. VI.

R. Dedekind[39] showed the importance of the result implied by Gauss and explicitly stated by Cayley that if we are given any bilinear substitution (8) we can write down three forms F, f, f' whose coefficients are determined by those of the substitution such that $F = ff'$. By an evident modification of Arndt's[11] remark, we can deduce a second substitution for which $f' = Ff$. Since composition is commutative, Arndt's method when applied to $F = f'f$ leads to a third substitution for which $f = Ff'$. Dedekind was apparently not familiar with the results due to Arndt and Cayley, since he approached the subject anew and developed it in a symmetrical and simple manner.

Employing three pairs of independent variables x_i, y_i and eight arbitrary constants α_j, β_j, and any permutation r, s, t, of 1, 2, 3, he wrote

$$(31) \qquad F_r = F_r(x_r, y_r) = A_r x_r^2 + B_r x_r y_r + C_r y_r^2$$
$$= (\beta_s x_r + \alpha_t y_r)(\beta_t x_r + \alpha_s y_r) - (\alpha_r x_r + \beta_0 y_r)(\alpha_0 x_r + \beta_r y_r).$$

where therefore

$$A_r = \beta_s\beta_t - \alpha_0\alpha_r, \qquad C_r = \alpha_s\alpha_t - \beta_0\beta_r, \qquad B_r = \alpha_s\beta_s + \alpha_t\beta_t - \alpha_r\beta_r - \alpha_0\beta_0.$$

He proved that $F_r(X_r, Y_r) = F_s F_t$ under the substitution

$$(32) \quad X_r = (\beta_r x_s + \alpha_t y_s)x_t + (\alpha_s x_s + \beta_0 y_s)y_t, \quad Y_r = -(\alpha_0 x_s + \beta_s y_s)x_t - (\beta_t x_s + \alpha_r y_s)y_t.$$

For, if we denote $\partial F_r/\partial x_r$ by $2u_r$, $\partial F_r/\partial y_r$ by $2v_r$, we see that

$$\alpha_0 X_r + \beta_r Y_r = -A_s x_s y_t - A_t y_s x_t - \tfrac{1}{2}(B_t + B_s) y_s y_t = -y_s u_t - u_s y_t,$$
$$\beta_t X_r + \alpha_s Y_r = -y_s v_t + u_s x_t, \qquad \beta_s X_r + \alpha_t Y_r = x_s u_t - v_s y_t,$$
$$\alpha_r X_r + \beta_0 Y_r = x_s v_t + v_s x_t,$$

whence by matrix multiplication (row of first by column of second),

$$\begin{pmatrix} x_s & v_s \\ -y_s & u_s \end{pmatrix}\begin{pmatrix} u_t & v_t \\ -y_t & x_t \end{pmatrix} = \begin{pmatrix} \beta_s X_r + \alpha_t Y_r, & \alpha_r X_r + \beta_0 Y_r \\ \alpha_0 X_r + \beta_r Y_r, & \beta_t X_r + \alpha_s Y_r \end{pmatrix}.$$

The determinant of the third matrix is $F_r(X_r, Y_r)$, while that of the first is F_s, and that of the second is F_t.

We can now readily prove that any three forms

$$f_1 = f_1(x_1, y_1) = a_1 x_1^2 + b_1 x_1 y_1 + c_1 y_1^2, \qquad f_2 = f_2(x_2, y_2), \qquad f_3 = f_3(x_3, y_3),$$

[39] Jour. für Math., 129, 1905, 1–34.

whose discriminants $\partial_1=b_1^2-4a_1c_1$, ∂_2, ∂_3 are each not zero and for which $f_1(X_1, Y_1)=f_2f_3$ under a bilinear substitution

$$(33)\quad X_1=\beta_1x_2x_3+a_2x_2y_3+a_3y_2x_3+\beta_0y_2y_3,\quad Y_1=-a_0x_2x_3-\beta_3x_2y_3-\beta_2y_2x_3-a_1y_2y_3,$$

are proportional to F_1, F_2, F_3 in (31). For greater symmetry, let $f_1=k_1f_2f_3$ where k_1 is a constant $\neq 0$. Since (33) is of the form (32) with $r=1$, $s=2$, $t=3$, when we regard x_2, y_2 as constants we have a binary substitution of determinant $-F_2$ which transforms $f_1(X_1, Y_1)$ into the product of f_3 by the constant k_1f_2, whence[40]

$$(k_1f_2)^2\partial_3=(-F_2)^2\partial_1,\qquad (k_1f_3)^2\partial_2=(-F_3)^2\partial_1,$$

the second being obtained similarly by regarding x_3, y_3 as constants. Hence $F_2=n_2f_2$, $F_3=n_3f_3$, where n_2 and n_3 are constants $\neq 0$. By multiplication,

$$k_1F_2F_3=n_2n_3f_1(X_1, Y_1).$$

But $F_2F_3=F_1(X_1, Y_1)$ by the former result. Since X_1, Y_1 may be made to take any assigned values x_1, y_1, we have $F_1(x_1, y_1)=n_1f_1(x_1, y_1)$, where $n_1=n_2n_3/k_1$. Hence

$$F_1=n_1f_1,\qquad F_2=n_2f_2,\qquad F_3=n_3f_3.$$

Conversely, these relations imply that each of the forms f_1, f_2, f_3 is transformed into the product of the remaining two multiplied by a constant. The six conclusions of Gauss (Art. 235) are now easily proved. The three bilinear substitutions (32) are such that

$$\begin{aligned}H&=y_1X_1-x_1Y_1=y_2X_2-x_2Y_2=y_3X_3-x_3Y_3\\&=a_0x_1x_2x_3+a_1x_1y_2y_3+a_2y_1x_2y_3+a_3y_1y_2x_3\\&\quad+\beta_0y_1y_2y_3+\beta_1y_1x_2x_3+\beta_2x_1y_2x_3+\beta_3x_1x_2y_3.\end{aligned}$$

A new derivation of $F_r=F_sF_t$ is made by a study of general trilinear forms in which, in contrast to the preceding, the a_j, β_j do not enter explicitly.

H. Weber[41] began with the preceding trilinear form H. Then

$$X_r=\frac{\partial H}{\partial y_r},\qquad Y_r=-\frac{\partial H}{\partial x_r}$$

have the values (32). Further,

$$\begin{vmatrix}\dfrac{\partial X_r}{\partial x_s} & \dfrac{\partial X_r}{\partial y_s}\\ \dfrac{\partial Y_r}{\partial x_s} & \dfrac{\partial Y_r}{\partial y_s}\end{vmatrix}=-F_t,$$

where F_r is given by (31). Assign to the a's, β's such values that each F_t is a product of two distinct linear factors, and introduce these six linear functions as new variables in H. If the new variables are given the initial notation, we have $A_r=C_r=0$, $B_r\neq 0$. A simple examination of the resulting conditions on the a's, β's, shows that H has the normal form $a_0x_1x_2x_3+\beta_0y_1y_2y_3$. Then F_t becomes $-a_0\beta_0x_ty_t$, and $f_t\equiv F_t(X_t, Y_t)$ becomes $a_0^2\beta_0^2x_ry_rx_sy_s$, whence $f_t=F_rF_s$. Also the discriminant D_t of F_t becomes $a_0^2\beta_0^2$, whence $D_1=D_2=D_3$. Although these relations were derived for the

[40] For $F_t=f_t$, we have $\partial_3=\partial_2$. Thus the three forms (31) have equal discriminants.
[41] Göttingen Nachrichten, 1907, 86–100. We interchange his f_s, F_s.

normal form, they hold also for the general form H, since F_t and D_t are (relative) covariants of H under any three binary linear transformations of x_i, y_i $(i=1, 2, 3)$. Several further known properties of composition are proved from the present standpoint.

T. Lalesco[42] called $(a_1, b_1, c_1)=a_1x^2+b_1xy+c_1y^2$ and (a_2, b_2, c_2) composable if they have the same determinant D and if $a_1, a_2, \frac{1}{2}(b_1+b_2)$ have no common divisor. Then they are equivalent if and only if there exist integers x, y such that

$$x^2-Dy^2=4a_1a_2, \qquad x+b_1y\equiv 0 \pmod{2a_1}, \qquad x-b_2y\equiv 0 \pmod{2a_2}.$$

It is then easily shown that, if f and F are composable forms equivalent to f' and F' respectively, the compound of f and F is equivalent to the compound of f' and F'.

Lalesco[43] recalled that, if m, n are represented properly by classes K_m, K_n of primitive forms of determinant D, without a square factor, then mn is represented by the composite class $K_m\cdot K_n$. But is this representation proper or improper? First, consider numbers prime to $2D$. By Dirichlet,[14] proper representations of aa_1 by (aa_1, b, c) can result from the composition of proper representations of the relatively prime numbers a and a_1 by (a, b, a_1c) and (a_1, b, ac). Let

$$m=a_1^{\alpha_1}\ldots a_k^{\alpha_k}b_1^{\beta_1}\ldots b_p^{\beta_p}, \qquad n=a_1^{\alpha'_1}\ldots a_k^{\alpha'_k}c_1^{\gamma_1}\ldots c_q^{\gamma_q}$$

and let K_λ and K_λ^{-1} be the two opposite classes which alone represent properly the prime λ. Then the general expression for the 2^{p+k} classes representing m properly is

$$K_m=K_{a_1}^{\pm\alpha_1}\ldots K_{a_k}^{\pm\alpha_k}K_{b_1}^{\pm\beta_1}\ldots K_{b_p}^{\pm\beta_p}.$$

Of the $2^{p+k}\cdot 2^{q+k}$ representations of mn obtained by composition of the proper representations of m and n, evidently only 2^{p+q+k} are proper, the condition for an improper one being that in the proper representations of m and n at least one of the common factors is represented by opposite classes.

A divisor d of $D=d\delta$ is represented properly only by the ambiguous class $dx^2-\delta y^2$. No power of d is represented properly by forms of determinant D.

F. Mertens[44] considered primitive positive forms

$$f=\left(a, \frac{b}{\sigma}, c\right)=ax^2+\frac{2b}{\sigma}xy+cy^2 \qquad (\sigma=1 \text{ or } 2)$$

for which $D=b^2-\sigma^2ac$ is negative. It is call *restricted* (*schlicht*) if $0\leqq b<\sigma a$. Let $\Sigma=\left(\begin{smallmatrix}\alpha & \beta\\ \gamma & \delta\end{smallmatrix}\right)$ be a given substitution with integral coefficients of determinant unity. Denote the g.c.d. of a and γ by ρ and write

$$\eta=\frac{\gamma}{\rho}, \qquad \xi=\frac{1}{\rho}\left(a\alpha+\frac{b+1-\sigma}{\sigma}\gamma\right).$$

Then ξ, η are relatively prime integers. We can choose integers ξ', η' such that $\left(\begin{smallmatrix}\xi & \xi'\\ \eta & \eta'\end{smallmatrix}\right)$ is of determinant unity and replaces the principal form $H_\sigma=(1, (\sigma-1)/\sigma, [(\sigma-1)^2-D]/\sigma^2)$ by a restricted form Θ. It is proved that the form by which Σ

[42] Nouv. Ann. Math., (4), 7, 1907, 145–150.
[43] Bull. Soc. Math. France, 35, 1907, 248–252.
[44] Sitzungsber. Akad. Wiss. Wien (Math.), 119, IIa, 1910, 241–7.

replaces f is a form parallel[45] to the compound of f and Θ, the compound being determined by the congruences of Gauss (Art. 243).

A. Speiser[46] gave a simple and pleasing derivation of the main facts concerning composition. Consider the bilinear substitution

$$(34)\qquad x_1=py_1z_1+p'y_1z_2+p''y_2z_1+p'''y_2z_2,\qquad x_2=qy_1z_1+q'y_1z_2+q''y_2z_1+q'''y_2z_2,$$

with the same coefficients as Gauss' substitution (8). Solving for y_1, y_2, we get

$$(35)\qquad y_1=\frac{Y_1}{\phi_3},\qquad y_2=\frac{Y_2}{\phi_3},\qquad \phi_3(z_1,z_2)=\begin{vmatrix} pz_1+p'z_2 & p''z_1+p'''z_2\\ qz_1+q'z_2 & q''z_1+q'''z_2\end{vmatrix},$$

$$(36)\ Y_1=q''x_1z_1+q'''x_1z_2-p''x_2z_1-p'''x_2z_2,\qquad Y_2=-qx_1z_1-q'x_1z_2+px_2z_1+p'x_2z_2.$$

Next, we solve (34) for z_1, z_2 and obtain

$$(37)\qquad z_1=\frac{Z_1}{\phi_2},\qquad z_2=\frac{Z_2}{\phi_2},\qquad \phi_2(y_1,y_2)=\begin{vmatrix} py_1+p''y_2 & p'y_1+p'''y_2\\ qy_1+q''y_2 & q'y_1+q'''y_2\end{vmatrix},$$

$$(38)\ Z_1=q'x_1y_1+q'''x_1y_2-p'x_2y_1-p'''x_2y_2,\ Z_2=-qx_1y_1-q''x_1y_2+px_2y_1+p''x_2y_2.$$

Solving the first two equations (35) for z_1/ϕ_3, z_2/ϕ_3, we get

$$z_1/\phi_3=-Z_1/\phi_1,\qquad \phi_1(x_1,x_2)=\begin{vmatrix} q''x_1-p''x_2 & q'''x_1-p'''x_2\\ -qx_1+px_2 & -q'x_1+p'x_2\end{vmatrix},$$

where Z_1 is given by (38). Replacing Z_1 by its value from (37_1), we get

$$(39)\qquad -\phi_1(x_1,x_2)=\phi_2(y_1,y_2)\cdot\phi_3(z_1,z_2),$$

which therefore becomes an identity under substitution (34). Hence if we are given any bilinear substitution (34), we can find three quadratic forms ϕ_i such that $-\phi_1(x_1,x_2)$ is transformed into the product of $\phi_2(y_1,y_2)$ and $\phi_3(z_1,z_2)$ by the given substitution. This result was stated explicitly by Cayley and implied by Gauss.

We now give a simple proof of Dedekind's theorem that if f_1, f_2, f_3 are three quadratic forms, no one of which is a perfect square, such that

$$(40)\qquad f_1(x_1,x_2)=f_2(y_1,y_2)\cdot f_3(z_1,z_2)$$

under the substitution (34), then $f_i=c_i\phi_i$, where c_i is a constant $\neq 0$. Since (40) will become an identity in x_1, x_2, z_1, z_2 under substitution (35)

$$\phi_3^2(z_1,z_2)\cdot f_1(x_1,x_2)=f_2(Y_1,Y_2)\cdot f_3(z_1,z_2).$$

Thus f_3, whose linear factors are distinct, must divide ϕ_3, whence $f_3=c_3\phi_3$. Next, under the substitution (37), (40) becomes

$$\phi_2^2(y_1,y_2)\cdot f_1(x_1,x_2)=f_2(y_1,y_2)\cdot f_3(Z_1,Z_2).$$

Hence f_2 must divide ϕ_2, $f_2=c_2\phi_2$. Thus (40) becomes

$$f_1(x_1,x_2)/c_2c_3=\phi_2(y_1,y_2)\cdot\phi_3(z_1,z_2).$$

Inserting the values of y_1, y_2 from (35) and cancelling a factor ϕ_3, we get

$$(41)\qquad \phi_2(Y_1,Y_2)=\frac{1}{c_2c_3}\cdot f_1(x_1\ x_2)\cdot\phi_3(z_1,z_2),$$

---

[45] f is *parallel* to $(a, b'/\sigma, c')$ if $b'\equiv b \pmod{\sigma a}$.
[46] Festschrift H. Weber, 1912, 375–395.

under the substitution (36). This implies that $f_1(x_1, x_2)$ divides $\phi_1(x_1, x_2)$, for the same reason that (40) implied that f_2 divides ϕ_2. Thus $f_1=c_1\phi_1$.

Since $f_i=c_i\phi_i$, a comparison of (39) and (40) gives $-c_1=c_2c_3$. Thus, by (41),

$$-\phi_2(Y_1, Y_2)=\phi_1(x_1, x_2)\cdot\phi_3(z_1, z_2)$$

under the substitution (36). Inserting the values of z_1, z_2 from (37) in (39), we see that

$$-\phi_3(Z_1, Z_2)=\phi_1(x_1, x_2)\cdot\phi_2(y_1, y_2)$$

under the substitution (38). Hence the ϕ's form a symmetrical triple, as noted by Dedekind.

The explicit expressions for ϕ_3, ϕ_2 are given by (22), (25) of Pepin.[30] That their discriminants are equal follows at once from the identity (15).

Given any integers $P, \ldots, U$ satisfying (15), it is proved by means of Gauss (Art. 279) and also by a second, rapid method that we can find integral solutions $p, \ldots, q'''$ of the six equations (9), in fact with $p'''=0$. Hence any two forms which have integral coefficients, equal discriminants, and relatively prime divisors can be identified with ϕ_2, ϕ_3 and hence be compounded under a bilinear substitution with integral coefficients. If M is the matrix of one such substitution, the matrix of every such substitution is given by the product VM, where V is a square matrix whose four elements are integers of determinant unity, and the resulting forms ϕ_1 compounded of ϕ_2, ϕ_3 are all equivalent (see the above report of Gauss, Arts. 239, 234). Hence any two forms with equal discriminants and relatively prime divisors can be compounded into all of the forms of a class of equivalent forms. It is proved in detail that composition of classes of forms obeys the associative law.

L. Aubry[47] gave an example (?) of distinct classes of forms of the same determinant which, under composition with the same properly primitive class, give the same class.

H. Brandt[48] gave a new theory of composition of forms ϕ, ϕ' with the same discriminant D and relatively prime divisors without using the bilinear substitution. At the same time he generalized the concept of composition without disturbing its main properties. Consider the principal form $\eta=x^2+\epsilon xy-dy^2$, where $\epsilon=0$ or 1, $\epsilon^2+4d=D$. Let $\phi=ax^2+bxy+cy^2$ be any form of discriminant D. If m is representable properly by ϕ, there exists a binary substitution of determinant m which replaces η by $m\phi$, and is called a *substitution producing* ϕ. It and a substitution producing ϕ' determine a third substitution producing Gauss' compound of ϕ, ϕ'. The generalization employs an arbitrary ground form χ of discriminant D instead of η, and obtains composition of ϕ, ϕ' relative to χ.

On the composition of quadratic forms in four variables, see papers 18, 23, 42, 44 of Ch. IX, and 5, 34, 44 of Ch. X. On the composition of higher forms, see papers 5, 18, 19 of Ch. XIV and those cited at the end of that chapter.

---

[47] L'intermédiaire des math., 20, 1913, 6–7.
[48] Jour. für Math., 150, 1919, 1–46.

CHAPTER IV.

ORDERS AND GENERA; THEIR COMPOSITION.

Classes of binary quadratic forms are separated into orders and genera by means of the values of certain arithmetical invariants, called characters in the case of genera. If we compound any form of one order with any one of another order, we always obtain a form of a unique third order, so that we may speak of composition of orders. The same is true of genera. Gauss relied upon ternary quadratic forms to prove the difficult point that exactly half of the notationally possible characters actually correspond to genera. More elementary proofs have been given by various writers. Forms transformable into each other by linear substitutions with rational coefficients are put into the same genus in the final papers of Pund and Speiser, as had been done for n-ary forms by Eisenstein in 1852 (Ch. XI).

C. F. Gauss[1] (Art. 226) wrote m for the g.c.d. of a, b, c and called the form (a, b, c) *primitive* (*ursprüngliche*) if $m=1$, but if $m>1$ spoke of it as being *derived* from the primitive form $(a/m, b/m, c/m)$.

If any class of forms of determinant D contains a primitive form, all forms of the class are primitive, and the class is called primitive. If a form F of a class K of determinant D is derived from a primitive form f of a class k of determinant D/m^2, then all forms of K are derived from forms of k, so that K is said to be derived from the primitive class k.

A primitive form (a, b, c) is called *properly* (*eigentlich*) or *improperly primitive*, according as a, c are not both or are both even; in the respective cases, the g.c.d. of a, $2b$, c is 1 or 2. The determinant of an improperly primitive form is $\equiv 1 \pmod 4$. Any class is called properly or improperly primitive according as one form (and hence all forms) of it is properly or improperly primitive.

According as the g.c.d. of a, $2b$, c is m or $2m$, (a, b, c) is derived from the properly or improperly primitive form $(a/m, b/m, c/m)$. A class is derived either from a properly or improperly primitive class.

Two classes are said to belong to the same *order* (*Ordnung*) if and only if, when (a, b, c) and (a', b', c') are representative forms of the two classes, not only a, b, c have the same g.c.d. as a', b', c', but also a, $2b$, c have the same g.c.d. as a', $2b'$, c'. Thus the properly primitive classes form one order and the improperly primitive classes another order.

If n^2 divides the determinant D, the classes derived from the properly primitive classes of the determinant D/n^2 form an order, and similarly for the improperly primitive classes. In case D has no square factor >1, there exist no derived orders, so that either the order of the properly primitive classes is the only order (when

[1] Disquisitiones Arithmeticae, 1801; Werke, 1, 1863; German transl. by H. Maser, 1889.

$D\equiv 2$ or 3 modulo 4), or the orders of the properly and improperly primitive classes are the only orders (when $D\equiv 1$ modulo 4). By applying the principles of the theory of combinations it is stated that we have the following general rule. Express D in the form $D'\cdot 2^{2\mu}a^{2\alpha}b^{2\beta}\ldots$, where D' has no square factor, and $a, b, \ldots$ are distinct odd primes; then the number of orders is

$$(\mu+1)(\alpha+1)(\beta+1)\ldots \quad \text{if } D'\equiv 2 \text{ or } 3 \pmod 4,$$
$$(\mu+2)(\alpha+1)(\beta+1)\ldots \quad \text{if } D'\equiv 1 \pmod 4.$$

Any (Art. 228) properly primitive form $F=ax^2+2bxy+cy^2$ represents infinitely many integers not divisible by a given prime p. For, if a is not divisible by p, and we take x not divisible by p, and y divisible by p, then F is not divisible by p. If both a and c are divisible by p, so that $2b$ is not, assign to x and y values not divisible by p.

The following results (Art. 229) form the basis of the definitions of characters and genera. If m, m' are any integers not divisible by the prime factor p of the determinant D of a primitive form F, and both are representable by F, then m, m' are both quadratic residues or both quadratic non-residues of p. For,

$$m=ag^2+2bgh+ch^2, \qquad m'=ag'^2+2bg'h'+ch'^2$$

imply

$$mm'=\{agg'+b(gh'+hg')+chh'\}^2-D(gh'-hg')^2,$$

whence mm' is congruent to a square modulo D and hence modulo p.

Next, if $D\equiv 3 \pmod 4$, the odd integers represented by the primitive form F are either all $\equiv 1$ or all $\equiv 3 \pmod 4$. For, as just proved, $mm'=P^2-DQ^2$. When m, m' are odd, one of P, Q is even and the other odd, whence $mm'\equiv 1 \pmod 4$.

If $D\equiv 2 \pmod 8$, the odd integers represented by F are either all $\equiv \pm 1$ or all $\equiv \pm 3 \pmod 8$. For, P is odd and $P^2\equiv 1 \pmod 8$. Evidently $Q^2\equiv 0$, 4 or 1 (mod 8). Hence $mm'\equiv \pm 1 \pmod 8$.

Similarly, if $D\equiv 6 \pmod 8$, the odd integers represented by F are either all $\equiv 1, 3$ or all $\equiv 5, 7 \pmod 8$.

If $D\equiv 0 \pmod 4$, the odd numbers represented by F are all $\equiv 1$ or all $\equiv 3$ (mod 4), since mm' is a quadratic residue of 4 and hence is $\equiv 1 \pmod 4$.

If $D\equiv 0 \pmod 8$, the odd numbers represented by F are all $\equiv 1$, or all $\equiv 3$, or all $\equiv 5$, or all $\equiv 7 \pmod 8$, since mm' is a quadratic residue of 8 and hence is $\equiv 1 \pmod 8$.

If (Art. 230) p is a prime factor of the determinant D of a primitive form F, and if only quadratic residues of p can be represented by F, we say that F has the (*special*) *character* Rp. If only quadratic non-residues can be represented, F has the character Np. Again, if F represents no odd integers except those $\equiv 1 \pmod 4$, F is said to have the character 1, 4. Similarly for the characters 3, 4; 1, 8; etc. [For simpler notations, see Dirichlet.[8]] The various characters of a primitive form (a, b, c) are evidently determined by a, c, which are represented by the form and are not both divisible by a prime factor p of D, and are not both even.

Since all numbers which are representable by one form F of a class are representable by every form of the class, we speak of the various characters of F as characters of the class. Opposite classes have the same characters.

The totality (Art. 231) of the special characters of a class is said to constitute the *total character* of the class. All classes which have the same total character form a *genus* (*Geschlecht, genre*). The principal form $(1, 0, -D)$ belongs to the principal class, which in turn belongs to the principal genus, all of whose characters are $+1$.

If D is divisible by 4, but not by 8, and if m is the number of distinct odd prime factors of D, there are 2^{m+1} total characters. For, there are two special characters 1, 4 and 3, 4 with respect to 4, while there are two special characters Rp and Np with respect to each odd prime factor p of D.

Similarly, if $D\equiv 0 \pmod 8$, there are 2^{m+2} total characters; if D is even, but not divisible by 4, there are 2^{m+1} total characters; if $D\equiv 1 \pmod 4$, there are 2^m; if $D\equiv 3 \pmod 4$, there are 2^{m+1} total characters. But it does not follow that there exist as many genera as possible total characters (see the report below on Arts. 261, 287).

If (Art. 245) f and g belong to the same order, and f' and g' to the same order, the form compounded of f, f' has the same determinant and belongs to the same order as the form compounded of g, g'. Hence we speak of *composition of orders*.

Given (Art. 246) two primitive forms f, f', whose compound is F, we may derive the genus to which F belongs from the genera to which f, f' belong. First, let f be properly primitive. The determinant D of F is then the g.c.d. of dm'^2 and d', where $m'=1$ or 2, according as f' is properly or improperly primitive. If p is an odd prime factor of D, it divides d and d'. Since the product of numbers represented by f, f' respectively is representable by F, if f and f' each have the character Rp, or both the character Np, then F has the character Rp; but if one of f, f' has the character Rp and the other the character Np, then F has the character Np. Similarly, F has the character 1, 4 or 3, 4 according as both f, f' have the same character (1, 4 or 3, 4) or one of them has the character 1, 4 and the other 3, 4. The investigation is made also for $D\equiv 0$ or $2 \pmod 8$, and also when both f, f' are improperly primitive.

It follows (Arts. 247–8) that if f and g are forms of the same genus, and f' and g' are forms of the same genus, the forms compounded of f, f' and of g, g' belong to the same genus. Hence we speak of *composition of genera*. The principal genus (and it alone) plays the rôle of unity in composition. The compound of two properly primitive forms of the same determinant belongs to the principal genus if and only if they belong to the same genus. When any primitive form is compounded with itself, the resulting form (arising by *duplication*) belongs to the principal genus.

If (Arts. 250–1) D is divisible by m^2, $(\pm m, 0, \mp D/m)$ is called the *simplest form* of divisor m and determinant D, derivable from a properly primitive form (by multiplication by m). If also $D/m^2\equiv 1 \pmod 4$,

$$\left(\pm 2m, \pm m, \pm \frac{m^2-D}{2m}\right)$$

is the simplest form of divisor m and determinant D, derivable from an improperly primitive form. Given a form F of divisor m, we can find a properly primitive form whose compound with the simplest form of divisor m is F.

There occur (Art. 252) equally many classes in the various genera of the same order of a given determinant. This is not true of genera of different orders.

The number (Art. 258) of all ambiguous, properly primitive, positive[2] classes of determinant D, not a square, is exactly half the number of all possible characters for determinant D.

Hence (Art. 261) for a given determinant D, not a square, half of all possible characters do not correspond to properly primitive genera (the genera to be positive if D is negative). By means of the reciprocity law for quadratic residues, it is determined (Arts. 263–4) which half of the characters do not correspond to genera.

In Art. 286, Gauss' theorems on ternary quadratic forms (see Ch. IX) are applied to prove that there exists (and to show how to find) a binary quadratic form by whose duplication arises any given form of the principal genus. It follows at once (Art. 287) that at least half and hence (Art. 261) exactly half of the possible characters for a given determinant actually correspond to properly primitive positive genera.

The median value of the number (Art. 301) of genera in a properly primitive (for negative determinant, positive) order of determinant $\pm D$ is stated to be $\alpha \log D+\beta$, where $\alpha=0.405\ldots$, $\beta=0.883\ldots$. Cf. Dirichlet.[4]

The number m (Art. 305) of forms in the period defined by a class in the principal genus is a divisor of the number n of classes. The nth power of any class is the principal class (Art. 306, I).

Miscellaneous remarks (Art. 307) on genera other than the principal genus are applied to compute all properly primitive classes for a regular determinant.

G. L. Dirichlet[3] expressed the results of Gauss (Art. 229) in the following convenient notations: If p is an odd prime divisor of D, the integers m (not divisible by p) which can be represented by the same properly primitive form F of determinant D are all such that Legendre's symbol (m/p) has the same value $+1$ or -1. If $D\equiv 3 \pmod 4$, for every odd integer m representable by F, $(-1)^{\frac{1}{2}(m-1)}$ has the same value. Similarly for $(-1)^{\frac{1}{8}(m^2-1)}$ when $D\equiv 2 \pmod 8$, $(-1)^{\frac{1}{2}(m-1)+\frac{1}{8}(m^2-1)}$ when $D\equiv 6 \pmod 8$, $(-1)^{\frac{1}{2}(m-1)}$ when $D\equiv 4 \pmod 8$, and both $(-1)^{\frac{1}{2}(m-1)}$ and $(-1)^{\frac{1}{8}(m^2-1)}$ when $D\equiv 0 \pmod 8$. Each of the expressions

$$\left(\frac{m}{p}\right),\qquad (-1)^{\frac{1}{2}(m-1)},\qquad (-1)^{\frac{1}{8}(m^2-1)},\ldots$$

is called a special character of the properly primitive form F.

If there are λ such special characters and hence 2^λ combinations of them, only half of them (apart from one exception noted below) exist and define genera. Indeed, there exists a relation between the special characters. For, if S^2 is the largest square dividing D, denote the quotient by P or $2P$, according as it is odd or even. When $P\neq\pm 1$, let $p, p', \ldots$ be the odd primes whose product is P. We can assign

[2] In any form (a, b, c) of negative determinaut, the outer coefficients a and c have like signs, and the same is true of any form equivalent to it. If a and c are positive, the form is called *positive* and the class to which it belongs is called a *positive class*. Similarly for negative forms and negative classes (Art. 225).

[3] Jour. für Math., 19, 1839, 335–340, 365–9; Werke, 1, 1889, 425–9, 456–460. Cf. Zahlentheorie, Suppl. IV, §§ 121–6, 1863, 1871, 1879, 1894 (with modifications by Dedekind, who remarked in a note to § 125 that a brief proof is possible by means of the theorem on the infinitude of primes in an arithmetical progression).

relatively prime values to x, y such that the value m of F is positive, odd, and prime to D. Then D is a quadratic residue of m. Hence

$$\left(\frac{P}{m}\right)=1 \text{ if } D=PS^2, \qquad \left(\frac{2P}{m}\right)=\left(\frac{2}{m}\right)\left(\frac{P}{m}\right)=1 \text{ if } D=2PS^2.$$

By the reciprocity law for Jacobi's symbols,

$$\left(\frac{P}{m}\right)=\left(\frac{m}{P}\right)(-1)^{\frac{1}{2}(P-1)\cdot\frac{1}{2}(m-1)}.$$

The final factor is $+1$ or $(-1)^{\frac{1}{2}(m-1)}$ according as $P\equiv 1$ or $P\equiv 3 \pmod 4$. Also replacing (m/P) by $(m/p)(m/p')\ldots$, and $(2/m)$ by $(-1)^{\frac{1}{8}(m^2-1)}$, we obtain the desired relation between the special characters, except in the case arising for $D=PS^2$, $P\equiv 1 \pmod 4$, when the indicated relation $(m/p)(m/p')\ldots=1$ involves no existing symbols on account of the absence of factors $p, p', \ldots$ of P, i. e., if $P=+1$, whence D is a positive square.

By use of infinite series it is proved (§ 6, V) that the $2^{\lambda-1}$ total characters satisfying the relation mentioned correspond to existing genera, whose number is therefore $2^{\lambda-1}$; and that there occur equally many classes in the various genera of the properly primitive order, or in the improperly primitive order if it exists for the given determinant not a square.

Dirichlet[4] stated that, in view of known theorems, the number of genera of a negative determinant $-n$ is $\phi(n)$ if $n=8h$ or $n=4h+1$, but is $\frac{1}{2}\phi(n)$ if $n=8h+4$ or $4h+2$ or $4h+3$, where $\phi(n)=2^\rho$, ρ being the number of distinct prime factors of n. The mean value of $\phi(s)$ is found, not when s takes all the values $1, \ldots, n$, but when s ranges over the numbers of one of the preceding five linear forms. Introducing the factor $\frac{1}{2}$ in the last three cases above and the factor 2 when the form is $4h+c$ (to reduce all five forms to $8h+k$), adding and dividing by 8, we obtain

$$\frac{4}{\pi^2}\left(\log_e n+\frac{12C'}{\pi^2}+2C-\tfrac{1}{6}\log_e 2\right)$$

as the median value of the number of genera of determinant $-n$, for n arbitrary and not limited to a special linear form. Here

$$C=0.5772156\ldots, \qquad C'=\sum_{s=2}^{\infty}(\log_e s)/s^2.$$

This median value, which holds also for positive determinants, agrees with the result of Gauss (Art. 301).

F. Arndt[5] gave an elementary proof of Gauss' theorem on the existence of genera, which states in effect that there exists a (properly) primitive form of any assigned determinant D which represents any given number m prime to D such that D is a quadratic residue of m.

Let $f=(a, b, c)$ be any given properly primitive positive form belonging to the principal genus and having a determinant D not divisible by a square. It is first proved that f represents a square h^2 prime to D. After a preliminary transformation,

---

[4] Abh. Akad. Wiss. Berlin, 1849, Math., 82–83; Werke, II, 65–66. Further details by Bachmann.[18]

[5] Jour. für Math., 56, 1859, 72–78.

we may assume that a is prime to $2D$. By $b^2-ac=D$, D is a quadratic residue of a, and a of D since f is in the principal genus. Write $a=\theta^2a'$, where a' has no square factor. Hence by the theorem of Legendre [this History, Vol. II, p. 365], $x^2-Dy^2=a'z^2$ has integral solutions without a common factor. Then z is prime to D. Since the properly primitive forms x^2-Dy^2 and f represent $a'z^2$ and a, respectively, their compound f represents $a'z^2\cdot a=h^2$, where $h=\theta a'z$ is prime to D.

Next, the duplication of a properly primitive form of determinant D gives f. Here we may replace f by a properly equivalent form $\phi=(h^2, l, n)$ of determinant D. According as h is odd or even, the duplication of the properly primitive form

$$(h,\ l,\ nh),\qquad \left(2h,\ l+h,\ l+\frac{h}{2}(n+1)\right)$$

of determinant D gives ϕ.

Use is made of the notation $D=PS^2$ or $2PS^2$ of Dirichlet[3] and his (elementary) derivation of the relation between the special characters. If $S=1$ the existence of all notationally possible genera was proved above. Also if $S>1$ there exists a properly primitive positive form $F=ax^2+2bxy+cy^2$ of determinant P or $2P$ whose characters satisfy the relation mentioned. It is shown that integral values can be assigned to x, y such that for the resulting number m represented by F the characters (m/r), (m/r'), ... take any prescribed values ±1, where $r, r', \ldots$ are the distinct prime factors of S not dividing P. In the proof we may assume that a is prime to $2D$. It is readily proved that we can solve the congruence

$$ax'^2+2bx'y'+cy'^2\equiv\rho \pmod r,$$

where ρ is any given integer not divisible by r, and similarly

$$ax''^2+2bx''y''+cy''^2\equiv\rho' \pmod{r'},$$

etc. As in the Chinese remainder problem [Vol. II, p. 57], we can determine integers x, y such that

$$x\equiv x' \pmod r,\quad x\equiv x'' \pmod{r'},\ \ldots,\quad y\equiv y' \pmod r,\quad y\equiv y'' \pmod{r'},\ \ldots.$$

For the resulting value m of F, (m/r), (m/r'), ... will have prescribed values ±1 since $\rho, \rho', \ldots$ were arbitrary.

H. J. S. Smith[6] gave a report on Gauss' work on genera.

A. Cayley[7] tabulated, for each D between -100 and $+100$ not a positive square, representatives of each class of forms of determinant D, their characters, the generators of the group of the class, and for positive determinants the periods of the reduced forms.

L. Kronecker[8] proved by means of analytic methods used by Dirichlet[3] the theorems of Gauss that each genus of properly primitive forms contains the same number of classes, and that all classes of the principal genus arise by duplication.

E. Schering[9] proved for properly primitive forms that the period numbers of the fundamental classes (Ch. I[25]) of the principal genus are odd. An ambiguous class

---

[6] Report British Assoc. for 1862, 520–6; Coll. Math. Papers, I, 251–262.
[7] Jour. für Math., 60, 1862, 357–369; Coll. Math. Papers, V, 141.
[8] Monatsber. Akad. Wiss. Berlin, 1864, 285–303.
[9] Abh. Gesell. Wiss. Göttingen, 14, 1869; Werke, I, 147–8.

arises by composition only from fundamental classes not belonging to the principal genus. If δ is the number of the latter classes, 2^δ is the number of ambiguous classes, as well as the number of genera.

R. Dedekind[10] gave an account of Gauss' proof that at most half of the notationally possible total characters correspond to existing genera, and, by means of ideas similar to those employed by Arndt,[5] proved that every class of the principal genus arises by duplication.

T. Pepin[11] proved that the characters of the class compounded of two classes are obtained by multiplying their corresponding characters. Gauss' theorem that every class of the principal genus P arises by duplication is proved (p. 45) by means of the fact that every form of P represents odd squares a^2 prime to $2D$. For, let a^2 be represented by $f=(a^2,\ b,\ (b^2-D)/a^2)$ of P. Then f arises by duplication of $(a, b, (b^2-D)/a)$.

A comparison is made (pp. 63–69) between the number of classes of each genus for the positive determinants D and Dp^2, when p is 2 or an odd prime. There is a single class in each genus for the determinants 3^t, $2\cdot 3^t$, $4\cdot 3^t$, 5^t, 7^t, 11^t, ..., where t is odd, and the determinant $3^{2k+1}p^{2a}$, if $p=12l+11$ and $6l+5$ are both primes.

H. Weber[12] proved Gauss' theorem that, if the number of independent special characters is λ, then exactly $2^{\lambda-1}$ genera exist. For, it is shown that the class-equation (p. 337), having its source in elliptic functions, decomposes into as many factors as there are genera, after the adjunction of certain square roots, and that each factor vanishes for the class-invariants of a genus.

H. Weber[13] gave a brief derivation of the main properties of characters and the number of genera, following Dirichlet,[3] but employing forms $ax^2+bxy+cy^2$.

J. A. de Séguier[14] gave a simplification (following Kronecker[8]) of the proof by Dirichlet[3] and proved the following theorem: Let R_d be the group of the primitive classes of discriminant $D=D'd^2$ (D' having the form of a discriminant) which when compounded with any class of divisor d and discriminant D leads to a definite class of divisor d and discriminant D. Then every character belonging to both D and D' has the value $+1$ in all classes of R_d, while one belonging to D and not to D' has the value $+1$ as often as the value -1.

P. Bachmann[15] gave an exposition of the work of Gauss[1] and Dirichlet.[3, 4]

D. Hilbert[16] introduced the symbol $\left(\frac{n,\ m}{w}\right)$ to have the value $+1$ if the rational integer n is congruent, with respect to any arbitrary power of the rational prime w as modulus, to the norm of an integral algebraic number of the field $k(\sqrt{m})$, where m is a rational integer not a square, and the value -1 in the contrary case. If

10 Dirichlet–Dedekind, Zahlentheorie, Suppl. X, §§ 152–3, 155, 158, ed. 2, 1871; ed. 3, 1879; ed. 4, 1894.
11 Atti Accad. Pont. Nuovi Lincei, 33, 1879–80, 36–72.
12 Elliptische Functionen und Algebraische Zahlen, 1891, 411–421.
13 Göttingen Nachr., 1893, 57–62, 147–9; Algebra, III, 1908, 380–7, 409.
14 Formes quadratiques et multiplication complexe, 1894, 135–153, 333–4.
15 Die Analytische Zahlentheorie, 1894, 233–271, 472-9.
16 Jahresbericht d. Deutschen Math.-Vereinigung, 4, 1894–5 (1897), 286–316. French transl., Ann. Fac. Sc. Toulouse, (3), 2, 1910, 260–286. Cf. Math. Annalen, 51, 1899, 12, 42; Acta Math., 26, 1902, 99. J. Sommer, Vorlesungen über Zahlentheorie, 1907, 127–164; French transl. by A. Lévy, 1911, 134–172.

$l_1, \ldots, l_t$ are the distinct prime divisors of the discriminant of the field k, and a is any rational integer, the set of t numbers (each $+1$ or -1)

$$(1) \qquad \left(\frac{a,\, m}{l_1}\right), \ldots, \quad \left(\frac{a,\, m}{l_t}\right)$$

is called the *system of characters* of a in k. If k is an imaginary field, the system of characters of an ideal $\mathfrak{a}$ of k is the set of numbers (1) with a replaced by the (positive) norm $n(\mathfrak{a})$ of $\mathfrak{a}$. Next, let k be a real field. If all the numbers

$$(2) \qquad \left(\frac{-1,\, m}{l_1}\right), \ldots, \quad \left(\frac{-1,\, m}{l_t}\right)$$

are $+1$, we define the system of characters of $\mathfrak{a}$ as above. But if one of the numbers (2), say the last one, is -1, we choose the sign of $n' = \pm n(\mathfrak{a})$ so that

$$\left(\frac{n',\, m}{l_t}\right) = +1,$$

and define the set of numbers

$$\left(\frac{n',\, m}{l_1}\right), \ldots \quad \left(\frac{n',\, m}{l_{t-1}}\right)$$

to be the system of characters of $\mathfrak{a}$.

All the ideals of a class have the same system of characters. All the classes of ideals which have the same system of characters are said to form a *genus*. The principal genus contains those classes whose characters are all $+1$; it contains the principal class. Multiplication of classes of ideals of two genera yields the classes of ideals of a genus, whose system of characters are the products of corresponding characters of the two genera. Hence the square of any class of ideals belongs to the principal genus. Every genus evidently contains equally many classes. By means of ideals, there is given (§§ 67–78) an arithmetical proof of Gauss' theorem that a set of $r = t$ or $t-1$ units ± 1 is a system of characters of a genus of $k(\sqrt{m})$ if and only if the product of the r units is $+1$; the number of genera is thus 2^{r-1}. The transcendental proof by Dirichlet is also given (§§ 79–82). To obtain a (1, 1) correspondence between classes of ideals and classes of quadratic forms, we must use the idea of narrow[17] equivalence of ideals (§ 83).

F. Mertens[18] proved Gauss' theorem that every class of the principal genus arises by duplication, using from the theory of ternary forms only Legendre's result that there exist integral solutions of $\xi^2 - D_0\eta^2 - A_0\zeta^2 = 0$ when the necessary congruencial conditions are satisfied.

Ch. de la Vallée Poussin[19] proved Gauss' theorem that all properly primitive forms of the principal genus arise by duplication. The proof is arithmetical and employs only principles of the theory of binary forms, making no use of ternary forms (as had Gauss, Arndt, Dedekind). Ch. II gives various theorems involving relations between the classes of forms which represent the same numbers.

---

17 Dedekind, Dirichlet's Zahlentheorie, eds. 2–4, 1871–1894, Suppl. XI.

18 Sitzungsber. Akad. Wiss. Wien (Math.), 104, IIa, 1895, 137–143. A report of the first part is given under Composition, and of the rest under Class Number.

19 Mém. couronnés et autres mém. Acad. Belgique, 53, 1895–6 mém. No. 3, 59 pp. Summary by P. Mansion, Bull. Acad. Belgique, (3), 30, 1895, 189–195.

Humbert[86] of Ch. XI proved that the hyperabelian curves associated with the classes of forms belonging to the same genus are of the same genus.

F. Mertens[20] gave a short proof, by use of binary forms only, of Gauss' theorem that every class of primitive forms of the principal genus arises by duplication. The proof is by induction on the *Stufe* $s=\rho+\pi$ of the determinant of the form, where ρ is the number of characters and π is the number of prime factors of D.

O. Pund[21] proved that two forms belong to the same genus if and only if they are transformable into each other by a substitution

$$\begin{pmatrix} a & b \\ c & d \end{pmatrix} = \begin{pmatrix} \alpha/\epsilon & \beta/\epsilon \\ \gamma/\epsilon & \delta/\epsilon \end{pmatrix}$$

with rational coefficients of determinant unity, where $\alpha, \beta, \gamma, \delta$ are integers without a common divisor >1 such that $\alpha\delta-\beta\gamma=\epsilon^2$. The conditions that this substitution shall transform $ax^2+2bxy+cy^2$ into $a'x^2+\ldots$ of the same determinant

$$d=b^2-ac=b'^2-a'c' \tag{3}$$

are shown at once, by setting $\eta=a\alpha+b\gamma$, to be equivalent to

$$aa'\epsilon^2+d\gamma^2=\eta^2 \tag{4}$$

$$\eta+b'\gamma=a'\delta, \qquad (b-b')\eta+(bb'-d)\gamma=-aa'\beta, \tag{5}$$

and hence to the existence of integral solutions ϵ, γ, η of (4) such that $\eta-b\gamma$ is divisible by a, and such that the left members of (5) are divisible by a' and aa' respectively. Replacing the given primitive forms by suitably chosen equivalent forms, we may assume that $a, a', 2d$ are relatively prime in pairs. Assume only that d is prime to aa'. Then necessary and sufficient conditions that (4) be solvable are that aa' be a quadratic residue of d and vice versa, and when $d<0$ that $aa'>0$. Then there are solutions ϵ, γ, η without a common divisor such that if $\xi^2\equiv d$, also $\xi\eta\equiv d\gamma \pmod{aa'}$. Determining ξ so that $\xi\equiv b \pmod{a}$, $\xi\equiv -b' \pmod{a'}$, we see that the divisibility conditions mentioned below (5) are satisfied.

The condition that aa' be a quadratic residue of d and that $aa'>0$ when $d<0$ are shown to require that the forms have the same characters.

A Speiser[22] regarded two forms as of the same genus if they have the same discriminant D and if they can be transformed into each other by a linear substitution with fractional coefficients of determinant unity such that their common denominator is prime to $2D$. Every form of one genus may be compounded with every form of another genus under a bilinear substitution with fractional coefficients whose common denominator is prime to $2D$ to give any preassigned form of the resulting genus. It is proved very simply that every form of the principal genus arises by duplication. That genera so defined are identical with those defined by characters is proved when D has no square factor and the principal form is x^2-Dy^2.

---

20 Jour. für Math., 129, 1905, 181–6.
21 Mitt. Math. Gesell. Hamburg, 4, 1905, 206–210. See Eisenstein,[8] Ch. XI.
22 H. Weber Festschrift, 1912, 392–5. See Pund.[21]

CHAPTER V.

IRREGULAR DETERMINANTS.

C. F. Gauss[1] called a determinant *regular* or *irregular,* according as all the classes of the principal genus do or do not form a single *period,* i. e., are all powers of a single one of these classes. If (§ VII) the principal genus contains the classes C, C' whose periods are composed of m, m' classes, it contains a class C'' whose period is composed of M classes, where M is the least common multiple of m, m'. For, if M is the product of two relatively prime divisors r, r' of m, m' respectively, $C''=C^{m/r}C'^{m'/r'}$. Thus the greatest number of classes (of the principal genus) in any period is divisible by the number in any other period. The quotient of the number n of all classes of the principal genus by the number in the greatest period is an integer called the *exponent of irregularity* e when the determinant is irregular.

The following special results are stated without proof (§ VIII). If the principal genus contains more than two ambiguous classes, the determinant is irregular and e is even. If only 1 or 2 ambiguous classes occur, either the determinant is regular or e is odd. All negative determinants of the type $-(216k+27)$, $k\neq 0$, are irregular and e is divisible by 3; the same is true of $-(1000k+75)$, $k\neq 0$, and $(-1000k+675)$. If e is divisible by a prime p, n is divisible by p^2. For negative determinants $-D$, the irregular ones occur more frequently as D increases. There are 13 irregular determinants $-D$ with $D<1000$, viz., 576, 580, 820, 884, 900, having $e=2$; 243, 307, 339, 459, 675, 755, 891, 974, having $e=3$. In the second thousand there are 13 with $e=2$ and 15 with $e=3$. In the third thousand there are 18 with $e=2$ and 19 with $e=3$. In the tenth thousand there are 31 with $e=2$ and 32 with $e=3$. Apparently the ratio of the frequency of the irregular negative determinants $-D$ to that of the regular approaches a constant as D increases. For positive determinants not a square, the irregular determinants are much rarer; there is an infinitude with e even, but none with e odd have been found.

Gauss[2] proved the following results: (I) The number of properly primitive classes of determinant D whose p^π power (p a prime) is the principal class is unity or a power of p. (II) If the number of the properly primitive classes of the principal genus is $a^\alpha b^\beta\ldots$, where $a, b, \ldots$ are distinct primes, there exist in this genus $a^\alpha, b^\beta, \ldots$ classes whose powers $a^\alpha, b^\beta, \ldots$, respectively, give the principal class.

Gauss[3] gave a table of the classes of binary quadratic forms which shows the number of genera (here often called *Ordo*), number of classes, and the exponent of

[1] Disquisitiones Arith., 1801, Art. 306, §§ VI–VIII; Werke, I, 1863, 371. Maser's transl., Untersuchungen, . . . , 1889, pp. 357–9, 450.

[2] Posth. MS. of 1801, Werke, 2, 1863, 266–8 (French); German transl. by H. Maser, Untersuchungen . . . , 1889, 653-4.

[3] Posth. MS., Werke, 2, 1863, 449–476. Corrections, pp. 498–9, by editor E. Schering.

irregularity for negative determinants for the following hundreds: 1, ..., 30, 43, 51, 61, 62, 63, 91, ..., 100, 117, 118, 119, 120 (also rearranged for the first, third, and tenth thousand), and for the first 800 determinants of each of the types $-(15n+7)$, $-(15n+13)$. Also for the positive determinants of the first, second, third, ninth, and tenth hundreds, and a few others.

E. E. Kummer[4] noted that Gauss' theorems (§ VII) hold also for all classes of ideal numbers formed from a λth root of unity, λ playing the same rôle as the determinant of the quadratic form.

A. Cayley[5] tabulated representatives of the classes, their characters, and the generators of the group of the classes, for Gauss' 13 negative irregular determinants $-D$, $D<1000$.

E. Schering[6] noted that Gauss[3] omitted the exponent of irregularity 3 for the determinant -972. Schering[7] proved Gauss'[1] statement (§ VIII) that if e is divisible by p, n is divisible by p^2.

T. Pepin[8] proved that the greatest number λ of classes in a period generated by one of the H classes of the principal genus is a divisor of $H=e\lambda$, e being the exponent of irregularity for the determinant D. If a is an odd divisor of t^2-D, every power of a can be represented properly by the principal form provided the exponent of a is divisible by 2λ, or simply by λ when there exists a single genus of properly primitive classes.

Every determinant $-243(3l+1)$ is irregular, its e being a multiple of 3. If d is irregular and e its exponent of irregularity, dm^2 is irregular and its exponent of irregularity is divisible by e, if m is a prime >2. The exponent of irregularity of $-243(24l+1)$, $l\neq 0$, is a multiple of 9, provided $-3(24l+1)$ is regular. If the number of ambiguous classes of the principal genus is $2^a>2$, the determinant is irregular and its e is divisible by 2^{a-1}. Several theorems of Gauss[1] (§ VIII) are proved.

J. Perott[9] employed Gauss'[2] two theorems to prove the following result on properly primitive forms x of determinant D of the principal genus. Let $\alpha-k$ be the least positive integer such that $x_a{}^{\alpha-k}$ is in the principal class for every x for which $x^{a\alpha}$ is in it, with a and α as by Gauss[2]. If $k>0$ there are at least a^2 forms x for which x^a is in the principal class, and conversely. Then D is an irregular determinant and all irregular determinants can be so found. The irregular determinants -468 and -931 were omitted by Gauss.[1, 3]

Perott[10] proved that, if p is any given odd prime, we can find a determinant $\Delta(p)$ whose exponent of irregularity is divisible by p. Let t_1, u_1 be the least positive solutions of $t_1^2-pu_1^2=1$. Let

$$t_p+u_p\sqrt{p}=(t_1+u_1\sqrt{p})^p.$$

Expanding the second member by the binomial theorem, we see that $u_p/(pu_1)$ and

[4] Bericht Akad. Wiss. Berlin, 1853, 194–200.
[5] Jour. für Math., 60, 1862, 370–2; Coll. Math. Papers, V, 154–6.
[6] Jour. für Math., 100, 1887, 447–8; Werke, I, 103–4 (letter to Kronecker, 1863).
[7] Abh. Gesell. Wiss. Göttingen, 14, 1869; Werke, I, 145–6.
[8] Atti Acad. Pont. Nuovi Lincei, 33, 1879–80, 53, 370–391.
[9] Jour. für Math., 95, 1883, 232–6.
[10] *Ibid.*, 96, 1884, 327–347.

t_p/t_1 are integers >1 and prime to $2pt_1u_1$. Let q_1 be the least prime which divides u_p but not $2pt_1u_1$. Let q_2 be the least prime which divides t_p but not $2pt_1u_1$. Let p^s be the highest power of p which divides u_1. If $s \gtreqless 0$, it is shown that $\Delta(p) = p^{2s+3} q_1^2 q_2^2$ has an exponent of irregularity divisible by p.

J. Perott[11] proved that, if q is an odd prime, t a positive odd integer $<q$, and k an integer either positive and $>(q^2-4t^2)/(8q)$, or zero and $>q^2-4t^2$, or negative and $>(3q^2-12t^2)/(32q)$, then the negative determinant $-8kq^3-3t^2q^2$ is irregular and its exponent of irregularity is divisible by 3. The case $q=3$, $t=1$, is due to Gauss[1] (§ VIII).

G. B. Mathews[12] proved a generalization of Gauss' remarks (§ VIII) on determinants of the types $-(216k+27)$, etc. Let $D=-(8km+3)m^2$, where k is an integer >0, and m is odd >1. The forms

$$f_1=(m^2,\ m,\ 8km+4), \qquad f_2=(4m^2,\ m,\ 2km+1)$$

are properly primitive and belong to the principal genus for the determinant D. By means of Arndt's formula for composition [Ch. III[16]], it can be shown that the duplicate of the class to which f_1 belongs is the opposite class, whence the triplicate of f_1 belongs to the principal class. Either f_1 or the neighboring form $(8km+4,\ -m,\ m^2)$ is reduced. Like remarks apply to f_2. The two reduced forms are not equivalent. Hence D is an irregular determinant.

L. I. Hewes[13] tabulated the classes, etc., in the notation used by Cayley,[5] for the irregular determinants -468 and -931 of Perott.[9]

A. M. Nash[14] tabulated in MS. irregular determinants up to $-20{,}000$. There is here printed a list of 56 errata, relating to irregularity, in Gauss'[3] table of negative determinants.

Th. Gosset[15] noted that Cayley omitted the irregular determinants -544, -547, -972 and gave for them the necessary additions to Cayley's table.

L. J. Mordell[16] noted that if k is of the form x^3-y^2 where x, y are such that k is positive, $\equiv 3 \pmod 8$, and has no square factor, and such that k is not of one of the forms $3a^2\pm1$, $3a^2\pm8$, then $-k$ is an irregular determinant whose exponent of irregularity is a multiple of 3. Hence -547 is irregular.

---

[11] Johns Hopkins Univ. Circular, 9, 1889–90, p. 30.
[12] Messenger Math., 20, 1891, 70–74.
[13] Bull. Amer. Math. Soc., 9, 1902–3, 141–2.
[14] *Ibid.*, 466–7.
[15] Messenger Math., 40, 1911, 135–7.
[16] *Ibid.*, 42, 1912–13, 124.

CHAPTER VI.*

NUMBER OF CLASSES OF BINARY QUADRATIC FORMS WITH INTEGRAL COEFFICIENTS.

INTRODUCTION.

Particular interest in the mere number of the classes of binary quadratic forms of a given determinant dates from the establishment by C. F. Gauss of the relation between the number h of properly primitive classes of the negative determinant $-D$ and the number of proper representations of D as the sum of three squares. Gauss himself found various expressions for h. G. L. Dirichlet elaborated Gauss' method exhaustively and rigorously.

L. Kronecker, by a study of elliptic modular equations, deduced recurrence formulas for class-number which have come to be called *class-number relations*. C. Hermite obtained many relations of the same general type by equating certain coefficients in two different expansions of pseudo-doubly periodic functions. Hermite's method was extended by K. Petr and G. Humbert to deduce all of Kronecker's relations as well as new and independent ones of the same general type. The method of Hermite was translated by J. Liouville into a purely arithmetical deduction of Kronecker's relations.

The modular function of F. Klein, which is invariant only under a certain congruencial sub-group of the group of unitary substitutions, was employed by A. Hurwitz and J. Gierster just as elliptic moduli had been employed by Kronecker and so the range of class-number relations was vastly extended.

Taking the suggestion from R. Dedekind in his investigation of the classes of ideals of the quadratic field of discriminant D, Kronecker departed from the tradition of Gauss and chose the representative form $ax^2+bxy+cy^2$, where b is indifferently odd or even, and regarded as primitive only forms in which the coefficients have no common divisor. Kronecker thus simplified Dirichlet's results and at the same time set up a relation in terms of elliptic theta functions between the class-number of two discriminants; so he referred the problem of the class-number of a positive discriminant to that of a negative discriminant.

By a study of quadratic residues, M. Lerch and others have curtailed the computation of the class-number. A. Hurwitz has accomplished the same object by approximating $h(p)$, p a prime, by a rapidly converging series and then applying a congruencial condition which selects the exact value of $h(p)$.

Reports are made on several independent methods of obtaining the asymptotic expression for the class-number, and also methods of obtaining the ratio between

* This chapter was written by G. H. Cresse.

the number of classes of different orders of the same determinant. The chief advances that have been made in recent years have been made by extending the method of Hermite.

We shall frequently avoid the explanation of an author's peculiar symbols by using the more current notation. Where there is no local indication to the contrary, $h(D)$ denotes the number of properly primitive, and $h'(D)$ the number of improperly primitive, classes of Gauss forms (a, b, c) of determinant $D=b^2-ac$. Referring to Gauss' forms, $F(D)$, $G(D)$, $E(D)$, though printed in italics, will have the meaning which L. Kronecker (p. 109) assigned to them when printed in Roman type. The class-number symbol $H(D)$ is defined as $G(D)-F(D)$. By $K(D)$ or $Cl D$, we denote the number of classes of primitive Kronecker forms of discriminant $D=b^2-4ac$. A determinant is *fundamental* if it is of the form P or $2P$; a discriminant is *fundamental* if it is of the form P, $4P=4(4n-1)$ or $8P$, where P is an odd number without a square divisor other than 1. The context will usually be depended on to show to what extent the Legendre symbol (P/Q) is generalized.

Reduced form and *equivalence* will have the meanings assigned by Gauss (cf. Ch. I). Among definite forms, only positive forms will be considered; and the leading coefficient of representative indefinite forms will be understood to be positive. Ordinarily, τ will be used to denote the number of automorphs for a form under consideration; but when $D>0$, $\tau=1$.

Some account will be given of the modular equations which lead to class-number relations. In reports of papers involving elliptic theta functions, the notations of the original authors will be adopted without giving definitions of the symbols. For the definitions and a comparison of the systems of theta-function notation, the reader is referred to the accompanying table. The different functions of the divisors of a number will be denoted by the symbols of Kronecker,[54] and without repeating the definition. A Gauss form will be called *odd* if it has at least one odd outer coefficient; otherwise it is an *even* form. These terms are not applied to Kronecker forms.

TABLES OF THETA-FUNCTIONS.

Jacobi Hermite	Kronecker	Klein–Fricke	Smith	Hermite Weber Mordell	Petr	Humbert Chapelon	Bell (Tannery)	Fricke
$\Theta_1(z)$ or Θ_1	$=\vartheta_3(x)$	$=\vartheta_3(x, r)$	$=\theta_{00}(z)$	$=\theta_{00}(v)$	$=\Theta_3(v)$	$=\Theta_1(x)$ or Θ_1	$=\vartheta_3(x)$	$=\vartheta_3(v, q)$
$\Theta(z)$ or Θ	$=\vartheta_0(x)$	$=\vartheta_0(x, r)$	$=\theta_{01}(z)$	$=\theta_{01}(v)$	$=\Theta_2(v)$	$=\Theta(x)$ or Θ	$=\vartheta_4(x)$	$=\vartheta_0(v, q)$
$H_1(z)$ or H_1	$=\vartheta_2(x)$	$=\vartheta_2(x, r)$	$=\theta_{10}(z)$	$=\theta_{10}(v)$	$=H_1(v)$	$=H_1(x)$ or H_1	$=\vartheta_2(x)$	$=\vartheta_2(v, q)$
$H(z)$ or H	$=\vartheta_1(x)$	$=\vartheta_1(x, r)$	$=\theta_{11}(z)$	$=\theta_{11}(v)$	$=H(v)$	$=H(x)$ or H	$=\vartheta_1(x)$	$=\vartheta_1(v, q)$

Here, $r=q^2$, $z=2Kx/\pi$, $v=x/\pi$, n is any, m is any odd, integer; and, according to Humbert.

$$\Theta_1(x)=\sum_{n=-\infty}^{\infty} q^{n^2}\cos 2nx,\quad \Theta(x)=\sum_{n=-\infty}^{\infty}(-1)^n q^{n^2}\cos 2nx,$$

$$H_1(x)=\sum_{n=-\infty}^{\infty} q^{m^2/4}\cos mx,\quad H(x)=\sum_{n=-\infty}^{\infty}(-1)^{\frac{1}{2}(m-1)} q^{m^2/4}\sin mx.$$

For $x=0$, the following systems of special symbols are represented in this chapter.

Humbert	Kronecker, Hurwitz, Hermite, Stieltjes	Weber, Mordell	Hermite, Smith	Humbert, Chapelon	Petr	Bell	Fricke
$\Theta_1(0)$	$=\vartheta_3(q)$ or $\theta_3(q)$	$=\theta_{00}$	$=\theta_1$	$=\theta_1$	$=\Theta_3$	$=\vartheta_3$	$=\vartheta_3$
$\Theta(0)$	$=\vartheta_0(q)$ or $\theta(q)$	$=\theta_{01}$	$=\theta$	$=\theta$	$=\Theta_2$	$=\vartheta_4$	$=\vartheta_0$
$H_1(0)$	$=\vartheta_2(q)$ or $\theta_2(q)$	$=\theta_{10}$	$=\eta$	$=\eta_1$	$=\Theta_1$	$=\vartheta_2$	$=\vartheta_2$
$H'(0)$	$=\vartheta_1'(q)$ or $\theta_1'(q)$	$=\pi\theta_{11}'$		$=\eta'$		$=\vartheta_1'$	

In connection with these tables, the following relations will need to be recalled:

$$q=e^{i\pi\omega},\ \omega=\tau=iK'/K;$$
$$\sqrt{\kappa}=\theta_2(q)/\theta_3(q),\ \sqrt{\kappa'}=\theta(q)/\theta_3(q),\ \theta_1'(q)=\theta(q)\theta_2(q)\theta_3(q);$$
$$\sqrt{2K/\pi}=\theta_3(q),\ \sqrt{2\kappa'K/\pi}=\theta(q),\ \sqrt{2\kappa K/\pi}=\theta_2(q);$$
$$k=\kappa=\phi^4(\omega),\ k'=\kappa'=\psi^4(\omega).$$

A. M. Legendre[1] excluded every reduced form ("quadratic divisor") whose determinant has a square divisor. Each reduced form $py^2+2qyz+2mz^2$ of determinant $-a=-(4n+1)$ has a *conjugate* reduced form $2py^2+2qyz+mz^2$; here p, q, m are all odd.

If a is of the form $8n+5$, one of p, m is of the form $4n+1$ and the other of the form $4n-1$. Hence the odd numbers represented by one of the quadratic forms are all of the form $4n+1$ and those represented by the conjugate form are of the form $4n+3$. Thus a form and its conjugate are not equivalent and the total number of reduced forms is even.

If $a=8n+1$, the number of reduced forms may be even or odd,[1] but is odd[2] if $a=8n+1$ is prime.

Legendre[3] counted (r, s, t) and $(r, -s, t)$ as the same form. Hence for $a=4n+1$, his number of forms is $\frac{1}{2}\{h(-a)-A\}$, where $h(-a)$, in the terminology of Gauss[4] (Art. 172), is the number of properly primitive classes and A is the number of ambiguous properly primitive classes plus the number of classes represented by forms of the type (r, s, r).

C. F. Gauss,[4] by the composition of classes, proved (Art. 252) that the different genera of the same order have the same number of classes (cf. Ch. IV). He[5] then set for himself the problem of finding an expression in terms of D for the number of classes in the principal genus of determinant D. He succeeded later[8] in finding an expression for the total number of primitive classes of the determinant and thus solved his former problem only incidentally.

---

1 Theorie des nombres, Paris, 1798, 267–8; ed. 2, 1808, 245–6; ed. 3, 1830, Vol. I, Part II, § XI (No. 217), pp. 287–8; German transl. by H. Maser, Zahlentheorie, I, 283.

2 *Ibid.*, Part IV, Prop. VIII, 1798, 449; ed. 2, 1808, 385; ed. 3, II, 1830, 55; Zahlentheorie, II, 56.

3 *Ibid.*, 1798, No. 48, p. 74; ed. 2, 1808, p. 65; ed. 3, I, p. 77; Zahlentheorie, I, p. 79.

4 Disquisitiones Arithmeticae, 1801; Werke, I, 1876; German transl. by H. Maser, Untersuchungen ueber Höhere Arithmetik, 1889; French transl. by A. C. M. Poullet-Delisle, Reserches Arithmetiques, 1807, 1910.

5 Werke, I, 466; Maser, 450; Supplement X to Art. 306. Cf. opening of Gauss'[8, 9] memoirs of 1834, 1837.

If (Art. 253) Q denotes the number of classes of the (positive) order O of determinant D, and if r denotes the number of properly primitive classes of determinant D which, being compounded with an arbitrary class K of the order O, produce a given arbitrary class L of the order O, then the number of properly primitive (positive) classes is rQ. We take both K and L to be the simplest form (Art. 250). It is proved (Arts. 254–6) by the composition of forms that the above r classes are included among certain r' primitive forms, r' being given by

$$r' = \tfrac{1}{2} A n \prod_a \left[1 - \left(\frac{D'}{a}\right)\frac{1}{a}\right],$$

in which (A, B, C) is the simplest form of order O, $D' = 4D/A^2$, and a ranges over the distinct odd divisors of A, while $n=2$ if D/A^2 is an integer, $n=1$ if $4D/A^2 \equiv 1 \pmod 8$, $n=3$ if $4D/A^2 \equiv 5 \pmod 8$.

Now $r=r'$ if D is a positive square or a negative number except in the cases $D=-A^2$ and $-\frac{3}{4}A^2$, in which cases $r=r'/2$ and $r'/3$ respectively. No general relation (Art. 256, IV, V) is found between r and r' for D positive and not a square.

The problem of finding the ratio of the number of classes of different orders of a determinant will be hereafter referred to as the Gauss Problem. It was solved completely by Dirichlet,[20, 93] Lipschitz,[41] Dedekind,[115] Pepin,[120, 137] Dedekind,[127a] Kronecker,[171] Weber,[220] Mertens,[237] Lerch,[277] Chatelain,[316] and de Séguier.[226]

If O is the improperly primitive order, the same method gives the following result (Art. 256, VI):

If $D \equiv 1 \pmod 8$, $r=1$; if $D<0$ and $\equiv 5 \pmod 8$, $r=3$ (except when $D=-3$ and then $r=1$); if $D>0$ and $\equiv 5 \pmod 8$, $r=1$ or 3, according as the three properly primitive forms

$$(1, 0, -D), \qquad (4, 1, \tfrac{1}{4}(1-D)), \qquad (4, 3, \tfrac{1}{4}(9-D))$$

belong to one or three different classes.

Gauss (Art. 302) gave the following expression for the asymptotic median number of the properly primitive classes of a negative determinant $-D$:

$$M(D) = m\sqrt{D} - \frac{2}{\pi^2}, \qquad m = \frac{2\pi}{7(1+\frac{1}{8}+\frac{1}{27}+\frac{1}{64}+\dots)}.$$

He later corrected[6] this formula to $m\sqrt{D}$.

His tables of genera and classes led him (Art. 303) to the conjecture[304] that the number of negative determinants which have a given class-number h is finite for every h (cf. Joubert,[60] Landau,[260] Lerch,[262] Dickson,[327] Rabinovitch[336a] and Nagel[336b]).

The asymptotic median value of $h(k^2)$ is $8k/\pi^2$ (Art. 304). He conjectured that the number of positive determinants which have genera of a single class is infinite. Dirichlet[40] proved that this is true. He stated (Art. 304) that, for a positive determinant D, the asymptotic median value of $h(D)\log(T+U\sqrt{D})$ is $m\sqrt{D}-n$, where T, U give the fundamental solution of $t^2-Du^2=1$ and[7] for m as above, while n is a constant as yet not evaluated (cf. Lipschitz[102]).

---

[6] Werke, II, 1876, 284; Maser's transl., 670. Cf. Lipschitz.[102]

[7] On the value of m, see Supplement referring to Art. 306 (X). Maser's transl., p. 450; Werke, I, 1863, 466.

C. F. Gauss[8] considered the lattice points within or on the boundary of an ellipse $ax^2+2bxy+cy^2=A$, where A is a positive integer. The area is $\pi A/\sqrt{D}$, where $-D=b^2-ac$. Hence as A increases indefinitely, the number of representations of all positive numbers $\leqq A$ by means of the definite form (a, b, c) bears to A a ratio which approaches $\pi/\sqrt{D}$ as a limit.

Hereafter[9] the determinant $-D$ has no square divisors, and the asymptotic number of representations of odd numbers $\leqq M$ by the complex C of representative properly primitive forms of determinant $-D$ is

$$\frac{\pi M}{2\sqrt{D}}h(-D).$$

To evaluate $h(-D)$, a second expression for this number of representations is found; but Gauss gives the deduction only in fragments. Thus if (n) denotes the number of representations of n by C and p is an odd prime, then[10]

1. $(pn)=(n)$, if p is a divisor of D;

2. $(pn)=(n)+(h)$, if $\left(\frac{-D}{p}\right)=1$;

3. $(pn)=-(n)+h$, if $\left(\frac{-D}{p}\right)=-1$,

where $n=hp^\mu$, μ arbitrary, h prime to p.

This implies in the three cases

1. $(h)=(ph)=(p^2h)=(p^3h)=\ldots;$

2. $(ph)=2(h)$, $(p^2h)=3(h)$, $(p^3h)=4(h)$, $\ldots;$

3. $(ph)=0$, $(p^2h)=(h)$, $(p^3h)=0$, $(p^4h)=(h)$, $\ldots.$

Hence the ratio of the mean number of representations by C of all odd numbers $\leqq M$ to the mean number of representations of those numbers after the highest possible power of p has been removed from each as a factor is, in each of the three cases,[11]

$$1/\left\{1-\left(\frac{-D}{p}\right)\frac{1}{p}\right\}.$$

A second odd prime divisor p' is similarly eliminated from the odd numbers $\leqq M$; and so on. Eventually the number of representations of the numbers is asymptotically $\frac{1}{2}\tau M$. Gauss, supposing $-D<-1$, takes the number τ of automorphs to be 2. (See Disq. Arith., Art. 179; Gauss[35] of Ch. I.) Hence the original number of representations is asymptotically[12]

$$M/\Pi\left(1-\left(\frac{-D}{p}\right)\frac{1}{p}\right).$$

[8] Posthumous paper presented to König. Gesells. der Wiss. Göttingen, 1834; Werke, II, 1876, 269–276; Untersuchungen über Höhere Arith., 1889, 655–661.

[9] Posthumous fragmentary paper presented to König. Gesells. der Wiss. Göthingen, 1837; Werke, II, 1876, 276–291; Untersuchungen über Höhere Arith., 1889, 662–677.

[10] Cf. remarks by R. Dedekind, Werke of Gauss, 1876, II, 293–294; Untersuchungen über Höhere Arith., 1889, 686.

[11] Cf. R. Dedekind, Werke of Gauss, II, 1876, 295–296; Untersuchungen, 1889, 687.

[12] Cf. remarks of R. Dedekind, Werke of Gauss, II, 1876, 296; Untersuchungen, 688.

And hence (Untersuchungen, 670, III; cf. Dirichlet,[19] (1))

$$h(-D)=\frac{2\sqrt{D}}{\pi}\Pi\left\{1-\left(\frac{-D}{p}\right)\frac{1}{p}\right\}^{-1}.$$

Gauss gives without proof five further forms of $h(-D)$ including

$$h(-D)=\Sigma\frac{\sqrt{D}}{N}\left(\frac{-D}{n}\right)\cot n\theta,$$

where $\theta=\pi/N$, $N=D$ *or* $4D$, n is odd >0 and $<D$. Cf. Lebesgue,[36] (1).

By considering the number of lattice points in a certain hyperbolic sector,[13] $h(D)$ is found to be, for $D>0$,

$$\frac{2\sqrt{D}(1\pm\frac{1}{3}\pm\frac{1}{5}\pm\ldots)}{\log(T+U\sqrt{D})}=\frac{\log\sin\frac{\theta}{2}\pm\log\sin\frac{3\theta}{2}\pm\log\sin\frac{5\theta}{2}\pm\ldots}{\log(T+U\sqrt{D})},$$

where the coefficient ±1 of $1/m$ and of $\log\sin m\theta/2$ is (D/m). Cf. Dirichlet,[23] (7), (8).

For a negative prime determinant, $-D=-(4n+1)$, $h(-D)$ is stated incorrectly to be $\alpha-\beta$, where α and β are respectively the number of quadratic residues and non-residues of D in the first quadrant of D. [This should be $2(\alpha-\beta)$; see Dirichlet,[23] formula (5).]

Extensive tables lead by induction to laws which state, in terms of the class-number of a prime determinant p, the distribution of quadratic residues of p in its octants and 12th intervals.

G. L. Dirichlet[14] obtained $h(-q)$, where q is a positive prime $=4n+3>3$. By replacing infinite sums by infinite products he obtained the lemma:

$$\sum\frac{2^{\mu}}{m^s}=\sum\frac{1}{n^s}\cdot\sum\left(\frac{n}{q}\right)\frac{1}{n^s}\Big/\sum\frac{1}{n^{2s}},$$

where n ranges in order over all positive odd integers prime to q, and m ranges over all positive numbers which have only prime divisors f such that $(f/q)=1$; while μ is the number of such distinct divisors of m; and s is arbitrary >1. Now

$$ax^2+2bxy+cy^2,\qquad a'x^2+2b'xy+c'y^2,\ldots$$

denotes a complete set of representative properly primitive (positive) forms of determinant $-q$. Then, by the lemma, since the number of representations of m by the forms is $2^{\mu+1}$ (cf. Dirichlet, Zahlentheorie, § 87), we have

$$(1)\quad 2\Sigma\frac{1}{n^s}\cdot\Sigma\left(\frac{n}{q}\right)\frac{1}{n^s}=\Sigma\frac{1}{(ax^2+2bxy+cy^2)^s}+\Sigma\frac{1}{(a'x^2+2b'xy+c'y^2)^s}+\ldots,$$

where x, y take every pair of values for which the values of the quadratic forms are

---

[13] Remarks of R. Dedekind, Gauss' Werke, II, 1876, 299; Maser's translation, 691. Cf. G. L. Dirichlet, Zahlentheorie, §98.

[14] Jour. für Math., 18, 1838, 259–274; Werke, I, 1889, 357–370.

prime[15] to $2q$. We let $s=1+\rho$, and let $\rho>0$ approach zero. The limit of the ratio of each double sum in the right member to

$$\frac{q-1}{2q\sqrt{q}}\frac{\pi}{\rho}$$

is found from the lattice points of an ellipse to be 1. But

$$\lim \Sigma \frac{1}{n^s} : \frac{q-1}{2q} \cdot \frac{1}{\rho} = 1.$$

Hence (cf. C. F. Gauss,[9] Werke, II, 1876, 285),

$$(2) \qquad h(-q) = \frac{2\sqrt{q}}{\pi} \sum_{n=1}^{\infty} \left(\frac{n}{q}\right)\frac{1}{n} \equiv \frac{2\sqrt{q}}{\pi} S.$$

To evaluate S, we consider

$$S/\left\{1-\left(\frac{2}{q}\right)\frac{1}{2}\right\} = \Sigma\left(\frac{n}{q}\right)\frac{1}{n},$$

where n now ranges in order over all integers $\overline{\gtrless} 1$. In the cyclotomic theory,[16]

$$\Sigma \sin\frac{2an\pi}{q} - \Sigma \sin\frac{2bn\pi}{q} = \left(\frac{n}{q}\right)\sqrt{q}.$$

Hence

$$\frac{\sqrt{q}}{1-\left(\frac{2}{q}\right)\frac{1}{2}} \cdot S = \sum_a \sum_n \frac{1}{n} \sin\frac{2an\pi}{q} - \sum_b \sum_n \frac{1}{n} \sin\frac{2bn\pi}{q},$$

where $(a/q)=1$, $(b/q)=-1$, and a, b are >0 and $<q$. Since (cf. W. E. Byerly, Fourier's Series, 1893, 39) $z=2b\pi/q$ is between 0 and 2π,

$$\frac{\pi-z}{2} = \Sigma \frac{1}{n} \sin nz;$$

and so[17]

$$(3) \qquad h = 2\left[1-\left(\frac{2}{q}\right)\frac{1}{2}\right]\frac{\Sigma b - \Sigma a}{q}.$$

Evaluating S itself by cyclotomic considerations, Dirichlet gives the result[18]

$$(4) \qquad h = A - B = 2A - \tfrac{1}{2}(q-1),$$

where A and B are respectively the number of quadratic residues and non-residues of q which are $<\frac{1}{2}q$. For $p=4n+1$, Dirichlet obtained

$$(5) \qquad h(-p) = 2(A-B) = 4A - \tfrac{1}{2}(p-1),$$

---

[15] This restriction is removed by G. Humbert, Comptes Rendus, Paris, 169, 1919, 360–361.

[16] Cf. C. F. Gauss, Werke, II, 1876, 12. G. L. Dirichlet, Zahlentheorie, §116.

[17] Stated empirically by C. G. J. Jacobi, Jour. für Math., 9, 1832, 189–192; detailed report in this History, Vol. I, 275–6; J. V. Pexider,[320] Archiv Math. Phys., (3), 14, 1909, 84–88, combined (3) with the known relation $\Sigma b + \Sigma a = \frac{1}{2}q\ (q-1)$ to express h in terms of Σa alone or Σb alone.

[18] G. B. Mathews, Proc. London Math. Soc., 31, 1899, 355–8, expressed $A-B$ in terms of the greatest integer function.

where A and B are the number of a's and b's respectively between 0 and $\frac{1}{2}p$; and without proof, he stated that

$$h(-pq)=(\Sigma b-\Sigma a)/pq \quad \text{or} \quad 3(\Sigma b-\Sigma a)/pq,$$

according as pq is $\equiv 7$ or 3 (mod 8), where a, b are positive integers $<pq$, and $(a/p)=(a/q)$, $(b/p)=-(b/q)$. For $h(q)$, the factor π in (2) must be replaced by $\log(T+U\sqrt{D})$; as the lattice points involved must now lie in a certain hyperbolic sector rather than an ellipse (cf. Gauss,[9] Dirichlet[19]).

G. L. Dirichlet[19] considered the four cases of a determinant: $D=P\cdot S^2$, $P\equiv 1$ and 3 (mod 4); $D=2P\cdot S^2$, $P\equiv 1$ and 3 (mod 4), where S^2 is the greatest square divisor of D. He defined δ and ϵ in the four cases as follows:

$$\delta=\epsilon=1, \qquad \delta=-1,\ \epsilon=1, \qquad \delta=1,\ \epsilon=-1, \qquad \delta=\epsilon=-1.$$

Employing the notation of his former memoir,[14] he found for all four cases, if m is representable,

$$\delta^{\frac{1}{2}(f-1)}\epsilon^{\frac{1}{8}(f^2-1)}\left(\frac{f}{P}\right)=1.$$

Consequently the generalization of (1) of the preceding memoir[14] is, for $D=-D_1<0$,

$$\Sigma\frac{1}{(ax^2+2bxy+cy^2)^s}+\Sigma\frac{1}{(a'x^2+2b'xy+c'y^2)^s}+\ldots=2\Sigma\frac{1}{n^s}\Sigma\delta^{\frac{1}{2}(n-1)}\epsilon^{\frac{1}{8}(n^2-1)}\left(\frac{n}{P}\right)\frac{1}{n^s},$$

where the restrictions on s, x, y, n are the same as for (1) in the preceding memoir.

A lemma shows that

$$\Sigma\frac{1}{n^s}=\frac{\phi(D_1)}{2D_1}\cdot\frac{1}{\rho} \text{ or } \frac{\phi(D_1)}{D_1}\ \frac{1}{\rho},$$

according as D is odd or even, where $s=1+\rho$ and ρ is indefinitely small, and ϕ is the Euler symbol. The study of lattice points in the ellipse $ax^2+2bxy+cy^2=N$ for very great N leads to

$$\frac{\pi}{2}\frac{\phi(D_1)}{\sqrt{D_1^3}}\cdot\frac{1}{\rho} \text{ or } \frac{\pi\phi(D_1)}{\sqrt{D_1^3}}\cdot\frac{1}{\rho}$$

as the asymptotic value of each of the h sums in the first member, according as D is odd or even. Hence for $D=-D_1<0$,

(1) $$h=\frac{2}{\pi}\sqrt{D_1}\Sigma\delta^{\frac{1}{2}(n-1)}\epsilon^{\frac{1}{8}(n^2-1)}\left(\frac{n}{P}\right)\frac{1}{n}.$$

Dirichlet obtained independently an analogous formula for the number h' of improperly primitive classes of determinant $D=-D_1<0$. For $D>0$, results analogous to all those for $D<0$, are obtained by considering all the representations of positive numbers $\leqq N$ by $ax^2+2bxy+cy^2$, where a is >0 and (x, y) are lattice points in the hyperbolic sector having $y>0$ and bounded by $y=0$, $U(ax+by)=Ty$, and $ax^2+2bxy+cy^2=N$. For $D>0$, these restrictions on a, x, y are hereafter understood in this chapter of the History.

---

19 Jour. für Math., 19, 1839, 324–369; 21, 1840, 1–12, 134–155; Werke, I, 1889, 411–496; Ostwald's Klassiker der exakten Wissenschaften, No. 91, 1897, with explanatory notes by R. Haussner.

Incidentally Dirichlet[20] stated for $D<-3$ the "fundamental equation of Dirichlet" (see Zahlentheorie, § 92, for the general statement):

$$(2)\quad \Sigma'\phi(ax^2+2bxy+cy^2)+\Sigma'\phi(a'x^2+2b'xy+c'y^2)+\dots$$
$$=2\Sigma\delta^{\frac{1}{2}(n-1)}\epsilon^{\frac{1}{8}(n^2-1)}\left(\frac{n}{P}\right)\phi(nn'),$$

where ϕ is an arbitrary function which gives absolute convergence in both members; the forms are a representative primitive system; x and y take all pairs of integral values (excepting $x=y=0$) in each form for which the value of the form is prime[21] to $2D$ if the form is properly primitive, but half of the value of the form is prime to $2D$ if the form is improperly primitive; the second member is a double sum as to n and n'. Kronecker[171] and Lerch[277] (Chapter I of his Prize Essay) used this identity to obtain a class-number formula.

Dirichlet noted from the results in his[19] former memoir that for $D<0$, $h=h'$ or $3h'$, according as $D\equiv 1$ or 5 (mod 8), except that $h=h'$ for $D=-3$. For $D>0$, if $D=8n+1$, $h=h'$; but, if $D=8n+5$, $h=h'$ or $3h'$, according as the fundamental solutions of $t^2-Du^2=4$ are odd or even. (Cf. Gauss,[4] Disq. Arith., Art. 256.)

Since the series in (1) may be written as

$$\Pi\left\{1-\delta^{\frac{1}{2}(n-1)}\epsilon^{\frac{1}{8}(n^2-1)}\left(\frac{n}{P}\right)\frac{1}{n}\right\}^{-1},$$

where n is a positive odd prime, and prime to D, it follows that if h and h' denote respectively the number of properly primitive classes of the two negative determinants D and $D'=D\cdot S^2$, D having no square divisor, then

$$\frac{h'}{h}=S\prod_r\left[1-\delta^{\frac{r-1}{2}}\epsilon^{\frac{r^2-1}{8}}\left(\frac{r}{P}\right)\frac{1}{r}\right],$$

where[22] r ranges over the odd prime positive divisors of S (except if $D=-1$, the ratio thus given should be divided by 2). The corresponding ratio is found for $D'>0$.

Dirichlet[23] hereafter took $S=1$ and, representing the series in (1) by V, found that for $D=\pm P\equiv 1$ (mod 4), for example,

$$\frac{V}{1-\left(\frac{2}{P}\right)\frac{1}{2}}=-\int_0^1\frac{\Sigma\left(\frac{n}{P}\right)x^{n-1}}{x^P-1}\,dx=-\frac{1}{P}\sum_m\sum_n\left(\frac{n}{P}\right)e^{\frac{n}{P}\cdot 2m\pi i}\int_0^1\frac{dx}{x-e^{2m\pi i/P}},$$

$n,\ m=0,1,2,\dots,P-1;\ P=|D|$.

---

[20] Jour. für Math., 21, 1840, 7; Werke, I, 1889, 467. The text is a report of Jour. für Math. 21, 1840, 1-12; Werke, I, 1889, 461-72.
[21] This restriction is removed by G. Humbert, Comptes Rendus, Paris, 169, 1919, 360-361.
[22] Cf. Disq. Arith., Art. 256, V; R. Lipschitz,[41] Jour. für Math., 53, 1857, 238.
[23] From this point, Jour für Math., 21, 1840, 134-155; Werke, I, 1889, 479-496. Cf. Zahlentheorie, §§-103-105.

The identity, in Gauss sums,

$$(3)\qquad \Sigma\left(\frac{n}{P}\right)e^{\frac{n}{P}\cdot 2m\pi i}=i^{\left(\frac{P-1}{2}\right)^2}\left(\frac{m}{P}\right)\sqrt{P}$$

now gives

$$\frac{V}{1-\left(\frac{2}{P}\right)\frac{1}{2}}=-\frac{i^{\left(\frac{P-1}{2}\right)^2}}{\sqrt{P}}\sum_{m=1}^{P}\left(\frac{m}{P}\right)\left(\log\sin\frac{m\pi}{P}-\frac{m\pi i}{P}\right);$$

whence

$$(4)\qquad \begin{cases} V=-\dfrac{1}{\sqrt{P}}\left(1-\left(\dfrac{2}{P}\right)\dfrac{1}{2}\right)\Sigma\left(\dfrac{m}{P}\right)\log\sin\dfrac{m\pi}{P}, \text{ if } P=4\mu+1,\\[2ex] V=\dfrac{-\pi}{\sqrt{P^3}}\left(1-\left(\dfrac{2}{P}\right)\dfrac{1}{2}\right)\Sigma\left(\dfrac{m}{P}\right)m, \text{ if } P=4\mu-1.\end{cases}$$

For $D=-P$, $P=4\mu-1$, the comparison of (1) and (3) gives

$$h(D)=\frac{2}{\pi}\Sigma\left(\frac{m}{P}\right)\Sigma\frac{1}{n}\sin n\frac{2m\pi}{P};$$

whence[24] finally by grouping quadratic residues and non-residues, we have:

$$h(D)=\Sigma\left(\frac{m'}{P}\right),\qquad 0<m'<\frac{P}{2}.$$

So Dirichlet[25] obtained his classic formulas for $D<0$:

$$(5)\qquad \begin{cases} D=-P,\quad P=4\mu+3,\ h(D)=\displaystyle\sum_{0}^{\frac{1}{2}P}\left(\frac{S}{P}\right);\\ D=-P,\quad P=4\mu+1,\ h(D)=2\displaystyle\sum_{0}^{\frac{1}{4}P}\left(\frac{S}{P}\right);\\ D=-2P,\ P=4\mu+3,\ h(D)=2\displaystyle\sum_{\frac{1}{8}P}^{\frac{3}{8}P}\left(\frac{S}{P}\right);\\ D=-2P,\ P=4\mu+1,\ h(D)=2\displaystyle\sum_{0}^{\frac{1}{8}P}\left(\frac{S}{P}\right)-2\sum_{\frac{3}{8}P}^{\frac{1}{2}P}\left(\frac{S}{P}\right).\end{cases}$$

From (1) and (4) and their analogues, he wrote also in the four cases of $D<0$:

$$(6)\quad -h(D)=\frac{1}{P}\left(2-\left(\frac{2}{P}\right)\right)\Sigma\left(\frac{S}{P}\right)\epsilon_1 S,\ \frac{1}{4P}\Sigma\left(\frac{S}{P}\right)\epsilon_2 S,\ \frac{1}{8P}\Sigma\left(\frac{S}{P}\right)\epsilon_3 S,\ \frac{1}{8P}\Sigma\left(\frac{S}{P}\right)\epsilon_4 S,$$

where $S=m$ ranges from 0 to P, $4P$, $8P$, $8P$ in the four respective cases, and

$$\epsilon_1=1,\ \epsilon_2=(-1)^{\frac{1}{2}(m-1)},\ \epsilon_3=(-1)^{\frac{1}{8}(m^2-1)},\ \epsilon_4=(-1)^{\frac{1}{2}(m-1)+\frac{1}{8}(m^2-1)}.$$

For $D>0$, the analogue of (1) is

$$(7)\qquad h(D)=\frac{2\sqrt{D}}{\log(T+U\sqrt{D})}\Sigma\delta^{\frac{1}{2}(n-1)}\epsilon^{\frac{1}{8}(n^2-1)}\left(\frac{n}{P}\right)\frac{1}{n},$$

[24] Fourier-Freeman, Theory of Heat, Cambridge, 1878, 243.
[25] Jour. für Math., 21, 1840, 152; Werke, I, 492–3; Zahlentheorie, §106.

where T, U are the fundamental solution of $t^2-Du^2=1$. Hence[26] from equations like (4), we obtain:

$$(8_1)\qquad D=P,\quad P=4\mu+1,\quad h(D)=\frac{2-\left(\frac{2}{P}\right)}{\log(T+U\sqrt{P})}\log\frac{\Pi\sin b\pi/P}{\Pi\sin a\pi/P},$$

where a and b range over the integers $<P$ and prime to P for which

$$(a/P)=+1,\qquad (b/P)=-1;$$

$$(8_2)\qquad D=P,\quad P=4\mu+3,\quad h(D)=\frac{1}{\log(T+U\sqrt{P})}\log\frac{\Pi\sin\frac{1}{4}b\pi/P}{\Pi\sin\frac{1}{4}a\pi/P},$$

where a and b range over the integers $m<4P$ and prime to $4P$ for which

$$(-1)^{\frac{1}{2}(m-1)}\left(\frac{m}{P}\right)=+1 \text{ or } -1, \text{ according as } m=a \text{ or } b;$$

$$(8_3)\qquad D=2P,\quad h(D)=\frac{1}{\log(T+U\sqrt{2P})}\log\frac{\Pi\sin\frac{1}{8}b\pi/P}{\Pi\sin\frac{1}{8}a\pi/P},$$

where a and b range over the integers $m<8P$ and prime to $8P$, for which

if $P\equiv 1 \pmod 4$, $(-1)^{\frac{1}{8}(m^2-1)}\left(\frac{m}{P}\right)=+1$ or -1, according as $m=a$ or b;

if $P\equiv 3 \pmod 4$, $(-1)^{\frac{1}{2}(m-1)+\frac{1}{8}(m^2-1)}\left(\frac{m}{P}\right)=+1$ or -1, according as $m=a$ or b.

If $D=P=4\mu+1>0$, (4) and (7) with cyclotomic considerations give[27]

$$(9)\qquad h(D)=\left[4-2\left(\frac{2}{P}\right)\right]\frac{\log\frac{1}{2}[Y(1)+Z(1)\sqrt{P}]}{\log(T+U\sqrt{P})},$$

where $\frac{1}{2}[Y(x)+Z(x)\sqrt{P}]\equiv\Pi(x-e^{2\pi bi/P})$.

Arndt[53] supplied formulas for the other three cases.

A. L. Cauchy[28] proved that if p is a prime of the form $4l+3$,

$$\frac{A-B}{2}\equiv -3B_{(p+1)/4} \text{ or } B_{(p+1)/4} \pmod p,$$

according as $p=8l+3$ or $8l+7$, where A is the number of quadratic residues and B that of the non-residues of p which are >0 and $<\frac{1}{2}p$, and B_k is the kth Bernoullian number. This implies, by G. L. Dirichlet,[23] (5), that

$$(1)\qquad h(-p)\equiv 2B_{(p+1)/4} \text{ or } -6B_{(p+1)/4} \pmod p,$$

according as $p=8l+7$ or $8l+3$ [cf. Friedmann and Tamarkine[321]].

Cauchy[29] obtained also the equivalent of the following for n free from square factors, and of the form $4x+3$:

$$(2)\qquad A-B=\left[2-\left(\frac{2}{n}\right)\right]\frac{\Sigma b-\Sigma a}{n}=\left[2-\left(\frac{2}{n}\right)\right]\frac{\Sigma b^2-\Sigma a^2}{n^2},$$

---

26 Jour. für Math., 21, 1840, 151; Werke, I, 492.

27 See this History, Vol. II, Ch. XII, 372 [117]; Cf. Dirichlet, Zahlentheorie, 1894, 279, § 107.

28 Mém. Institut de France, 17, 1840, 445; Oeuvres, (1), III, 172. Bull. Sc. Math., Phys., Chim. (ed., Férussac), 1831.

29 Mém. Institut de France, 17, 1840, 697; Oeuvres, (1), III, 388. Comptes Rendus, Paris, 10, 1840, 451.

where A, B are the number of quadratic residues and non-residues of n, which are $<\frac{1}{2}n$, while a, b are >0 and $<n$, $(a/n)=1$, $(b/n)=-1$: and similar formulas for $n=4x+1$. Hence, for $n=4x+3$,

$$h(-n)=\left[2-\left(\frac{2}{n}\right)\right]\frac{\Sigma b^2-\Sigma a^2}{n^2}$$

is called Cauchy's class-number formula.[30]

M. A. Stern[31] found that when P is a prime $8m+7$, or $8m+3$ respectively,

$$\prod_a \cot\frac{2\pi a}{P}=\pm(-1)^N\frac{1}{\sqrt{P}},$$

where a ranges over all positive integers $<P$ prime to P such that $(a/P)=1$, and N denotes the number of quadratic divisors of determinant $-P$. This formula has been made to include the case $P=4m+1$ by Lerch.[323]

G. Eisenstein[32] proposed the problem: If $D>0$ is $\equiv 5 \pmod 8$, to determine *a priori* whether $p^2-Dq^2=4$ can be solved in odd or even integers p, q; that is[33] to furnish a criterion to determine whether the number of properly primitive classes of determinant D is 1 or 3 times the number of improperly primitive classes of the same determinant. He also proposed the problem[34]: To find a criterion to determine whether the number of properly primitive classes of a determinant D is divisible by 3; and if this is the case, a criterion to determine those classes which can be obtained by triplication[35] of other classes.

V. A. Lebesgue[36] employed the notation of Dirichlet[23] and, in his four cases, set $p=P$, $4P$, $8P$, $8P$, and $f(x)=\Sigma\epsilon_i(a/p)x^a$, summed over all the positive integers $a<p$, for $i=1, 2, 3, 4$. Then

$$V'=\int_0^1\frac{f(x)\,dx}{x(1-x^p)}$$

is the sum of integrals (for the various values of a), with proper signs prefixed,

$$\int_0^1\frac{x^{a-1}dx}{1-x^p}=-\frac{1}{p}\sum_{m=1}^{p}\cos m\frac{2a\pi}{p}\log\sin\frac{m\pi}{p}+\frac{\pi}{2p}\cot\frac{a\pi}{p}.$$

For a negative determinant, the terms involving the logarithm cancel each other and then, by the theory of Gauss[37] sums, V' reduces to[38]

(1) $$V'=\frac{\pi}{p}\Sigma\cot\frac{A\pi}{p},\quad \epsilon_i\left(\frac{A}{p}\right)=1,\quad 1\leqq A<p.$$

H. W. Erler[39] developed a hint by Gauss (Disq. Arith., Art. 256, § V, third case) that there is a remarkable relation between the totality B of properly primitive forms

---

[30] Cf. T. Pepin,[120] Annales sc. de l'Ecole Norm. Sup., (2), 3, 1874, 205; M. Lerch,[277] Acta Math., 29, 1905, 381.

[31] Jour. de Math., (1), 5, 1840, 216–7. This is proved by means of C. G. J. Jacobi's result in this History, Vol. I, 275–6.

[32] Jour. für Math., 27, 1844, 86.

[33] Cf. G. L. Dirichlet, Zahlentheorie, 1894, Art. 99; Dirichlet.[23]

[34] Jour. für Math., 27, 1844, 87.

[35] Cf. C. F. Gauss, Disq. Arith., Art. 249; Maser's translation, 1889, 261; Werke, I, 1876, 272.

[36] Jour. de Math., 15, 1850, 227–232.

[37] Disq. Arith., Art. 356.

[38] Cf. C. F. Gauss, memoir of 1837, Werke, II, 1876, 286; Untersuchungen, 1889, 671.

[39] Eine Zahlentheoretische Abhandlung, Progr. Zullichau, 1855, p. 18.

of determinant D which represent A^2 and the least solution t_1, u_1 of $t^2-Du^2=A^2$. Erler considered the case in which A^2 divides D, whence A divides t_1. Write $\tau_1=t/A$, $D'=D/A^2$, whence $t_1^2-D'u_1^2=1$. Find the period of the solution of the latter for modulus A. From each pair of simultaneous values of τ_1, u_1, we can derive one and only one from the set B which is equivalent to the principal form. The terms of every later period give the same forms in the same sequence as those of the first period. In case bisection of the period is possible, the terms of the second half are the same as in the first half. The forms obtained from the terms of the first half (or from the entire first period, if bisection is impossible) are distinct.

G. L. Dirichlet[40] recalled (see Dirichlet,[23] (8)) that for a positive determinant $D'=D\cdot S^2$,

$$h(D')=h(D)\frac{\log(T+U\sqrt{D})}{\log(T'+U'\sqrt{D'})}\cdot S\cdot R\equiv\frac{h(D)\cdot S\cdot R}{N},$$

in which R is independent of $\alpha_1, \alpha_2, \ldots, \alpha_k$ in $S=p_1^{\alpha_1}\cdot p_2^{\alpha_2}\cdots p_k^{\alpha_k}$, where the p's are distinct primes. By the theory of the Pell equation it is found (see this History, Vol. II, p. 377, Dirichlet[184]) that if each α increases indefinitely, S/N is eventually a constant. Hence for every D, there is an infinitude of determinants $D'=DS^2$ for which $h(D')=h(D)$. And a proper choice of D and the primes $p_1, p_2, \ldots, p_k$ leads to an infinite sequence of determinants D' for which the number of genera coincides with the value of $h(D')$. This establishes the conjecture of C. F. Gauss (Disq. Arith.,[4] Art. 304) that there is an infinitude of determinants which have genera of a single class.

R. Lipschitz[41] called the linear substitutions

$$\begin{pmatrix}\alpha, & \beta\\ \gamma, & \delta\end{pmatrix},\quad \begin{pmatrix}\mathrm{A} & \mathrm{B}\\ \Gamma & \Delta\end{pmatrix}$$

equivalent if $\alpha, \ldots, \Delta$ are integers and if integers $\alpha', \beta', \gamma', \delta'$ exist such that

$$\begin{pmatrix}\alpha & \beta\\ \gamma & \delta\end{pmatrix}\begin{pmatrix}\alpha' & \beta'\\ \gamma' & \delta'\end{pmatrix}=\begin{pmatrix}\mathrm{A} & \mathrm{B}\\ \Gamma & \Delta\end{pmatrix},\quad \alpha'\delta'-\beta'\gamma'=1.$$

Every substitution of odd prime order p is equivalent to one of the $p+1$ non-equivalent substitutions:

$$(1)\qquad \begin{pmatrix}1 & 0\\ 0 & p\end{pmatrix},\ \begin{pmatrix}0 & -p\\ 1 & 0\end{pmatrix},\ \begin{pmatrix}1 & -p\\ 1 & 0\end{pmatrix},\ \begin{pmatrix}2 & -p\\ 1 & 0\end{pmatrix},\ldots,\ \begin{pmatrix}p-1 & -p\\ 1 & 0\end{pmatrix}.$$

Let (a, b, c), a properly primitive form of determinant D, be transformed by (1) into $p+1$ forms (a', b', c'). Then $D'=D\cdot p^2$. The coefficients of every form (a', b', c') satisfy the system of equations

$$ap^2=a'\delta^2-2b'\delta\gamma+c'\gamma^2,$$
$$bp^2=-a'\delta\beta+b'(\alpha\delta+\beta\gamma)-c'\gamma\alpha,$$
$$cp^2=a'\beta^2-2b'\beta\alpha+c'\alpha^2.$$

---

40 Bericht. Acad. Berlin, 1855, 493–495; Jour. de Math., (2), 1, 1856, 76–79; Jour. für Math., 53, 1857, 127–129; Werke, II, 191–194.

41 Jour. für Math., 53, 1857, 238–259. See H. J. S. Smith, Report Brit. Assoc., 1862, § 113; Coll. Math. Papers, I, 246–9; also G. B. Mathews, Theory of Numbers, 1892, 159–170.

Hence (a', b', c') has no other divisor than p, and the condition that p be a divisor is

$$aa' = a(a\alpha^2 + 2b\alpha\gamma + c\gamma^2) = (a\alpha + b\gamma)^2 - D\gamma^2 \equiv 0 \pmod{p}.$$

Now, a may be assumed relatively prime to p. The number of solutions of this congruence is the number of substitutions in (1) which do not lead to properly primitive forms (a', b', c'). This number is 2, 0, 1 according as $(D/p) = 1, -1, 0$. Hence the number of properly primitive forms (a', b', c') is $p - (D/p)$.

If $\binom{\alpha\ \beta}{\gamma\ \delta}$ be one substitution of (1) which carries (a, b, c) over into a particular (a', b', c'), then all the substitutions in (1) which effect the transformation are $\binom{\mathrm{A}\ \mathrm{B}}{\Gamma\ \Delta}$, in which (Gauss, Disq. Arith., Art. 162; report in Ch. I)

$$\mathrm{A} = \alpha t - (\alpha b + \gamma c)u, \qquad \Gamma = \gamma t + (a\alpha + \gamma b)u,$$
$$\mathrm{B} = \beta t - (\beta b + \delta c)u, \qquad \Delta = \delta t + (\beta a + \delta b)u,$$

t, u ranging over σ pairs of integers which satisfy $t^2 - Du^2 = 1$, where σ is the smallest value of i for which u_i is a multiple of p in

$$(T + U\sqrt{D})^i = t_i + u_i\sqrt{D}.$$

If $u_\sigma = pu'$, $t_\sigma = t'$, then

$$(T + U\sqrt{D})^\sigma = t' + u'\sqrt{p^2 \cdot D}, \qquad \sigma = \frac{\log(T' + U'\sqrt{p^2 D})}{\log(T + U\sqrt{D})},$$

where T, U is the fundamental solution of $t^2 - Du^2 = 1$, and T', U', of $t^2 - Dp^2u^2 = 1$. Since only one form (a, b, c) can be carried over into a particular form (a', b', c') by (1), Dirichlet's[42] ratio $h(S^2D)/h(D)$ follows at once.[43] Similarly, Lipschitz obtained the ratio of the number of improperly primitive classes to the number of properly primitive classes for the same determinant.[44]

L. Kronecker[45] stated that if n denotes a positive odd number >3 and κ denotes the modulus of an elliptic function, then the number of different values of κ^2 which admit of *complex multiplication* by $\sqrt{-n}$ [i. e., for which $sn^2(u\sqrt{-n}, \kappa)$ is rationally expressible in terms of $sn^2(u, \kappa)$ and κ] is six times[46] the number of classes of quadratic forms of determinant $-n$. These values of κ^2 are the sole roots of an algebraic equation with integral coefficients, which splits into as many integral factors as there are orders of binary quadratic forms of determinant $-n$. To each order corresponds one factor whose degree is six times the number of classes belonging to that order. The two following recursion formulas[47] and one immediately deducible from them are given. Let $n \equiv 3 \pmod{4}$; let $F(m)$ be the number of properly primitive classes of $-m$ plus the number of classes derived from them;

[42] Jour. für Math., 21, 1840, 12. See Dirichlet.[20]
[43] For details, see G. B. Mathews,[218] Theory of Numbers, Cambridge, 1892, 159–166; also H. J. S. Smith's Report.[79]
[44] For details see G. B. Mathews,[218] Theory of Numbers, 1892, 166–169.
[45] Monatsber. Akad. Wiss. Berlin, Oct., 1857, 455–460. French trans., Jour. de Math., (2), 3, 1858, 265–270.
[46] Cf. H. J. S. Smith, Report Brit. Assoc., 35, 1865, top of p. 335; Coll. Math. Papers, I, 305.
[47] Cf. L. Kronecker, Jour. für Math., 57, 1860, 249.

$\phi(n)$ be the sum of the divisors of n which are $>\sqrt{n}$; $\psi(n)$ be the sum of the other divisors. Then

$$\text{(I)} \qquad 2F(n)+4F(n-2^2)+4F(n-4^2)+\ldots=\phi(n)-\psi(n),$$

$$\text{(II)} \qquad 4F(n-1^2)+4F(n-3^2)+4F(n-5^2)+\ldots=\phi(n)+\psi(n),$$

where, in the left members, $n-i^2>0$.

Using the *absolute invariant* j instead of κ^2, H. Weber[48] has deduced in detail a similar relation which these two imply.[214]

C. Hermite[49] set $u=\phi(\omega)=\kappa^{\frac{1}{4}}$, κ being the ordinary modulus in elliptic functions, and found that the algebraic discriminant of the standard modular equation for transformations of prime order n,

$$\Pi\left[\phi^8(\omega)-\phi^8\left(\frac{\delta_1\omega+16m}{\delta_2}\right)\right]=0,\ \delta_1\delta_2=n,\ m=0,\ 1,\ 2,\ \ldots,\ \delta_2-1,$$

is of the form

$$u^{n+1}(1-u^8)^{n+(2/n)}\theta^2(u),$$

where $\theta(u)=a_0+a_1u^8+a_2u^{16}+\ldots+a_\nu u^{8\nu}$ is a reciprocal polynomial with no multiple roots and $\theta(u)$ is relatively prime to u and $1-u^8$; moreover,

$$\nu=\frac{n^2-1}{8}-\frac{1}{2}\left[n+\left(\frac{2}{n}\right)\right].$$

By means of the condition for equality of two roots[50] of the modular equation, he set up a correspondence between these equal roots and the roots of certain quadratic equations of determinant $-\Delta$ and so proved the following theorem.[51] Let

$$\Delta'=(8\delta-3n)(n-2\delta)>0,\ \Delta''=8\delta(n-8\delta)>0,\ \Delta'''=\delta(n-16\delta)>0.$$

Then

$$\nu=2\Sigma F(\Delta')+2\Sigma F(\Delta'')+6\Sigma G(\Delta''').$$

(Cf. H. J. S. Smith, Report Brit. Assoc., 1865; Coll. Math. Papers, I, 344–5.)

Those roots $x=\phi(\omega)$ of $\theta(u)=0$ are now segregated which correspond to the roots ω of a representative system of properly primitive forms of a given negative determinant $-\Delta$; similarly for a system of improperly primitive forms. If the representative form (A, B, C) of each properly primitive class is chosen with C even, A uneven, then to the roots ω of the equations $A\omega^2+2B\omega+C=0$ correspond values of $u^8=\phi^8(\omega)$ which are the principal roots of a reciprocal equation $F\ (x, \Delta)=0$ with integral

---

[48] Elliptische Functionen und Algebraische Zahlen, Braunschweig, 1891, 393–401; Algebra, III, Braunschweig, 1908, 423–426. For the same theory see also Klein-Fricke, Elliptischen Modulfunctionen, Leipzig, 1892, II, 160–184.

[49] Comptes Rendus, Paris, 48, 1859, 940–948, 1079–1084, 1096–1105; 49, 1859, 16–24, 110–118, 141–144. Oeuvres, II, 1908, 38–82. Reprint, Paris, 1859, Sur la théorie des équations modulaires et la résolution de l'équation du cinquième degré, 29–68.

[50] Cf. C. Hermite, Sur la théorie des équations modulaires, 1859, 4; Comptes Rendus, Paris, 46, 1859, 511; Oeuvres, II, 1908, 8. Cf. also H. J. S. Smith, Report Brit. Assoc., 1865, 330; Coll. Math. Papers, I, 299. For properties of the discriminant of the modular equation, see L. Koenigsberger, Vorlesungen über die Theorie der Elliptischen Functionen, Leipzig, 1874, Part II, 154–6.

[51] For an equivalent result see Kronecker.[124]

coefficients and of degree the double of the number of those classes. Moreover, $F(x, \Delta)$ can be decomposed into factors of the form

$$(x+1)^4+a_1x(x-1)^2, \text{ if } \Delta=\begin{cases}2n-1,\ 2n-9,\ 2n-25,\ldots;\\ 2n,\ 2n-4,\ 2n-16,\ldots.\end{cases}$$

This illustrates the rule that, excepting $\Delta=1$, 2, the number of properly primitive classes of $-\Delta$ is even if Δ is $\equiv 1$ or 2 (mod 4).

In a theorem analogous to the preceding and concerning improperly primitive classes, $\mathcal{F}(x, \Delta)=0$ is a reciprocal equation with integral coefficients and of degree 2 or 6 times the number of those classes, according as $\Delta\equiv -1$ or 3 (mod 8); and $\mathcal{F}(x,\Delta)$ can be decomposed into factors of the form $(x^2-x+1)^3+a(x^2-x)^2$, if $\Delta=4n-1$, $4n-9$, $4n-25,\ldots$.

For a few small determinants the class-number is exhibited as by the following example. After Jacobi, the modular equations of orders 3 and 5 respectively are

$$(q-l)^4=64(1-q^2)(1-l^2)(3+ql),$$
$$(q-l)^6=256(1-q^2)(1-l^2)[16ql(9-ql)^2+9(45-ql)(q-l)^2]$$

where $q=1-2\kappa^2$, $l=1-2\lambda^2$. These equations combined with

$$u^8=\frac{1}{1-v^8}=x, \text{ where } u^8=\kappa^2,\ v^8=\lambda^2,$$

give, respectively,

$$[x^2-x+1][(x^2-x+1)^3+2^7(x^2-x)^2]\equiv\mathcal{F}(x,3)\cdot\mathcal{F}(x,11)=0,$$
$$[(x^2-x+1)^3+2^7(x^2-x)^2][(x^2-x+1)^3+2^7\cdot 3^3(x^2-x)^2]$$
$$\equiv\mathcal{F}(x,11)\cdot\mathcal{F}(x,19)=0.$$

The common factor of the two left members must be identical with $\mathcal{F}(x, 11)$. Then the numbers of improperly primitive classes of determinants -3, -11, -19 are one-sixth of the degrees of the expressions in brackets in the left members of the last two equations. P. Joubert's[52] modification of this method is given for determinants -15, -23, -31.

F. Arndt[53] wrote

$$\phi(x)=\prod_a(x-e^{2a\pi i/P}),\qquad \psi(x)=\prod_b(x-e^{2b\pi i/P})$$

and, in the three cases which Dirichlet had omitted (see Dirichlet,[23] (9)), obtained the following:

$$\text{(II)}\qquad D=P,\ P=4\mu+3,\ (T+U\sqrt{P})^h=\mp\psi(i)^4,$$

where $\mp$ means $-$ or $+$ according to P is or is not prime;

$$\left.\begin{array}{ll}\text{(III)} & D=2P,\ P=4\mu+1,\ (T+U\sqrt{2P})^h=\psi(x)^4\cdot\phi(-x)^4\\ \text{(IV)} & D=2P,\ P=4\mu+3,\ (T+U\sqrt{2P})^h=\psi(x)^4\cdot\psi(x^3)^4\end{array}\right\},\ x=\sqrt{\tfrac{1}{2}}(1+i).$$

[52] Cf. Joubert,[62] Comptes Rendus, Paris, 50, 1860, 911.
[53] Jour. für Math., 56, 1859, 100.

L. Kronecker[54] published without demonstration eight class-number recursion formulas derived from singular moduli in the theory of elliptic functions.[55] They are algebraically-arithmetically independent of each other; and any other formula of this type derived from an elliptic modular equation[49] is a linear combination of Kronecker's eight. He employed the following permanent[56] notations.

n is any positive integer; m any positive uneven integer; r any positive integer $8k-1$; $s=8k+1>0$.

$G(n)$ is the number of classes of determinant $-n$; $F(n)$ is the number of uneven classes.

$X(n)$ is the sum of the odd divisors of n; $\Phi(n)$ is the sum of all divisors.

$\Psi(n)$ is the sum of the divisors of n which are $>\sqrt{n}$ minus the sum of those which are $<\sqrt{n}$.

$\Phi'(n)$ is the sum of the divisors of the form $8k\pm1$ minus the sum of the divisors of the form $8k\pm3$.

$\Psi'(n)$ is the sum both of the divisors of the form $8k\pm1$ which are $>\sqrt{n}$ and of the divisors of the form $8k\pm3$ which are $<\sqrt{n}$ minus the sum both of the divisors of the form $8k\pm1$ which are $<\sqrt{n}$ and of the divisors of the form $8k\pm3$ which are $>\sqrt{n}$.

$\phi(n)$ is the number of divisors of n which are of the form $4k+1$ minus the number of those of the form $4k-1$.

$\psi(n)$ is the number of divisors of n which are of the form $3k+1$ minus the number of those of the form $3k-1$.

$\phi'(n)$ is half the number of solutions of $n=x^2+64y^2$; and $\psi'(n)$ is half the number of solutions of $n=x^2+3\cdot64y^2$, in which positive, negative, and zero values of x and y are counted for both equations.

$$\text{(I)}\quad F(4n)+2F(4n-1^2)+2F(4n-2^2)+2F(4n-3^2)+\dots =2X(n)+\Phi(n)+\Psi(n),$$

$$\text{(II)}\quad F(2m)+2F(2m-1^2)+2F(2m-2^2)+2F(2m-3^2)+\dots =2\Phi(m)+\phi(m),$$

$$\text{(III)}\quad F(2m)-2F(2m-1^2)+2F(2m-2^2)-2F(2m-3^2)+\dots=-\phi(m),$$

$$\text{(IV)}\quad 3G(m)+6G(m-1^2)+6G(m-2^2)+6G(m-3^2)+\dots =\Phi(m)+3\Psi(m)+3\phi(m)+2\psi(m),$$

$$\text{(V)}\quad 2F(m)+4F(m-1^2)+4F(m-2^2)+4F(m-3^2)+\dots =\Phi(m)+\Psi(m)+\phi(m),$$

$$\text{(VI)}\quad 2F(m)-4F(m-1^2)+4F(m-2^2)-4F(m-3^2)+\dots =(-1)^{\frac{m-1}{2}}[\Phi(m)-\Psi(m)]+\phi(m),$$

$$\text{(VII)}\quad 2F(r)-4F(r-4^2)+4F(r-8^2)-4F(r-12^2)+\dots =(-1)^{\frac{1}{8}(r-7)}[\Phi'(r)-\Psi'(r)],$$

$$\text{(VIII)}\quad 4\Sigma(-1)^{\frac{s-k^2}{16}}\left[2F\left(\frac{s-k^2}{16}\right)-3G\left(\frac{s-k^2}{16}\right)\right] =(-1)^{\frac{1}{8}(s-1)}[\Phi'(s)-\Psi'(s)]+\phi(s)+4\psi(s)-4\phi'(s)-8\psi'(s).$$

---

[54] Jour. für Math., 57, 1860, 248–255; Jour. de Math., (2), 5, 1860, 289–299.

[55] Demonstrated by the same method by H. J. S. Smith, Report Brit. Assoc., 1865, 349–359; Coll. Math. Papers, I, 325–37.[100]

[56] Later in the report of this paper will be noted the historical modification of Kronecker's F and G printed in Roman type.

In all recursion formulas (except those of G. Humbert[355]) of this chapter, the determinants are $\leqq 0$. In the above 8 formulas, $F(0)=0$, $G(0)=\frac{1}{2}$. The functions $\phi(n)$, $\psi(n)$, $\phi'(n)$, $\psi'(n)$ are removed hereafter from the formulas by replacing italic letters F and G throughout by Roman letters F and G, which agree respectively with the earlier symbols except that $\mathrm{F}(0)=0$, $\mathrm{G}(0)=-\frac{1}{12}$, and except that classes (1, 0, 1), (2, 1, 2) and classes derived from them are each counted as $\frac{1}{2}$ and $\frac{1}{3}$ of a class respectively. Later writers have commonly adopted these conventions but have not insisted on printing the symbols in Roman type.

The following also result[57] from the theory of elliptic functions:

$$\mathrm{F}(4n)=F(4n), \text{ for all } n;$$
$$\mathrm{F}(4n)=2\mathrm{F}(n),\ \mathrm{G}(4n)=\mathrm{F}(4n)+\mathrm{G}(n), \text{ for all } n;$$
$$\mathrm{G}(n)=\mathrm{F}(n), \text{ if } n\equiv 1 \text{ or } 2 \pmod 4;$$
$$3\mathrm{G}(n)=[5-(-1)^{\frac{1}{4}(n-3)}]\mathrm{F}(n), \text{ if } n\equiv 3 \pmod 4.$$

By means of these relations, Kronecker obtained from the original eight formulas the following[58]:

$$\text{(IX)}\quad \mathrm{F}(n)+\mathrm{F}(n-2)+\mathrm{F}(n-6)+\mathrm{F}(n-12)+\mathrm{F}(n-20)+\ldots=\tfrac{1}{8}\Psi(4n+1),$$
$$\text{(X)}\quad \mathrm{E}(n)+2\mathrm{E}(n-1)+2\mathrm{E}(n-4)+2\mathrm{E}(n-9)+\ldots=\tfrac{2}{3}[2+(-1)^n]X(n),$$

where $\mathrm{E}(n)=2\mathrm{F}(n)-\mathrm{G}(n)$. But

$$\Sigma(2\pm 1)X(n)q^n=\Sigma\frac{q^n}{(1\pm q^n)^2},\quad n=1, 2, \ldots,$$

the plus or minus sign being taken on both sides according as n is even or odd. Hence formula (X) is equivalent to the important formula

$$\text{(XI)}\qquad 12\Sigma \mathrm{E}(n)q^n=\theta_3^3(q),\quad \theta_3(q)=\sum_{n=-\infty}^{+\infty} q^{n^2},\quad q=e^{\pi i\omega},$$

which implies that the number of representations[59] of n as the sum of three squares is $12\mathrm{E}(n)$. (Cf. this History, Vol. II, 265.)

By (VI) and (VII), Kronecker calculated $F(m)$ for m uneven from 1 to 10,000.

P. Joubert,[60] referring to a conjecture of Gauss,[61] proved that if n is a fixed prime and $\Delta>0$ grows through a range of values which are quadratic residues of n, then the number of classes in a genus of the forms of determinant $-\Delta$ has a lower limit for the range.

P. Joubert[62] considered the principal root ω of $P\omega^2+Q\omega+R=0$. If ω furnishes a root $\phi^2(\omega)$ of the modular equation for transformations of order 2^μ, μ arbitrary, he found that just two values of $\phi^2(\omega)$ are furnished as roots by all the forms (P, Q, R) of a given improperly primitive class which have third coefficients a

57 For the means of immediate arithmetical deduction, see Lipschitz[41] and H. J. S. Smith, Report Brit. Assoc., 1862, 514–519; Coll. Math. Papers, I, 246–51.

58 See H. J. S. Smith, Report Brit. Assoc., 1865, 348; Coll. Math. Papers, I, 323.

59 Cf. C. F. Gauss,[4] Disq. Arith., Arts 291–2. For a report, see this History, Vol. II, 262; while on pp. 263, 265, 269, are reports on papers by Dirichlet, Kronecker and Hermite giving applications of class-number to sums of three squares.

60 Comptes Rendus, Paris, 50, 1860, 832–837.

61 Disq. Arith.,[4] Art. 303.

62 Comptes Rendus, Paris, 50, 1860, 907–912.

multiple of 16. If (A, B, C) is a form of this kind and of negative determinant $-\Delta=(S^2-2^{\mu+2})/T^2$, in which S, T are odd, it is equivalent to $(2^\mu A, B, C/2^\mu)$, and these two forms give the same value of $\phi^2(\omega)$. Consequently, if in the ordinary modular equation we set $u^2=v^2=x$, the resulting equation $f(x)=0$ has a degree which is double the number of representative improperly primitive forms (A, B, C) of negative determinant $-\Delta$; and $f(x)$ can be decomposed into polynomial factors each of degree the double of the number of the improperly primitive classes of the corresponding determinant $-\Delta$.

For example, if $\mu=1$, the only possible determinant is -7. The modular equation for transformations of order 2 is $v^4=2u^2/(1+u^4)$, and becomes $x^2+x+2=0$. Therefore there is a single improperly primitive class of determinant -7. For somewhat larger values of determinant $-(8k-1)$, Hermite's[49] device is used for identifying common factors which belong to the same Δ and which occur in the left members of $f(x)=0$ for neighboring values of $n=2^\mu$.

In the modular equation $F(\lambda, \kappa)=0$ for transformations of odd prime order n, Joubert wrote $\lambda=2x/(1+x^2)$, $\kappa=x^2$, and obtained $f(x)=0$ in which $f(x)$ is a product of polynomials which have the same characteristic properties as in the former case. If ω is such that $\phi^2(\omega)=\sqrt{\kappa}$ is a root of $F(\lambda, \kappa)=0$, then ω is the principal root of an equation

$$A\omega^2+2B\omega+C=0,$$

where (A, B, C) is improperly primitive and the negative determinant $-\Delta$ has Δ equal to one of the numbers $8n-1^2, 8n-3^2, 8n-5^2, \ldots$. Moreover, C is divisible by 16 and again there are therefore just two values of $\phi^2(\omega)$ for each improperly primitive class; and the roots $\phi^2(\omega)$ lead to forms (A, B, C) which just exhaust the classes of negative determinants $-(8n-\sigma^2)$. Hence the aggregate number of improperly primitive classes of the sequence of determinants is read off as in the following example. Let $n=3$; then $\Delta=23, 15$,

$$F(\lambda, \kappa)=\lambda^4-4\lambda^3(4\kappa^3-3\kappa)+6\lambda^2\kappa^2+4\lambda(3\kappa^3-4\kappa)+\kappa^4,$$
$$f(x)\equiv(x^4+4x^3+5x^2+2x+4)(x^6-x^5+9x^4+13x^3+18x^2+16x+8).$$

Since $15=2\cdot8-1$, the first factor in $f(x)$ has already been associated with $\Delta=15$ by the use of $n=2^\mu=2$. The number of improperly primitive classes of determinant -23 may be read off as half the degree of the second factor and also as the index of its constant term regarded as a power of 2.

Joubert[63] illustrated his method by many examples.

Joubert,[64] in the modular equations for transformations of odd order $n=p^\alpha q^\beta r^\gamma$ (p, q, r different primes), and with the roots

$$v^8=\phi^8\left(\frac{g\omega+16m}{g_1}\right),$$

added to the usual conditions the restriction that g and g_1 be relatively prime. In the modular equation $f(x, y)=0$, he took $y=1/x$. Now $f(x, 1/y)$ is of degree

$$2N=2p^{\alpha-1}q^{\beta-1}r^{\gamma-1}(p+1)(q+1)(r+1)$$

[63] Comptes Rendus, Paris, 50, 1860, 940–4.
[64] *Ibid.*, 1040–1045.

and has $2\mathcal{N}+\phi\sqrt{n}$ or $2\mathcal{N}$ roots equal to unity, according as n is or is not a square, where

$$\mathcal{N}=\sum_{d}\sum_{\gamma}\gamma\cdot\phi(d),$$

d^2 ranging over the square divisors of n (n omitted if it is a perfect square), while $\gamma\gamma_1=n/d^2$, $\gamma<\gamma_1$, γ and γ_1 being relatively prime. Excluding the unity roots, he established a correspondence between the roots of $f(x, 1/x)=0$ and the roots of certain quadratic equations, and obtained the following formula when n is or is not a square respectively:

$$F(n)+2F(n-1^2)+2F(n-2^2)+2F(n-3^2)+\ldots=N-\mathcal{N} \text{ or } N-\mathcal{N}-\tfrac{1}{2}\phi(\sqrt{n}).$$

where $F(D)$ denotes the number of odd classes of determinant $-D$ which have all their divisors prime to n. If, however, a form is involved which is derived from (1, 0, 1), the right member in each case should be diminished by the number of proper decompositions of n into the sum of two squares. Numerous[65] other class-number relations in the modified F and a similarly modified G are obtained. Tables[66] verify the formulas in F. The interdependence of Joubert's and Kronecker's[54] class-number relations has been discussed by H. J. S. Smith.[67]

H. J. S. Smith[68] reproduced the principal parts of the researches of Gauss[4] and Dirichlet[19, 20, 23] on the class-number of binary quadratic forms. For $D>0$ and $\equiv 1 \pmod 4$, he wrote

$$h(D)=\left(\frac{2}{D}\right)\frac{2}{\log(T+U\sqrt{D})}\sum\left(\frac{D}{m}\right)\log\tan\frac{m\pi}{2D},$$

where m is positive, odd, prime to D, and $<D$. (Cf. Berger,[166] (3).)

C. Hermite[69] began with the factorization

$$\frac{H^2(z)\Theta_1(z)}{\Theta^2(z)}=\frac{H(z)\Theta_1(z)}{\Theta(z)}\cdot\frac{H(z)}{\Theta(z)}$$

and expanded each factor after C. G. J. Jacobi,[70] setting $z=2Kx/\pi$. In the product of the two expansions, the term independent of x is*

$$\text{(1)}\qquad \mathcal{A}=\sum\frac{\sqrt{q^{2n+1}}}{1-q^{2n+1}}\cdot q^{\frac{1}{4}(2n+1)^2-a^2}\equiv\sum F(N)q^{N/4},$$

where in the first sum, $a=0, \pm1, \pm2, \ldots \pm n$; while in the second sum, N ranges over all positive numbers $\equiv 3 \pmod 4$ which can be represented by (I), and hence by each of the three identically equal expressions

(I) $(2n+1)(2n+4b+3)-4a^2$,
(II) $(2n+1)(4n+4b+4-4a)-(2n+1-2a)^2$,
(III) $(2n+1)(4n+4b+4+4a)-(2n+1+2a)^2$.

* The expansion of the first fraction in (1) is Σq^k, $k=\frac{1}{2}(2n+1)+(2n+1)b$, $b\geqq 0$.

[65] Comptes Rendus, Paris, 50, 1860, 1095–1100.
[66] *Ibid.*, 1147–1148.
[67] Report Brit. Assoc., 35, 1865, 364; Coll. Math. Papers, I, 343–4.
[68] Report Brit. Assoc., 1861, 324–340; Coll. Math. Papers, I, 1894, 163–228.
[69] Comptes Rendus, Paris, 53, 1861, 214–228; Jour. de Math., (2), 7, 1862, 25–44; Oeuvres, II, 1908, 109–124.
[70] Fundamenta Nova Funct. Ellipticarum, 1829, §§ 40–42; Werke, I, 1881, 159–170.

Thus $F(N)$ denotes the number of ways in which N can be represented by any one of the expressions (I), (II), (III). We represent N by (I), (II), or (III), according as

$$|a|<\tfrac{1}{4}(2n+1); \quad a \geqq \tfrac{1}{4}(2n+1); \quad a<0, \text{ but } |a| \geqq \tfrac{1}{4}(2n+1).$$

Now (I), (II), (III) are respectively the negatives of the determinants of the quadratic forms

$$(2n+1,\ 2a,\ 2n+4b+3),$$
$$(2n+1,\ 2n+1-2a,\ 4n+4b+4-4a),$$
$$(2n+1,\ 2n+1+2a,\ 4n+4b+4+4a).$$

Thus we have $F(N)$ forms which are reduced. Moreover, the $F(N)$ forms exhaust the reduced uneven forms of determinant $-N$. For, those of the first type constitute all uneven reduced forms of determinant $-N$ which have an even middle coefficient. Those of the second and third types constitute all forms (p, q, r) of determinant $-N$ in which p and q are uneven, $p>2q$, $r>2q>0$. Hence, since (p, q, r) is here never equivalent to $(p, -q, r)$, the number of forms of the three types together is $F(N)$, in the class-number sense.[54]

A second factorization yields[70]

$$(2) \qquad \frac{K}{2\pi}\sqrt{\frac{2kK}{\pi}\cdot\frac{H^2(z)}{\Theta(z)}\cdot\frac{\Theta_1(z)}{\Theta(z)}}$$
$$=\mathscr{A}\Theta_1(z)-\Sigma \cos 2nx\cdot q^{n^2}(\sqrt[4]{q^{-1}}+3\sqrt[4]{q^{-3}}+\ldots+(2n-1)\sqrt[4]{q^{-(2n-1)^2}}).$$

For $x=0$, the first member vanishes and the terms under the summation sign are of the type

$$q^{N/4}\cdot\tfrac{1}{2}(\Sigma d'-\Sigma d),$$

where $N\equiv 3 \pmod 4$, d' is any divisor $>\sqrt{N}$ of N and d is any divisor $<\sqrt{N}$. In Kronecker's[54] symbols, we get, by (1) and (2),

$$\Theta_1(0)\Sigma F(N)q^{N/4}=\tfrac{1}{2}\Sigma\Psi(N)q^{N/4}.$$

Or, since $\Theta_1(0)=1+2q+2q^4+2q^9+2q^{16}+\ldots$,

$$(3) \qquad F(N)+2F(N-2^2)+2F(N-4^2)+\ldots=\tfrac{1}{2}\Psi(N).$$

In Kronecker's[54] formulas this is (V)+(VI).

A third factorization combined with the first yields the following:

$$(4) \qquad F(4n-1)+F(4n-3^2)+F(4n-5^2)+\ldots=\Phi_1(n)-\Psi_1(n),$$

where $\Phi_1(n)$ denotes the sum of the divisors of n whose conjugates are odd, and $\Psi_1(n)$ denotes the sum of all the divisors $<\sqrt{n}$ and of different parity from their conjugates. Similarly,

$$(5) \quad F(N)-2F(N-2^2)+2F(N-4^2)-\ldots+2(-1)^kF(N-4k^2)\ldots$$
$$=(-1)^{\frac{1}{4}(N-3)}\Psi_2(N)=(-1)^{\frac{1}{4}(N-3)}\cdot\tfrac{1}{2}(\Phi(N)-\Psi(N)),\ N\equiv 3 \pmod 4,$$

where $\Psi_2(n)$ denotes the sum of the divisors of n which are $<\sqrt{n}$. Hermite's three class-number relations above are all derivable from Kronecker's[71] eight.

[71] See H. J. S. Smith, Rep. Brit. Assoc., 35, 1865, 364; Coll. Math. Papers, I, 1894, 343.

Since N in (1) is of the form $4n+3$, (1) implies

$$(6) \qquad \tfrac{1}{2}(\mathscr{A}-\epsilon\mathscr{A}_1)=\sum_0^\infty F(8n+3)q^{\frac{1}{4}(8n+3)},$$

where $\epsilon=(1+i)/\sqrt{2}$, $\epsilon^4=-1$, and $\mathscr{A}_1$ is the result of replacing q by $-q$ in $\mathscr{A}$. Another expression for $\frac{1}{2}(\mathscr{A}-\epsilon\mathscr{A}_1)$ is found by means of the integral of the product quoted at the beginning of this report; comparison of it with (6) gives

$$(7) \qquad \Sigma F(8n+3)q^{\frac{1}{4}(8n+3)}=(\sqrt[4]{q}+\sqrt[4]{q^9}+\sqrt[4]{q^{25}}+\ldots)^3.$$

This result is implicitly included[72] in Kronecker's[54] (XI) and can be deduced from it by elementary algebra.[73] When the coefficients of equal powers of q are equated in the two members, this formula implies that the number of odd classes of determinant $-(8n+3)$ is the number of positive solutions of

$$8n+3=x^2+y^2+z^2.$$

L. Kronecker[74] referring to his[54] earlier memoir, multiplied formulas (I), (II), (V) respectively by q^{4n}, q^{2m}, $\frac{1}{2}q^m$, added the results, and summed for all values of n and m, and obtained

$$(1) \qquad \Sigma F(n)q^n=\tfrac{1}{2}\sqrt{\frac{\pi}{2K}}\,\Sigma\frac{n}{q^n-q^{-n}}(q^{n^2+n}-2+q^{n^2-n}).$$

Similarly from formulas (I), (III), (VI), he obtained

$$(2) \qquad \Sigma F(n)q^n=\tfrac{1}{2}\sqrt{\frac{\pi}{2k'K}}\,\Sigma n(-q)^{n^2}\cdot\frac{q^n-q^{-n}}{q^n+q^{-n}}.$$

Now (1) and (2) imply the following three formulas[75]:

$$\Sigma F(2m)q^{\frac{1}{2}m}=\frac{kK}{\pi}\cdot\sqrt{\frac{K}{2\pi}}, \qquad \Sigma F(4n+1)q^n=\frac{1}{q^{\frac{1}{4}}}\,\frac{K}{\pi}\cdot\sqrt{\frac{kK}{2\pi}},$$

$$\Sigma F(8n+3)q^{2n}=\frac{1}{q^{\frac{3}{4}}}\cdot\frac{kK}{2\pi}\sqrt{\frac{kK}{2\pi}};$$

and these imply Kronecker's[54] (IV).

By means of an expansion[76] of $\sin^2 \mathrm{am}\, 2Kx/\pi$ in terms of cosines of multiples of x, (1) takes the form

$$(3) \qquad \Sigma F(n)q^n=\frac{1}{q^{\frac{1}{4}}}\cdot\frac{k^2K}{2\pi^2}\sqrt{\frac{K}{2\pi}}\int_0^\pi \sin^2 \mathrm{am}\,\frac{2Kx}{\pi}\cdot\theta_2(x)\cos x\,dx.$$

"From (3), all the formulas[54] (I)–(VIII) can be deduced." Other such relations are indicated by means of theta-functions, although the eight formulas "are algebraically-arithmetically independent."

---

[72] Jour. für Math., 57, 1860, 253.
[73] Cf. L. J. Mordell, Messenger Math., 45, 1915, 79.
[74] Monatsber. Akad. Wiss. Berlin, 1862, 302–311. French transl., Annales Sc. Ecole Norm. Sup., 3, 1866, 287–294.
[75] Cf. C. Hermite,[69] Comptes Rendus, Paris, 53, 1861, 226.
[76] Cf. C. G. J. Jacobi, Fundamenta Nova, 1829, 110, (1), Werke, I, 1881, 166.

Kronecker stated that he had obtained arithmetical deductions of certain of his class-number relations by following the plan of Jacobi[77] who had first found by equating coefficients in two expansions, the number of expressions for n as the sum of four squares and had later translated the analytic method into an arithmetical one.[78] The following theorem, which Kronecker deduced from his formula (V), was offered as a suggestion for a means of deducing his class-number relations arithmetically: Let p be any odd prime and let

$$a_1z^2+2b_1z+c_1\equiv 0, \qquad a_2z^2+2b_2z+c_2\equiv 0\ldots \pmod p$$

be a succession of congruences corresponding to reduced forms of determinants, $-p$, $-(p-1^2)$, $-(p-2^2)$, ... respectively (with b taken negative in the reduced form if $a=c$); then the number of roots of the congruences is

$$F(p)+2F(p-1^2)+2F(p-2^2)+2F(p-3^2)+\ldots;$$

that is to say, by formula (V), the number is $p+1$ or p according as p is $\equiv 1$ or 3 (mod 4).

H. J. S. Smith[79] gave an account of Lipschitz's[41] method of obtaining the ratio of $h(D\cdot S^2)$ to $h(D)$.

C. Hermite[80] gave a list of expansions of quotients obtained from theta-functions and showed how the products and quotients of theta-functions lead to class-number relations (cf. Hermite[69]). This list of doubly periodic functions of the third kind has been extended by C. Biehler,[81] P. Appell,[81a] Petr,[252, 258] Humbert,[293] and E. T. Bell.[82] Finally, Hermite deduced Kronecker's[54] relation (XI).

Hermite[83] generalized a theorem of Legendre (this History, Vol. I, 115, (5)) into the *Lemma*: If $m=a^\alpha b^\beta c^\gamma\ldots k^\kappa$, where $a, b, c, \ldots, k$ are μ different primes, then the number of integers which are less than or equal to x and relatively prime to m is

$$\Phi(x)=E(x)-\Sigma E\left(\frac{x}{a}\right)+\Sigma E\left(\frac{x}{ab}\right)-\Sigma E\left(\frac{x}{abc}\right)+\ldots\pm E\left(\frac{x}{abc\ldots k}\right),$$

with the convention $\Phi(x)=E(x)$ if $m=1$. It follows that

$$\text{(1)} \qquad \Phi(x)=\frac{x}{m}\phi(m)+2^{\mu-1}\epsilon, \qquad -1<\epsilon<+1.$$

Now $F(n)$ is defined by $F(n)=\Sigma_{i=1}^{i=n}f(i)$, where $f(i)=0$ if i is not a divisor of n or if i is a divisor of n but is not prime to m; also $F(n)=0$, if m and n are not relatively prime. Then, by definition,

$$\sum_{k=1}^{m} F(k)=\sum_{i=1}^{n} f(i)\Phi\left(\frac{n}{i}\right).$$

[77] Fundamenta Nova, 1829, Art. 66; Werke, I, 1881, 239.
[78] Jour für Math., 12, 1834, 167–172; Werke, VI, 1891, 245–251.
[79] Report Brit. Assoc., 1862, § 113; Coll. Math. Papers, I, 1904, 246–9.
[80] Comptes Rendus, Paris, 55, 1862, 11, 85; Jour. de Math., (2), 9, 1864, 145–159; Oeuvres, II, 1908, 241–254.
[81] Thesis, Paris, 1879.
[81a] Annales de l'École Normale, (3), 1, 1884, 135–164; 2, 1885, 9–36.
[82] Messenger Math., 49, 1919, 84.
[83] Comptes Rendus, Paris, 55, 1862, 684–692. Oeuvres, II, 1908, 255–263.

Now (cf. Dirichlet,[93] (1)), if $D=S^2D_0$, where D_0 is a fundamental determinant, and if n is any positive uneven integer relatively prime to D, then for $f(i)=(D_0/i)$ and D uneven, for example, the formula

$$(2)\qquad F(n)=k\sum_{i=1}^{n}\left(\frac{D_0}{i}\right)\Phi\left(\frac{n}{i}\right),\quad k=2 \text{ if } D<-3,\ k=1 \text{ if } D>0,\ m=2|D|,$$

gives the sum of the number of representations of integers from 1 to n which are uneven and relatively prime to D by the representative properly primitive forms of determinant D with the usual restriction[84] on x and y in case $D>0$.

Hermite omits the rather difficult proof that the term containing ϵ in (1) is negligible[85] for n very great and concludes from (1) and (2) that, for n very great,

$$F(n)=k\sum_{i=1}^{n}\left(\frac{D_0}{i}\right)\frac{n}{2i(-1)^{k+1}D}\,\phi((-1)^{k+1}\cdot 2D).$$

C. F. Gauss[86] and G. L. Dirichlet[87] had found geometrically the asymptotic mean number of such representations furnished by each form for n large. A comparison yields the class-number (Dirichlet,[19] (1)).

J. Liouville[88] stated that the number[89] of solutions of $yz+zx+xy=n$ in positive odd integers with $y+z\equiv 2 \pmod 4$, $n\equiv 3 \pmod 4$, is $F(n)$.

J. Liouville[90] obtained an arithmetical deduction of a Kronecker[54] recursion formula in the form

$$F(2m-1^2)+F(2m-3^2)+F(2m-5^2)+\ldots=\tfrac{1}{2}[\zeta_1(m)+\rho(m)],$$

where m is an arbitrary uneven integer, $\zeta_1(m)$ represents the sum of the divisors of m, and $\rho(m)$ is the excess of the number of divisors of m which are $\equiv 1 \pmod 4$ over the number of divisors $\equiv 3 \pmod 4$.

Lemma 1. Let any uneven integer m be subjected to the two types of partitions

$$(1)\qquad m=2m'^2+d''\delta'',\qquad 2m=m_1^2+d_2\delta_2+2^{a_3+1}d_3\delta_3,$$

where m_1, d_2, d_3, δ_2, δ_3 are positive uneven integers; $a_3>0$; while m' is any positive, negative, or zero integer. Then, if $f(x)$ is an even function,

$$(2)\qquad \begin{aligned}&\tfrac{1}{2}\Sigma[d''f(2m')-f(2m')-2f(2m'+2)-2f(2m'+4)-\ldots-2f(2m'+\delta''-1)]\\ &\qquad=\Sigma\left[f\left(\frac{d_2+\delta_2}{2}-d_3\right)-f\left(\frac{d_2+\delta_2}{2}+d_3\right)\right].\end{aligned}$$

Now take $f(x)$ so that $f(0)=1$, $f(x)=0$ if $x\neq 0$. Then the only partitions of the second type (1) which furnish terms in the right member of (2) are those in which $d_3=\frac{1}{2}(d_2+\delta_2)$. Hence the right member of (2) has for its value the number of solutions of

$$2m-m_1^2=d_2\delta_2+2^{a_3}(d_2+\delta_2)\delta_3.$$

---

[84] G. L. Dirichlet,[19] Zahlentheorie, Art. 90, ed. 4, 1894, 225 and 226.

[85] Cf. T. Pepin, Annales Sc. de l'Ecole Norm. Sup., (2), 3, 1874, 165; M. Lerch, Acta Math., 29, 1905, 360.

[86] Werke, II, 1876, 281 (Gauss[4]).

[87] Jour. für Math.,[19] 19, 1839, 360 and 364.

[88] Jour. de Math., (2), 7, 1862, 44.

[89] Cf. Bell,[370] and Mordell.[372]

[90] Jour. de Math., (2), 7, 1862, 44–48.

We set $d_2+\delta_2=2u$, $d_2-\delta_2=4z$. Hence $u>2z$. Keeping m_1 fixed, Liouville followed the method of Hermite[69] and obtained the result that the number of solutions of $2m-m_1^2=u(u+4d)-4z^2$ is

$$\sum_{m_1} F(2m-m_1^2)-\tfrac{1}{2}\sum_{m_1}\zeta(2m-m_1^2)-\tfrac{1}{2}\sum_{s_1}\omega(2m-s_1^2),$$

in which $\zeta(n)$ denotes the number of divisors of n, $\omega(n)=1$ or 0 according as n is or not a perfect square. Hence $\Sigma\omega(2m-s_1^2)=\rho(m)$.

Now in the first member of (2), the summation of the first two terms in the bracket is equal to $\zeta_1(m)-\zeta(m)$. Furthermore the expression in (2):

$$f(2m'+2)+f(2m'+4)+\ldots+f(2m'+\delta''-1)$$

will have the value 1 for each pair of values $m'<0$, $2m'+\delta''>0$ and the value 0 for all other values m' and $2m'+\delta''$. Let A denote the number of pairs of values $m'<0$, $2m'+\delta''>0$ in the partition (1_1). We have now proved that

$$(3)\qquad \sum_{m_1} F(2m-m_1^2)-\tfrac{1}{2}\sum_{m_1}\zeta(2m-m_1^2)-\tfrac{1}{2}\rho(m)=\tfrac{1}{2}\{\zeta_1(m)-\zeta(m)\}-A,$$

Lemma 2. Let any uneven integer M be subjected to the two types of partitions

$$M=2M'^2+D''\Delta'',\qquad 2M=M_1^2+D_2\Delta_2,$$

where M_1, D_2, Δ_2, D'', Δ'' are positive odd integers, while M' is any integer. Then, if $f_1(x)$ is an uneven function,

$$(4)\qquad \Sigma f_1(D''+2M')=\Sigma f_1\left(\frac{D_2+\Delta_2}{2}\right).$$

To evaluate A, we identify m and M and specialize $f_1(x)$ so that $f_1(x)=1$ if $x>0$, $f_1(x)=0$ if $x=0$, $f_1(x)=-1$ if $x<0$. Since the number of solutions of $M=2M'^2+D''\Delta''$ with $M'>0$ is equal to the number with $M'<0$, the left member of (4) is composed of the following four parts:

$$\begin{aligned} A&=\Sigma f_1(D''+2M'),\quad M'<0,\quad D''+2M'>0;\\ -B&=\Sigma f_1(D''+2M'),\quad M'<0,\quad D''+2M'<0;\\ A+B&=\Sigma f_1(D''+2M'),\quad M'>0,\quad D''+2M'>0;\\ \zeta(m)&=\Sigma f_1(D''+2M'),\quad M'=0,\quad D''+2M'>0.\end{aligned}$$

Hence (4) implies that

$$2A+\zeta(m)=\Sigma f_1\left(\frac{D_2+\Delta_2}{2}\right)=\Sigma\zeta(2m-m_1^2).$$

Thus (3) becomes[91]

$$\sum_{m_1} F(2m-m_1^2)=\tfrac{1}{2}[\zeta_1(m)+\rho(m)].$$

This result has been established in detail by Bachmann[91] and Meissner.[292] From the same two lemmas, H. J. S. Smith[92] obtains a different form of the right member, for the case m odd.

---

[91] Cf. P. Bachmann, Niedere Zahlentheorie, Leipzig, II, 1910, 423–433.
[92] Report Brit. Assoc., 35, 1865, 366; Coll. Math. Papers, I, 1894, 346–350.

Hermite's discovery[69] of the relation between the number of classes of determinant N and the number of certain decompositions of N, also enabled Liouville to announce that formulas exist analogous to those of Kronecker,[54] but in which the successive negative determinants are respectively $2s^2-n$, $3s^2-n$, $4s^2-n$, ..., where n is fixed and s has a sequence of values.[92a]

G. L. Dirichlet[93] reproduced in a text-book the theory of his memoirs[14, 19, 20, 23] of 1838, 1839, 1840. Continuing his former notation, he obtained (Arts. 105–110) new expressions for

$$N=-\frac{i^{\left(\frac{P-1}{2}\right)^2}}{8\sqrt{P}}\sum_r j^r(1+\delta i^{3r})[1+\epsilon(-1)^r]\int_0^1\sum_s\left(\frac{s}{P}\right)\frac{dx}{x-j^r\theta^s},$$

$j=e^{\pi i/4}$, $\theta=e^{2\pi i/P}$, while s ranges over a complete set of incongruent numbers (mod P) prime to P. The result[94] is, for $D>0$, $D\equiv 1 \pmod 4$, for example,

$$N\cdot 2\sqrt{P}=-\left\{1-\left(\frac{2}{P}\right)\tfrac{1}{2}\right\}\log\{F(1)^2\},\qquad F(x)=\Pi(x-\theta^s)^{(s/P)}.$$

Thence in the notation of the Pellian equation, for example,

$$(1)\quad D=P\equiv 1 \pmod 8,\quad (T+U\sqrt{P})^{h(D)}=(t+u\sqrt{D})^l,\qquad l=\left[2-\left(\frac{2}{P}\right)\right](2-\kappa),$$

where $\kappa=1$ or 0 according as P is prime or composite, and t, u are positive integers satisfying $t^2-Du^2=1$. From five such relations, Dirichlet points out divisibility properties of $h(D)$; e. g., if $D\equiv 1 \pmod 4$, $h(D)$ is odd or even according as P is prime or composite.

Incidentally (Art. 91), Dirichlet proved that the number of representations of a number σn by a system of primitive forms of determinant D is

$$(2)\qquad \tau\Sigma(D/\delta)$$

where $\sigma=1$ or 2 according as the forms are proper or improper, n is prime to $2D$, and δ ranges over the divisors of n.

This formula has been used by Hermite,[83] Pepin,[120] Poincaré[271] to evaluate the class-number.

V. Schemmel[95] denoted by p an arbitrary positive odd number which has no square divisors. By the use of Gauss sums he set up such identities as the following, when $p=4n+3$:

$$(1)\qquad \sum_1^{p-1}\left(\frac{m}{p}\right)\sin m\alpha=\frac{1}{2\sqrt{p}}\sum_1^{p-1}\left(\frac{m}{p}\right)\frac{\sin p\alpha\,\sin 2m\pi/p}{\cos 2m\pi/p-\cos\alpha},$$

where α is an arbitrary real number. He took $\alpha=\pi/2$ in both members, then

$$A-B-C+D=-\frac{1}{2\sqrt{p}}\sum_1^{p-1}\left(\frac{m}{p}\right)\tan\frac{2m\pi}{p},$$

[92a] Cf. Liouville,[107, 109] Gierster [145], Stieltjes,[154, 162] Hurwitz,[167, 184] Petr [258], Humbert [293], Chapelon [340].

[93] Vorlesungen über Zahlentheorie, Braunchweig, 1863, 1871, 1879, 1894, Ch. V.

[94] Cf. G. L. Dirichlet,[28] Jour. für Math., 21, 1840, 154; Werke, I, 1889, 495; Arndt.[53]

[95] De multitudine formarum secundi gradus disquisitiones, Diss., Breslau, 1863, 19 pp.

where A, B, C, D are the number of positive quadratic residues which are $<p$ and of the respective forms $4n+1$, $4n+2$, $4n+3$, $4n+4$. Whence,[96]

$$(2)\qquad h(-p)=-\frac{1}{2\sqrt{p}}\sum_1^{p-1}\left(\frac{m}{p}\right)\tan\frac{2m\pi}{p}.$$

After differentiating both members of (1) with respect to a, he took $a=0$. The result is[23]

$$\sum_1^{p-1}\left(\frac{m}{p}\right)m=-\frac{p}{2\sqrt{p}}\sum_1^{p-1}\left(\frac{m}{p}\right)\cot\frac{m\pi}{p},$$

whence follows Lebesgue's[36] class-number formula (1):

$$(3)\qquad p=4n+3,\qquad h(-p)=\frac{2-\left(\frac{2}{p}\right)}{2\sqrt{p}}\sum_1^{p-1}\left(\frac{m}{p}\right)\cot\frac{m\pi}{p}.$$

Similarly to (3) are obtained

$$(4)\qquad p=4n+1,\qquad h(-p)=\frac{1}{2\sqrt{p}}\sum_1^{p-1}\left(\frac{m}{p}\right)\sec\frac{2m\pi}{p};$$

$$p=4n+3,\qquad h(-2p)=-\frac{1}{\sqrt{p}}\sum_1^{p-1}\left(\frac{m}{p}\right)\frac{\sin 2m\pi/p}{\cos 4m\pi/p};$$

$$p=4n+1,\qquad h(-2p)=\frac{1}{\sqrt{p}}\sum_1^{p-1}\left(\frac{m}{p}\right)\frac{\cos 2m\pi/p}{\cos 4m\pi/p}.$$

Schemmel, without discussing convergence, decomposed an infinite series by the identity

$$\sum_n\left(\frac{n}{p}\right)\cos na=\lim_{\kappa=\infty}\sum_1^{p-1}\left(\frac{m}{p}\right)\{\cos ma+\cos(p+m)a+\ldots+(\kappa p+m)a\},$$

where $p\equiv 3 \pmod 4$, and n is positive and relatively prime to p. After transforming the right member, he integrated both members between the limits 0 and $\frac{1}{2}\pi$, with Dirichlet's[23] formula (8_2) as the final result:

$$(5)\qquad h(p)=\frac{2}{\log(T+U\sqrt{p})}\log\Pi\frac{\cos(b\pi/p+\pi/4)}{\cos(a\pi/p+\pi/4)}.$$

Employing the usual cyclotomic notation,[53]

$$\psi(x)=\Pi(x-e^{2a\pi i/p}),\qquad \phi(x)=\Pi(x-e^{2b\pi i/p}),$$

Schemmel found that, for $p=4n+3>0$,

$$(6)\qquad \frac{\psi'(1)}{\psi(1)}-\frac{\phi'(1)}{\phi(1)}\equiv\Sigma\frac{1}{1-e^{2a\pi i/p}}-\Sigma\frac{1}{1-e^{2b\pi i/p}}=\frac{i}{2}\sum_1^{p-1}\left(\frac{m}{p}\right)\cot\frac{m\pi}{p},$$

which by (3) gives a new class-number formula for $-p$ (see H. Holden[280]). He noted that, for $p=4n+3>0$,

$$\frac{\phi(i)}{\psi(i)}=\Pi\frac{\cos\left(\frac{b\pi}{p}+\frac{\pi}{4}\right)e^{\Sigma b\pi/p}}{\cos\left(\frac{a\pi}{p}+\frac{\pi}{4}\right)e^{\Sigma a\pi/p}};\quad e^{(\Sigma b-\Sigma a)\pi/p}=1 \text{ or } -1,$$

---

[96] G. L. Dirichlet,[23] Jour. für Math., 21, 1840, 152; Zahlentheorie, Art. 104, ed. 4, 1894, 264.

according as p is composite or prime. Hence by (5), if we set

$$F(x)=\log \pm \frac{\phi(x)}{\psi(x)},$$

we have, for $p=4n+3>0$,

$$h(p)=\frac{2}{\log(T+U\sqrt{p})}F(i).$$

Similarly for $p=4n+3>0$,

$$\frac{\psi'(i)}{\psi(i)}-\frac{\phi'(i)}{\phi(i)}=-\frac{1}{2}\sum_{1}^{p-1}\left(\frac{m}{p}\right)\tan\frac{2m\pi}{p},\qquad h(-p)=\frac{1}{\sqrt{p}}F'(i).$$

Moreover, if $p=4n+3>0$,

$$h(2p)=\frac{2}{\log(T+U\sqrt{2p})}[F(\omega^3)-F(-\omega^3)];$$
$$h(-2p)=-\frac{i+1}{2\sqrt{p}}[F'(\omega^3)+F'(-\omega^3)],$$

and, if $p=4n+1$,

$$h(2p)=\frac{4}{\log(T+U\sqrt{2p})}[F(\omega)-F(-\omega)],\quad h(-2p)=\frac{i-1}{2\sqrt{p}}[F'(\omega)-F'(-\omega)],$$

where, in the last four formulas, $\omega=(1+i)/\sqrt{2}$.

L. Kronecker[97] obtained, more simply than had G. L. Dirichlet,[98] the fundamental equation (2) of Dirichlet,[20] and specialized it in the form

$$(1)\qquad \tau\Sigma\left(\frac{D}{n}\right)(nn')^{-1-\rho}=\sum_{a,b,c}\sum_{x,y}(ax^2+2bxy+cy^2)^{-1-\rho}.$$

For a particular (a, b, c), the sum

$$\sum_{x=s}^{\infty}(ax^2+2bxy+cy^2)^{-1-\rho}\equiv\sum_{x=s}^{\infty}\phi(x,y)$$

lies between the two values

$$\pm\phi(hy,y)+\int_{hy}^{\infty}\phi(x,y)\,dx,$$

if $hy<s<hy+1$. Hence

$$\lim_{\rho=0}\rho\sum_{y=1}^{\infty}\sum_{x=1}^{\infty}\phi(x,y)=\lim_{\rho=0}\rho\sum_{y=1}^{\infty}\int_{hy}^{\infty}\phi(x,y)\,dx,$$

where h is taken so that $ah^2+2bh+c\neq 0$. When we set $ax+by=zy$, this limit is given by

$$\lim_{\rho=0}\tfrac{1}{2}\int_{ah+b}^{\infty}\frac{dz}{z^2-D}=\frac{1}{4\sqrt{D}}\log\frac{t+u\sqrt{D}}{t-u\sqrt{D}},\ ah+b=\frac{t}{u},\ D>0$$
$$=\frac{\pi}{\sqrt{-D}}\qquad\qquad ,\ D<0$$

Hence, when we exclude[99] from the final sum (1) those terms for which the form takes values not prime to P, (1) implies, for $\rho=0$,

$$\rho\Sigma\left(\frac{D}{n}\right)(nn')^{-1-\rho}=\tfrac{1}{2}\cdot\frac{1}{\sqrt{D}}\log(t_1+u_1\sqrt{D})\cdot h(D)\Pi\left(1-\frac{1}{p}\right)\left[1-\left(\frac{D}{p}\right)\frac{1}{p}\right],$$

---

[97] Monatsber. Akad. Wiss. Berlin, 1864, 285–295.
[98] Jour. für Math., 21, 1840, 7; Werke, I, 1889, 467.
[99] Cf. R. Dedekind, Remarks on Gauss' Untersuchungen über höhere Arithmetik, Berlin, 1889, 685–686; Gauss' Werke, II, 293–4.

where p ranges over the distinct prime divisions of P, and t_1, u_1 are fundamental solutions of $t^2-Du^2=1$. For $D>0$, this proves that $h(D)$ is finite, since the left member is a definite number.

H. J. S. Smith[100] discussed the researches of Kronecker,[54, 74] Hermite,[49, 69, 80] Joubert,[62, 64] and Liouville[90] in class-number relations. He found proofs of Kronecker's class-number relations[64] by means of the complex multiplication of elliptic functions. The details are based on the methods used by Joubert and Hermite. L. Kronecker[101] has commended the report for its mastery and insight.

For instance, formula (V) of Kronecker is proved by putting $x=\kappa^2$ and $1-x=\lambda^2$ in the ordinary modular equation $f_8(\kappa^2, \lambda^2)=0$ for transformations of uneven order m. The right member of the desired formula is found as the order of the infinity of $f_8(x, 1-x)$ as x increases without limit. The left member is the aggregate multiplicity of the roots of $f_8(x, 1-x)=0$.

R. Lipschitz[102] developed a general theory of asymptotic expansions for number-theoretic functions and found that, in the special case of the number of properly primitive classes, the asymptotic expression is

$$h(-m)=\frac{2\pi}{7\Sigma s^{-3}}m^{\frac{3}{2}},\quad s=1, 2, 3, \ldots;\ m>0.$$

This agrees with C. F. Gauss[103] since

$$\frac{7}{2}\left(1+\frac{1}{2^3}+\frac{1}{3^3}+\cdots\right)=4\left(1+\frac{1}{3^3}+\frac{1}{5^3}+\cdots\right).$$

And asymptotically,

$$h(m)=\frac{24\log 2}{7\Sigma s^{-3}}m^{\frac{3}{2}},\quad s=1, 2, 3, \ldots;\ m>0.$$

The method of Lipschitz is illustrated by C. Hermite.[104]

J. Liouville[105] stated without proof that if a and a' denote respectively the [odd] minimum and second [odd] minimum of the forms of a properly primitive class of determinant $-k=-(8n+3)<0$, then

$$\sum_{Cl} a(a'-a)=\tfrac{2}{3}k\cdot h(-k).$$

He discussed as examples the cases $k=3$, 11, 19, 27. The theorem has been proved arithmetically by Humbert.[293]

Liouville[106] let m be an arbitrary number of the form $8n+3$, whence the only reduced ambiguous forms of negative determinant $-(m-4\sigma^2)$ are $(d, 0, \delta)$, where $d\delta=m+4\sigma^2$ and $d\leqq\sqrt{m-4\sigma^2}$. Hence the d's are the values of the minima of the uneven ambiguous classes of determinant $-(m-4\sigma^2)$. And hence, if n_1 [and n_2] denotes the number of ambiguous classes of determinant $-m$ whose minima are $\equiv 1$

---

100 Report Brit. Assoc., 35, 1865, 322–375; Collected Papers, I, 1894, 289–358.
101 Sitzungsber. Akad. Wiss. Berlin, 1875, 234.
102 Sitzungsber. Akad. Berlin, 1865, 174–185. Reproduced by P. Bachmann, Zahlentheorie, Leipzig, II, 1894, 438–459.
103 Werke, II, 1876, 284; Untersuchungen,[9] Berlin, 1889, 670.
104 Bull. des Sc. Math.,[204] (2), 10, I, 1886, 29; Oeuvres, IV, 220–222.
105 Jour. de Math. (2), 11, 1866, 191–192.
106 Jour. de Math. (2), 11, 1866, 221–224.

(mod 4) [and $\equiv 3$ (mod 4)], and if p_1 [and p_2] denotes the number of uneven ambiguous classes of determinants $-(m-4\sigma^2)$ excluding $\sigma=0$, whose minima are $\equiv 1$ (mod 4) [and $\equiv 3$ (mod 4)], then in the notation of this History, Vol. II, p. 265 (Liouville[33a]),

$$n_1-n_2+2(p_1-p_2)=\rho'(m)+2\rho'(m-4\cdot 1^2)+2\rho'(m-4\cdot 2^2)+\ldots.$$

By the theorem there stated, it follows from Hermite[69] that

$$F(m)=n_1-n_2+2(p_1-p_2).$$

Liouville[107] stated that he had obtained the following results arithmetically. He generalized Hermite's[69] formula (4) both to

$$(1)\qquad \sum_i F(2^{a+2}m-i^2)=2^a\Sigma d-\Sigma D,\quad i>0,$$

in which i and m are odd; and to

$$(2)\qquad \sum_i i^2F(2^{a+2}m-i^2)=2^am(2^a\Sigma d-\Sigma D)-\tfrac{1}{3}\Sigma D^3,\quad i>0$$

where a is an integer $\gtreqless 0$, d denotes a divisor of m; and D is a divisor of 2^am which is of opposite party to its conjugate divisor. By the nature of their second members, these formulas represent what Humbert[293] has called the second type of Liouville's formulas.

For $m=d\delta=12g+7$ or $12g+11$, he gave

$$(3)\qquad \sum_i F(2m-3i^2)=\frac{1}{8}\left[3+\left(\frac{m}{3}\right)\right]\sum_d\left(\frac{3}{\delta}\right)d,$$

where $i=1, 3, 5\ldots.$ He stated that if m denotes an odd positive number prime to 5; and a, β are given positive numbers or zero, and $m=d\delta$, then

$$(4)\qquad \sum_i F(8\cdot 2^a5^\beta m-5i^2)=2^{a-2}\left[5^{\beta+1}-(-1)^a\left(\frac{m}{5}\right)\right]\sum_d\left(\frac{5}{\delta}\right)d,$$

where $i=1, 3, 5, \ldots, m=d\delta$. A special case of this relation is proved by Chapelon[340] as his formula (3) below.

If m is a positive integer of the form $24g+11$, then

$$F(m)+2\Sigma F(m-48\cdot s^2)=\tfrac{1}{4}\sum_d\left(\frac{3}{\delta}\right)d,\quad s>0.$$

Finally, if $m=4g+3$ and $t=g-s$, then

$$\sum_{s=0}^{g}(8s+3)F(8t+3)=\tfrac{1}{8}\Sigma(-1)^{(\delta-1)/2}d^2.$$

The right members here characterize what Humbert has called the first type of Liouville's formulas. G. Humbert[108] has deduced formulas of this type, by C. Hermite's method, from elliptic function theory.

---

[107] Comptes Rendus, Paris, 62, 1866, 1350; Jour. de Math., (2), 12, 1867, 98–103.
[108] Jour. de Math.,[293] (6), 3, 1907, 366–368, 446–447.

Liouville[109] by replacing n by $3m$ in Hermite's[69] formula (4), decomposed it into two class-number relations

$$\Sigma' F(12m-i^2)=\zeta_1(3m)-\zeta_1(m), \quad i\not\equiv 0 \pmod 3,\ i\equiv 0 \pmod 2$$
$$\Sigma F(12m-9i^2)=\zeta_1(m), \quad i\equiv 0 \pmod 3,$$

where $\zeta_1(n)$ is the sum of the divisors of n; and m is odd.

Liouville[110] announced without proof the relation[111]

$$\Sigma\left(\frac{-1}{i}\right)iF(4m-i^2)=\Sigma(a^2-4b^2),$$

where $i=1, 3, 5, 7, \ldots$; a is positive and uneven; and a, b range over the integral solutions of $m=a^2+4b^2$; m odd.

Stieltjes[160] and G. Humbert[112] have each given a proof by Hermite's method of equating coefficients in expansions of doubly periodic functions of the third kind.

Liouville[113] stated for $m\equiv 5 \pmod{12}$ that

$$\Sigma F\left(\frac{2m-i^2}{3}\right)=\tfrac{1}{4}\Sigma\left(\frac{3}{\delta}\right)d, \tag{1}$$

where $i=1, 5, 7, 11, 13, 17, \ldots$ is relatively prime to 6; $m=\delta d$. For[114] m odd and relatively prime to 5,

$$F(10m)+2\Sigma F(10m-25t^2)=2\Phi(m),$$

where $t=1, 2, 3, \ldots$; $\Phi(m)$ denotes the sum of the divisors of m.

R. Dedekind,[115] by the composition of classes, solved completely the Gauss[4] problem, obtaining the results of Dirichlet.[20]

R. Götting,[116] to evaluate Dirichlet's[14] formula (4) for $h(-p)$, p a prime of the form $4n+3$, proved that

$$\sum_{a=0}^{\frac{1}{2}p}\left(\frac{a}{p}\right)=-\frac{p-1}{2}+2\sum_{j=0}^{\frac{1}{4}(p-3)}\sigma_j-2\sum_{j=1}^{\frac{1}{4}(p-3)}\rho_j,$$

$$\sum_{j=0}^{\frac{1}{4}(p-3)}\sigma_j=\frac{1}{p}\sum_{a=0}^{\frac{1}{2}p}\left(\frac{a}{p}\right)a+\frac{p^2+1}{12},$$

where $$\sigma_j=\left[\sqrt{pj+\frac{p}{2}}\right], \quad \rho_j=[\sqrt{pj}].$$

Hence if $$p=8n+7, \quad \Sigma\sigma_j=\tfrac{1}{12}(p^2-1);$$

if $$p=8n+3, \quad \Sigma\sigma_j+2\Sigma\rho_j=\tfrac{1}{4}(p^2-1).$$

He obtained numerous formulas for computing $\Sigma(a'/p)$.

---

109 Jour. de Math., (2), 13, 1868, 1–4.
110 Jour. de Math. (2), 14, 1869, 1–6.
111 Cf. * T. Pepin, Memoire della Pontifica Accad. Nuovi Lincei, 5, 1889, 131–151.
112 Jour. de Math.,[293] (6), 3, 1907, 367, Art. 30.
113 Jour. de Math., (2), 14, 1869, 7.
114 *Ibid.*, 260–262. Proved on p. 171 of Chapelon's[340] Thesis.
115 Supplement X to G. L. Dirichlet's Zahlentheorie, ed. 3, 1871; ed. 4, 1894, §§ 150–151.
116 Ueber Klassenzahl quadratischen Formen. Sub-title: Ueber den Werth des Ausdrucks $\Sigma(a'/p)$ wenn p eine Primzahl von der Form $4n+3$ und a' jede ganze Zahl zwischen 0 und $\frac{1}{2}p$ bedeutet. Prog., Torgau, 1871, 20 pp.

F. Mertens[117] denoted by $\psi(s, x)$ the number of positive classes of negative determinants 1, 2, 3, ..., x which have reduced forms with middle coefficient $\pm s$; by $\chi(s, x)$ the number of these classes which are even. By a study of the coefficients of reduced forms, it is found that the number of uneven classes of negative determinants 1, 2, 3, ..., x is[118]

$$F(x) = \sum_{0}^{\sqrt{x/3}} [\psi(s, x) - \chi(s, x)],$$

where, except for terms of the order of x,

$$\sum_{0}^{\sqrt{x/3}} \psi(s, x) = \frac{2\pi}{9} x^{\frac{3}{2}}, \qquad \sum_{0}^{\sqrt{x/3}} \chi(s, x) = \frac{\pi}{18} x^{\frac{3}{2}}.$$

If we set $f(N) = \Sigma_1^N h(-n)$, we have

$$\begin{aligned} F(x) &= f(x) + f(x/3^2) + f(x/5^2) + f(x/7^2) + f(x/9^2) + \dots \\ F(x/3^2) &= \qquad f(x/3^2) \qquad\qquad\qquad + f(x/9^2) + \dots \\ F(x/5^2) &= \qquad\qquad f(x/5^2) \qquad\qquad\qquad + \dots \\ &\cdots\cdots\cdots\cdots \end{aligned}$$

and we solve for $f(x)$ by multiplying the respective equations by $\mu(1)$, $\mu(3)$, $\mu(5)$, ..., where $\mu(n)$ is the Moebius function (this History, Vol I, Ch. XIX). Thus

$$f(x) = \sum_{n=1}^{x} \mu(n) F(x/n^2).$$

But

$$F\left(\frac{x}{n^2}\right) = \frac{\pi x^{\frac{3}{2}}}{6n^3} + O\left(\frac{x}{n^2}\right),$$

where $Of(x)$ denotes a function of the order of $f(x)$, or more exactly a function whose quotient by $f(x)$ remains numerically less than a fixed finite value for all sufficiently large values of x.

Hence, when terms of the order of x are neglected,

$$f(x) = \frac{\pi}{6} x^{\frac{3}{2}} \sum_{1}^{\infty} \frac{\mu(n)}{n^3} = \frac{\pi}{6} x^{\frac{3}{2}} \left(1 - \frac{1}{3^3}\right)\left(1 - \frac{1}{5^3}\right)\left(1 - \frac{1}{7^3}\right)\left(1 - \frac{1}{11^3}\right) \dots$$

Then, asymptotically,

$$\sum_{n=1}^{N} h(-n) = \frac{4\pi}{21 S_3} N^{\frac{3}{2}}, \qquad S_3 = 1 + \frac{1}{2^3} + \frac{1}{3^3} + \frac{1}{4^3} + \dots$$

And therefore the asymptotic median class number is[119] $2\pi\sqrt{N}/(7S_3)$.

T. Pepin[120] let Σm be the total number of representations of numbers n relatively prime to a given number Δ, $0 \leqq n \leqq M$, M being an arbitrary positive integer, by a system of properly primitive forms of negative determinant D. He also let Σm be the total number of representations of numbers $2n$, n relatively prime to Δ,

---

[117] Jour. für Math., 77, 1874, 312-319. Reproduced by P. Bachmann, Zahlentheorie, Leipzig, II, 1894, 459.

[118] Cf. C. F. Gauss, Disq. Arith., Art. 171.

[119] Cf. C. F. Gauss, Disq. Arith., Art. 302; Werke, II, 1876, 284. Cf. R. Lipschitz,[102] Sitzungsber. Akad., Berlin, 1865, 174–185.

[120] Annales Sc. de l'Ecole Norm. Sup., (2), 3, 1874, 165–208.

$0 \leqq n \leqq M$, by a system of improperly primitive forms of determinant D. In every representation, let

$$x=\alpha x_1+\gamma, \quad y=\beta y_1+\delta, \quad \gamma<\alpha, \quad \delta<\beta, \quad \alpha,\ \beta,\ \gamma,\ \delta \text{ each} \gtreqless 0;$$

and in each of the two cases above, let K, K' be respectively the number of pairs of values γ, δ possible for given α, β. Then[121]

$$\frac{Kh(D)\pi M}{\alpha\beta}+M\epsilon=\frac{K'h'(D)\pi M}{\alpha\beta}+M\eta=\Sigma m,$$

where the limits of $M\epsilon$ and $M\eta$ for $M=\infty$ are finite.

A comparison of K and K' for $\alpha=\beta=\Delta=2$ gives Dirichlet's[20] ratio h/h'. The corresponding result is obtained for the other orders and for the positive determinant.

Pepin avoids the convergence difficulty of Hermite[83] and obtains Dirichlet's[23] classic closed expression (5) for $h(D)$, $D<0$, by extending a theorem of Dirichlet[93] (2), to give

$$\Sigma m=\kappa\sum_n\sum_i (D_0/i),$$

in which κ is the automorph factor 2, 4 or 6; D_0 is a fundamental determinant, $D=D_0S^2$; i ranges over all divisors of n, while n ranges over all odd numbers $\leqq M$; and (D_0/i) is the Jacobi-Legendre symbol.

Pepin translating certain results of A. Cauchy[122] on the location of quadratic residues, found in Dirichlet's[23] notation

$$(1) \qquad h(-n)=\left[2-\left(\frac{2}{n}\right)\right]\frac{\Sigma b-\Sigma a}{n}=\left[2-\left(\frac{2}{n}\right)\right]\frac{\Sigma b^2-\Sigma a^2}{n^2},$$

where $-n=-(4\mu+3)$ is a fundamental negative determinant. This latter class-number formula, called Cauchy's, has been simply deduced by M. Lerch, Acta Math., 29, 1905, 381. Other results of Cauchy[123] give, in terms of Bernoullian numbers,

$$h(-n)\equiv 2B_{(n+1)/4} \text{ if } n=8l+7; \quad \equiv -6B_{(n+1)/4} \text{ if } n=8l+3,$$

modulo n a prime. And without proof Pepin states, for $n>0$, that

$$h(-n)=\left[2-\left(\frac{2}{n}\right)\right]\left\{\tfrac{1}{3}[2l+1][4l+1]-2\sum_{i=1}^{l} i[\sqrt{in}]\right\}, \quad l=\frac{n-3}{4}$$

L. Kronecker[124] obtained from his[54] eight classic relations new ones, as, for example, by combining (IV), (V), (VI), the following:

$$\sum_h(-1)^hF(n-4h^2)=\tfrac{1}{2}(-1)^{\frac{1}{4}(n-3)}\{\Phi(n)+\Psi(n)\}, \quad n\equiv 3 \pmod 4, \quad h\gtrless 0$$

By means of[125]

$$(1) \qquad 4\sum_0^\infty F(4n+2)q^{n+\frac{1}{2}}=\theta_2^2(q)\theta_3(q),$$

---

[121] C. F. Gauss, Werke, II, 1876, 280; Untersuchungen über höhere Arithmetik, Berlin, 1889, 666.

[122] Mém. Institut de France,[29] 17, 1840, 697; Oeuvres, (1), III, 388.

[123] Mém. Institut de France, 17, 1840, 445 (Cauchy[28]); Oeuvres, (1), III, 172.

[124] Monatsber. Akad. Wiss. Berlin, 1875, 223-236.

[125] Cf. Monatsber. Akad. Wiss. Berlin,[74] 1862, 309.

he obtained formulas for

$$\sum_h G\left(\frac{s-h^2}{16}\right), \quad \sum_h F\left(\frac{s-h^2}{16}\right), \quad s\equiv 1 \pmod 8.$$

He obtained two analogues[156] of (1), and stated that, in his[54] classic relations, $\frac{1}{4}(IV)-\frac{15}{32}(V)+\frac{3}{32}(VI)-\frac{1}{8}(VIII)$ is, when m is the square of a prime, equivalent to Hermite's[49] first class-number relation.

R. Dedekind[126] supplied the details of Gauss'[9] fragmentary deduction of formulas for $h(D)$ and $h(-D)$. He also[127] deduced and complemented Gauss'[9] set of theorems which state, in terms of the class-number of the determinant $-p$, the distribution of quadratic residues and non-residues of p in octants and 12th intervals of p, where p is an odd prime.

Dedekind,[127a] in a study of ideals, obtained results which he translated[373] immediately into the solution of the Gauss Problem.[4]

Dedekind[128] extended the notion of equivalence in modular function theory by removing the condition[129] that β and γ be even in the unitary substitution $\binom{\gamma\ \delta}{\alpha\ \beta}$. Each point ω in the upper half of the complex plane is equivalent to just one point ω_0, called a reduced point, in a fundamental triangle defined as lying above the circle $x^2+y^2=1$ and between the lines $x=\pm\frac{1}{2}$ and including only the right half of the boundary (cf. Smith[95] of Ch. I). The function, called the *valence* of ω,

$$v=val(\omega)=\frac{4}{27}\frac{(k+\rho)^3(k+\rho^2)^3}{k^2(1-k)^2}, \quad k=\kappa^2, \tag{1}$$

where ρ is an imaginary cube root of unity, is invariant[130] under the general unitary substitution. Dedekind's v is $-4/27$ times C. Hermite's[131] a. Let

$$v_n=val\left(\frac{C+D\omega}{A+B\omega}\right), \qquad \begin{vmatrix} C & D \\ A & B \end{vmatrix}=-n,$$

where A, B, C, D are integers without common divisor. Then v_n ranges exactly over the values

$$val\left(\frac{c+d\omega}{a}\right),$$

where a, c, d are integers $\gtreqless 0$ and $ad=n$; moreover, if e is the g.c.d. of a and d, then c ranges over those of the numbers $0, 1, 2, \ldots, a$ which are relatively prime to e. Hence the number of distinct values of v_n is

$$\nu=\sum_a \frac{a}{e}\,\phi(e)=n\Pi\left(1+\frac{1}{p}\right), \tag{2}$$

where p ranges over the distinct prime divisors of n.

---

[126] Remark on Disq. Arith., in Gauss's Werke, II, 1876, 293–296; Untersuchungen über Hohere Arithmetik, 1889, 686–688.

[127] Gauss's Werke, II, 1876, 301–303; Untersuchungen, 1889, 693–695.

[127a] Über die Anzahl der Ideal-classen in der verschiedenen Ordnungen eines endlichen Körpers. Festschrift zur Saecularfeier des Geburstages von Carl Frederich Gauss, Braunschweig, 1877, 55 pp.

[128] Jour. für Math., 83, 1877, 265–292.

[129] Cf. H. J. S. Smith,[100] Rep. Brit. Assoc., 35, 1865, 330; Coll. Math. Papers, I, 299.

[130] Cf. C. F. Gauss, Werke, III, 1876, 386.

[131] Oeuvres, II, 1908, 58 (Hermite [49]).

Dedekind discussed the equations whose roots are the ν values of v_n.

H. J. S. Smith[132] called the totality of those indefinite forms which are equivalent with respect to his normal substitution (Smith[95] of Ch. I) a *subaltern class.* He found that if σ denotes 2 or 1, according as U is even or uneven in $T^2-NU^2=1$, the circles of each properly primitive class of determinant N are divided into 3σ subaltern classes which in sets of σ satisfy the respective conditions

$$\text{(A)}\ a\equiv c\equiv 1 \pmod 2;\qquad \text{(B)}\ a\equiv 0,\ c\equiv 1 \pmod 2;\qquad \text{(C)}\ a\equiv 1,\ c\equiv 0 \pmod 2.$$

Since the circle $[a, b, c]$ corresponds to both (a, b, c) and $(-a, -b, -c)$, the number of subaltern classes of properly primitive circles of determinant N is $H=\frac{3}{2}\sigma h(N)$. There is a similar relation for the improperly primitive circles.

Now $\omega=x+iy$, representing a point in the fundamental region Σ, is inserted in

$$\phi^8(\omega)=\tfrac{1}{2}+X+iY,\qquad \psi^8(\omega)=\tfrac{1}{2}-X-iY,$$

where $\phi^8(\omega)$, $\psi^8(\omega)$ are Hermite's[49] symbols in elliptic function theory. Then if the circle $[a, b, c]$ satisfy (A), for example, the arcs within Σ of all and only circles (completely) equivalent to $[a, b, c]$ are transformed by the modular equation $F(k^2, \lambda^2)=0$ of order N into a certain algebraic curve, an interlaced lemmiscatic spiral. Hence all the circles of determinant N that satisfy (A) go over into a *modular curve* consisting of $\frac{1}{3}H$ distinct algebraic branches. This is called by F. Klein the *Smith-curve.*[133]

The number of improperly primitive subaltern classes of determinant N (not a square) is just the number of branches of a modular curve which is derived as the preceding from circles of determinant N, in which $a\equiv c\equiv 0 \pmod 2$.

F. Klein[134] called Dedekind's[128] v the absolute invariant J and, instead of v_n, he wrote J'. The equation, $\Pi(J-J')=0$ is called the transformation equation of order n. He gave an account of its Galois group, fundamental polygon, and Riemann surface. Simplest forms of Galois resolvents are found for $n=2, 3, 4, 5$. For example, the simplest resolvent for $n=5$ is the icosahedron equation.

Define $\eta(\omega)$ as a modular function if it is invariant under a subgroup of the group of unitary substitutions $\binom{\alpha\ \beta}{\gamma\ \delta}$. Then ω_1 and ω_2 are *relatively equivalent* if $\eta(\omega_1)=\eta(\omega_2)$. A subgroup $\binom{\alpha\ \beta}{\gamma\ \delta}$ is said to be of *grade* (*stufe*) q if

$$\begin{pmatrix}\alpha & \beta\\ \gamma & \delta\end{pmatrix}\equiv\begin{pmatrix}a & b\\ c & d\end{pmatrix} \pmod q,$$

where a, b, c, d are constants. Klein ascribed the grade q to any modular function which is invariant under only (that is, *belongs* to) such a subgroup. The subgroup

$$\begin{pmatrix}\alpha & \beta\\ \gamma & \delta\end{pmatrix}\equiv\begin{pmatrix}1 & 0\\ 0 & 1\end{pmatrix} \pmod q$$

is called the *principal* subgroup; and it is found that the icosahedron irrationality belongs to this subgroup if $q=5$. This result for the case of $n=5$ is extended to all

---

132 Atti della R. Accad. Lincei, fis. math. nat. (3), 1, 1877, 134–149; Coll. Math. Papers, II, 1894. 224–239; Abstract, Transunti, (3), 1, 68-69.

133 Elliptische Modulfunctionen,[217] II, 1892, 167 and 205.

134 Math. Annalen, 14, 1879, 111–162.

odd primes n. A modular function which belongs to the principal subgroup is called a principal modular function.

If n is an odd prime, the simplest Galois resolvent is of order $\frac{1}{2}n(n^2-1)$ and its Riemann surface is equivalent to $\frac{1}{2}n(n^2-1)$ triangles in the modular division of the plane. These triangles are chosen so as to form a polygon; and the surface of the resolvent is formed from the polygon by joining the points in the boundary which are relatively equivalent. The genus of the surface is

$$p=\tfrac{1}{24}(n-3)(n-5)(n+2).$$

Klein hereafter ascribes the p of the surface to η itself. Hence if a principal modular function η has $q=3$ or 5 then $p=0$; but if $q=7$, then $p=3$. It follows that if q is an odd prime, J is a rational function of η if and only if $q=3$ or 5. It is found similarly that if $q=2$ or 4, J is a rational function of η.

The modular equation of prime order $n\neq 5$ and of grade 5 is written as

$$\text{(1)} \qquad \Pi[\eta(\omega)-\eta(\omega')]=0,$$

where $\eta(\omega)$ is the icosahedron function, and the $n+1$ relatively non-equivalent representatives ω' are displayed in detail.

J. Gierster[135] wrote a set of eight class-number relations which he stated he had found from the icosahedron equation (Klein,[134] (1)) by the method of L. Kronecker[136] and Smith.[100] For example,

$$3\Sigma H(4n-k_1^2)=\Phi(n), \qquad n\equiv\pm 1 \pmod 5,$$

where, as always hereafter, $H(m)$ denotes the number of even classes of determinant $-m$ with the usual conventions[54]; k_1 ranges over positive quadratic residues of 5 which are $\leq \sqrt{4n}$.

A combination of these eight relations gives

$$\text{(A)} \qquad \Sigma H(4n-k^2)=\Phi(n)+\Psi(n),$$

which may be expressed in terms of Kronecker's[54] original eight:

$$\text{I}(n)-\text{II}(m) \text{ or } \text{I}(n)-\tfrac{2}{3}\text{I}(m)+\tfrac{1}{3}\text{IV}(m)-\tfrac{1}{3}\text{V}(m),$$

according as μ is odd or even in $n=2^\mu m$, where m is odd.

T. Pepin[137] completed the solution of Gauss'[4] problem. He accomplished this by finding the number of properly primitive classes of determinant $S^2\cdot D$ which when compounded with $(S, 0, -D\cdot S)$ reproduce that class. Similarly he found the ratio between the number of properly and improperly primitive classes of the same determinant.

F. Klein[138] emphasized the importance of the study of the modular functions (cf. Klein[134]) which are invariants of subgroups of finite index (i. e., subgroups whose substitutions are in $(1, k)$ correspondence with those of the modular group) and in particular those in which the subgroups are at once (a) congruence sub-

[135] Göttingen Nach., 1879, 277–81; Math. Annalen, 17, 1880, 71-3.
[136] Monatsber. Akad. Wiss. Berlin, 1875, 235.
[137] Atti Accad. Pont. Nuovi Lincei, 33, 1879–80, 356–370.
[138] Math. Annalen, 17, 1880, 62–70.

groups, (b) invariant subgroups, and (c) of genus zero. In the last case, a (1, 1) correspondence can be set up between the points of the fundamental polygon of the sub-group in the ω plane and the points of the complex plane by means of the equation $J=f(\eta)$ of genus zero where $\eta(\omega)$ is called a *haupt modul.* But if the genus p is >0, $\eta(\omega)$ must be replaced by a *system* of modular functions $M_1(\omega)$, $M_2(\omega)$, ... Klein and after him A. Hurwitz and J. Giester always chose $M_i(\omega)$ so that

$$M_i\left(\frac{\alpha\omega+\beta}{\gamma\omega+\delta}\right),\ \alpha\delta-\beta\gamma=1,$$

for all values of i, is a linear combination of $M_1(\omega)$, $M_2(\omega)$,.... The representatives ω' are $(A\omega+B)/D$, with $AD=n$, $0\leqq B<D$, B having no factor common to A and D. The analogue of the vanishing of $\Pi[\eta(\omega)-\eta(\omega')]$ in the modular equation[134] for the case $p=0$, is for the case $p>0$ the coincidence of the values of $M_1(\omega)$, $M_2(\omega)$, ... with those of $M_1(\omega')$, $M_2(\omega')$, ... respectively. This analogue of the modular equation is called the *modular correspondence* and it is said to be *grade* q if the M's are of grade q.

J. Gierster[139] stated that all of F. Kronecker's[54] eight class-number relations are obtainable as formulas of grades 2, 4, 8, 16. From F. Klein's[140] correspondence of order n and grade $q>2$, Gierster obtained $r=\frac{1}{2}q(q^2-1)$ correspondences by means of the unitary substitutions. He also considered the case where A, B, D have a common factor, i. e., the reducible correspondence. The number of coincidences of a reducible correspondence at points ω in the fundamental polygon[134] for q can be determined arithmetically in terms of class-number and algebraically in terms of the divisors of n. Excluding the coincidences which occur at the vertices, in the real axis, of the fundamental polygon, he gave briefly the chief material for the arithmetical determination. This he[145] made complete later.

If a given congruence subgroup G is not invariant, Gierster indicated a method of finding the number of coincidences of a correspondence for G in terms of the number of coincidences of the r reducible correspondences for the largest invariant subgroup under G and hence in terms of a class-number aggregate (cf. Gierster[148] for details).

He here stated (but later[141] proved) a full set of class-number relations of grade 7 (failing to evaluate just one arithmetical function $\xi(n)$ which occurs in several of the relations). These relations for the case when n is relatively prime to 7 were derived in detail later by Gierster[148] and A. Hurwitz[142] by different methods, Gierster employing modular functions which belong to other than invariant congruence subgroups.

A. Hurwitz[143] denoted by D any positive or negative integer which has no square factor other than 1, and wrote

$$F(s,D)=\left[1-(-1)^{\frac{1}{8}(D^2-1)}\frac{1}{2^s}\right]^{-1}\Sigma\left(\frac{D}{n}\right)\frac{1}{n^s},\ \text{if } D\equiv 1 \pmod 4,$$

$$F(s,D)=\Sigma\left(\frac{D}{n}\right)\frac{1}{n^s}\quad\text{in all other cases,}$$

139 Sitzungsber. Münchener Akad., 1880, 147-63; Math. Annalen, 17, 1880, 74-82.
140 Math. Annalen, 17, 1880, 68 (Klein 138).
141 *Ibid.*, 22, 1883, 190-210 (Giester 148).
142 *Ibid.*, 25, 1885, 183-196 (Hurwitz 184).
143 Zeitschrift Math. Phys., 27, 1882, 86-101.

where the summation extends over all integers $n>0$ prime to $2D$. (Cf. Dirichlet,[19] (1).) He proved the following four theorems:

(I) The functions $F(s, D)$ are everywhere one-valued functions of the complex variable s.

(II) Every function $F(s, D)$, except $F(s, 1)$, has a finite value for every finite value of s.

(III) For every finite value of s, the function $F(s, 1)$ has a finite value except when $s=1$. Then $F(s, 1)$ becomes infinite in such a way that

$$\lim_{s\to 1}[(s-1)F(s, 1)]=1.$$

(IV) If $D>0$,

$$F(1-s,\ D)=\left(\frac{2\pi}{\kappa D}\right)^{1-s}\frac{\Gamma(s)}{\pi}\sqrt{\kappa D}\cos\tfrac{1}{2}s\pi\cdot F(s, D)$$

$$=\left(\frac{\kappa D}{\pi}\right)^{s-\frac{1}{2}}\frac{\Gamma\left(\frac{s}{2}\right)}{\Gamma\left(\frac{1-s}{2}\right)}F(s, D);$$

if $D<0$,

$$F(1-s,\ D)=\left(\frac{2\pi}{-\kappa D}\right)^{1-s}\frac{\Gamma(s)}{\pi}\sqrt{-\kappa D}\sin\tfrac{1}{2}s\pi\cdot F(s, D)$$

$$=\left(\frac{-\kappa D}{2\pi}\right)^{s-\frac{1}{2}}\frac{\Gamma\left(\frac{s}{2}+\frac{1}{2}\right)}{\Gamma\left(\frac{1-s}{2}+\frac{1}{2}\right)}\cdot F(s, D),$$

where $\kappa=1$ if $D\equiv 1 \pmod 4$, $\kappa=4$ in all other cases. These four results are extended to $D=D'\cdot S^2$ by the use of Dirichlet's identity

$$\Sigma\left(\frac{D}{n}\right)\frac{1}{n^s}=\Sigma\left(\frac{D'}{n'}\right)\frac{1}{n'^s}\Pi\left[1-\left(\frac{D'}{r}\right)\frac{1}{r^s}\right],$$

where n' ranges over all positive integers prime to $2D'$, and r ranges over all prime numbers which are divisors of D' but not of D (cf. Dirichlet, Zahlentheorie, § 100).

The memoir ends with an ingenious proof of the three following theorems:

$$\text{If } D>0 \text{ and } D\neq 1,\ F(s, D)=0,$$

for $s=0$ and for all negative even integral values of s. If $D<0$, $F(s, D)=0$ for all negative odd integral values of s.

$$F(s, D)\cdot\Gamma\left(\frac{s}{2}\right)\cdot\left(\frac{\kappa D}{\pi}\right)^{s/2}(D>0),\quad F(s, D)\Gamma\left(\frac{s+1}{2}\right)\cdot\left(\frac{-\kappa D}{\pi}\right)^{s/2}(D<0)$$

are not altered in value when s is replaced by $1-s$.

L. Kronecker[144] proved six of his[54] eight classic relations by means of a formula for the class-number of bilinear forms and a correspondence between classes of bilinear forms and classes of quadratic forms (Kronecker[14] of Ch. XVII).

Two quadratic forms are completely equivalent if and only if one is transformed into the other by a unitary substitution congruent to the identity (mod 2). (For

[144] Abhand. Akad. Wiss. Berlin, 1883, II, No. 2; Werke, II, 1897, 425–490.

more details, see Kronecker[113] of Ch. I.) Whence $12G(n)$ and $12F(n)$ are the number of classes and of odd classes respectively of determinant $-n$ under this new definition of equivalence. Two bilinear forms are likewise completely equivalent if they are transformed into each other by cogredient substitutions of the above kind. Then the number of representative bilinear forms $Ax_1y_1+Bx_1y_2-Cx_2y_1+Dx_2y_2$ having a determinant $\Delta=AD+BC$ is $12(G(n)-F(n))$ or $12G(n)$, according as $B+C$ is odd or even where $n=-\Delta+\frac{1}{4}(B+C)^2$ is the determinant of the quadratic form $(A, \frac{1}{2}(B-C), D)$. But since $G(4n)-F(4n)=G(n)$, the number of classes of bilinear forms of determinant Δ is

$$12\sum_h[G(4\Delta-h^2)-F(4\Delta-h^2)], \quad -2\sqrt{\Delta}<h<2\sqrt{\Delta}.$$

And there are $12\Sigma F(\Delta-h^2)$ classes of those bilinear forms of determinant Δ, for which at least one of the outer coefficients A and D is odd and the sum of the middle coefficients B and C is even.

The class-number of bilinear forms is now obtained in terms of $\Phi(\Delta)$, $\Psi(\Delta)$ and $X(\Delta)$. This gives immediately such class-number relations as

$$\sum_h[G(4\Delta-h^2)-F(4\Delta-h^2)]=\Phi(\Delta)+\Psi(\Delta);$$

and so (I)–(VI) of Kronecker.[54]

J. Gierster[145] gave a serviceable introductory account of the modular equation $f(J', J)\equiv\Pi(J-J')=0$ and of the congruencial modular equation, and also of the congruencial modular correspondence. He determined (p. 11) the location and order of the branch-points of the Riemann surface of the transformed congruencial modular function $\mu(\omega')$ as a function of $\mu(\omega)$, for the case q a prime, n prime to q, and $\mu(\omega)$ belonging to the unitary sub-group,

$$(1) \qquad \begin{pmatrix}\alpha & \beta\\ \gamma & \delta\end{pmatrix}\equiv\begin{pmatrix}1 & 0\\ 0 & 1\end{pmatrix} \pmod{q}.$$

From the condition that ω furnish a root of the reducible modular equation[134] $f(J', J)=0$, namely, that integers a, b, c, d exist such that

$$(2) \qquad \omega=\frac{a\omega+b}{c\omega+d}, \quad ad-bc=n,$$

he established (p. 17) a correspondence between the roots of $f(J', J)=0$ and the roots of certain quadratic equations $P\omega^2+Q\omega+R=0$ of all discriminants $-\Delta=(d+a)^2-4n<0$. Whence the number of zeros of $f(J', J)$ in the fundamental triangle is

$$\Sigma H(4n-\kappa^2), \qquad \kappa=0, \pm1, \pm2, \ldots, \qquad \kappa^2<4n.$$

To study the infinities of $f(J, J')$ in the fundamental triangle, Gierster (after Dedekind[128]) took $\omega'=(A\omega+B)/D$, noted the initial terms in the expansion of J and J' in powers of $q=e^{\pi i\omega}$, and found that

$$(J-J')^{D/T}=\text{const.}\ J^{g/T}$$

in the neighborhood of $\omega=i\infty$; in which g is the greater of A and D, and T is the

[145] Math. Annalen, 21, 1883, 1–50. Cf. Gierster[139]; Klein-Fricke, Elliptische Modulfunctionen, II, 160–235.

g.c.d. of A and D. Whence, taking into account the number of values of B, he arrived at the class-number relation

$$\Sigma H(4n-\kappa^2)=\Phi(n)+\Psi(n), \qquad \kappa=0, \pm 1, \pm 2, \ldots.$$

The result also follows from the Chasles correspondence principle.[146]

The irreducible correspondence[139] is now studied (p. 29) between $\mu_i(\omega)$ and $\mu_i(\omega')$, where the $\mu_i(\omega)$ are a system of functions invariant only of the subgroup of unitary substitutions (1), and ω' ranges over a complete set of relatively non-equivalent representatives

$$\frac{a\omega+b}{c\omega+d}, \quad ad-bc=n,$$

where n is prime to q, and a, b, c, d have fixed residues (mod q). Now ω in the fundamental polygon[134] furnishes a finite coincidence if and only if there exist integers a, b, c, d satisfying (2). Hence the condition is that ω be the vanishing point for some form $P\omega^2+Q\omega+R$, for which

$$(3) \qquad \pm P\equiv \pm c, \qquad \pm Q\equiv \pm(d-a), \qquad \pm R\equiv \mp b \pmod q.$$

For an arbitrary reduced form $P_0\omega^2+Q_0\omega+R_0$, let g be the number of equivalent forms $P_\nu\omega^2+Q_\nu\omega+R_\nu$ which have roots in the fundamental polygon and which satisfy both (3) and

$$(4) \qquad (P_\nu, Q_\nu, R_\nu)\binom{\alpha\ \beta}{\gamma\ \delta}=(P_0, Q_0, R_0).$$

In the particular case, $b\equiv c\equiv 0$, $d\equiv a\equiv \sqrt{n}$, we have $a+d\equiv 2\sqrt{n}$, $0\equiv P_0\equiv Q_0\equiv R_0 \equiv \Delta=4n-(a+d)^2 \pmod q$; and (3) and (4) impose no condition on $\alpha, \beta, \gamma, \delta$. Hence (Klein[134]), $g=\frac{1}{2}q(q^2-1)$ and the number of finite coincidences is

$$\tfrac{1}{2}q(q^2-1)\Sigma H'\left(\frac{4n-l^2}{q^2}\right),$$

where l ranges over the positive and negative integers for which $4n-l^2$ is positive and divisible by q^2, while $H'(m)$ is the number of classes of forms of discriminant $-m$ which have no divisor which is a divisor of n. The number of finite coincidences of the reducible correspondence of order n is therefore

$$z_\infty=q(q^2-1)\Sigma H\left(\frac{4n-\kappa^2}{q^2}\right),$$

where $\kappa\equiv\kappa_{2\sqrt{n}}$ ranges over the positive integers $\leqq 2\sqrt{n}$ which are $\equiv \pm 2\sqrt{n} \pmod q$.

Gierster now finds for the reducible correspondence the number of infinite coincidences in the fundamental polygon. For the above particular case, this is

$$\sigma_\infty=z_\infty+(q^2-1)U_{\sqrt{n}},$$

where U_i denotes the sum of the divisors of n which are $<\sqrt{n}$ and $\equiv \pm i \pmod q$, provided that, if $\sqrt{n}$ is an integer $\equiv \pm i \pmod q$ then $\frac{1}{2}\sqrt{n}$ is to be added to the sum. He evaluated σ in many further cases.

---

146 M. Chasles, Comptes Rendus Paris, 58, 1864, 1775. A. Cayley, On the Correspondence of Two Points on a Curve, Proc. London Math. Soc., 1, 1865–6, Pt. VII; Coll. Math. Papers, VI, 9–13.

For q's such that[134] $p=0$, the σ's are evaluated also by the principle of Chasles.[146] And so for $q=3$ and 5, twelve exhaustive class-number relations are written such as (for our particular case above):

$$q=5,\quad \left(\frac{n}{5}\right)=1,\quad 60\Sigma H\left(\frac{4n-\kappa_{2\sqrt{n}}^2}{25}\right)=\Phi(n)-12U_{\sqrt{n}}.$$

J. Gierster[147] tabulated congruence sub-groups of prime grade q of the modular group and calculated their genus (Klein[134]) for $q\leqq 13$.

Gierster[148] continued his[145] investigation but now replaced his former invariant subgroup of grade q by any one not invariant. There the total number of coincidences in the correspondences was expressed as a sum σ of class-numbers. Here the analogues of the σ's are found to be mere linear combinations of the former σ's. Employing congruence groups of grade 7, 11, 13 and genus[134, 138] zero, he deduced class-number relations[149] including for example

$$4\Sigma H(4n-\kappa_s^2)=\Phi(n),\quad q=7,\quad s=4\sqrt{-n},\quad (n/7)=-1.$$

A. Berger[150] employed an odd prime p, integers m, n and put

$$U_n=(-1)^{[\sqrt{n}]}=4[\tfrac{1}{2}\sqrt{n}]-2[\sqrt{n}]+1,\qquad 0\leqq n<p^2,$$
$$S_m=\sum_{k=}^{p} U_{m+4(k-1)p},\qquad 0\leqq m<4p,$$

where $[x]$ denotes the largest integer $\leqq x$. Various expressions for S_m are found. For example, if $p\equiv 1 \pmod 4$,

$$S_m=\epsilon+2\sum_{k>(m-p)/4}^{k\leqq m/4}\left(\frac{k}{p}\right), \tag{1}$$

where $\epsilon=+1$ if $m\equiv 0 \pmod 4$, $\epsilon=-1$ if $m\equiv 1$, 2, or 3 $\pmod 4$. Write

$$L_r=\sum_{k>(r-1)p/8}^{k<rp/8}\left(\frac{k}{p}\right).$$

Let K_1 be the number of properly primitive classes of determinant $-p$, and K_2 that of determinant $-2p$. A study of L_r and Dirichlet's[23] formula (5) give

$$K_1=2(L_1+L_2),\qquad K_2=2(L_1-L_4),\qquad \text{if } p\equiv 1 \pmod 4;$$
$$L_1=L_8=\tfrac{1}{4}(K_1+K_2),\quad L_2=L_4=L_5=L_7=\tfrac{1}{4}(K_1-K_2),\quad \text{if } p\equiv 1 \pmod 8.$$

Whence, for $p\equiv 1 \pmod 8$, he found by (1) such relations as

$$S_0=1+K_1,\qquad S_p=-1+K_1,\qquad S_{2p}=-1-K_1,$$
$$S_{3p}=-1-K_1,\qquad S_{(p-1)/2}=1+K_1+K_2,\qquad S_{(3p-1)/2}=-1-K_1.$$

Similar relations are obtained for $p\equiv 3, 5, 7 \pmod 8$.

---

147 Math. Annalen, 22, 1883, 177–189.

148 *Ibid.*, 190–210.

149 Notations of Gierster[145] (2), or more fully in Math. Ann., 22, 1883, 43–50.

150 Nova Acta Reg. Soc. Sc. Upsaliensis, (3), 11, 1883, No. 7, 22 pp. For some details of the proof of (1), see Fortschritte Math., 14, 1882, 143, where the denotation of (1) is incorrectly given.

Berger wrote $Q(x)$ for the largest square $\leqq x$ and deduced eight theorems like the following: Among the p squares

$$(2)\qquad Q(0),\quad Q(4p),\quad Q(8p),\ \ldots,\ Q\{4(p-1)p\},$$

there are $\frac{1}{2}(p+1+K_1)$, $\frac{1}{2}(p+1)$, $\frac{1}{2}(p+1+K_1)$, or $\frac{1}{2}(p+1-2K_1)$ even numbers, according as $p\equiv 1, 3, 5$, or $7 \pmod 8$. Since K_1 and K_2 are positive, the squares (2) include at least $\frac{1}{2}(p+5)$, $\frac{1}{2}(p+1)$, $\frac{1}{2}(p+3)$, or $\frac{1}{2}(p-1)$ even numbers in the respective cases.

C. Hermite[151] communicated to Stieltjes and Kronecker the fact that if $F(D)$ denotes the number of uneven classes of determinant $-D$, then (cf. Hermite,[164] (2))

$$F(3)+F(7)+\ldots+F(4n-1)=\Sigma E\left(\frac{n-\nu^2}{2\nu+1}\right)+2\Sigma E\left(\frac{n-\nu^2-2\nu}{2\nu+3}\right)$$
$$+2\Sigma E\left(\frac{n-\nu^2-4\nu}{2\nu+5}\right)+\ldots+2\Sigma E\left(\frac{n-\nu^2-2k\nu}{2\nu+2k+1}\right);$$

in which $n-\nu^2-2k\nu \gtreqless 2\nu+2k+1$.

Hermite[152] stated Oct. 24, 1883, that if $F(N)$ denotes the number of properly primitive [he meant uneven] classes of determinant $-N$ and $\psi(n)=\Sigma(-1)^{(d-1)/2}$, where d ranges over all divisors $<\sqrt{n}$ of n, then

$$F(3)+F(11)+F(19)+\ldots+F(n)=\psi(3)+\psi(11)+\ldots+\psi(n)$$
$$+2\Sigma\psi(k)E(\tfrac{1}{4}\sqrt{n-k})+2\Sigma\psi(l)E(\tfrac{1}{4}\sqrt{n-l}+\tfrac{1}{2});$$

$k=3, 11, 19, \ldots, n$; $l=7, 15, 23, \ldots, n-4$.

T. J. Stieltjes[153] observed that this result is equivalent to

$$F(n)=\psi(n)+2\psi(n-4\cdot 1^2)+2\psi(n-4\cdot 2^2)+2\psi(n-4\cdot 3^2)+\ldots;$$

and this is equivalent to an earlier result of J. Liouville, Jour. de Math., (2), 7, 1862, 43–44. [For, by definition, Liouville's $\rho'(n)$ is Hermite's $\psi(n)$; see this History, Vol. II, Ch. VII, 265, note 33a.]

Stieltjes[154] let $F(n)$ denote generally the number of classes of determinant $-n$ with positive outer coefficients, but in case $n=8k+3$ with even forms excluded. Then he found, when $n\equiv 5 \pmod 8$, that $\frac{1}{2}F(n)$ is the number of solutions of $n=x^2+2y^2+2z^2$, x, y, z each >0 and uneven. Consequently setting $\phi(n)=\Sigma(2/d_1)d$, $dd_1=n$, he found that

$$F(n)+2F(n-8\cdot 1^2)+2F(n-8\cdot 2^2)+\ldots=\tfrac{1}{2}\phi(n),\quad n\equiv 3 \text{ or } 4 \pmod 8;$$
$$F(n-2\cdot 1^2)+F(n-2\cdot 3^2)+F(n-2\cdot 5^2)+\ldots=\tfrac{1}{4}\phi(n),\quad n\equiv 5 \text{ or } 7 \pmod 8.$$

On Nov. 15, 1883, Stieltjes[155] observed that the former of the last two theorems is a corollary to Gauss, Disq. Arith., Art. 292. For $i=1, 2, 3, 5$ or 6, he found that

$$\sum_{n=1}^{N} F(8n+i)=\frac{\pi}{48}N^{\frac{3}{2}},$$

[151] Aug., 1883, Correspondance d'Hermite et Stieltjes, Paris, I, 1905, 26.
[152] Correspondance d'Hermite et Stieltjes, Paris, I, 1905, 43.
[153] *Ibid.*, 45; Oct. 28, 1883.
[154] Correspondance d'Hermite et Stieltjes, Paris, 1905, I, 50–52, Nov. 12, 1883.
[155] *Ibid.*, 52–54.

asymptotically (cf. Gauss,[4] Disq. Arith., Art. 302, Mertens,[117] Gegenbauer,[199] Lipschitz[102]).

Stieltjes,[156] by the use of the two Kronecker[124] formulas,

$$4\sum_0^\infty F(4n+1)q^{n+\frac{1}{4}}=\theta_2(q)\theta_3^2(q),\qquad 8\sum_0^\infty F(8n+3)q^{2n+\frac{3}{4}}=\theta_2^3(q),$$

obtained the following three results: Let

$$\Phi(n)=\Sigma(2/d')d,\ dd'=n;\quad \Psi(n)=\Sigma(-2/d),$$

whence $2\Psi(n)$ is the total number of representations of n by x^2+2y^2; then

$$\left.\begin{array}{ll} n\equiv 1 \pmod 8, & \Sigma F(n-8r^2)=\frac{1}{2}\Phi(n)+\frac{1}{2}\Psi(n) \\ n\equiv 3,\ 5 \pmod 8, & \Sigma F(n-8r^2)=\frac{1}{2}\Phi(n)\end{array}\right\} r=0,\ \pm1,\ \pm2,\ \ldots$$

$$n\equiv 3,\ 5,\ 7 \pmod 8,\quad \Sigma F(n-2s^2)=\tfrac{1}{4}\Phi(n)+\tfrac{1}{4}\Psi(n),\quad (s=1,\ 3,\ 5,\ \ldots).$$

Stieltjes[157] stated that he had deduced Liouville's[110] class-number relation of 1869 and other similar formulas both by arithmetical methods and by the theory of elliptic functions. For example, for $N>0$,

$$2\Sigma(-1)^{\frac{1}{2}(s-1)}sF(4N-2s^2)=(-1)^{\frac{1}{2}N(N-1)}\Sigma(x^2-2y^2),\quad s=1,\ 3,\ 5,\ \ldots,$$

summed for all integral solutions of $x^2+2y^2=N$. This he[158] later proved in detail. For $N>0$,

$$\Sigma(-1)^{\frac{1}{2}(s-1)}sF(16N-3s^2)=\Sigma(x^2-3y^2),\quad s=1,\ 3,\ 5,\ \ldots,$$

summed for all integral solutions of different parity of $x^2+3y^2=N$. The method of verifying this formula was indicated[159] later.

Stieltjes[160] obtained from classic expansions the expansion

(1) $$\theta(q)\theta_2^4(q)\theta_3(q)=16\Sigma(x^2-y^2)q^{x^2+y^2},\quad x=1,\ 3,\ 5,\ 7,\ \ldots,\quad y=0,\ \pm2,\ \pm4,\ \ldots.$$

But

(2) $$\theta(q)\theta_2(q)\theta_3(q)=2(q^{1/4}+3q^{9/4}+5q^{25/4}+\ldots);$$

and (cf. Hermite,[69] (7))

(3) $$\theta_2^3(q)=8\sum_0^\infty F(8n+3)q^{\frac{1}{4}(8n+3)}.$$

A comparison of (1), (2), (3) gives at once a Liouville[110] class-number relation. Stieltjes added three new relations of the same type; e. g., for $N=8k+1$,

$$2\Sigma(-1)^{\frac{1}{2}(s-1)+\frac{1}{8}(s^2-1)}sF(2N-s^2)=\Sigma(-1)^y(x^2-8y^2),$$

summed for all integral solutions of $x^2+8y^2=N$ in which $x>0$ and uneven.

Stieltjes[161] stated, for the Kronecker[54] symbol $F(n)$, that

(1) $$F(np^{2k})=[p^k+p^{k-1}+\ldots-\left(\frac{-n}{p}\right)(p^{k-1}+p^{k-2}+\ldots)]F(n).$$

[156] Correspondance d'Hermite et Stieltjes, Paris, I, 1905, 54, Nov. 24, 1883.
[157] Comptes Rendus, Paris, 97, 1883, 1358–1359; Oeuvres, I, 1914, 324–5.
[158] Correspondance d'Hermite et Stieltjes, Paris, I, 1905, 63, Nov. 27, 1883.
[159] *Ibid.*, 69–70, Dec. 8, 1883.
[160] Comptes Rendus, Paris, 97, 1883, 1415–1418; Oeuvres, I, 1914, 326–8.
[161] Correspondance d' Hermite et Stieltjes, Paris, I, 1905, 81, 85–87, letter to Hermite. Jan. 6, 1884.

He gave[162] a proof depending on the fact that $F(n)=\rho\Sigma h(n/d)$, where d ranges over the odd square divisors of n; $\rho=\frac{1}{2}$ or 1, according as n is or is not an uneven square; $h(m)$ denotes the number of properly primitive classes of determinant $-m$.

Stieltjes[162] put

$$\psi(n)=\Sigma(-1)^{(d_1^2-1)/8}d \quad (dd_1=n); \quad \chi(n)=\Sigma x,$$

where x ranges over the solutions of $n=x^2-2y^2>0$, $x>0$, $|y|<\frac{1}{2}x$; and stated that, when n is odd,

$$2\Sigma(-1)^r F(n-2r^2)=(-1)^{(n-1)/2}\chi(n), \qquad r=0, \pm1, \pm2, \ldots;$$
$$2\Sigma F(n-2r^2)=2\psi(n)-\chi(n), \qquad r=0, \pm1, \pm2, \ldots.$$

These and two similar formulas he was unable to deduce by equating coefficients of powers of q in expansions. This was later done for formulas which include these as special cases by Petr,[258] Humbert,[293] and Mordell.[352]

C. Hermite[163] imparted to Stieltjes in advance the outline of the deduction of Hermite's[164] formula (1).

Hermite,[164] by the same study of the conditions on the coefficients of reduced forms as he employed[69] in 1861, found that

$$\theta_2^3(q)=24\Sigma[(N)+2f(N)]q^{N/4}-16\epsilon,$$

where (N) denotes the number of ambiguous, and $f(N)$ the number of unambiguous, even classes of determinant $-N$; while $\epsilon=1$ or 0, according as N is or is not the treble of a square. For the case $N\equiv3 \pmod 8$ a comparison of this with his earlier result[69] $\theta_2^3(q)=8\Sigma F(N)q^{N/4}$, where $F(N)$ is the number of uneven classes of $-N$, gives at once the ratio between the number of classes of the two primitive orders (cf. Gauss,[4] Disq. Arith., Art. 256, VI).

Kronecker's[124] formula (1) implies that

$$(1) \quad \frac{\theta_2^2(q)\theta_3(q)}{1-q}=4(\sum_0^\infty q^n)\sum_0^\infty F(4n+2)q^{n+\frac{1}{2}}=4\sum_n[F(2)+F(6)+\ldots+F(4n+2)]q^{n+\frac{1}{2}}.$$

But obviously

$$\theta_2^2(q)\equiv[2\sum_0^\infty q^{\frac{1}{4}(2n+1)^2}]^2=\sum_0^\infty f(8c+2)q^{2c+\frac{1}{2}},$$

where $f(n)$ denotes the number of solutions of $x^2+y^2=n$. Moreover,

$$\theta_3(q)=1+2\sum_1^\infty q^{n^2}.$$

Therefore, in the identity

$$\frac{\theta_2^2(q)\theta_3(q)}{1-q}=\frac{\theta_2^2(q)}{1-q}+2\frac{(\Sigma q^{n^2})\theta_2^2(q)}{1-q},$$

the first term of the right member is $\Sigma f(8c+2)q^{n+\frac{1}{2}}$, summed for $n=0, 1, 2, \ldots$;

[162] Correspondance d'Hermite et Stieltjes, Paris, I, 1905, 82–85, Jan. 15, 1884.
[163] *Ibid.*, I, 88–89, Feb. 28, 1884.
[164] Bull. de l'Acad. des Sc. St. Petersburg, 29, 1884, 325–352; Acta Math., 5, 1884–5, 297–330; Oeuvres, IV, 1917, 138–168.

$c=0, 1, 2, \ldots, [\frac{1}{2}n]$; the second term, by a lemma on the Legendre greatest-integer symbol, is

$$2\Sigma f(8c+2)\cdot[\sqrt{n-2c}]\cdot q^{n+\frac{1}{2}},$$

summed for $n=0, 1, 2, \ldots, c=0, 1, 2, \ldots, [\frac{1}{2}(n-1)]$. Hence a comparison with (1) gives

$$4[F(2)+F(6)+\ldots+F(4n+2)]= \overset{n}{\underset{c_1=0}{\Sigma}} f(8c_1+2)+2\Sigma f(8c+2)\cdot[\sqrt{n-2c}].$$

By the use of Jacobi's expansion formula:

$$\theta_2^2(q)=4\sqrt{q}\,\frac{1+q^2}{1-q^2}-4\sqrt{q^9}\,\frac{1+q^6}{1-q^6}+4\sqrt{q^{25}}\,\frac{1+q^{10}}{1-q^{10}}-\ldots,$$

Hermite found similarly other expressions for $F(2)+F(6)+\ldots+F(4n+2)$, such as

(1) $$\Sigma(-1)^{\frac{1}{2}(a-1)}+2\Sigma(-1)^{\frac{1}{2}(a-1)}\left[\frac{4n+2-a^2-b^2}{4a}\right],$$

where a and b range over all odd positive integers satisfying

$$4n+2-a^2-b^2 \gtreqless 0.$$

By means of two other formulas of Kronecker, Hermite evaluated similarly

$$F(1)+F(5)+\ldots+F(4n+1), \qquad F(3)+F(11)+\ldots+F(8n+3).$$

He announced without proof that

(2) $$F(3)+F(7)+\ldots+F(4n+3)=2\Sigma\left[\frac{n+1-c^2-2cc'}{2c+2c'+1}\right],$$

$c>0$, $c'>0$ and satisfying $(c+1)(2c+2c'+1)\leqq n+1$, counting half of each term in which $c'=0$.

T. J. Stieltjes[165] stated that by the theory of elliptic functions he obtained the theorem: If d range over the odd divisors of n and

$$\psi(n)=\Sigma(-1)^{\frac{1}{2}(d-1)+\frac{1}{8}(d^2-1)}=\Sigma\left(\frac{-2}{d}\right), \quad \psi(0)=\tfrac{1}{2},$$

then, for $n\equiv 2 \pmod 4$, in Kronecker's[54] notation,

$$F(n)=\tfrac{1}{2}\Sigma\psi(n-2r^2)=\Sigma\psi(n-8r^2), \quad r=0, \pm1, \pm2, \ldots.$$

Thence he verified his[161] earlier theorem (1) for the cases $n=k^2$ and $n=2k^2$ by the method used by Hurwitz in finding the number of decompositions of a square into the sum of five squares (see this History, Vol. II, 311).

A. Berger,[166] to evaluate Dirichlet's[14] series (2), namely,

$$V=\overset{\infty}{\underset{k=1}{\Sigma}}\left(\frac{\Delta}{k}\right)\frac{1}{k},$$

[165] Comptes Rendus, Paris, 98, 1884, 663–664; Oeuvres, I, 1914, 360–1.
[166] Nova Acta Regiae Soc. Sc. Upsaliensis, (3), 12, 1884–5, No. 7, 31 pp.

Δ being a fundamental discriminant, started from Kronecker's[171] identity (4a) in the form

$$\sum_{h=1}^{\epsilon\Delta-1}\left(\frac{\Delta}{h}\right)e^{2hk\pi i/(\epsilon\Delta)}=(\sqrt{\Delta})\left(\frac{\Delta}{k}\right),$$

where ϵ is the sign of Δ, and $k>0$. By separating the real from the imaginary and by a study of quadratic residues and non-residues, he obtained

$$(1)\qquad \Delta<0,\ \sum_{h=1}^{-\Delta/2}\left(\frac{\Delta}{h}\right)\sin\frac{2hk\pi}{-\Delta}=-\tfrac{1}{2}\sqrt{-\Delta}\cdot\left(\frac{\Delta}{k}\right),\quad k>0.$$

Since (cf. Dirichlet[14])

$$(2)\qquad \sum_{n=1}^{\infty}\frac{\sin nu}{n}=\frac{\pi-u}{2},\quad 0<u<2\pi,$$

we get, by dividing (1) by k and summing, Dirichlet's[23] formula (6) for $\Delta<0$.

Similarly by the use of the identity

$$-\log\left(2\sin\frac{u}{2}\right)=\sum_{n=1}^{\infty}\frac{\cos nu}{n},$$

Berger obtained Dirichlet's[23] closed formula (8), for $\Delta>0$.

To obtain Dirichlet's[23] second closed form, Berger took, for $\Delta<0$ (cf. Dirichlet, Zahlentheorie, § 89, ed. 4, p. 224)

$$V=\frac{1}{r}\Pi\left[1-\left(\frac{\Delta}{p}\right)\frac{1}{p}\right]^{-1}=\frac{1}{r}\sum_{k=1}^{\infty}\left(\frac{\Delta}{2k-1}\right)\frac{1}{2k-1},$$

where[171] $r=1-\frac{1}{2}(\Delta/2)$, and p ranges over all odd positive primes. By means of (1), this becomes

$$r\sum_{k=1}^{\infty}\left(\frac{\Delta}{k}\right)\frac{1}{k}=\frac{-2}{\sqrt{-\Delta}}\sum_{h=1}^{h<-\Delta/2}\left(\frac{\Delta}{h}\right)\sum_{k=1}^{\infty}\frac{1}{2k-1}\sin 2h(2k-1)\pi/\Delta.$$

But (2) implies that the final factor is $\pi/4$. Hence we get Dirichlet's[23] classic formula (5). By parallel procedure, Berger obtained, for $\Delta>0$,

$$(3)\qquad V=\frac{2}{\left\{2-\left(\frac{\Delta}{2}\right)\right\}\sqrt{\Delta}}\sum_{h=1}^{h<\Delta/2}\left(\frac{\Delta}{h}\right)\log\cot\frac{h\pi}{\Delta}.$$

Cf. Dirichlet,[23] (8).

A. Hurwitz[167] gave without proof[168] thirteen class-number relations of the 11th grade which he had deduced by the method which he had used to obtain relations of the 7th grade.[169]

For example,

$$6\sum_{k}H(4n-\kappa^2)=\Phi(n)+\psi_1(n)+\psi_2(n)-\psi_3(n),\quad \left(\frac{n}{11}\right)=1,$$

where κ ranges over all positive integers whose square is $\equiv n \pmod{11}$; while

---

[167] Berichte Sächs. Gesells., Math-Phys. Classe, 36, 1884, 193–197.
[168] For proof, see F. Klein and R. Fricke, Vorlesungen über Elliptischen Functionen,[217] Leipzig, II, 1892, 663–664.
[169] Math. Annalen,[184] 25, 1885, 157–196.

$\psi_1(n)=\frac{1}{2}\Sigma x$, where x ranges over those solutions of $4n=x^2+11y^2$ in which x and y are not 0, and $(x/11)=-1$;

$$\psi_2(n)=\tfrac{1}{8}[5Z(n)-12\Phi(n)],\qquad \psi_3(n)=\tfrac{1}{8}[3Z_0(n)-Z(n)],$$

in which $Z(n)$ denotes the number of solutions of $4n=x^2+11y^2+z^2+11u^2$ for which $x+y$ is even; $Z_0(n)$, the number for which one of $x, z, x-z, x+z$ is divisible by 11.

By eliminating ψ_2 and ψ_3 from his set, Hurwitz obtained a new set which he showed to include J. Gierster's[170] class-number relations of grade 11.

L. Kronecker,[171] unlike Gauss, studied quadratic forms $ax^2+bxy+cy^2$ in which b may be even or uneven. He defined primitive forms as those in which a, b, c have no common factor. He denoted by $K(D)$ the number of primitive classes of discriminant $D=b^2-4ac$. He put

$$H(D)=\sum_{h=1}^{\infty}\left(\frac{D}{h}\right)\frac{1}{h},\qquad \left(\frac{D}{h}\right)=\left(\frac{2^g}{D}\right)\left(\frac{D}{h'}\right),$$

if $h=2^g h'$, h' uneven, in which the symbols of the last right member are the Jacobi-Legendre signs.

Dirichlet's[20] fundamental formula (2) is specialized as follows:

$$(1)\qquad \tau\sum_{h,k}\left(\frac{Q^2}{h}\right)\left(\frac{D}{k}\right)F(hk)=\sum_{a,b,c}\ \sum_{m,n}\left(\frac{Q^2}{m}\right)F(am^2+bmn+cn^2),$$

where h, k range over all positive integers; m, n over all integers not both zero; a, b, c over the coefficients of a system of representative forms (a, b, c) of the primitive classes of the discriminant $D=D_0\cdot Q^2$ (D_0 being fundamental); $a>0$ is relatively prime to Q; and b and c are divisible by all the prime divisors of Q; $F(x)$ is any function for which the series in each member is convergent.

By Dirichlet's methods (Zahlentheorie, Arts. 93–98) are obtained the following results:

$$(2)\qquad \tau H(D)=\frac{2\pi}{\sqrt{-D}}K(D),\quad D<0;\qquad H(D)=\frac{K(D)}{2\sqrt{D}}\log\frac{T+U\sqrt{D}}{T-U\sqrt{D}},\quad D>0.$$

These are combined into one formula

$$H(D)=K(D)\int_{T/U}^{\infty}\frac{dz}{z^2-D},$$

where T, U denote that fundamental solution of $T^2-DU^2=1$ or 4 for which T/U is the greater. This is equivalent to

$$(3)\qquad H(D)=\frac{K(D)}{\sqrt{D}}\log E(D),\qquad E(D)=\frac{1}{r}(T+U\sqrt{D}),\ r=1\text{ or }2.$$

But (cf. Dirichlet, Zahlentheorie, Art. 100),

$$H(D)=H(D_0)\Pi\left(1-\left(\frac{D_0}{q}\right)\frac{1}{q}\right),$$

[170] Math. Annalen,[148] 22, 1883, 203–206.
[171] Sitzungsber. Akad. Wiss. Berlin, 1885, II, 768–780.

q ranging over the prime divisors of Q. Hence,

$$(4)\qquad \frac{K(D)}{K(D_0)}=Q\Pi\left\{1-\left(\frac{D_0}{q}\right)\frac{1}{q}\right\}\frac{\log E(D_0)}{\log E(D)}.$$

In the light of the identity (p. 780)

$$(4a)\qquad \left(\frac{D_0}{r}\right)=\frac{1}{\sqrt{D_0}}\sum_k\left(\frac{D_0}{k}\right)e^{2rk\pi i/|D_0|},\quad k=1,3,5,\ldots,2|D_0|-1;\ r>0,$$

(2) implies

$$(5)\qquad \begin{cases} K(D_0)=\dfrac{\tau}{2D_0}\displaystyle\sum_{k=1}^{-D_0+1}\left(\frac{D_0}{k}\right)k, & D_0<0,\\[2ex] K(D_0)\log E(D_0)=-\displaystyle\sum_{k=1}^{D_0-1}\left(\frac{D_0}{k}\right)\log(1-e^{2k\pi i/D_0}), & D_0>0.\end{cases}$$

H. Weber[172] and J. de Séguier[173] have modified the above identity (4a) so as to be true also for $D_0\equiv 0 \pmod 4$, which is not the case in Kronecker's form of it. De Séguier has given the deduction in full of (5) and has shown that (5_2) holds also for $D_0<0$. Dirichlet[174] at this point needed to treat eight cases instead of Kronecker's two and de Séguier's one.

Kronecker[175] had defined the function $\theta(\zeta,\omega)$ by

$$\theta(\zeta,\omega)=\sum_\nu e^{\frac{1}{4}(\nu^2\omega+4\nu\zeta-2\nu)\pi i},\quad \nu=\pm1,\ \pm3,\ \pm5,\ldots,$$

and the function Λ by

$$\Lambda(\sigma,\tau,\omega_1,\omega_2)=(4\pi^2)^{\frac{1}{3}}e^{\tau^2(\omega_1+\omega_2)\pi i}\cdot\frac{\theta(\sigma+\tau\omega_1,\omega_1)\theta(\sigma-\tau\omega_2,\omega_2)}{[\theta'(0,\omega_1)\theta'(0,\omega_2)]^{\frac{1}{3}}},$$

in which σ, τ are arbitrary complex numbers; ω_1, ω_2 are any complex numbers such that $\omega_1 i$ and $\omega_2 i$ have negative real parts. He[176] found that if ω_1 and $-\omega_2$ are the roots of $a+bw+cw^2=0$, where $b^2-4ac=-\Delta$ is a negative discriminant, then

$$(6)\qquad \log\Lambda(\sigma,\tau,\omega_1,\omega_2)=\frac{-\sqrt{\Delta}}{2\pi}\lim_{\rho=0}\sum_{m,n}\frac{e^{2(m\sigma+n\tau)\pi i}}{(am^2+bmn+cn^2)^{1+\rho}}$$

and therefore Λ is a class invariant. Relation (6) was afterward developed by Kronecker[177] into what J. de Séguier[178] has called Kronecker's second fundamental formula.

For D_1, D_2 two arbitrary conjugate divisors of $D=D_1\cdot D_2=D_0\cdot Q^2$ (1) is found to imply what J. de Séguier[179] has called Kronecker's first fundamental formula, namely,[180]

$$\tau\sum_{h=1}^{\infty}\sum_{k=1}^{\infty}\left(\frac{D_1Q^2}{h}\right)\left(\frac{D_2Q^2}{k}\right)F(hk)$$
$$=\tfrac{1}{2}\sum_{a,b,c}\left[\left(\frac{D_1}{A}\right)+\left(\frac{D_2}{A}\right)\right]\sum_{m,n}\left(\frac{Q^2}{m}\right)F(am^2+bmn+cn^2),$$

172 Götting. Nachr., 1893, 51–52.
173 Formes quadratiques et multiplication complexe,[226] Berlin, 1894, 32.
174 Zahlentheorie, Art. 105, ed. 4, 1894, 274–5.
175 Sitzungsber. Akad. Wiss. Berlin, 1883, I, 497–498.
176 *Ibid.*, 528.
177 Sitzungsber. Akad. Wiss. Berlin, 1889, I, 134, formula (16); 205, formula (18).[213]
178 Formes quadratiques et multiplication complexe, 1894, 218, formula (3).[226]
179 *Ibid.*, 133, formula (6).[226]
180 L. Kronecker, Sitzungsber. Akad. Wiss. Berlin, 1885, II, 779.

with ranges of summation as in (1), while $2am/n+b \gtreqless U/T$ and $n>0$, if D is >0; A is an arbitrary number relatively prime to $2D$ and representable by (a, b, c). An elegant demonstration has been given by H. Weber.[181]

Take $Q=1, D_1<0, D_2>0, F(x)=x^{-1-\rho}$. When (6) is applied to the right member, the result, when $\rho=0$, is

$$(7) \qquad \frac{\tau\sqrt{\Delta}}{2\pi}H(D_1)H(D_2)=\sum_{a,b,c}\left(\frac{D_1}{a}\right)\log c[\theta'(0,\omega_1)\theta'(0,\omega_2)]^{-\frac{2}{3}}, \quad \Delta=-D.$$

This formula refers the problem of the class-number of a positive discriminant to that of a negative discriminant. For the purposes of calculation, this formula has been improved by J. de Séguier.[182]

L. Kronecker[183] considered solutions (U, V) of $U^2+DV^2=4p$, where $p\equiv 1 \pmod{D}$, D a prime$=4n+3>0$. If $x^p=1$, $\alpha^D=1$, $x\neq 1$, $\alpha\neq 1$, and g is a primitive root of p, then

$$\prod_a(x+\alpha^a x^g+\alpha^{2a}x^{g^2}+\ldots+\alpha^{(p-2)a}x^{g^{p-2}})=\tfrac{1}{2}u+\tfrac{1}{2}v\sqrt{-D},$$

where a ranges over the incongruent quadratic residues of D, and u and v are integers. Whence finally he stated that U and V are determined from

$$\frac{u+v\sqrt{-D}}{u-v\sqrt{-D}}=\left(\frac{U+V\sqrt{-D}}{U-V\sqrt{-D}}\right)^{\frac{\Sigma b-\Sigma a}{D}},$$

Cf. Dirichlet's[23] formula (6).

A. Hurwitz[184] stated that his[185] modular equations of the 8th grade[134] yield those class-number relations which L. Kronecker[124] had given in Monatsber., Berlin, 1875, 230–233. He modified Gierster's[145] deduction of the class-number relation of the first grade by showing that a modular function $f(J, J')$ has as many poles as zeros in the fundamental polygon.

For genus[138] $p>0$, Hurwitz employed a system of normalized integrals $j_1(\omega)$, $j_2(\omega), \ldots, j_p(\omega)$ of the first kind on the Riemann surface formed from the fundamental polygon for the largest invariant sub-group of grade q. For arbitrary constants e_r the θ functions[186] of j_r have the property

$$\theta[j_r(T(\omega))-e_r]=\theta[j_r(\omega)-e_r]e^k, \qquad k\equiv\sum_{r=1}^{p}2t_r(j_r(\omega)-e_r)+C_t,$$

where T is an arbitrary unit substitution $\equiv \binom{1\ 0}{0\ 1} \pmod{q}$; while $t_1, t_2, t_3, \ldots, t_p, C_t$ depend only on T. Constants c_r are so chosen that

$$\theta[j_r(\omega)-j_r(\Omega)-c_r]=\theta[j_r(\Omega)-j_r(\omega)+c_r],$$

and $\theta=0$ when and only when the zero regarded as a value of ω (and Ω) is relatively

---

181 Reproduced by de Séguier, Formes quadratiques, 332–334.
182 Formes quadratiques et multiplication complexe, Berlin, 1894, 314, (25).
183 Göttingen Gelehrte Anzeigen; Nachrichten Konigl. Gesells. Wiss., 1885, 368–370, letter to Dirichlet.
184 Math. Annalen, 25, 1885, 157–196.
185 Göttingen Nachr., 1883, 350.
186 Cf. B. Riemann: Jour. für Math., 65, 1866, 120; Werke, 1892, 105; Oeuvres, 1898, Mém. XI. 207; C. Neumann, Theorie der Abel'schen Integrale, Leipzig, 1884, Chaps. XII, XIII.

equivalent to $\Omega, \omega_1, \omega_2, \ldots, \omega_{p-1}$ (and $\omega, \omega_p, \omega_{p+1}, \ldots, \omega_{2p-2}$), where $\omega_1, \omega_2, \ldots, \omega_{2p-2}$ are constants chosen almost[187] arbitrarily; moreover, that zero is of the first order.

The transformations $R_1(\omega)$, $R_2(\omega)$, ... are a system of representative substitutions[188] of order n and are

$$\Omega \equiv \frac{a\omega+b}{c\omega+d} \pmod q,$$

where a, b, c, d are fixed for all R's.

Consider the function

$$\Phi(\omega) = \prod_i \theta[j_r(\omega) - j_r(R_i(\omega)) - c_r],$$

where if n is a square, we omit the representative

$$\equiv \frac{\sqrt{n}\omega}{\sqrt{n}} \pmod q,$$

which is relatively[134] equivalent to ω. Aside from the zero values which are due to the choice of $\omega_1, \omega_2, \ldots, \omega_{2p-2}$, and aside from the rational points ω, the theory of the zeros[189] of a θ-function shows that, since $\Phi(\omega)$ is reproduced except for a finite exponential factor under the substitution $T(\omega)$, $\Phi(\omega)$ vanishes in the fundamental polygon as many times as there are identities

$$\omega_0 = \frac{a'\omega_0+b'}{c'\omega_0+d'}, \qquad a'd'-b'c'=n, \qquad \begin{pmatrix} a'b' \\ c'd' \end{pmatrix} \equiv \begin{pmatrix} ab \\ cd \end{pmatrix} \pmod q.$$

From this point Hurwitz treats the θ-functions as Gierster[145] had treated the factors $\eta(\omega)-\eta(\omega')$ of the modular equation and his determination of Gierster's σ differs only in details from Gierster's determination.

To complete Gierster's nine class-number relations[190] of the 7th grade for $n \not\equiv 0 \pmod 7$ and without recourse to non-invariant subgroups, Hurwitz, after F. Klein,[191] put

$$z_1(\omega) = \Sigma(-1)^\nu q^{\frac{1}{28}[7(2\nu+1)+8]^2}, \qquad z_2(\omega) = \Sigma(-1)^\nu q^{\frac{1}{28}[7(2\nu+1)+1]^2},$$
$$z_4(\omega) = \Sigma(-1)^\nu q^{\frac{1}{28}[7(2\nu+1)+2]^2}.$$

Three normalized integrals of the first kind and of grade 7 are

$$I_r(\omega) = -\frac{1}{7}\int_0^q z_r \theta_1'(0|q)\frac{dq}{q} = \Sigma \frac{\psi_r(m)}{m} q^{2m/7}, \quad r=1, 2, 4;$$

summed for values of $m \equiv r \pmod 7$, where necessarily $\psi_r(m) = \frac{1}{2}\Sigma\alpha$, the summation extending over all positive and negative integer solutions α, β of $4m = \alpha^2 + 7\beta^2$, $m \equiv r \pmod 7$, $(\alpha/7)=1$. Now $I_r(\omega)$ has the property

$$\sum_{i=1}^{\Phi(n)} I_r(R_i(\omega)) = \text{const., or } \psi(n) I_r(S(\omega)) + \text{const.},$$

according as $(n/7) = -1$ or $+1$, while

$$S(\omega) \equiv \frac{a\omega+b}{c\omega+d} \pmod 7,$$

---

[187] Cf. H. Poincaré and E. Picard, Comptes Rendus, Paris, 97, 1883, 1284.
[188] F. Klein, Math. Annalen, 14, 1879, 161.
[189] B. Riemann, Jour. für Math., 65, 1866, 161–172; Werke, 1892, 212–224.
[190] Math. Annalen, 17, 1880, 82; 22, 1883, 201–202.
[191] *Ibid.*, 17, 1880, 569.

and $\psi(n)=\frac{1}{2}\Sigma a$, the summation extending over all positive and negative integer solutions a, β of $4n=a^2+7\beta^2$, $(a/7)=1$. Let this property of the integrals I_r be possessed by the integrals j_1, j_2, j_4. Hurwitz put

$$\Phi(\omega', \omega)=\prod_{i=1}^{\Phi(n)} \theta[j_r(\omega')-j_r(R_i(\omega))-c_r],$$

$$\Psi(\omega', \omega)=\begin{cases}\theta[j_r(\omega')-j_r(S(\omega))-c_r]^{-\psi(n)}, & \text{if } (n/7)=1,\\ 1 & \text{, if } (n/7)=-1,\end{cases}$$

$$\psi(\omega', \omega)=\Phi(\omega', \omega)\cdot\Psi(\omega', \omega), \qquad F(\omega', \omega)=\frac{\psi(\omega', \omega)}{\psi(\omega_0', \omega)\cdot\psi(\omega', \omega_0)};$$

where ω_0', ω_0 are arbitrary fixed values of ω with positive imaginary parts. Then $F(\omega', \omega)$ is invariant under $T(\omega')$ and hence as a function of ω and of ω' is an algebraic function belongs to the Riemann surface of the 7th grade. $F(\omega', \omega)=0$ expresses algebraically the modular correspondence[192] of grade q and order n.

$F(\omega', \omega)$ is an algebraic function which belongs to the surface and has as many zeros as poles in the fundamental polygon. Hence

$$(1) \qquad \sigma-k\cdot\psi(n)=2\Phi(n)-2\psi(n) \text{ if } \begin{pmatrix}a & b\\ c & d\end{pmatrix}\neq\begin{pmatrix}\sqrt{n} & 0\\ 0 & \sqrt{n}\end{pmatrix},$$

where k is the number of zeros of $\theta[j_r(\omega)-j_r(S(\omega))-c_r]$ in the fundamental polygon, and σ has the value given by Gierster.[145]

Similarly

$$(2) \qquad \sigma=2\Phi(n)-6\psi(n)+\eta, \text{ if } \begin{pmatrix}a & b\\ c & d\end{pmatrix}=\begin{pmatrix}\sqrt{n} & 0\\ 0 & \sqrt{n}\end{pmatrix},$$

where $\eta=4$ or 0 according as n is or is not a square.

From (1) and (2) and the relation[145]

$$2(U_\rho+U_{2\rho}+U_{4\rho})=\Phi(n)-\psi(n), \qquad \rho=\sqrt{n\left(\frac{n}{7}\right)},$$

Gierster's[193] class-number relations of grade 7 follow at once; for, Gierster's[139] $\xi(n)$ is Hurwitz's $-2\psi(n)$.

A. Hurwitz[194] generalized completely his[184] deduction of the class-number relations of grade 7 to grade q, where q is a prime >5; and showed that the right member of these relations is $2\Phi(n)$ plus a simple linear combination of coefficients $\psi(n)$ which occur in an expansion of Abelian integrals of the first kind and of grade q. That is, if $\sigma(n)$ be determined in terms of class-number as by Gierster[145] and Hurwitz,[184]

$$\sigma(n)-2\Phi(n)-\eta=h_1\psi_1(n)+h_2\psi_2(n)+\ldots+h_\mu\psi_\mu(n),$$

where $\eta=2(p-1)$ or 0 according as n is or is not a square; and $h_1, h_2, \ldots, h_\mu$ are independent of n. Klein and Fricke[217] have since shown for $q=7$, 11, how the h's may be simply evaluated when the ψ's are known.

---

192 Cf. A. Hurwitz, Göttingen Nachr., 1883, 359.
193 Math. Annalen, 22, 1883, 199–203 (Gierster 148).
194 Berichte Königl. Sächs. Gesells., Leipzig, 37, 1885, 222–240.

E. Pfeiffer[195] wrote $H(n)$ for the number of classes of forms of negative determinant $-n$, and sharpened Merten's[117] asymptotic expression for the sum $\Sigma H(n)$ to the equivalent of

$$\sum_{n=1}^{x} H(n)=\frac{2}{9}\pi x^{\frac{3}{2}}-\frac{x}{2}+O(x^{\frac{5}{6}+\epsilon}),$$

where the order[117] only of the last term is indicated and ϵ is a small positive quantity. Pfeiffer, in a discussion which lacks rigor, indicated a method of proof (see Landau[330] and Hermite[204]).

L. Gegenbauer[196] denoted by $f(n)$ the number of representations of n as the sum of two squares, and deduced from four of Kronecker's formulas like[124] (1) four formulas similar to and including the following:

$$12\sum_{x=1}^{n} E(x)=f_2(n)+2\sum_{x=1}^{[\sqrt{n}]} f_2(n-x^2),$$

where[54]

$$E(n)=2F(n)-G(n),\quad f_2(r)=\sum_{x=1}^{r} f(x)=\sum_{x=0}^{[\sqrt{r}]}[\sqrt{r-x^2}].$$

His earlier result[197]

$$\sum_{x=0}^{[\sqrt{m/a}]}[\sqrt{m-ax^2}]=\frac{\pi m}{4\sqrt{a}}+O(\sqrt{m})$$

transforms this into

$$\sum_{x=1}^{n} E(x)=\tfrac{1}{9}\pi n^{3/2}+O(n).$$

(For the notation O, see F. Mertens.[117]) The other analogous results are

$$\lim_{n=\infty}\sum_{x=0}^{n} F(4x+a)/n^{3/2}=\tfrac{1}{3}\pi,\qquad \lim_{n=\infty}\sum_{x=0}^{n} F(8x+3)/n^{3/2}=\tfrac{1}{3}\pi\sqrt{2},$$

where $a=1$ or 2. Hence the asymptotic median number in the three cases is $\frac{1}{2}\pi\sqrt{n}$, $\frac{1}{2}\pi\sqrt{n}$, $\pi\sqrt{n/2}$. These four results combined with those of Gauss[198] and Mertens[117] give the asymptotic median number of odd classes as

$$\pi\sqrt{n}\left\{\frac{1}{12}+\frac{1}{7\zeta(3)}\right\},\qquad \zeta(3)\equiv 1+\frac{1}{2^3}+\frac{1}{3^3}+\frac{1}{4^3}+\cdots.$$

Gegenbauer[199] derived from four of Kronecker's[200] and four of Hurwitz's[202] formulas, twelve class-number relations with more elegance than he[196] or Hermite[164] had derived three of the same formulas. For example, from the following formula of Hurwitz,[202]

$$4\sum_{n=0}^{\infty} F(8n+1)q^{2n+\frac{1}{4}}=\theta_2(q)\theta_3^2(q^2),$$

---

195 Jahresbericht der Pfeiffer'schen Lehr-und Erziehungs-Anstalt, Jena, 1885-1886, 1–21.
196 Sitzungsber. Akad. Wiss. Wien, Math-Natur., 92, II, 1885, 1307–1316.
197 *Ibid.*, 384.
198 Disq. Arith.,[4] Art. 302; Werke, II, 1876, 284.
199 Sitzungsber. Akad. Wiss. Wien., Math.-Natur., 93, II, 1886, 54–61.
200 Monatsber. Akad. Wiss. Berlin,[124] 1875, 229.

it follows that $4F(8n+1)$ is the number of integral (positive, negative or zero) solutions of

$$8n+1=8x^2+8y^2+(2z-1)^2.$$

Put $x^2+y^2=k$ and solve for z. For a fixed k, the number of integer values of z as n ranges from 1 to N is therefore

$$[\tfrac{1}{2}\sqrt{8N+1-8k}+\tfrac{1}{2}].$$

Hence

$$2\sum_{x=0}^{n} F(8x+1)=\sum_{x=0}^{n} f(x)[\tfrac{1}{2}\sqrt{8n+1-8x}+\tfrac{1}{2}],$$

where $f(x)$ denotes the number of representations of x as the sum of two squares. The symbol $f(x)$ is decomposed so that the last formula becomes

$$\sum_{x=0}^{n} F(8x+1)=\tfrac{1}{3}\pi\sqrt{2}n^{3/2}+O(n),$$

with O as in Mertens.[117] As in the previous case,[196] Gegenbauer now finds that the asymptotic median number of odd classes of the determinant $-(8n+i)$, $i=1, 2, 3, 5$, or 6 is $\pi\sqrt{n/2}$.

Gegenbauer[201] without giving proofs supplemented his earlier list[199] of 12 class-number relations with 20 others which are easily deduced by processes analogous to those used before[199] and which include the following three types:

$$\sum_{x=0}^{n} F(8x+3)[\tfrac{1}{2}\sqrt{8n+1-8x}+\tfrac{1}{2}]=\sum_{x=0}^{n}\psi_1(2x+1),$$

in which [presumably] $\psi_1(n)$ denotes the number of representations of $4n$ as the sum of four uneven squares, where the order of terms is regarded, but $(-a)^2$ is regarded as the same as $(+a)^2$.

$$\sum_{x=0}^{n} F(16x+14)=2\sum_{x=0}^{n}\rho(8x+5)[\tfrac{1}{2}\sqrt{4n+1-4x}+\tfrac{1}{2}],$$

$\rho(m)=\Sigma(-2/d_1)$, d_1 ranging over the odd divisors of n.

$$\sum_{x=0}^{n} F(8x+6)=2\sum_{y}(-1)^{y-1}+4\sum_{y,z}(-1)^{y-1}\left[\frac{2(n-2y^2+2y)-z^2+z}{8y-4}\right],$$

$y\gtreqless 1$; $z\gtreqless 1$; $2n-4y^2+4y-z^2+z\gtreqless 0$.

A. Hurwitz[202] employed four formulas of Kronecker[203] all of the same type and including

(1) $$4\Sigma F(4n+2)q^n=q^{-\frac{1}{2}}\theta_2^2(q)\theta_3(q),$$

(2) $$4\Sigma F(4n+1)q^n=\tfrac{1}{2}q^{-\frac{1}{4}}\theta_2(q^{\frac{1}{4}})\cdot\theta_3^2(q).$$

He enlarged the list to 12 such formulas by simple methods, for example by replacing q by $-q$ in (1), adding the result to (1), and then using the relation

$$\theta_2^2(q)=2\theta_2(q^2)\theta_3(q^2).$$

---

201 Sitzungsber. Akad. Wiss. Wien, Math-Natur., 93, II, 1886, 288–290.
202 Jour. für. Math., 99, 1886, 165–168; letter to Kronecker, 1885.
203 Monatsber. Akad. Wiss. Berlin,[124] 1875, 229–230.

The result in this case is

$$(5)\qquad 2\Sigma F(8n+2)q^n = q^{\frac{1}{4}}\theta_2(q)\theta_3(q)\theta_3(q^2).$$

Seven class-number relations are obtained similarly to the following. We multiply (2) by $\theta_2(iq^{\frac{1}{2}})$. The relation

$$\theta_2(iq)\theta_3^2(q^2) = \theta_1'(iq)$$

now gives

$$4\theta_2(iq^{\frac{1}{2}})\cdot\Sigma F(4n+1)\cdot q^{n+\frac{1}{4}} = \theta_2(q)\theta_1'(iq^{\frac{1}{2}});$$

and the equating of coefficients here gives

$$\sum_h (-1)^{(h^2-1)/8} F\left(\frac{m-h^2}{2}\right) = \tfrac{1}{4}\Omega_2(m), \qquad m\equiv 3 \pmod 8)$$

in which h is uneven and positive, $\Omega_2(m)=\Sigma(-2/\nu)\nu$, where ν ranges over all positive uneven numbers satisfying $m=\nu^2+2n^2$.

C. Hermite[204] represented the totality of reduced unambiguous quadratic forms of negative determinant and positive middle coefficient by $(2s+r,\ s,\ 2s+r+t)$, $r, s, t=1, 2, 3, \ldots$. Hence in

$$S = 2\Sigma q^{(2s+r)(2s+r+t)-s^2}$$

the coefficient of q^N is the number of unambiguous classes of determinant $-N$. And if we put $n=2s+r$, we get

$$S = 2\Sigma\frac{q^{n^2+n-s^2}}{1-q^n}, \qquad n=3, 4, 5, \ldots;\quad s=1, 2, 3, \ldots, \left[\frac{n-1}{2}\right].$$

The number of ambiguous forms $(A,\ 0,\ C)$, $A\leqq C$, of determinant $-N$ is the number of factorizations $N=n(n+i)$, where n is a positive integer and $i \geqq 0$. This implies that the number of ambiguous forms of this type is the coefficient of q^N in the doubly infinite sum

$$S_1 = \sum_{n,i} q^{n(n+i)} = \Sigma\frac{q^{n^2}}{1-q^n}.$$

Similarly the number of ambiguous reduced forms of the type $(2B,\ B,\ C)$ and $(A,\ B,\ A)$ of determinant $-N$ is the coefficient of q^N in the expansion of

$$S_2 = \Sigma\frac{q^{n^2+2n}}{1-q^{2n}}, \qquad n=1, 2, 3, \ldots.$$

This gives[205]

$$S_1+S_2 = \Sigma\frac{q^m}{1-q^{2m}} + \Sigma\frac{q^{2n}}{1-q^{2n}}, \qquad n=1, 2, 3, \ldots;\quad m=1, 3, 5, 7, \ldots.$$

Hence, if $H(n)$ denotes the number of classes of determinant $-n$,

$$\Sigma H(n)q^n = \Sigma\frac{q^{n^2}}{1-q^n} + \Sigma\frac{q^{n^2+2n}}{1-q^{2n}} + 2\Sigma\frac{q^{n^2+n-s^2}}{1-q^n}.$$

[204] Bull. des sc. math., 10, I, 1886, 23–30; Oeuvres, IV, 1917, 215–222.

[205] Cf. C. G. J. Jacobi, Fundamenta Nova, 1829, Art. 65, p. 187; Werke, I, 1881, 239 (transformation of C. Clausen).

We divide each member by $1-q$ and expand according to increasing powers of q. Then the coefficient of q^N in the left member is $U=H(1)+H(2)+\ldots+H(N)$. By the use of the identity[206]

$$\text{(1)}\qquad \frac{q^b}{(1-q)(1-q^a)}=\Sigma E\left(\frac{N+a-b}{a}\right)q^N,$$

the coefficient of q^N in the second member becomes

$$U=\Sigma E\left(\frac{N+n-n^2}{n}\right)+\Sigma E\left(\frac{N-n^2}{2n}\right)+2\Sigma E\left(\frac{N+s^2-n^2}{n}\right).$$

Neglecting quantities of the order of $E(\sqrt{N})=\nu$, we get

$$\Sigma E\left(\frac{N+n-n^2}{n}\right)=\Sigma\frac{N+n-n^2}{n}=N\left(1+\tfrac{1}{2}+\tfrac{1}{3}+\ldots+\frac{1}{\nu}\right)-\frac{\nu^2-\nu}{2}$$
$$=N(\tfrac{1}{2}\log N+C)-\tfrac{1}{2}N;$$

$$\Sigma E\left(\frac{N-n^2}{2n}\right)=\Sigma\frac{N-n^2}{2n}=\frac{N}{2}\left(1+\tfrac{1}{2}+\tfrac{1}{3}+\ldots+\frac{1}{\nu}\right)-\frac{\nu^2-\nu}{4}$$
$$=\tfrac{1}{2}N(\tfrac{1}{2}\log N+C)-\tfrac{1}{4}N,$$

where C is the Euler constant.[330] In short,

$$U=\tfrac{3}{4}N\log N+2\Sigma\frac{N+s^2-n^2}{n}.$$

Geometric[207] considerations give the approximate value of the last term as

$$2\iint\frac{N+x^2-y^2}{y}\,dxdy,\quad x,y>0,$$

where the limits of integration are given by the relations $y>2x$, $N+x^2-y^2>0$. Hence for N very great, $U=\frac{2}{9}\pi N^{\frac{3}{2}}$. Cf. Pfeiffer,[195] Landau.[330]

L. Gegenbauer,[208] employing the same notation as had G. L. Dirichlet[209] and the same restrictions, obtained by new methods the results of Dirichlet, that the mean number of representations of a single positive integer by a system of representative forms of fundamental discriminant Δ is

$$\tau\sum_{x=1}^{\infty}\left(\frac{\Delta}{x}\right)\frac{1}{x},\ \text{if}\ \Delta\gtrless 0;\qquad 2\pi K(\Delta)/\sqrt{-\Delta},\ \text{if}\ \Delta<0,$$

where $K(\Delta)$ is the number of classes of negative discriminant Δ. For example, in the first case, the identity

$$\sum_{x=1}^{n}\left[\frac{n}{x}\right]\left(\frac{\Delta}{x}\right)=\sum_{x,\,y=1}^{n}\epsilon\left(\frac{n}{x\,y}\right)\left(\frac{\Delta}{x}\right)=\sum_{r=1}^{n}\epsilon\left(\frac{n}{r}\right)\sum_{d}\left(\frac{\Delta}{d}\right),$$

in which, presumably, $\epsilon(x)=0$ or 1 according as $x<1$ or $\geqq 1$; and the last summation extends over divisors of r, implies that

$$\sum_{x=1}^{n}\tau\sum_{d}\left(\frac{\Delta}{d}\right)=\tau\sum_{x=1}^{n}\left[\frac{n}{x}\right]\left(\frac{\Delta}{x}\right),$$

[206] C. Hermite, Acta Math., 5, 1884-5, 311; Oeuvres, IV, 1917, 152.
[207] Cf. R. Lipschitz,[102] Sitzungsber. Akad. Wiss. Berlin, 1865, 174-175.
[208] Sitzungsber. Akad. Wiss. Wien, 96, II, 1887, 476-488.
[209] Zahlentheorie, Braunchweig, 1894, 229; Dirichlet.[19]

where d ranges over the divisors of x and $\tau\Sigma(\Delta/d)$ is Dirichlet's[93] expression (2) for the number of representations of x by a system of representative forms of determinant Δ. Hence

$$\sum_{x=1}^{n}\tau\sum_{d}\left(\frac{\Delta}{d}\right)=\tau n\sum_{x=1}^{\infty}\left(\frac{\Delta}{x}\right)\frac{1}{x}-\tau n\sum_{x=[\sqrt{n}]+1}^{\infty}\left(\frac{\Delta}{x}\right)\frac{1}{x}-\tau\sum_{x=1}^{[\sqrt{n}]}\epsilon_x\left(\frac{\Delta}{x}\right)+\tau\sum_{x=[\sqrt{n}]+1}^{n}\left[\frac{n}{x}\right]\left(\frac{\Delta}{x}\right),$$

where $0\leqq\epsilon_x<1$, and each of the last three terms remains finite when n becomes infinite.

Gegenbauer[210] defined a certain function by

$$\chi_k(n)=n^k\sum_d\frac{\mu(d)}{d^k}\left(\frac{\Delta}{d}\right),$$

in which (Δ/d) is the Jacobi-Legendre symbol, d ranges over the divisors of n, and $\mu(x)$ is the Moebius function (this History, Vol. I, Ch. XIX). Then

$$\sum_{x=1}^{n}\left[\frac{n}{x}\right]\left(\frac{\Delta}{x}\right)\chi_k(x)=\sum_{x=1}^{n}\left(\frac{\Delta}{x}\right)x^k,$$

if Δ is prime to $1, 2, 3, \ldots, n$. This relation combined with Kronecker's[171] formulas (2) and (5) gives the number of classes of a prime discriminant Δ. That is,

$$K(\Delta)=\frac{\tau}{2\Delta}\sum_{x=1}^{|\Delta|-1}\left[\frac{|\Delta|-1}{x}\right]\left(\frac{\Delta}{x}\right)\chi_1(x),$$
$$K(\Delta)=\frac{\tau}{2\left(2-\left(\frac{\Delta}{2}\right)\right)}\sum_{x=1}^{[\frac{1}{2}|\Delta|]}\left[\frac{|\Delta|}{2x}\right]\left(\frac{\Delta}{x}\right)\chi_0(x).\qquad \Delta<0$$

For example, if $\Delta=-7$, $\chi_0(1)=1$, $\chi_0(2)=0$, $\chi_0(3)=2$, $\chi_1(1)=1$, $\chi_1(2)=1$, $\chi_1(3)=4$, $\chi_1(4)=2$, $\chi_1(5)=6$, $\chi_1(6)=4$. Therefore $K(-7)=1$.

C. Hermite[211] employed an earlier result[69]

$$\sum\frac{q^{\frac{1}{4}(a^2+2a)-c^2}}{1-q^a}=\Sigma F(n)q^{\frac{n}{4}},\quad n=4m-1,\quad \begin{cases}a=1, 3, 5, \ldots\\ c=0, \pm1, \pm2, \ldots \pm\left(\frac{a-1}{2}\right),\end{cases}$$

for $a=2c'+1$, divided by $q^{\frac{3}{4}}$, then applied his[204] identity (1), and equated coefficients of q^{m-1} and obtained

$$\sum_{r=1}^{m}F(4r-1)=2\Sigma E\left(\frac{m-dd'}{d+d'+1}\right),$$

where d, d' are of the same parity; $d'\gtreqless d$; $m\gtreqless(d+1)(d'+1)$; and the coefficient 2 is to be replaced by 1 if $d=d'$. But when in mathematical induction $m-1$ is replaced by m, the right member of the last equation is increased by double the number of solutions of

$$\frac{m-dd'}{d+d'+1}=c,\text{ or } 4m-1=4(c+d)(c+d')-(2c-1)^2,$$

in which $c=1, 2, \ldots, m$; $d\equiv d' \pmod 2$, $d'>d$; while, if $d'=d$, each solution is counted $\frac{1}{2}$. This gives the value of $F(4m-1)$.

---

210 Sitzungsber. Akad. Wiss. Wien (Math.), 96, II, 1887, 607–613.
211 Jour. für Math., 100, 1887, 51–65; Oeuvres, IV, 1917, 223–239.

Hermite equated the coefficients of certain powers of q in two expansions of $H_1^3(0)$ and found that, for $m\equiv 3 \pmod 8$, the number of odd classes of the negative determinant $-m$ is $\Sigma\Phi(m-b^2)$, in which $b=0, \pm 2, \pm 4, \ldots; b^2<m$; and $\Phi(m)=\Sigma(-1)^{\frac{1}{2}(d'+1)}$, d' ranging over the divisors of m which are $>\sqrt{m}$ and $\equiv 3 \pmod 4$.

P. Nazimow[212] gave an account of the use[54, 145] of modular equations, and of Hermite's[69] method of equating coefficients in the theta-function expansions, to obtain class-number relations.

X. Stouff[128] of Ch. I extended Dirichlet's[19] determination of the class-number when the quadratic forms and the definition of equivalence both relate to a fixed set of integers called modules.

L. Kronecker[213] let $ax^2+bxy+cy^2$ be a representative form of negative discriminant $D=-\Delta=b^2-4ac$; put $a=a_0\sqrt{\Delta}$ and (Cf. Kronecker[175])

$$\Lambda'(0, 0, \omega_1, \omega_2)=\frac{4\pi^2}{c_0}\left(\frac{\theta'(0,\omega_1)}{2\pi}\cdot\frac{\theta'(0,\omega_2)}{2\pi}\right)^{\frac{2}{3}},$$
$$\omega_1=\frac{-b+i\sqrt{\Delta}}{2c}, \qquad \omega_2=\frac{b+i\sqrt{\Delta}}{2c}.$$

He obtained the fundamental formula

$$\lim_{\rho=0}\left[-\frac{1}{\rho}+\frac{1}{2\pi}\sum_{m,n}\frac{(\sqrt{\Delta})^{1+\rho}}{(am^2+bmn+cn^2)^{1+\rho}}\right]=\log 4\pi^2+2C-\log\Lambda'(0,0,\omega_1,\omega_2),$$

where C is a constant independent of D, a, b, c. When each member of this identity is summed for the $K(D_0)$ representative forms of fundamental discriminant D_0, the result enables Kronecker[171] to evaluate the ratio $H'(-\Delta_0)/H(-\Delta_0)$ in terms of $K(D_0)$, where

$$H(-\Delta_0)=\sum_{k=1}^{\infty}\left(\frac{-\Delta_0}{k}\right)\frac{1}{k}, \qquad H'(-\Delta_0)=\sum_{k=1}^{\infty}\left(\frac{-\Delta_0}{k}\right)\frac{\log k}{k}.$$

This is called Kronecker's limit ratio.

H. Weber[214] denoted by ω the principal root of a reduced quadratic form of determinant $-m$, and denoted by $j(\omega)$ the product of F. Klein's[134] class-invariant J by 1728. The *class equation*

$$(1) \qquad \Pi[u-j(\omega)]=0,$$

in which ω ranges over the principal roots of a representative system of primitive quadratic forms of determinant $-m$, he expressed by

$$(2) \quad H_m(u)=0, \quad \text{or} \quad (3) \quad H'_m(u)=0,$$

according as the forms are of proper or improper order. By applying transformations of the second order to ω, he set up a correspondence between the roots of (2) and (3). This correspondence is 1 to 1, if $m\equiv -1 \pmod 8$; 3 to 1, if $m\equiv 3 \pmod 8$, except when $m=3$. Whence he obtained Dirichlet's[20] ratio between $h(D)$ and $h'(D)$, $D<0$.

212 On the applications of the theory of elliptic functions to the theory of numbers, 1885, (Russian). Summary in Annales Sc. de l'Ecole Norm. Sup., (3), 5, 1888, 23-48, 147-176 (French).
213 Sitzungsber. Akad. Berlin, 1889, I, 199-220.
214 Elliptische Functionen und Algebraische Zahlen, 1891, 338-344.

Weber[215] gave the name (cf. Dedekind's[128] valence equation (1)) *invariant equation* to

$$(4) \qquad \Pi\left[j(\omega)-j\left(\frac{c+d\omega}{a+b\omega}\right)\right]=0,$$

of order $ad-bc=n$, in which the g.c.d. of a, b, c, d is 1, and

$$\omega'=\frac{c+d\omega}{a+b\omega}$$

is a complete set of non-equivalent representatives. He observed that, if ω furnishes a root $j(\omega)$ of (4), then ω must be the principal root of a quadratic form

$$(5) \qquad A\omega^2+B\omega+C, \qquad B^2-4AC=D,$$

where, for a positive integer x,

$$b=Ax, \qquad c=-Cx, \qquad a-d=Bx;$$

and if we set $a+d=y$, we must have

$$(6) \qquad 4n=y^2-Dx^2.$$

Conversely, for each of the k representations of $-D$ in the form

$$-D=\frac{4n-y^2}{x^2},$$

there are $Cl(D)=h'(D)$ forms (5) each of whose principal roots furnishes one root of (4). Hence (4) can be written (cf. Weber's Algebra, III, 1908, 421)

$$(7) \qquad CH'^{k_1}_{D_1}(u)H'^{k_2}_{D_2}(u)\ldots=0, \qquad u=j(\omega).$$

If $j(\omega)$ is a root of (4), expansion of the left member in powers of $q=e^{\pi i\omega}$ shows that the degree of (4) in $j(\omega)$ is

$$2\Sigma\frac{\partial}{e}\phi(e)+\phi(\sqrt{n}) \text{ or } 2\Sigma\frac{\partial}{e}\phi(e),$$

according as n is or is not a square (cf. Dedekind,[128] (2)) where $\partial>\sqrt{n}$ is a divisor of n. The degree of (7) in $j(\omega)$ is $\Sigma h'(D_i)k_i$, summed for $i=1, 2, 3, \ldots$.

For brevity, (4) is written $F_n(u, u)=0$. The simplest case of deducing a class-number relation of L. Kronecker's type[48] is presented by equating two valuations of the highest degree of $u=j(\omega)$ in the reducible invariant equation

$$F_{n'_0}(u,u)\cdot F_{n'_1}(u,u)\cdot F_{n'_2}(u,u)\ldots=0,$$

where $n'_0, n'_1, n'_2, \ldots$ are derived from n in every possible way by removing square divisors including 1, but excluding n when n is square. The relation is

$$K(n)+2K(n-1)+2K(n-4)+\ldots+2K'(4n-1)+2K'(4n-9)+\ldots$$
$$=2\Sigma\partial \text{ or } 2\Sigma\partial+\sqrt{n}+\tfrac{1}{6},$$

according as n is not or is a square. Here $K(m)$ denotes the number of classes of

[215] Elliptische Functionen und Algebraische Zahlen, 1891, 393–401.

determinant $-m$, and $K'(m)$ denotes the number of classes of determinant $-m$ derived from improperly primitive classes. Finally, $\Sigma\partial$ is the sum of the divisors of n which are $>\sqrt{n}$.

J. Hacks[216] considered the negative prime determinant $-q$, where $q=4n+3$; he put

$$S=\sum_{s=1}^{\frac{1}{2}(q-1)}\left[\frac{s^2}{q}\right],\quad S'=\sum_{s=1}^{\frac{1}{2}(q-1)}\left[\frac{2s^2}{q}\right],$$

and found that the number of properly primitive classes of determinant $-q$ is $h=\frac{1}{2}(q-1)-2S'+4S$. This is given the two following modified forms

$$h=\frac{q-1}{2}-2\sum_{1}^{\frac{1}{2}(q-3)}(-1)^s[\sqrt{\tfrac{1}{2}q\cdot s}],\qquad h=\frac{q-1}{2}+2\sum_{s=1}^{\frac{1}{2}(q-3)}\sum_{1}^{[2s^2/q]}(-1)^t;$$

and finally is reduced to Dirichlet's[23] formula (6).

F. Klein and R. Fricke[217] reproduced the theory of modular functions of Dedekind[128] and Klein,[134, 138] also (Vol. II, pp. 160–235, 519–666) the application by Gierster,[135, 139, 145, 147] Hurwitz,[167, 184, 194] and Weber[214] of that theory to the deduction of class-number relations of negative determinants. They gave (Vol. II, p. 234) the relations of grade 3 which come from the tetrahedron equation and (Vol. II, pp. 231–233) the relations of grade 5 that come from the icosahedron equation. Their formulas (1_1) p. 231, and (7), p. 233, should all have their right members divided by 2. They reproduced (Vol. II, pp. 165-73, 204-7) the theory of the relation between modular equations and Smith's[132] reduced forms of positive determinant.

In connection with Hurwitz's[194] general class-number relation of prime grade $q>5$ and relatively prime to n, Klein and Fricke constructed a table of values of ψ_i and χ_i for $n\leqq 43$. A sample of the table follows (p. 616):

n	ψ_1	ψ_2	ψ_3	n	χ_1	χ_2
1	-1	1	0	2	0	-1
3	1	1	-1	6	1	0
4	2	0	1	7	-1	0

For $q=11$ and $(n/q)=-1$, Hurwitz's first general formula becomes

$$6\sum_{3\sqrt{-n}} H(4n-\kappa^2)=2\Phi(n)+t_1\chi_1+t_2\chi_2,$$

where κ is positive or negative and $\kappa^2\equiv(3\sqrt{-n})^2 \pmod q$. Hence, by the table for $n=2$, $n=6$,

$$12H(4)=6-t_2,\qquad 12H(23)=24+t_1.$$

But it is known that $H(4)=\frac{1}{2}$; $H(23)=3$. Therefore $t_1=12$; $t_2=0$.

G. B. Mathews[218] reproduced in outline the researches of G. L. Dirichlet[93] on the number of properly primitive classes of a given determinant; and those of Lipschitz[219] on the ratio of the numbers of classes of different orders of the same determinant.

---

216 Acta Math., 14, 1890-1, 321–328.
217 Elliptische Modulfunctionen, I, 1890, 163–416; II, 1892, 37–159.
218 Theory of Numbers, Cambridge, 1892, 230–256.
219 *Ibid.*, 159–170.

H. Weber,[220] by arithmetical processes, obtained L. Kronecker's[221] expression for the number h of primitive classes of forms $ax^2+bxy+cy^2$ of discriminant D. For $D=Q^2\cdot\Delta$, Δ a fundamental discriminant, he obtained by Dirichlet's[20] methods, Kronecker's[171] ratio (4) of the class-number of D and of Δ. By the use of Gauss sums, he transformed the former result for $Q=1$ into

$$(1)\qquad h=\frac{\tau}{2\Delta}\sum_s(\Delta, s)s,\qquad \Delta<0,$$

$$(2)\qquad h\log\tfrac{1}{2}(T+U\sqrt{\Delta})=-\sum_s(\Delta, s)\log\sin s\pi/\Delta,\qquad \Delta>0,$$

in which[222] (Δ, s) is the generalized symbol (Δ/s) of Kronecker[171]; and $0<s<\pm\Delta$.

By Dirichlet's methods, he obtained the analogue of Dirichlet's[23] formulas (5). See Lerch,[240] (4). By use of the Gauss function

$$\Psi(u)=\lim_{m=\infty}\left(\log m-\frac{1}{u+1}-\frac{1}{u+2}-\cdots\frac{1}{u+m}\right),$$

the formulas written above become

$$(3)\qquad h=\frac{-\tau}{2\sqrt{-\Delta}}\sum_\nu(\Delta, \nu)\cot\frac{\pi\nu}{\Delta},\quad \Delta<0;$$

$$(4)\qquad h\log\tfrac{1}{2}(T+U\sqrt{\Delta})=\frac{-1}{\sqrt{\Delta}}\sum_\nu(\Delta,\nu)\left[\psi\left(\frac{-\nu}{\Delta}\right)+\psi\left(-1+\frac{\nu}{\Delta}\right)\right],\quad \Delta>0,$$

$0<\nu<\pm\Delta/2$ (cf. Lebesgue,[36] (1)).

For $\Delta=-m<0$ and uneven, (3) is equivalent (cf. M. Lerch,[238] (1)) to

$$(5)\qquad h=\frac{\tau}{2\sqrt{m}}\sum_\nu\cot\frac{\pi\nu^2}{m}.$$

Weber transformed (p. 264) his formula (2) above by cyclotomic considerations[223] and observed that $h(\Delta)$ is odd if Δ is an odd prime or 8, and even in all other cases. (Cf. Dirichlet, Zahlentheorie, 1894, §§ 107–109.)

P. Bachmann[224] reproduced (pp. 89–145, 188–227) a great part of the class-number theory of Gauss[4, 9] Dirichlet,[93] and (pp. 228-231) Schemmel[95]; and also (pp. 437–65) the researches of Lipschitz[102] and Mertens[117] on the asymptotic value of $h(D)$.

J. de Séguier[225] showed that Kronecker's[171] formula (5_2) is valid for $D_0<0$, if in the right member, D_0 be replaced by $|D_0|$. This proof is reproduced in his[226] treatise.

J. de Séguier[226] wrote a treatise on binary quadratic forms from Kronecker's[171] later point of view making special reference[227] to two fundamental formulas of

[220] Göttingen Nachr., 1893, 138–147, 263–4.
[221] Sitzungsber.[171] Akad. Wiss. Berlin, 1885, II, 771.
[222] Cf. H. Weber, Algebra, III, 1908, § 85, pp. 322–328.
[223] Cf. Dirichlet,[93] (1); Arndt.[53]
[224] Zahlentheorie, II, Die Analytische Zahlentheorie, Leipzig, 1894.
[225] Comptes Rendus, Paris, 118, 1894, 1407–9.
[226] Formes quadratiques et multiplication complexe; deux formules fondamentales d'après Kronecker, Berlin, 1894.
[227] *Ibid.*, 133, formula (6); p. 218, formula (3).

Kronecker.[228] He extended (p. 32) Kronecker's[171] identity (4a) in Gauss sums (cf. H. Weber, Gött. Nachr., 1893, 51) to the form

$$\sum_{s=1}^{|D_0|-1}\left(\frac{D_0}{s}\right)e^{2sh\pi i/|D_0|}=\left(\frac{D_0}{h}\right)\sqrt{D_0}(-1)^{\frac{1}{4}(\operatorname{sgn} h-1)(\operatorname{sgn} D_0-1)},\quad h\neq 0,$$

where $\operatorname{sgn} x=+1$ or -1, according as x is $>$ or $x<0$, while D_0 is a fundamental discriminant.

Then, whether D is positive or negative, it follows at once in Kronecker's[171] notation that the number of primitive classes is given by

$$(1)\quad K(D_0)\log E(D_0)=\sqrt{D_0}H(D_0)=\sqrt{D_0}\sum_{n=1}^{\infty}\left(\frac{D_0}{n}\right)\frac{1}{n}$$
$$=\sum_{k=1}^{|D_0|-1}\left(\frac{D_0}{k}\right)\sum_{n=1}^{\infty}\frac{e^{2nk\pi i/|D_0|}}{n}=-\sum_{k=1}^{|D_0|-1}\left(\frac{D_0}{k}\right)\log\left(1-e^{2k\pi i/|D_0|}\right),$$

in which $E(D_0)$ is a fundamental unit; and, if $z=re^{i\theta}$, then $\log z=\log r+i\theta$, $-\pi<\theta<\pi$ (pp. 118–126). For $D_0>0$, this formula is Kronecker's[171] (5_2). Elsewhere de Séguier[225] repeated briefly his own deduction of (1).

By noting that

$$\log\left(1-e^{2k\pi i/|D_0|}\right)=\log 2\sin k\pi/|D_0|+i(\tfrac{1}{2}\pi-k\pi/|D_0|),$$

he obtained from (1) two distinct formulas; one being Kronecker's[171] (5_1) and the other (p. 127) being Weber's[220] (2),

$$K(D_0)=-\frac{1}{\log E(D_0)}\sum_{k=1}^{D_0}\left(\frac{D_0}{k}\right)\log\sin\frac{k\pi}{D_0},\quad D_0>0.$$

By a study of groups of classes in respect to composition of classes, de Séguier (pp. 77–96) obtained the ratio of $Cl(D\cdot S^2)$ to $Cl(D)$. Cf. Gauss,[4] Arts. 254–256.

Denoting the Moebius function (see this History, Vol I, Ch. XIX) by ϵ_n, de Séguier found (p. 116) that for any function F which insures convergence in each member of the following formula, we have

$$\sum_{m=1}^{\infty}\left(\frac{Q^2}{m}\right)F(m)=\sum_{d|Q}\epsilon_d\sum_{n=1}^{\infty}F(nd).$$

If a, b, c are arbitrary constants (eventually integers) and F is taken such that $F(xy)=F(x)\cdot F(y)$, we have

$$\sum_{m,n}\left(\frac{Q^2}{m}\right)F(am^2+bmn+cn^2)=\sum_{d|Q}\epsilon_d\sum_{m,n}F(d)F\left(adm^2+bmn+\frac{c}{d}n^2\right),$$

$m, n=0, \pm 1, \pm 2, \ldots, \pm\infty$, except $m=n=0$. Let $F(u)$ be $\rho/u^{1+\rho}$. Since, for such a function,

$$\lim_{\rho\to 0}\sum_{m,n}F(am^2+bmn+cn^2)$$

---

[228] L. Kronecker,[171, 213] Sitzungsber. Akad. Wiss. Berlin, 1885, II, 779; 1889, I, 205.

depends only on b^2-4ac, we have

$$\lim_{\rho=0} \rho \sum_{m,n} \left(\frac{Q^2}{m}\right)(am^2+bmn+cn^2)^{-1-\rho}=\frac{\phi(Q)}{Q}\lim_{\rho=0}\rho\Sigma(adm^2+bmn+\frac{c}{d}n^2)^{-1-\rho}.$$

But[171]

$$\tau H(D)=K(D)\lim_{\rho=0}\rho\sum_{m,n}(am^2+bmn+cn^2)^{-1-\rho}.$$

Hence we have, for $D<0$,

$$(4) \qquad \tau_{D_0Q^2}\frac{H(D_0Q^2)}{K(D_0Q^2)}\frac{\phi(Q)}{Q}=\Sigma\frac{\epsilon_d}{d^2}\frac{H(D_0d'^2)}{K(D_0d'^2)}\tau_{D_0d'^2} \quad (dd'=Q).$$

To this formula is applied the following lemma due to Kronecker[229]: Let $f(n)$, $g(n)$ be two arbitrary functions of n and let $h(n)=\Sigma f(d)g(d')$ $(dd'=n)$, and let g have the property $g(mn)=g(m)g(n)$, $g(1)=1$; then

$$f(n)=\Sigma\epsilon_d g(d)\cdot h(d') \quad (dd'=n).$$

Hence we deduce from (4) the new relation (p. 128)

$$\tau_{D_0Q^2}\frac{H(D_0Q^2)}{K(D_0Q^2)}=Q^{-1}\sum_{dd'=Q}\tau_{D_0d^2}\frac{\phi(d)}{d'}\frac{H(D_0d^2)}{K(D_0d^2)},\quad D<0.$$

For discriminants $D_1<0$, $D_2>0$, de Séguier gave the following approximation formula (p. 314):

$$K(D_1)\log E(D_1)=\frac{-\tau_{D_2}}{2K(D_2)}\Sigma\left(\frac{D_1}{A}\right)\left(\frac{\pi\sqrt{-D}}{6a}+\log a\right),$$

the summation extending over a system of primitive forms (a, b, c) of discriminant $D=D_1\cdot D_2$; while A is an arbitrary number representable by (a, b, c) and relatively prime to $2D$.

M. Lerch,[230] in the case of Kronecker's forms of negative fundamental discriminant $-\Delta\equiv 5 \pmod 8$, gave to Dirichlet's[20] equation (2) the form

$$\sum_{a,b,c}\sum_{m,n}{}' F(am^2+bmn+cn^2)=\sum_{h,k}(-\Delta/h)F(hk),$$

$m, n=0, \pm 1, \pm 2, \ldots$, except $m=n=0$; $h, k=1, 2, 3, \ldots$. He took

$$F(x)=(-1)^x e^{-x\pi/\sqrt{\Delta}}$$

and obtained

$$(1) \qquad \sum_{a,b,c}\sum_{m,n}{}'(-1)^{mn+m+n}e^{-\pi(am^2+bmn+cn^2)/\sqrt{\Delta}}=\sum_{h,k}\left(\frac{-\Delta}{h}\right)(-1)^{hk}e^{-hk\pi/\sqrt{\Delta}}.$$

But by taking $\sigma=\tau=0$ in Kronecker's[231] fundamental formula, it is seen that the left member of (1) would vanish if it contained the terms with $m=n=0$. Hence the left member of (1) is $-Cl(-\Delta)$, and (1) can be written

$$Cl(-\Delta)=\sum_{h=1}^{\infty}\left(\frac{h}{\Delta}\right)\frac{H}{1+H},\qquad H\equiv(-1)^{h-1}e^{-h\pi/\sqrt{\Delta}}.$$

---

229 De Séguier's Formes quadratiques, 114; L. Kronecker, Sitzungsber. Akad. Wiss. Berlin, 1886, II, 708.

230 Comptes Rendus, Paris, 121, 1895, 879.

231 Sitzungsber. Akad. Wiss. Berlin, 1883, I, 505.[175]

By expressing the right member in terms of a θ-function,[175] we obtain

$$\sum_{\rho=1}^{\Delta-1}\left(\frac{\rho}{\Delta}\right)\frac{\theta_1'\left(\frac{\rho}{\Delta}\middle|\frac{-2}{\Delta+i\sqrt{\Delta}}\right)}{\theta_1\left(\frac{\rho}{\Delta}\middle|\frac{-2}{\Delta+i\sqrt{\Delta}}\right)}=\gamma Cl(-\Delta)\cdot(\Delta+i\sqrt{\Delta})\pi i,$$

$$\gamma=\tfrac{5}{3}\text{ if }\Delta=3;\qquad \gamma=1\text{ if }\Delta>3.$$

G. Osborn,[232] from Dirichlet's[23] formulas (6) and his own elementary theorems[233] on the distribution of quadratic residues, drew the immediate conclusion that the number of properly primitive classes of determinant $-N$, N a prime, is

$$\tfrac{1}{2}(N-1)-\frac{2}{N}\Sigma(R),\qquad N=8m-1>0,$$

but is 3 times that number if $N=8m+3>0$, where $\Sigma(R)$ is the sum of the quadratic residues of N between 0 and N.

*R. Götting[234] found transformations of the more complicated of Dirichlet's[23] closed expressions for class-numbers of negative determinants.

A. Hurwitz[235] denoted by $h(D)$ the number of classes of properly primitive positive forms of negative determinant $-D$. Let p be a prime $\equiv 3\pmod 4$ and write $p'=\frac{1}{2}(p-1)$. Since $(s/p)\equiv s^{p'}\pmod p$, Dirichlet's[25] result (5_1) implies

$$h(p)\equiv 1^{p'}+2^{p'}+\ldots+p'^{p'}\pmod p.$$

The right member is the coefficient of

$$(1)\qquad (-1)^{\frac{1}{2}(p'-1)}x^{p'}/p'!$$

in the expansion of

$$\phi(x)=\sin x+\sin 2x+\ldots+\sin p'x=\frac{\cos\frac{1}{2}x-\cos\frac{1}{2}px}{2\sin\frac{1}{2}x}.$$

This numerator is congruent to $\cos\frac{1}{2}x-1$ modulo p, and by applying a theorem on the congruence of infinite series, we get

$$\phi(x)\equiv\frac{\cos\frac{1}{2}x-1}{2\sin\frac{1}{2}x}=\frac{-2\sin^2\frac{1}{4}x}{4\sin\frac{1}{4}x\cos\frac{1}{4}x}\equiv-\tfrac{1}{2}\tan\tfrac{1}{4}x\pmod p.$$

But when x is replaced by $4x$, (1) is multiplied by $4^{p'}$ or $2^{p-1}\equiv 1\pmod p$. Hence $h(p)$ is congruent modulo p to the coefficient of (1) in the expansion of $-\frac{1}{2}\tan x$. When $p\equiv 1\pmod 4$, we employ the expansion of $\frac{1}{2}\sec x$. Other such theorems give $h(2p)$.

The same result of Dirichlet is used to prove that if $q\equiv 1\pmod 4$ and q has no square factor >1, and if

$$\frac{1}{\cos qx}\left\{\left(\frac{1}{q}\right)\sin x-\left(\frac{3}{q}\right)\sin 3x+\left(\frac{5}{q}\right)\sin 5x-\ldots-\left(\frac{q-2}{q}\right)\sin(q-2)x\right\}$$
$$=c_1x+c_2x^3/3!+c_3x^5/5!+\ldots,$$

---

232 Messenger Math., 25, 1895, 157.
233 *Ibid.*, 45.
234 Program No. 257 of the Gymnasium of Turgau, 1895.
235 Acta Math., 19, 1895, 351–384.

and if $p\equiv 3 \pmod 4$ is a prime not dividing q, then

$$h(pq)\equiv(-1)^{\frac{1}{4}(p+1)}c_{\frac{1}{4}(p+1)} \pmod p.$$

There are analogous theorems for $h(pq)$ and $h(2pq)$ for all combinations of residues 1 and 3 (mod 4) of p and q.

To obtain a lower bound for the number of times that 2 may occur as a divisor of h, the number of genera of the properly primitive order is calculated.[236] If $h_g(D)$ denote the number of classes in a properly primitive genus of determinant $-D$, the parities of $h_g(pq)$ and $h_g(2pq)$ depend only on the values of (p/q) and p (mod 8) and q (mod 8), and are shown in tables.

By combining the two theories of this memoir one obtains, for special q, results such as the following:

If $p\equiv 3 \pmod 4$, $h(5p)$ is the least positive residue modulo $2p$ of $(-1)^{\frac{1}{4}(p+1)}c_{\frac{1}{4}(p+1)}$, where $c_1, c_2, \ldots$ are the coefficients in the expansion

$$\frac{\sin x+\sin 3x}{\cos 5x}=c_1x+c_2\frac{x^3}{3!}+\ldots+c_n\frac{x^{2n-1}}{(2n-1)!}+\ldots.$$

F. Mertens[237] completed the solution of Gauss' problem (Disq. Arith.[4], Art. 256) to find by the composition of forms the ratio of the number of the properly primitive classes of the determinant $S^2\cdot D$ to that of D. He modified Gauss' procedure by taking *schlicht* forms (Mertens[37] of Ch. III) as the representatives of classes and by means of them found for any determinant the number of primitive classes which when compounded with an arbitrary class of order S would produce an arbitrary class of order S (Mertens[37] of Ch. III).

M. Lerch[238] rediscovered Lebesgue's[36] class-number formula (1) above, and wrote it for the case $\Delta=p=4m+3$, a prime:

$$2\Sigma'\cot\frac{k\pi}{p}=\frac{4\sqrt{p}}{\tau}Cl(-p),\quad k=1,2,\ldots,p-1,\ \left(\frac{k}{p}\right)=1.$$

By replacing k by $a^2-p[a^2/p]$, he obtained Weber's formula[220] (5):

$$(1)\qquad \frac{2\sqrt{p}}{\tau}Cl(-p)=\sum_{a=1}^{\frac{1}{2}(p-1)}\cot\frac{a^2\pi}{p}.$$

He found for $\Delta=4p$, $p=4m+1$, a prime >1,

$$(2)\qquad \sqrt{p}Cl(-4p)=\sum_\nu\frac{1}{\sin\nu^2\pi/(2p)}-\frac{p-1}{2}\ (\nu=1,3,5,\ldots,p-2).$$

For $\Delta=8p$, Lerch derived more complicated formulas which are analogous to (1) and (2).

L. Gegenbauer[239] in a paper on determinants of m dimensions and order n, stated the following theorem. If for $k=1,\ldots,n$ in turn in a non-vanishing determinant of even order m, we replace, in the sequence of elements which belong to any particular

---

[236] C. F. Gauss, Disq. Arith., Art. 252; G. L. Dirichlet, Zahlentheorie, Supplement IV, ed. 4, 1894, 313–330.
[237] Sitzungsber. Akad. Wiss. Wien, 104, IIa, 1895, 103–137.
[238] Sitzungsber. Böhm. Gesells. Wiss., Prague, 1897, No. 43, 16 pp.
[239] Denkschrift Akad. Wiss. Wien, Math.-Natur., 57, 1890, 735–52.

τth index, the elements which belong to the σth index k, by the corresponding elements respectively which have the σth index $k^2+k+\Delta$, where $-\Delta$ is a negative fundamental discriminant and where all the indices are taken modulo n; and if we divide each of the resulting determinants by the original, the product of $\sqrt{\Delta}$ by the sum of the quotients has mean value, $G(-\Delta)$, when n becomes infinite (p. 749). Three similar theorems include a case of n finite.

M. Lerch[240] employed

$$E^*(x)=x-\tfrac{1}{2}+\sum_{\nu=1}^{\infty}\frac{\sin 2\nu x\pi}{\nu\pi}.$$

Then $E^*(x)=[x]$ if $x>0$ is fractional, but $=[x]-\frac{1}{2}$ if x is an integer. In the initial equation, x is replaced by $x+am/\Delta$, where $-\Delta$ is a negative fundamental discriminant; each member is then multiplied by $(-\Delta/a)$ and summed for $a=1, 2, 3, \ldots, \Delta-1$. Since [a misprint is corrected here],

$$(1)\qquad \sum_{a=1}^{\Delta-1}\left(\frac{-\Delta}{a}\right)=0,\quad \left(\frac{-\Delta}{\Delta-a}\right)=-\left(\frac{-\Delta}{a}\right),$$

it follows from the theory of Gauss' sums (cf. G. L. Dirichlet, Zahlentheorie, Art. 116, ed. 4, 1894, p. 303) that

$$\sum_{a=1}^{\Delta-1}\left(\frac{-\Delta}{a}\right)E^*\left(x+\frac{am}{\Delta}\right)=\frac{m}{\Delta}\sum_{a=1}^{\Delta-1}\left(\frac{-\Delta}{a}\right)a+\left(\frac{-\Delta}{m}\right)\frac{\sqrt{\Delta}}{\pi}\sum_{\nu=1}^{\infty}\left(\frac{-\Delta}{\nu}\right)\frac{\cos 2\nu x\pi}{\nu}.$$

Then by Kronecker's[171] formula (5_1) we have

$$(2)\qquad \sum_{a=1}^{\Delta-1}\left(\frac{-\Delta}{a}\right)E^*\left(x+\frac{am}{\Delta}\right)+\frac{2m}{\tau}Cl(-\Delta)=\left(\frac{-\Delta}{m}\right)\frac{\sqrt{\Delta}}{\pi}\sum_{\nu=1}^{\infty}\left(\frac{-\Delta}{\nu}\right)\frac{\cos 2\nu x\pi}{\nu}.$$

By comparing this result with the case $m=1$, we have for $x=0$,

$$(3)\qquad \frac{2}{\tau}\left[m-\left(\frac{-\Delta}{m}\right)\right]Cl(-\Delta)=-\sum_{a=1}^{\Delta-1}\left(\frac{-\Delta}{a}\right)E^*\left(\frac{am}{\Delta}\right).$$

For m not divisible by Δ, $E^*(am/\Delta)$ is equal to $[am/\Delta]$. Taking $m=2$ and applying (1), we get[241]

$$(4)\qquad \frac{2}{\tau}\left[2-\left(\frac{2}{\Delta}\right)\right]Cl(-\Delta)=\sum_{a=1}^{\frac{1}{2}(\Delta-1)}\left(\frac{-\Delta}{a}\right).$$

Hereafter we take $\Delta>4$, i. e., $\tau=2$. Then, for $m=4$, we have

$$(5)\qquad \left[4-\left(\frac{4}{\Delta}\right)\right]Cl(-\Delta)=-\sum_{a=1}^{\Delta-1}\left(\frac{-\Delta}{a}\right)\cdot\left[\frac{4a}{\Delta}\right].$$

When we put $S(a,\ldots,b)$ for $\Sigma_a^b(-\Delta/a)$, formula (5) is reduced by means of (1) to

$$3S\left(0,\ldots,\frac{\Delta}{4}\right)+S\left(\frac{\Delta}{4},\ldots,\frac{\Delta}{2}\right)=\left[4-\left(\frac{4}{\Delta}\right)\right]Cl(-\Delta).$$

But (4) is equivalent to

$$S\left(0,\ldots,\frac{\Delta}{4}\right)+S\left(\frac{\Delta}{4},\ldots,\frac{\Delta}{2}\right)=\left[2-\left(\frac{2}{\Delta}\right)\right]Cl(-\Delta).$$

---

[240] Bull. des sc. math. (2), 21, I, 1897, 290–304.
[241] Cf. H. Weber,[220] Göttingen Nachr., 1893, 145.

By combining the last two formulas we obtain the two serviceable ones

$$(6)\qquad \sum_{a=1}^{[\Delta/4]}\left(\frac{-\Delta}{a}\right)=\tfrac{1}{2}\left[2+\left(\frac{2}{\Delta}\right)-\left(\frac{4}{\Delta}\right)\right]Cl(-\Delta),$$

$$(7)\qquad \sum_{a=[\Delta/4]+1}^{\frac{1}{2}(\Delta-1)}\left(\frac{-\Delta}{a}\right)=\tfrac{1}{2}\left[2-3\left(\frac{2}{\Delta}\right)+\left(\frac{4}{\Delta}\right)\right]Cl(-\Delta).$$

A still more expeditious formula is obtained by taking $m=3$ in (3), whence

$$\sum_{a=1}^{[\Delta/3]}\left(\frac{-\Delta}{a}\right)=\tfrac{1}{2}\left[3-\left(\frac{\Delta}{3}\right)\right]Cl(-\Delta);$$

and this relation combined with (6) yields

$$\sum_{a=[\Delta/4]+1}^{[\Delta/3]}\left(\frac{-\Delta}{a}\right)=\tfrac{1}{2}\left[1-\left(\frac{\Delta}{2}\right)+\left(\frac{\Delta}{3}\right)+\left(\frac{\Delta}{4}\right)\right]Cl(-\Delta).$$

For $m=1$, (2) becomes

$$(8)\qquad \frac{2}{\tau}Cl(-\Delta)=\frac{\sqrt{\Delta}}{\pi}\sum_{\nu=1}^{\infty}\left(\frac{-\Delta}{\nu}\right)\frac{\cos 2\nu x\pi}{\nu}\quad\left(0\leqq x<\frac{1}{\Delta}\right).$$

This is a generalization of Dirichlet's[19] formula (1) and it holds for $-\Delta$ not a fundamental discriminant. Lerch showed that (8) is valid for any negative discriminant when $0\leqq x<1/\Delta$ by reducing it from Dirichlet's[19] formula (1). By simply integrating (8), he deduced

$$Cl(-\Delta)=\frac{\tau\Delta\sqrt{\Delta}}{4\pi^2}\sum_{\nu=1}^{\infty}\left(\frac{-\Delta}{\nu}\right)\frac{\sin 2\nu\pi/\Delta}{\nu^2},$$

$$Cl(-\Delta)=\frac{\tau\Delta^2\sqrt{\Delta}}{2\pi^3}\sum_{\nu=1}^{\infty}\left(\frac{-\Delta}{\nu}\right)\frac{\sin^2 \nu\pi/\Delta}{\nu^3}.$$

M. Lerch[242] applied to Kronecker forms $ax^2+bxy+cy^2$ the unit substitution and for a given value of $b^2-4ac=D<0$ studied the number of principal roots ω of reduced forms which would lie in the fundamental region.[128] By arithmetical methods he obtained cumbersome formulas, involving the Legendre symbol $E(x)$, for $\Sigma F(4k)$ and $\Sigma F(4k-1)$, summed for $k=1, 2, \ldots, n$, where $F(\Delta)$ denotes the number of classes of discriminant $-\Delta$. He identified these results with the concise ones of Hermite[211] which had been obtained from elliptic functions for forms $ax^2+2bxy+cy^2$.

Lerch[243] in an expository article, deduced for negative and positive discriminants Dirichlet's[19] class-number formulas (1) in which enters $P(D)=\Sigma_1^\infty(D/h)/h$. For an arbitrary discriminant D, where $|D|=\Delta$, he found by logarithmic differentiation of the ordinary Γ-function that

$$P(D)=-\frac{1}{\Delta}\sum_{k=1}^{\infty}\left(\frac{D}{k}\right)\Gamma'\left(\frac{k}{\Delta}\right)\Big/\Gamma\left(\frac{k}{\Delta}\right).$$

---

[242] Rozpravy české Akad., Prague, 7, 1898, No. 4, 16 pp. (Bohemian).

[243] Rozpravy české Akad., Prague, 7, 1898, No. 5, 51 pp. (Bohemian); resumé in French, Bull. de l'Acad. des Sc. Bohème, 5, 1898, 33–36.

To this he applied the identity:

$$\Gamma'\left(\frac{k}{\Delta}\right)\Big/\Gamma\left(\frac{k}{\Delta}\right)-\Gamma'(1)=-\log 2\Delta-\frac{\pi}{2}\cot\frac{k\pi}{\Delta}+\sum_{a=1}^{\Delta-1}\cos\frac{2ak\pi}{\Delta}\log\sin\frac{a\pi}{\Delta}.$$

For the fundamental discriminant D_0, this furnishes familiar formulas including, e. g., for $D_0>0$, Weber's[220] formula (1).

Lerch[244] repeated the deduction of his[240] formula (8) and established the validity of the formula for a non-fundamental discriminant D for the interval $0\leqq x< 1/(\Delta_0 Q')$, where $D=\Delta_0\cdot Q$ and Q' is the product of the distinct factors of Q.

Lerch[245] transformed the Gauss sum

$$\sum_{a=0}^{n-1} e^{2a^2m\pi i/n}$$

as it occurs in class-number formulas (cf. G. L. Dirichlet, Zahlentheorie, Arts. 103, 115) and so obtained finally

$$\text{(1)}\qquad \sum_{a=1}^{n-1}\left\{\frac{a^2m}{n}-E\left(\frac{a^2m}{n}\right)\right\}=\frac{n-q}{2}-\sum_{d}\left(\frac{m}{d}\right)\frac{2}{\tau_d}Cl(-d),$$

where m, n are relatively prime positive integers, n is uneven and q^2 its greatest square divisor, while d ranges over the divisors of n which are $\equiv 3 \pmod 4$. Lerch has since[274] repeated the deduction in detail. From (1) follows [74]

$$\text{(2)}\ \sum_{a=1}^{n-1}\left\{\tfrac{1}{4}+\frac{a^2m}{n}-E\left(\tfrac{1}{4}+\frac{a^2m}{n}\right)\right\}$$
$$=\frac{2n-1}{4}-(-1)^{\frac{n-1}{2}}\sum_{d_1}\left(\frac{m}{d_1}\right)\frac{2}{\tau_{4d_1}}Cl(-4d_1)-\sum_{d_3}\left(\frac{m}{d_3}\right)\frac{1-(2/d_3)}{\tau_{d_3}}Cl(-d_3),$$

in which d_1 and d_3 range over the divisors of n such that $d_1\equiv 1$, $d_3\equiv 3 \pmod 4$.

J. de Séguier[246] in a paper primarily on certain infinite series and on genera simplified his results by substituting the class-number for its known value. He found, for example (p. 114), if $F(x)$ is an arbitrary function which insures convergence, then

$$\frac{1}{K(D_0Q^2)}\sum_{a,\,b,\,c}^{D_0Q^2}\left(\frac{D_1}{A}\right)F(am^2+bmn+cn^2)$$
$$=\sum_{d}^{Q}\frac{O(D_1,d)\tau(D_0d^2)}{K(D_0d^2)}\sum_{hk}\left(\frac{D_1d^2}{h}\right)\left(\frac{D_0D_1^{-1}d^4}{k}\right)F(d'^2hk),$$

where $K(m)$ is the number of properly primitive classes of discriminant m; A is representable by $am^2+bmn+cn^2$; $D=D_1D_2=D_0Q^2$, D_0 being fundamental; and $O(D_1, d)$ is the number of classes of discriminant D_1 and of order d, where $dd'=Q$.

*J. S. Aladow[247] evaluated in four separate cases the number G of classes of odd binary quadratic forms of prime negative determinant $-p$:

---

[244] Rozpravy české Akad., Prague, 7, 1898, No. 6; French resumé in Bull. de l'Acad. des Sc. Bohème, 5 1898, 36–37.

[245] Rozpravy české Akad., Prague, 7, 1898 No. 7 (Bohemian). French resumé in Bull de l'Acad. des Sc. Bohème, Prague, 5, 1898, 37–38.

[246] Jour. de Math. (5), 5, 1899, 55–115.

[247] St. Petersburg Math. Gesells., 1899, 103–5 (Russian).

(i) If $p\equiv 7 \pmod 8$, G equals the difference between the number of quadratic residues and non-residues $\leqq \frac{1}{8}\{p-3-2(3/p)\}$.

(ii) If $p\equiv 3 \pmod 8$, G equals the difference between the number of quadratic residues in the sequence

$$\tfrac{1}{4}(p+1),\quad \tfrac{1}{4}(p+5),\ldots,\quad \tfrac{1}{8}\{2p-3-(3/p)\},$$

and the number in the sequence

$$\tfrac{1}{8}\{p+3-2(3/p)\},\ldots,\quad \tfrac{1}{4}(p-3).$$

(iii) If $p\equiv 5 \pmod 8$, G equals twice the difference between the number of quadratic residues and non-residues in the sequence

$$\tfrac{1}{8}\{p+3+2(3/p)\},\quad \tfrac{1}{8}\{p+9+2(3/p)\},\ldots,\quad \tfrac{1}{4}(p-1).$$

(iv) If $p\equiv 1 \pmod 8$, G equals twice the sum of the difference between the number of quadratic residues and non-residues in the sequence

$$\tfrac{1}{4}(p+3),\ldots,\quad \tfrac{1}{8}\{2p-3+(3/p)\}$$

and the corresponding difference in the sequence

$$\tfrac{1}{8}\{p+3+2(3/p)\},\quad \tfrac{1}{8}\{p+9+2(3/p)\},\ldots,\quad \tfrac{1}{4}(p-1).$$

R. Dedekind,[248] in a long investigation of ideals in a real cubic field, proved the following result. If at least one of the integers a, b, ab is divisible by no square, and if we write $k=3ab$ or $k=ab$, according as a^2-b^2 is not or is divisible by 9, then the number of all non-equivalent, positive, primitive forms $Ax^2+Bxy+Cy^2$ of discriminant $D\equiv B^2-4AC=-3k^2$ is a multiple $3K$ of 3. For primes $p\equiv 1 \pmod B$, p not dividing D, K of the forms represent all and only such primes p of which ab^2 is a cubic residue, while the remaining $2K$ forms represent all and only such primes p of which ab^2 is a cubic non-residue.

D. N. Lehmer[249] calls any point in the cartesian plane a *totient point* if its two co-ordinates are integers and relatively prime. He wrote

$$P_{(m,k)}=\prod_{i=1}^{r}\frac{p_i-1}{p_i^{a_i-1}(p_i^{m+1}-1)},\qquad k=\prod_{i=1}^{r}p_i^{a_i}.$$

The number of totient points[250] in the ellipse $ax^2+2bxy+cy^2=N$, $b^2-4ac=D=-\Delta$, is

$$(1)\qquad \frac{12N}{\pi}\sqrt{\Delta}P_{(1,2\Delta)};$$

and in the hyperbolic sector, always taken[251] in this connection, the number is

$$(2)\qquad \frac{6}{\pi^2}\sqrt{D}P_{(1,2D)}N\log(T+U\sqrt{D}),$$

N being very great in both cases. Noting now Dirichlet's[93] formula (2) for the

[248] Jour. für Math., 121, 1900, 95.
[249] Amer. Jour. Math., 22, 1900, 293–335. Cf. Lehmer,[218] Ch. V, Vol. I. of this History.
[250] Cf. G. L. Dirichlet,[20] Zahlentheorie, Art. 95.
[251] Cf. *ibid.*,[19] Art. 98, ed. 4, 1894, 246

number of representations of a given number by a system of quadratic forms of determinant D, he finds the class-number, for example, for $D=-\Delta<0$,

$$h(D)=\epsilon\frac{\pi}{12}\frac{1}{\sqrt{\Delta}}P_{(1,\,2\Delta)}\lim_{N=\infty}\frac{1}{N}\sum_{x=1}^{N}2^{\nu(x)}\Theta_s(x),$$

in which ϵ is the number of solutions of $t^2-Du^2=1$; x is any positive number relatively prime to $2D$, $\nu(x)$ is the number of distinct prime factors of x; $\Theta_s(x)=1$ or 0, according as each prime divisor does or does not have D as a quadratic residue.

K. Petr,[252] by the use of five functions A (=Hermite's[69] $\mathscr{A}$), B, C, D, E, all analogous to Hermite's[69] $\mathscr{A}$, deduced all of Kronecker's[54] eight classic relations.

For example, from expansions by C. Jordan (Cours d'analyse, II, 1894, 409–411), he obtained

$$\text{(1)}\qquad \Theta_3\Theta_1^2\frac{\Theta^2(v)\Theta_1(v)}{\Theta_2^2(v)}=C\cdot\Theta_1(v)-8\sum_1^{\infty}\cos(2n+1)\pi v\cdot q^{(n+\frac{1}{2})^2}\sum_1^{n}kq^{-k^2}.$$

Also C is the coefficient[253] of $2q^{\frac{1}{4}}\cos\pi v$ in the product of the right member of

$$\text{(2)}\qquad \Theta_1\frac{\Theta(v)\Theta_1(v)}{\Theta_2(v)}=2\sum_1^{\infty}\sin 2n\pi v q^{n^2}\cdot\{2q^{-\frac{1}{4}}+2q^{-\frac{9}{4}}+\ldots+2q^{-(n-\frac{1}{2})^2}\}$$

by the right member of[254]

$$\text{(3)}\qquad \Theta_1\Theta_3\frac{\Theta(v)}{\Theta_2(v)}=\frac{4q^{\frac{1}{2}}\sin\pi v}{1-q}+\frac{4q^{3/2}\sin 3\pi v}{1-q^3}+\frac{4q^{5/2}\sin 5\pi v}{1-q^5}+\ldots.$$

But in that product, the coefficient of $\cos\pi v$ is a power series in q in which the coefficient of $q^{N+\frac{1}{4}}$ is 8 times the combined number of solutions of

$$n^2-(k+\tfrac{1}{2})^2+(n-\tfrac{1}{2})(2l+1)=N+\tfrac{1}{4},$$
$$n^2-(k+\tfrac{1}{2})^2+(n+\tfrac{1}{2})(2l+1)=N+\tfrac{1}{4},$$

where n and l are positive integers, l taking also the value zero; $k=0, 1, 2, \ldots, n-1$. But these equations can be written[255] in the forms

$$\text{(4)}\qquad \begin{cases}(n-k+1)(n+k)+(n-k-1)(l+1)+(n+k)(l+1)=N,\\ (n-k)(n+k+1)+(n-k)(l)+(n+k+1)(l)=N;\end{cases}$$

and the left members may be regarded as the discriminants $N=ab+bc+ca$ of reduced Selling[255a] quadratic forms $a(y-t)^2+b(t-x)^2+c(x-y)^2$, in which a, b, c do not agree in parity. Since there is a correspondence between such Selling forms of discriminant N and odd classes of Gauss forms of determinant $-N$, we have

$$\text{(5)}\qquad C=8\sum F(n)q^n.$$

The identity (Fundamenta Nova, § 41)

$$\Theta_1^2\Theta_3^2\frac{\Theta^2(v)}{\Theta_2^2(v)}=8\sum\frac{nq^n}{1-q^{2n}}-8\sum\frac{nq^n\cos 2\pi nv}{1-q^{2n}}$$

---

252 Rozpravy české Akad., Prague, 9, 1900, No. 38 (Bohemian); Abstract,[261] Bull. Internat. de l'Acad. des Sc. de Bohème, Prague, 7, 1903, 180–187 (German).
253 Cf. P. Appell, Annales de l'Ecole Norm. Sup. (3), 1, 1884, 135; 2, 1885, 9.
254 Cf. C. G. J. Jacobi, Fundamenta Nova 1829, p. 101, (19); Werke I, 1881, 157.
255 Cf. J. Liouville,[88] Jour. de Math. (2), 7, 1862, 44; Bell,[370] and Mordell.[372]
255a E. Selling, Jour. für Math., 77, 1874, 143.

is multiplied member by member with Jacobi's expansion formula for $\Theta_1(v)$. In the resulting left member, the coefficient of $\cos \pi v$ is $C \cdot 2q^{\frac{1}{4}}\Theta_3$. When this coefficient is equated to the coefficient of $\cos \pi v$ in the resulting right member, a comparison with (5) yields the relation:

$$\text{(I)} \qquad F(n)+2F(n-1^2)+2F(n-2^2)+\ldots=\Sigma d_\lambda-\Sigma d_1,$$

where d_λ denotes a divisor of n which has an odd conjugate and d_1 denotes a divisor of n which is $\leqq \sqrt{n}$ and which agrees with its conjugate in parity.

He also found the classic formula[54] for the number of solutions of $x^2+y^2+z^2=n$.

To obtain a class-number relation of Liouville's[256] second type, Petr expands in powers of v each member of an identity of the same general type as (1) above. Coefficients of v^2 are equated, with the result that

$$1^2F(8n-1^2)+3^2F(8n-3^2)+5^2F(8n-5^2)+\ldots$$
$$=2n\Sigma d_\lambda-2n\Sigma(d_{1l}+d_{1\lambda})-\Sigma(d_{1l}^3+d_{1\lambda}^3),$$

where, the d's are the divisors of $2n$; $d_1<\sqrt{2n}$; d_l is odd; d_λ has an odd conjugate; and the subscripts of d retain their significance when they are compounded.

To obtain a class-number relation of Liouville's[257] first type, each member of an identity of the same general type as (1) above is expanded in the neighborhood of $v=\frac{1}{2}$. Equating coefficients of v, Petr then obtains

$$H(8n-1^2)-3H(8n-3^2)+5H(8n-5^2)-\ldots=\Sigma(-1)^{\frac{1}{2}(d'_1+d'_2+1)}d_1'^2,$$

where d_1' is a divisor of $2n$ such that its conjugate d_2' is of different parity, and $d_1'<\sqrt{2n}$.

K. Petr,[258] employing the same notation as[252] in 1900, multiplied member by member the identity

$$\Theta_1^2\Theta_2\Theta_3^2\frac{\Theta(v)}{\Theta_2^2(v)}=8\Sigma(-1)^n(n+k)q^{n^2-\frac{1}{4}+k(2n+1)}\sin(2k-1)\pi v,$$
$$n=0,1,2,3,\ldots;\quad k=1,2,3,\ldots.$$

by the formula for transformation of order 2

$$\Theta(v)\Theta_1(v)/\Theta_2(0,2\tau)=\Theta(2v,2\tau).$$

In the resulting left member, the coefficient of $q^{\frac{1}{4}}\cos\pi v$ is $16\Sigma F(n)q^n\Theta_2(0,2\tau)$; in the right member it is 8 times the sum of

$$(-1)^{n+k-1}(n+2k)q^{(n+2k)^2-2k^2}, \qquad (-1)^{n+k-1}(n+2k-1)q^{(n+2k-1)^2-2(k-1)^2},$$

for $n=0,1,2,3,\ldots;k=1,2,3,\ldots.$ Hence

$$\text{(1)} \qquad \Sigma(-1)^\nu F(n-2\cdot\nu^2)=\Sigma(-1)^{x+y-1}x,$$

where x and y are the integer solution of $x^2-2y^2=n$, $x \gtreqless 2y$, $y \gtreqless 0$; while, as also in

[256] J. Liouville,[107] Jour. de Math., (2) 12, 1867, 99. Cf. G. Humbert,[293] Jour. de Math., (6), 3, 1907, 369–373, formulas (40)—(44), as numbered in the original memoir.

[257] Cf. J. Liouville,[107] Jour. de Math. (2), 14, 1866, 1; also G. Humbert,[293] *ibid.* (6), 3, 1907, 366–369, formulas (35), (36).

[258] Rozpravy české Akad., Prague, 10, 1901, No. 40 (Bohemian). Abstract, Bull. Internat. de l' Acad. des Sc. de Bohème Prague, 7, 1903 180–187, (German).

(2), ν ranges over all integers, positive, negative, or zero. In the summation x receives an extra coefficient $\frac{1}{2}$ if one of the inequalities becomes an equality. Similarly,

(2) $\Sigma(-1)^{\nu}F(8n-1-8\nu^2)=\Sigma(-1)^{\frac{1}{2}(x+y)}y,\quad x^2-2y^2=8n-1,\quad x>2y,\quad y>0.$

These are the first published class-number relations which are obtained from elliptic function theory and which involve an indefinite quadratic form, e. g., x^2-2y^2.

By means of the elementary relation

$$\pi\Theta_1\Theta_2\Theta_3=2\pi\Sigma(-1)^n(2n+1)q^{(n+\frac{1}{2})^2},\qquad n=0,1,2,3,\ldots$$

and the relation

$$\Theta_1^2\Theta_3=4\Sigma F(4n+2)q^{\frac{1}{4}(4n+2)}$$

the identity $\Theta_1^2\Theta_3\cdot\Theta_2=\Theta_1\Theta_2\Theta_3\cdot\Theta_1$ yields

$$F(4n+2)-2F(4n+2-4\cdot 1^2)+2F(4n+2-4\cdot 2^2)+\ldots=\Sigma(-1)^{\frac{1}{2}(x-1)}x,$$

$x,y>0$, $x^2+y^2=4n+2$; which is of the type of Hurwitz.[202]

A transformation formula of order 3 in a treatment similar to the above yields five such relations as

$$F(4n+3)-H(4n+3)-2[F(4n+3-3\cdot 1^2)-H(4n+3-3\cdot 1^2)]+\ldots=\Sigma(-1)^{\frac{1}{2}(x+y-1)}y,$$

$x^2+3y^2=4n+3$, $x\gtreqless 0$, $y>0$; and

(3) $$F(4n)-2F(4n-3\cdot 1^2)+2F(4n-3\cdot 2^2)-\ldots=-2\Sigma x,$$

$x^2-3y^2=2n$, $y\gtreqless 0$, $x\gtreqless 3y$.

From transformations of order 5, Petr obtained three relations including,

(4) $$F(8n)-2F(8n-5\cdot 1^2)+2F(8n-5\cdot 2^2)-\ldots=-4\Sigma x$$

$x^2-5y^2=2n$, $y\gtreqless 0$, $5y\leqq x$.

M. Lerch[259] wrote

$$R(\omega,s)=\sum_{\nu=0}^{\infty}\frac{1}{(\omega+\nu)^s},\quad K(a,b,c;s)=\Sigma'(am^2+bmn+cn^2)^{-s},$$

where ω is an arbitrary constant; $m,n=0,\pm1,\pm2,\ldots$, except $m=n=0$; (a,b,c), a positive form of negative discriminant $-\Delta$; a,b,c real. From Dirichlet's[20] fundamental equation (2), it follows that the relation

(1) $$\sum_{a,b,c}K(a,b,c;s)=\tau\Delta^{-s}R(1,s)\sum_{r=1}^{\Delta-1}\left(\frac{-\Delta}{r}\right)R\left(\frac{r}{\Delta},s\right)$$

is valid over the complex s-plane, if (a,b,c) ranges over a system of representative primitive positive forms of discriminant $-\Delta$, which is now supposed to be fundamental.

---

[259] Comptes Rendus, Paris, 135, 1902, 1314–1315.

Employ the Maclaurin developments in powers of s,

$$(2)\qquad R(\omega, s) = (\tfrac{1}{2}-\omega) + \log\frac{\Gamma(\omega)}{\sqrt{2\pi}}\cdot s + \dots;$$

$$(3)\quad K(a, b, c; s) = -1 - 2s\log\left[\frac{2\pi}{\sqrt{c}}H\left(\frac{-b+i\sqrt{\Delta}}{2c}\right)H\left(\frac{b+i\sqrt{\Delta}}{2c}\right)\right] + \dots,$$

where

$$H(\omega) = e^{\omega\pi i/12}\prod_1^\infty(1-e^{2n\omega\pi i}).$$

When substitution is made of (2) and (3) in (1), Lerch compares the terms which are independent of s and obtains Kronecker's[171] class-number formula (5).

E. Laudau[260] showed that every negative determinant <-7 has more than one properly primitive reduced form (cf. the conjecture of Gauss,[4] Disq. Arith., Art. 303) by proving that if $-\Delta = b^2 - ac$ is <-7, there is always another such form in addition to $(1, 0, \Delta)$. If there is no properly primitive reduced form $(a, 0, c)$ other than $(1, 0, \Delta)$, then Δ has no distinct factors, but must be of the form p^λ, p a prime.

(I) If $p=2$, and $\lambda \geqq 4$, there is the additional properly primitive reduced form $(4, 2, 2^{\lambda-2}+1)$.

(II) If p is an odd prime and if there is no reduced properly primitive form with $b=1$, then $\Delta+1$ cannot be expressed as $a\cdot c$, where one of the factors is uneven and >2. Hence $\Delta+1=2^\nu$. When $\nu \geqq 6$, there is an additional properly primitive reduced form $(8, 3, 2^{\nu-3}+1)$.

Landau now tested the few remaining admissible Δ's and found none which are >7 and have a single class.

K. Petr[261] gave in German an abstract of his two long Bohemian papers,[252, 258] including eleven class-number relations of the second paper. He indicated completely a method of expanding $\Theta_1^3\Theta_1(0, 5\tau)$, which leads to new expressions[340] for the number of solutions of $x^2+y^2+z^2+5\omega^2=n$ and hence to generalizations of Petr's[258] relation (4).

M. Lerch,[262] in order to find the negative discriminants $-\Delta$ for which $Cl(-\Delta)=1$, wrote $-\Delta = -\Delta_0 Q^2$, where Δ_0 is fundamental and q ranges over the distinct factors of $Q=Q'\Pi q$. Then the equation to be satisfied is (Kronecker,[171] (4))

$$Cl(-\Delta) = \frac{2}{\tau_0}Q'\prod_q\left\{q-\left(\frac{-\Delta_0}{q}\right)\right\}Cl(-\Delta_0) = 1.$$

If $\Delta_0=4$, then $\tau_0=4$, $Q=1$ or 2.

If $\Delta_0=3$, then $\tau_0=6$, $Q=1$, 2 or 3.

If $\Delta_0>4$, then $\tau_0=2$. Here $Cl(-\Delta)$ can be uneven only for $Q'=1$ and Δ_0 prime, or for $\Delta_0=8$. The case $\Delta_0=8$ is excluded if $Q\neq 1$. If Δ_0 is a prime, $Cl(-\Delta)>1$ unless $Q=q=2$, $(2/\Delta_0)=1$, i. e., $\Delta_0=8k-1$. But if $k\geqq 2$, $(1, 1, 2k)$ and $(2, 1, k)$ are non-equivalent reduced forms of discriminant $-\Delta_0$.

---

260 Math. Annalen, 56, 1902, 671–676.
261 Bull. Internat. de l'Acad. des Sc. de Bohème, Prague, 7, 1903, 180–187.
262 Math. Annalen, 57, 1903, 569–570.

Hence $Cl(-\Delta)=1$ for $\Delta=4, 8; 3, 12, 27; 8, 7, 28$. Any further solution Δ must be a prime $\equiv 3 \pmod 8$. But it is undecided whether there are such solutions other than 11, 19, 43, 67, 163.

Lerch[263] wrote $\psi(x)$ for $\Gamma'(x)/\Gamma(x)$ and observed that Dirichlet's[23] formula (7) for the number of positive classes of a positive fundamental discriminant D gives the relation

$$\sum_{h=1}^{D-1}\left(\frac{D}{h}\right)\psi\left(\frac{h}{D}\right)=-\sqrt{D}\,Cl(D)\log E(D).$$

From this ψ is eliminated by means of

$$-C-\log 4\pi+\log a-\psi(x)-\psi(1-x)$$
$$=\sum_{m=-\infty}^{\infty}\int_{1/a}^{\infty}a^{-(x+m)^2 z\pi}\frac{dz}{\sqrt{z}}+2\sum_{n=1}^{\infty}\cos 2nx\pi\int_a^{\infty}e^{-n^2 z\pi}\frac{dz}{z},$$

where C is the Euler constant[330] and a an arbitrary positive constant. The final result is that $Cl(D)$ is determined uniquely by

$$\frac{S-2(P_r+Q_r)}{\log E(D)}-2<Cl(D)<\frac{S-2(P_r+Q_r)}{\log E(D)},$$

in which, to a close approximation,

$$S=\tfrac{1}{2}\sqrt{D}(\log D+.046181)-\tfrac{1}{2}\log D+.023090,$$

$$P_r=\sum_{\beta\leqq r}\frac{2}{\beta\sqrt{\pi/D}}\int_{\beta\sqrt{D/\pi}}^{\infty}e^{-x^2}dx,\qquad Q_r=\tfrac{1}{2}\sum_{\beta\leqq r}\int_{\beta^2\pi/D}^{\infty}e^{-x}\frac{dx}{x},$$

while r is chosen sufficiently large to insure a unique determination of $Cl(D)$. For example, if $D=9817$, $\log E(D)=222$, $S=450.5$, whence $Cl(D)<450/222$. We need not compute P_r and Q_r since $Cl(D)$ is uneven (Dirichlet[93]) and hence is 1.

J. W. L. Glaisher[264] called a number s a *positive*, a *negative* or a *non-prime* with respect to a given number P, according as the Jacobi-Legendre symbol $(s/P)=+1$, -1, or 0. He denoted by a_r, b_r, λ_r, respectively, the number of positives, negatives and non-primes in the r-th octant of P. For example, if $P=8k+1$ is without a square factor, Dirichlet's[23] formulas (5) for the number of properly primitive classes of determinant $-P$ and $-2P$, respectively,

$$h'=2(a_1-b_1+a_2-b_2),\qquad h''=2(a_1-b_1-a_4+b_4)$$

become[265]

$$h'=4(\alpha_1+\alpha_2)-\tfrac{1}{2}(P-1),\qquad h''=4(\alpha_1-\alpha_2),$$

where $\alpha_r=a_r+\frac{1}{2}\lambda_r$. Similarly for other types of P. Obvious congruencial properties (mod 8) of h' and h'' are deduced from all of these formulas.

Again h' and h'' are expressed in terms of $\beta_r=b_r+\frac{1}{2}\lambda_r$ $(r=1, 2, 3, 4)$. Next, l_r and μ_r are used to denote respectively the number of positives and non-primes $<P$

---

[263] Jour. de Math. (5), 9, 1903, 377–401; Prace mat. fiz. Warsaw, 15, 1904, 91–113 (Polish).
[264] Quar. Jour. Math., 34, 1903, 1–27.
[265] Glaisher, Quar. Jour. Math., 34, 1903, 178–204.

which are of the form $8k+r$, while $L_r=l_r+\frac{1}{2}\mu_r$. A table (p. 13) transforms the preceding formulas into results such as

$$P=8k+1, \qquad h'=2(L_1-L_3), \qquad h''=4(L_1-L_7).$$

If Q_r denotes the number of uneven positives in the rth quadrant plus $\frac{1}{2}\lambda_r$, we have, for example,

$$P=8k+1, \qquad h'=2(Q_4-Q_2), \qquad h''=4(Q_4-Q_1).$$

L. C. Karpinski[266] gave details of R. Dedekind's[267] brief proofs of his theorems which state the distribution of quadratic residues of a positive uneven number P in octants and 12th intervals of P in terms of the class-number of $-P$, $-2P$ and $-3P$. He added to Dedekind's notation the symbols C_5 and C_6, which denote the number of properly primitive classes of determinant $-5P$ and $-6P$, respectively, and, by an argument precisely parallel to that of Dedekind, obtained for all positive uneven numbers P which have no square divisor, the distribution of quadratic residues in the 24th intervals of P as linear functions of C_1, C_2, C_3, C_6. He put $S_r^t=\Sigma(s_r/P)$, where t is a positive integer, and s_r ranges over the integers x for which

$$(r-1)P/t<x<rP/t.$$

He deduced such relations as the following: If $P\equiv 23 \pmod{24}$,

$$(1) \qquad S_1^6=-S_6^6=C_1, \qquad S_2^6=S_3^6=S_4^6=S_5^6=0.$$

If $P\equiv 1$, 5 or 17 (mod 24), C_3 is a multiple of 6. For $P\equiv 3 \pmod 4$,

$$C_1=S_1^{10}+S_2^{10}+S_3^{10}+S_4^{10}+S_5^{10}, \qquad C_5=2S_2^{10}+4S_3^{10}+2S_4^{10}.$$

Cf. Dirichlet,[23] (5). Three other relations among S_r^{10} which arise from familiar properties of quadratic residues lead to a complete determination of S_r^{10} as linear functions of C_1 and C_5 for $r=1, 2, 3, \ldots, 10$.

E. Landau[268] studied the identity

$$\sum_{n=1}^{\infty}\left(\frac{D}{n}\right)\frac{1}{n^s}=\frac{\Gamma(1-s)}{\pi}\left(\frac{2\pi}{\Delta}\right)^s\sqrt{\Delta}\cos\tfrac{1}{2}s\pi\sum_{n=1}^{\infty}\left(\frac{D}{n}\right)\frac{1}{n^{1-s}}, \quad D<0,$$

which is valid for a real s, $0<s<1$. The limit of the right member for $s=0$ is

$$\frac{\sqrt{\Delta}}{\pi}\sum_{n=1}^{\infty}\left(\frac{D}{n}\right)\frac{1}{n}.$$

The customary evaluation of the divergent left member for $s=0$ would give (Dirichlet,[20] (1) above) the erroneous result $h=\Sigma_1^{\infty}(D/n)$. A similar study is made of the limit for $s=0$ of the ratio

$$\sum_{n=1}^{\infty}\left(\frac{D}{n}\right)\frac{\log n}{n^s}\div\sum_{n=1}^{\infty}\left(\frac{D}{n}\right)\frac{1}{n^s},$$

which for $s=1$ is Kronecker's[213] limit ratio.

---

266 Thesis, Strassburg, 1903, 21 pp.; reprinted, Jour. für Math., 127, 1904, 1–19.
267 Werke of Gauss, II, 1863, 301–3; Maser's German translation of Disq. Arith., 1889, Remarks by Dedekind, 693–695.
268 Jour. für Math., 125, 1903, 130–132, 161–182.

*M. Lerch[269] denoted by g an arbitrary primitive root of a prime $p=2m+1$, and put

$$F_n(x)=\sum_{\nu=1}^{p=1} a^{\text{ind}\,\nu} x^\nu,$$

where a is an integer of index $m=p-1-n$ referred to the primitive root g as base. C. G. J. Jacobi[270] had found the relation

$$F_m(1+y)\equiv -Y_m/m!\pmod p,$$

where Y_m is the sum of the terms in $y^m, y^{m-1}, \ldots, y^{p-1}$ in the Maclaurin's expansion of $[\log(1+y)]^n$. Thus

$$\sum_{\nu=1}^{p-1}\left(\frac{\nu}{p}\right)x^\nu\equiv -\frac{1}{m!}Y_m(x-1)\pmod p.$$

Hence if c_j is the coefficient of y^j in $Y_m(y)$, and if we set

$$-\frac{1}{m!}\sum_{\nu=m}^{2m} c_\nu(i-1)^\nu=A+iB,$$

then $A\equiv B\equiv H\pmod p$, in which H is the number of positive quadratic forms of discriminant $-4p$.

H. Poincaré[271] wrote

$$F(q)=\sum_{m,n} q^{am^2+2bmn+cn^2},\quad q=e^{-t},$$

where (a, b, c) is a fixed representative properly primitive form of negative determinant $-p$ and the summation is taken over every pair of integers m, n, for which the value of (a, b, c) is prime to $2p$ except $m=n=0$. $F(q)$ is regarded as a special case of the Abelian function

$$(A)\qquad \Theta(x,y)=\Sigma e^{i(mx+ny)}q^{am^2+2bmn+cn^2}.$$

The theory of the flow of heat is used to show that if k, k' each range over all integral values, $\Theta(x, y)$ may be written

$$(B)\qquad \Theta(x,y)=\sum_{k,k'}\frac{2\pi}{Et}e^{-P},\quad -\tfrac{1}{4}E^2=b^2-ac=-p,$$

$$P=\frac{1}{E^2t}[a(y-2k'\pi)^2-2b(y-2k'\pi)(x-2k\pi)+c(x-2k\pi)^2].$$

Now for $x=y=0$ and t small, $\Theta(x, y)$ is asymptotically $F(1)$. Hence, in the neighborhood of $q=1$,

$$(1)\qquad F(q)=\frac{2\pi}{Et}$$

and is therefore independent of the choice of (a, b, c) of determinant $-p$.

But, for p a prime $\equiv 3\pmod 4$, we have (cf. Dirichlet's[20] formula (2))

$$(2)\qquad \sum_{(a,b,c)}\sum_{m,n} q^P=2\Sigma\left(\frac{n}{p}\right)q^{nn'},$$

---

[269] Bull. Int. de l'Acad. des Sc. de Cracovie, 1904, 57–70 (French).

[270] Monatsber. Akad. Wiss. Berlin, 1837, 127; Jour. für Math., 30, 1846, 166; Werke, Berlin, VI, 1891, 254–258.

[271] Jour. für Math., 129, 1905, 120–129.

where now $P=am^2+2bmn+cn^2$ ranges over a system of properly primitive forms of determinant $-p$, and m, n take all pairs of integral and zero values for which P is prime to $2p$; in the second member, n, n' range over every pair of odd positive integers each prime to P. By a simple transformation of each member, (2) can be written

$$(3) \qquad \sum_{(a,b,c)} F(q) - \sum_{(a,b,c)} F(-q) = 4\Sigma\left(\frac{n}{q}\right)\frac{q^n}{1-q^{2n}}.$$

But from (A), it follows that $F(q)=\Theta(0,0)$, $F(-q)=\Theta(\pi,\pi)$; and hence from (B), it follows that

$$F(q)=\Sigma\frac{2\pi}{Et}e^{-P}, \quad F(-q)=\Sigma\frac{2\pi}{Et}e^{-P}, \quad P=\frac{\pi^2}{E^2t}(a\mu^2-2b\mu\nu+c\nu^2),$$

where μ, ν are even integers in the case of $F(q)$, and odd integers in the case of $F(-q)$. Since for t small, all terms of the left member of (3) except those having $\mu=\nu=0$ are to be neglected, the left member becomes

$$\frac{\pi}{t\sqrt{p}}h(-p).$$

Moreover

$$\lim_{t=0}\frac{tq^n}{1-q^{2n}}=\frac{1}{2n}.$$

Hence[272] (3) becomes Dirichlet's[14] formula (2). Equation (3) is also transformed to give Dirichlet's[23] closed form (5) for $h(-p)$.

A. Hurwitz[273] by the substitution

$$u=\frac{\alpha_1x+\beta_1y+\gamma_1z}{\alpha x+\beta y+\gamma z}, \quad v=\frac{\alpha_2x+\beta_2y+\gamma_2z}{\alpha x+\beta y+\gamma z}$$

transformed the Cartesian area $\iint du\,dv$ of a plane region G into what he called the generalized area of G with respect to the form $\alpha x+\beta y+\gamma z$. Such a generalized area of the conic $xy-z^2=0$ is

$$(1) \qquad 2\pi/(\sqrt{4\alpha\gamma-\beta^2})^3.$$

For points on the conic, we put $x=r^2$, $y=rs$, $z=s^2$, and consider points $(x, y, z)=(r, s)=(-r, -s)$, r and s being relatively prime integers. An *elementary triangle* is one having as its three vertices the points

$$(2) \qquad (r, s), \quad (r_1, s_1), \quad (r+r_1, s+s_1), \quad rs_1-r_1s=\pm 1.$$

All such possible triangles in the aggregate cover the conic simply six times and their total area is

$$(3) \qquad \tfrac{1}{2}\Sigma\{\alpha r^2+\beta rs+\gamma s^2)(\alpha r_1^2+\beta r_1s_1+\gamma s_1^2)\,[\alpha(r+r_1)^2+\beta(r+r_1)(s+s_1)+\gamma(s+s_1)^2]\}^{-1}$$

summed for the solutions r, s, r_1, s_1 of $rs_1-r_1s=\pm 1$.

[272] Cf. G. L. Dirichlet, Zahlentheorie, Art. 97.
[273] Jour. für Math., 129, 1905, 187–213.

But if the Gauss form $au^2+\beta uv+\gamma v^2$ be subjected to all the unitary substitutions, it goes over into $\alpha' u'^2+\beta' u'v'+\gamma' v'^2$, where α', β', γ' have values such that (3) can be written as $\Sigma 8/\{\alpha'\gamma'(\alpha'+\beta'+\gamma')\}$, where $(\alpha', \beta'/2, \gamma')$ ranges over all forms equivalent to $(\alpha, \beta/2, \gamma)$. Hence by comparison of (1) and (3) we have

$$\frac{3\pi}{2D\sqrt{D}}\cdot h=\sum_{a,b,c}\frac{1}{ac(a+2b+c)},$$

where (a, b, c) ranges over all positive forms of determinant D.

By modifying his definition of generalized area Hurwitz obtained for the right member a more rapidly convergent series.

M. Lerch,[274] by use of his[240] trigonometric formula for E^*, showed by means of Gauss sums that

$$S\equiv\sum_{a=0}^{n-1}E^*\left(x+\frac{a^2m}{n}\right)=\sum_{a=0}^{n-1}\left(x+\frac{a^2m}{n}\right)-\frac{n}{2}+S_1,$$

in which S_1 is the imaginary part of

$$\sum_{\nu=1}^{\infty}\frac{1}{\nu\pi}\left(\frac{m\nu'}{d'_\nu}\right)d_\nu\sqrt{d'_\nu}\cdot i^{\frac{1}{4}(d'_\nu-1)^2}e^{2d_\nu\nu' x\pi i},$$

where d_ν is the g.c.d. of n and ν, and $d'_\nu=n/d_\nu$, $\nu'=\nu/d_\nu$. Then, if we put $d'_\nu=d$ $d_\nu=d'$, and also

$$\Phi(z,d)=\sqrt{d}\sum_{\nu=1}^{\infty}\left(\frac{\nu}{d}\right)\frac{\cos 2\nu z\pi}{\nu\pi},\text{ if } d\equiv-1 \pmod 4;$$

$$=\sqrt{d}\sum_{\nu=1}^{\infty}\left(\frac{\nu}{d}\right)\frac{\sin 2\nu z\pi}{\nu\pi},\text{ if } d\equiv+1 \pmod 4;$$

we find

$$S_1=\sum_{d|m}\left(\frac{m}{d}\right)\Phi(d'x,d).$$

Hence we get the chief formula of this memoir:

$$\text{(1)}\qquad \sum_{a=0}^{n-1}\left\{E^*\left(x+\frac{a^2m}{n}\right)-\left(x+\frac{a^2m}{n}\right)\right\}=-\frac{n}{2}+\sum_{d}\left(\frac{m}{d}\right)\Phi(d'x,d).$$

But by Kronecker,[171] (2), $\Phi(0,\Delta)=2\tau^{-1}Cl(-\Delta)$, where $Cl(-\Delta)$ denotes the number of primitive positive classes of discriminant $-\Delta$. And for $x=0$, m positive and relatively prime to n, (1) becomes[275] Lerch's formula[245] (1).

For $x=\frac{1}{2}$, (1) becomes[276]

$$\sum_{a=1}^{n-1}\left[\frac{a^2m}{n}-E\left(\frac{a^2m}{n}+\frac{1}{2}\right)\right]=\sum_{d}\left(\frac{m}{d}\right)\left\{1-\left(\frac{2}{d}\right)\right\}\frac{2}{\tau_d}Cl(-d),$$

d ranging over the divisors $4k+3$ of n. Similar results are obtained by taking $x=\frac{1}{4}$ (cf. Lerch[245] (2)) and $x=1$.

---

[274] Annali di Mat. (3), 11, 1905, 79–91.

[275] Cf. Lerch, Rozpravy české Akad., Prague, 7, 1898, No. 7; also Bull. de l'Acad. des Sc. Bohème, Prague, 1898, 6 pp.

[276] Reproduced by Lerch in his Prize Essay,[278] Acta Math., 30, 1906, 242, formula (40).

Lerch observed that the sum A of the quadratic residues of an odd number n, which are prime to n, >0 and $<n$, is given by

$$2^\omega A=\sum_{\nu=1}^{n}\left[1+\left(\frac{\nu}{p_1}\right)\right]\left[1+\left(\frac{\nu}{p_2}\right)\right]\cdots\left[1+\left(\frac{\nu}{p_\omega}\right)\right]\left(\frac{n^2}{\nu}\right)\nu,$$

where $p_1, p_2, \ldots, p_\omega$ are the distinct prime divisors of n. Hence

$$2^\omega A=\sum_{d}\sum_{\nu=1}^{n}\left(\frac{\nu}{d}\right)\left(\frac{n^2}{\nu}\right)\nu,$$

where d ranges over those divisors of n which have no square factor. By means of the Moebius function (this History, Vol. I, Ch. XIX), he transformed this into

$$2^\omega A=\tfrac{1}{2}n\phi(n)-n\sum_{d}\frac{2}{\tau_d}Cl(-d)M_d(n),$$

where d ranges over those divisors $\equiv -1 \pmod 4$ of n which have no square factor and

$$M_d(n)=\Pi\{1-(p/d)\},$$

where p ranges over the distinct prime factors of $d'=n/d$.

M. Lerch[277] in a prize essay wrote an expository introduction on class-number from the later view-point of L. Kronecker[171]; and stated without proof that if $\mathscr{R}(x)=x-[x]$ and

$$\mathscr{R}(m, n)=\sum_{\rho=1}^{m-1}\rho\mathscr{R}\left(\frac{n\rho}{m}\right)$$

and if $-\Delta_1$ and $-\Delta_2$ are two negative fundamental discriminants, and $D=\Delta_1\Delta_2$; moreover, if for an arbitrary positive integer τ, t and u be defined by

$$\left(\frac{T+U\sqrt{D}}{2}\right)^\tau=\frac{t+u\sqrt{D}}{2},$$

then

$$\frac{2\tau u}{\tau_1\tau_2}Cl(-\Delta_1)Cl(-\Delta_2)=\sum_{a,b,c}\left(\frac{-\Delta_1}{(a,b,c)}\right)\left[\frac{1}{a}\mathscr{R}\left(au, \frac{bu-t}{2}\right)+\frac{t}{12a}-\frac{au^2}{4}\right],$$

where (a, b, c) ranges over a complete system of representative forms of discriminant D, $a>0$.

In Ch. I, use is made of Dirichlet's[20] fundamental formula (2) to make rigorous Hermite's[83] deduction of Dirichlet's[23] classic class-number formula (5). By new methods he obtained the familiar evaluations of the class-number that are due to Dirichlet,[23] Kronecker,[171] Lebesgue,[36] and Cauchy,[29] and established anew Kronecker's[171] ratio (4) of $Cl(D_0\cdot Q^2)$ to $Cl(D_0)$.

He found that if D_i are fundamental discriminants ($i=1, 2, 3, \ldots, r$), and $|D_i|=\Delta_i$, and if 2ν of the determinants are negative, then

$$(1)\quad Cl(D_1D_2\ldots D_r)\log E(D_1D_2\ldots D_r)$$
$$=(-1)^{\nu+1}\sum_{h_1, h_2, \ldots, h_r}{}'\left(\frac{D_1}{h_1}\right)\left(\frac{D_2}{h_2}\right)\cdots\left(\frac{D_r}{h_r}\right)\log\sin\left(\frac{h_1}{\Delta_1}+\frac{h_2}{\Delta_2}+\ldots+\frac{h_r}{\Delta_r}\right)\pi,$$

---

[277] Full notes of the Essay were published in Acta Math., 29, 1905, 334–424; 30, 1906, 203–293; Mém. sav. étr., Paris, 1906, 244 pp.

where $0<h_i<\Delta_i$, the term containing $\log 0$ is to be suppressed, and $E(D)=\frac{1}{2}(T+U\sqrt{D})$. By taking $r=2$, $D_1=-\Delta$, $D_2=-4$, we obtain one of the corollaries:

$$Cl(4\Delta)\log E(4\Delta)=2\sum_{h=1}^{\frac{1}{2}(\Delta-1)}\left(\frac{-\Delta}{h}\right)\log\frac{1+\tan h\pi/\Delta}{1-\tan h\pi/\Delta}.$$

In Ch. II, Lerch extended his[240] methods of 1897 and obtained new formulas including the following comprehensive formula, suitable for computation:

$$\sum_{a=1}^{[D/2]}\left(\frac{D}{a}\right)\sum_{\nu=1}^{[a\Delta/D]}\left(\frac{-\Delta}{\nu}\right)=-\tfrac{1}{2}Cl(\Delta D),$$

where $-\Delta$, D are fundamental discriminants, and Δ, $D>0$. Also,

$$\frac{2}{\tau}Cl(-\Delta)=\frac{\sqrt{\Delta}}{\pi}\sum_{\nu=1}^{\infty}\left(\frac{-\Delta}{\nu}\right)\frac{1}{\nu}\frac{\Gamma(\xi)^2}{\Gamma(\xi+2\nu x/\Delta)\Gamma(\xi-2\nu x/\Delta)},\quad \xi\gtreqless\tfrac{1}{2},\quad 0<x<1.$$

In Ch.[278] III, the identity in cyclotomic theory[279]

$$\log\frac{A(x)}{B(x)}=-\sqrt{D}\operatorname{sgn} D\sum_{\mu=1}^{\infty}\left(\frac{D}{\mu}\right)\frac{x^\mu}{\mu},\quad |x|<1,$$

where $D>0$ is a fundamental discriminant, for the limiting value $x=1$, gives, by Kronecker,[171] (2) above, the formula,

$$Cl(D)\log E(D)=\log\frac{Y(1)+\sqrt{D}Z(1)}{Y(1)-\sqrt{D}Z(1)}=2\log\frac{|Y(1)+\sqrt{D}Z(1)|}{\sqrt{4F(1)}}.$$

Suppose D is prime and >3; if in the known identity $Y^2(1)-DZ^2(1)=F(1)=4D$, we put $Y(1)=Dz$, $Z(1)=y$, we get $y^2-Dz^2=-4$; and hence y and z do not satisfy the equation $t^2-Du^2=4$. Hence

$$\log\frac{|y|+|z|\sqrt{D}}{2}\div\log\frac{T+U\sqrt{D}}{2}$$

is not an integer. Therefore $Cl(D)$ is odd. Similarly, it is proved that if D is >8 and composite, $Cl(D)$ is even (cf. G. L. Dirichlet,[93] Zahlentheorie, near the end of each of the articles 108, 109, 110). Congruences (mod 2) are given for $Cl(-8m)$, m a prime.

Lerch showed (Acta Math., pp. 231–233) how to obtain $Y(x, D_1D_2)$ and $Z(x, D_1D_2)$ from the cyclotomic polynomials for D_1 and D_2, and thence found for D_1, D_2 fundamental and >0,

$$Cl(D_1D_2)\log E(D_1D_2)=\sum_{h=1}^{\Delta-1}\left(\frac{D_2}{h}\right)\log\frac{Y(g,D_1)+\sqrt{D_1}Z(g,D_1)}{Y(g,D_1)-\sqrt{D_1}Z(g,D_1)},\quad g\equiv e^{2h\pi i/\Delta_2}.$$

Lerch obtains the following as a new type of formula analogous to Gauss sums:

$$(1)\qquad \sum_{\nu=1}^{m-1}\cot\frac{\nu^2\pi}{m}=4m\sum_{\delta}\frac{1}{\tau_\delta\sqrt{\delta}}Cl(-\delta),$$

where $-m$ is a negative, fundamental, odd discriminant, and δ ranges over the divisors of m which have the form $4k+3$ (Acta Math., 1906, 248).

[278] Chapters III, IV appear in Acta Math., 30, 1906, 203–293.
[279] Cf. G. L. Dirichlet, Zahlentheorie, Art. 105, for notation.

To express $Cl(D)$, where D is fundamental, negative, and uneven (Acta Math., 1906, pp. 260–279), as the root of congruences (mod 4, 8, 16, ...), Lerch put $D=D_1D_2D_3\ldots D_m$, where the D_i are relatively prime discriminants, and put $\Delta=|D|$, $\Delta_i=|D_i|$. All possible products $D'=D_{r_1}D_{r_2}\ldots D_{r_a}$ and their complementary products $Q'=D_{r_{a+1}}D_{r_{a+2}}\ldots D_{r_m}$ are formed and Δ' is written for $|D'|$; also, we let

$$F(D')=\frac{2}{\tau'}Cl(D'),\text{ if } D'<0;\ =0 \text{ if } D'>0,$$

$(D', Q')=\Pi[1-(D'/q)]$, q ranging over the distinct divisors of Q'; and $(D', 1)=1$. Then

$$\tfrac{1}{2}\phi(\Delta)-\frac{2^m}{\Delta}\Sigma^* s=\sum_{D'}(D', Q')F(D'), \tag{2}$$

where $\Sigma^* s$ denotes the number of those of the integers $s=1, 2, \ldots, \Delta$ which satisfy $(D_i/s)=1$ for all D_i simultaneously.

For example, when $m=2$, $D_1=-p$, $D_2=+q$, p and q being primes, $p\equiv 3$, $q\equiv 1 \pmod 4$, then the last formula becomes

$$\tfrac{1}{2}(p-1)(q-1)-\frac{4}{pq}\sum_{1}^{pq}{}^* s=Cl(-pq)+\left[1-\left(\frac{q}{p}\right)\right]\frac{2}{\tau_p}Cl(-p).$$

Since $\frac{1}{2}(p-1)(q-1)$ is $\equiv 0 \pmod 4$ and $Cl(-p)\equiv 1 \pmod 2$, we have

$$Cl(-pq)\equiv 1-(q/p) \pmod 4.$$

Lerch also obtained congruences for $Cl(-pqr)$ modulis 8 and 16.

In Ch. IV, a complicated Kronecker relation in exponentials applied to Lebesgue's[36] class-number formula (1) gives finally the following result:

$$\begin{aligned}\frac{4\pi}{\tau_1\tau_2}&Cl(-\Delta_1)\cdot Cl(-\Delta_2)\\ &=\sqrt{\Delta_1}\sum_{m=1}^{\infty}\sum_{n=1}^{\infty}\left(\frac{-\Delta_1}{m}\right)\left(\frac{-\Delta_2}{n}\right)e^{2mn\omega\pi i}\,\frac{e^{2mu\pi i}+e^{-2mu\pi i}}{2m}\\ &\qquad+\sqrt{\Delta_2}\sum_{m=1}^{\infty}\sum_{n=1}^{\infty}\left(\frac{-\Delta_1}{m}\right)\left(\frac{-\Delta_2}{n}\right)e^{s}(e^{t}+e^{-t})/(2n),\end{aligned}$$

in which $s=-2mn\pi i/(\Delta_2\omega)$, $t=nu\pi i/(\Delta_2\omega)$, while u, ω are complex variables, the imaginary part of ω is real and, in the complex plane, u is in the interior of the parallelogram with vertices at $0, 1, 1+\omega, \omega$. Lerch specializes the formula in several ways. For example, for $\Delta_1=\Delta_2$, $u=0$, $\omega=i$, it becomes

$$[Cl(-\Delta)]^2=\frac{\tau^2\sqrt{\Delta}}{2\pi}\sum_{n=1}^{\infty}\left(\frac{-\Delta}{n}\right)\frac{\Theta_1(n)}{n}e^{-2n\pi/\Delta},$$

where $\Theta_1(k)$ is the sum of the divisors of k.

H. Holden,[280] in the usual notations[281] for the cyclotomic polynomial, wrote

$$4X_p=4X_1X_2=Y^2+pZ^2 \qquad H=h(-p)/[2-(2/p)].$$

[280] Messenger Math., 35 1906, 73–80 (first paper).
[281] Gauss, Disq. Arith., Art. 357.

For p a prime of the form $4n+3>0$, he found, putting $h=h(-p)$,

$$(1), (2)\quad \left(\frac{dY}{dx}\right)_{x=1}=(-1)^{\frac{1}{2}(h-1)}\left(\frac{2}{p}\right)pH,\qquad \left(\frac{d^2Y}{dx^2}\right)_{x=1}=(-1)^{\frac{1}{2}(h-1)}\left(\frac{2}{p}\right)\frac{p(p-3)}{2}\cdot H,$$

$$(3), (4)\quad \left(\frac{dZ}{dx}\right)_{x=-1}=-h,\qquad \left(\frac{d^2Z}{dx^2}\right)_{x=-1}=\frac{p-3}{2}h,$$

$$(5), (6)\quad \Sigma\frac{1}{1-r^\alpha}-\Sigma\frac{1}{1-r^\beta}=i\sqrt{p}H,\qquad \Sigma\frac{1}{1+r^\alpha}-\Sigma\frac{1}{1+r^\beta}=\left(\frac{2}{p}\right)i\sqrt{p}h,$$

$$\alpha,\ \beta>0,\ <p,\ (\alpha/p)=1,\ (\beta/p)=-1,\ r=e^{2\pi i/p}.$$

The fifth formula had been obtained in a different way by V. Schemmel.[282] The fifth and sixth are true also when $p=4n+3$ is a product of distinct primes.

Holden,[283] by a study of the quadratic residues and non-residues, transformed the Schemmel-Holden formula (5) above into

$$(7)\quad \left[q-\left(\frac{q}{p}\right)\right]H=(q-1)\sum_{a=0}^{p/q}\left(\frac{a}{p}\right)+(q-2)\sum_{p/q}^{2p/q}\left(\frac{a}{p}\right)+\ldots+\sum_{(q-2)p/q}^{(q-1)p/q}\left(\frac{a}{p}\right),$$

$$(8)\quad \left[q-\left(\frac{q}{p}\right)\right]H=(q-1)\sum_{a=0}^{p/q}\left(\frac{a}{p}\right)+(q-3)\sum_{p/q}^{2p/q}\left(\frac{a}{p}\right)+(q-5)\sum_{2p/q}^{3p/q}\left(\frac{a}{p}\right)+\ldots,$$

where q is any positive integer relatively prime to p; and the last series terminates with the last possible positive coefficient. If $q=3$ and $q=4$, (8) becomes

$$\left[3-\left(\frac{3}{p}\right)\right]H=2\sum_{a=0}^{p/3}\left(\frac{a}{p}\right),\qquad 3H=3\sum_{0}^{p/4}\left(\frac{a}{p}\right)+\sum_{p/4}^{p/2}\left(\frac{a}{p}\right),$$

the latter[284] being Dirichlet's[23] formula (5_1); for, the first or second term of the second member vanishes according as $p=8n+3$ or $8n+7$.

When $q=2$, 3, 6 successively, (8) becomes three equations which yield

$$H=2\sum_{0}^{p/6}\left(\frac{a}{p}\right)\Big/\left\{1+\left(\frac{2}{p}\right)+\left(\frac{3}{p}\right)-\left(\frac{6}{p}\right)\right\},$$

and which also lead simply to expressions[285] for

$$a_r^{(6)}=\sum_{(r-1)p/6}^{rp/6}\left(\frac{n}{p}\right),\quad 1\leqq r\leqq 6,$$

in terms of H. When $q=2$, 4, 8 successively, (8) leads to linear expressions for H in terms of the distribution of quadratic residues and non-residues in the first four of the octants of p.

When $q=p-1$, (7) yields Dirichlet's[23] formula (6_1).

By taking $q=2$, 3, 4, 6, 12 in (8), a table is constructed which shows an upper bound for h when $p\equiv 7$, 11, 19 or 23 (mod 24), as $h\leqq(p+5)/12$ if $p\equiv 7$.

---

[282] Dissertation, Breslau, 1863, 15: Schemmel,[95] (6).
[283] Messenger Math., 35, 1906, 102–110 (second paper).
[284] Cf. Zahlentheorie, Art. 106, ed. 4, 1894, 276.
[285] Cf. Remarks by R. Dedekind[127] in Maser's German translation of Disq. Arith., 693–695. Cf. L. C. Karpinski,[266] Jour. für Math., 127, 1904, 1–19.

Dirichlet's[23] formula (6_1) is transformed into

$$H=\tfrac{1}{6}(p-1)(p-2)-2\sum_{k=1}^{\frac{1}{4}(p-3)}[\sqrt{kp}], \quad p\equiv 3 \pmod 4).$$

Holden[286] multiplied each member of (7) by (p/q). The result for $q=4$ or $q=2$, $p\equiv 3 \pmod 4$, is reduced to

$$3H=3\Sigma\left(\frac{4n}{p}\right)+\Sigma\left(\frac{4n+1}{p}\right), \quad \left[2\left(\frac{2}{p}\right)-1\right]H=\Sigma\left(\frac{4n}{p}\right)-\Sigma\left(\frac{4n+1}{p}\right).$$

Hence

$$h=\Sigma(-1)^{\frac{1}{2}(n-1)}\left(\frac{n}{p}\right), \quad p\equiv 3 \pmod 4,$$

n odd, $n<p$. He found eight similar expressions for h including the cases of determinants $-D$, where $D=4m+3$, $2(4m+1)$, $2(4m+3)$ is a product of distinct primes.

Holden[287] for the case $p=4n+3$, a prime, put

$$\left(\omega^m, \frac{1}{1+r}\right)=\frac{1}{1+r}+\frac{\omega^m}{1+r^g}+\frac{\omega^{2m}}{1+r^{g^2}}+\dots+\frac{\omega^{(p-2)m}}{1+r^{g^{p-2}}},$$

where ω is a primitive root of $x^{p-1}=1$, g is a primitive root of $x^{p-1}\equiv 1 \pmod p$, and r is a root of $x^p=1$. Then (6) becomes:

$$\left(\omega^{\frac{1}{2}(p-1)}, \frac{1}{1+r}\right)=\left(\frac{2}{p}\right)i\sqrt{p}\cdot h(-p).$$

A study of the new symbol gives

$$h^2=\sum_{\mu=1}^{\frac{1}{2}(p-1)}\left(\frac{\mu}{p}\right)[\tfrac{1}{2}(p-1)-2\lambda_\mu], \quad bh=\Sigma\left(\frac{\mu}{p}\right)\lambda_\mu,$$

where λ_μ is the number of positive integral solutions $k, l\leqq \frac{1}{2}(p-1)$ for a given μ of the congruence $k\mu+l\equiv 0 \pmod p$, and b is the number of quadratic non-residues $<\frac{1}{2}p$ of p. Similarly,

$$h^2=-\sum_{\mu=1}^{p-1}(\mu/p)\lambda_\mu.$$

Holden[288] in a treatment similar to his first paper[280] obtained from his own transformation[289]

$$\frac{x^p-1}{x-1}=S^2-(-1)^{\frac{1}{2}(p-1)}pxT^2$$

of the cyclotomic polynomial, six expressions for h. For example, if p is prime,

$$\left(\frac{d^2S}{dx^2}\right)_{x=1}=(-1)^{\frac{1}{2}(h-1)}\tfrac{1}{4}p(p-3)H\left\{1-\left(\frac{2}{p}\right)\right\}, \quad \left(\frac{dT}{dx}\right)_{x=1}=-\left(\frac{2}{p}\right)\frac{h}{2},$$

according as $p\equiv 3$ or $p\equiv 1 \pmod 4$.

---

286 Messenger Math., 35, 1906, 110–117 (third paper).
287 Messenger Math., (2), 36, 1907, 37–45.
288 *Ibid.*, 36, 1907, 69–75 (fourth paper).
289 Quar. Jour. Math., 34, 1903, 235.

Holden[290] removed the restriction of his second paper[283] that q be relatively prime to p. He put $p=nP$, $q=nQ$, where P and Q are relatively prime, and found that, if $p=4m+3$ is free from square factors, then for any positive integer n,

$$\tfrac{1}{2}h\left[1+\left(\frac{n}{p}\right)\right]=a_1+a_3+a_5+\ldots,\qquad \tfrac{1}{2}h\left[1-\left(\frac{n}{p}\right)\right]=a_2+a_4+a_6+\ldots,$$

where a_r $(0<r\leqq n)$ is the sum of the quadratic characters of the integers between $(r-1)p/(2n)+1$ and $rp/(2n)$. As above,[283] he found that $\frac{1}{6}(p-3)$ is an upper bound of h for $p\equiv 3$ or 15 (mod 24).

Holden,[291] by a modification of his second paper,[283] obtained, when $p=4n+1$ is a product of distinct primes, Dirichlet's[23] formula (5); also writing

$$a_r=\sum_{n=(r-1)p/q}^{rp/q}\left(\frac{n}{p}\right),$$

with q prime to p, he found in the respective cases $q=8$, $q=12$,

$$a_1+a_4=\left(\frac{2}{p}\right)\frac{h}{2},\quad a_1-a_6=\tfrac{1}{2}\left[1+\left(\frac{3}{p}\right)\right]h.$$

In particular,

$$\begin{array}{lll} q=8, & p=\ 8n+\ 1, & h=a_1-a_3; \qquad p=8n+5,\quad h=a_2-a_4: \\ q=12, & p=24n+\ 1, & h=a_1-a_6=a_1-a_2=-2(a_3+2a_4); \\ & p=24n+\ 5, & h=2a_3=-2(a_4+a_5); \\ & p=24n+13, & \tfrac{1}{2}h=a_1=a_3=-a_2=-a_6; \\ & p=24n+17, & h=2(a_1-a_4)=-2(a_3+2a_4). \end{array}$$

E. Meissner[292] supplied the details of the arithmetical proof by Liouville[90] of a class-number relation of the Kronecker type.

G. Humbert,[293] following Hermite,[69] wrote

$$\mathscr{A}=\sum_0^\infty q^{\frac{1}{4}(4N+3)}f(4N+3),\qquad 4N+3=(2m+1)(2m+4\rho+3)-4\mu^2,$$

$$\mu=0,\ \pm 1,\ \pm 2,\ \ldots;\quad m,\ \rho=0,\ 1,\ 2,\ \ldots,$$

and recalled that the exponent of q has a chosen value as often as there are quadratic forms

$$\phi=(2m+1)x^2+4\mu xy+(2m+4\rho+3)y^2\equiv ax^2+2bxy+cy^2$$

satisfying the conditions $c>a$, $|b|<a$, a and c uneven, b even. By means of the modular division of the complex plane, he set up a (1, 1) correspondence between the principal roots of these forms and those of the reduced uneven forms of determinant $-(4N+3)$. Hence $f(4N+3)=F(4N+3)$.

Similarly Humbert employed $\mathscr{B}$ and $\mathscr{C}$ to mean the same as $\frac{1}{4}C$ and $\frac{1}{4}+\frac{1}{4}D$ in the notations of Petr.[252]

---

290 Messenger Math., 36, 1907, 75–77 (Addition to second paper 283).
291 Messenger Math., 36, 1907, 126–134 (fifth paper).
292 Vierteljahrs. Naturfors. Gesells. Zürich, 52, 1907, 208–216.
293 Jour. de Math., (6), 3, 1907, 337–449.

A new class-number relation analogous to Kronecker's[54] (VIII) is deduced by equating coefficients of $q^{N+\frac{1}{8}}$ in the identity

$$2e^{-i\pi/8}\eta_1(i\sqrt{q})\,q^{\frac{1}{8}}\sum_0 q^{\frac{1}{8}(8\nu+7)}F(8\nu+7)$$
$$=4\sum_0 \frac{mq^m}{1+q^{2m}}\left[q^{\frac{1}{8}(2m+1)^2}(-1)^{\frac{1}{2}m(m-1)}+q^{\frac{1}{8}(2m-1)^2}(-1)^{\frac{1}{2}(m-1)(m-2)}\right].$$

The result is

$$\sum_{x\geqq 0}\left(\frac{2}{2x+1}\right)F[8N-(2x+1)^2]=-\Sigma\delta\left(\frac{2}{\delta_1+\delta}\right),$$

where $2N=\delta\delta_1$, $\delta<\delta_1$, δ and δ_1 positive and of different parity.

Similar treatment leads to relations of the Kronecker-Hurwitz type[294] such as

$$\sum_{m\geqq 0}(-1)^{\frac{1}{2}m(m+1)}F[8N+4-(2m+1)^2]=\Sigma(-1)^{\frac{1}{2}(a-1)}a,$$

a ranging over the solutions of $2N+1=a^2+2b^2$, $a>0$.

Four class-number relations of Liouville's[107] first type are obtained, including two of Petr,[295] and also

$$8(-1)^N[\sum_{m\geqq 0}(-1)^m(2m+1)F(4N+1-(2m+1)^2]=-2\Sigma\left(\frac{-1}{d}\right)d^2+\Sigma(a^2-4b^2),$$

in which $4N+1=a^2+4b^2$; $a>0$; $4N+1=dd'$, $d\leqq d'$; the term in which $d=d'$ is divided by 2.

New deductions of five of Petr's[296] class-number relations of Liouville's[297] second type are given (pp. 369–371).

Like Petr,[258] by recourse to transformations of order 2 of theta functions, but independently, Humbert obtained class-number relations involving the forms x^2-2y^2, including Humbert's (57), which is a slight modification of Petr's[258] (1) above, and including Humbert's (52), which is Petr's[258] (2) above.

A geometric discussion, analogous to the one above in which Humbert evaluated $\mathscr{A}$, now shows (pp. 385–8) that for a negative determinant $-M$, $M\equiv 3 \pmod 8$, there is a (3, 1) correspondence between the proper and improper reduced forms. The corresponding well-known relation (Dirichlet[20]) is similarly established for $M\equiv 7 \pmod 8$.

To prove a theorem of Liouville,[105] Humbert finds (pp. 391–2) in Liouville's notation that, for a determinant $-(8M+3)$,

$$\Sigma a(a'-a)=2\Sigma(2m_1m_1'+2m_1m_1''+2m_1'm_1''-m_1^2-m_1'^2-m_1''^2),$$

where a and $a'\geqq a$ are the two odd minima of any odd class, while m_1, m', m_1'' denote the first uneven minima of the three odd classes corresponding to a single even class, and where summation on the right is taken over the even classes. But the right summand equals $8M+3$, whence

$$\Sigma a(a'-a)=\tfrac{2}{3}(8M+3)F(8M+3).$$

---

[294] L. Kronecker,[124] Monatsber. Akad. Wiss. Berlin, 1875, 230; A. Hurwitz,[202] Jour. für Math., 99, 1886, 167–168.

[295] Cf.[252] Rozpravy české Akad., Prague, 9, 1900, Mem. 38. In Humbert's memoir the two are (35), (36).

[296] Rozpravy české Akad.,[256] Prague, 9, 1900, Mem. 38.

[297] Jour. de Math.,[107] (2), 12, 1867, 99. The five are numbered (40)–(44) by Humbert.

To obtain class-number relations in terms of minima of classes, Humbert equated the coefficients of $q^{N+\frac{3}{4}}$ in the identity

$$\mathscr{A}\theta = \sum_{1} q^{m^2} a_m [1-2q^{2m}+\ldots+2(-1)^\rho q^{2\rho m}+\ldots],$$

where

$$a_m = q^{-1/4} - 3q^{-9/4} + \ldots + (-1)^{m-1}(2m-1)q^{-(2m-1)^2/4}, \quad \theta = \sum_{-\infty}^{+\infty} (-1)^m q^{m^2}.$$

The coefficient in the first member is

$$\sum_{x,\,y} (-1)^{x+y} F(4N+3-4x^2-4y^2).$$

In the second member, $4N+3=4m^2-(2\mu-1)^2+8m\rho$, $(m \gtreqless 1,\ \rho>0,\ 1 \leqq \mu \leqq m)$; and the coefficient is $\Sigma(-1)^{\mu+\rho-1}2(2\mu-1)$. When

$$4N+3=(2m+2\rho-2\mu+1)(2m+2\rho+2\mu-1)-4\rho^2$$

is identified with $ac-b^2$ the negative of the discriminant of form (a, b, c); a and c uneven; $c>a$; $a>b$; $b \gtreqless 0$; the latter coefficient is

$$2\Sigma(-1)^{\frac{1}{4}(c-a+2b-2)}\tfrac{1}{2}(c-a)=\tfrac{1}{2}(-1)^N\Sigma(\mu_2-\mu_1)(-1)^{\frac{1}{2}(\mu_1-1)},$$

where the summation on the right is over the proper classes of determinant $-(4N+3)$, and μ_1, μ_2 $(\mu_1 \leqq \mu_2)$ are the two uneven minima of a class.

Similarly, from $\mathscr{A}\eta_1\theta$ Humbert obtained

$$4\Sigma(-1)^x F[4N-4x^2-(2y+1)^2]=2\Sigma\mu(-1)^{\frac{1}{2}(\mu_1+\mu_2+2)},$$

summed over all pairs of integers x, y, where μ is the even minimum, μ_1, μ_2 the odd minima of an odd class of determinant $-4N$.

By equating the coefficients of q^N in the identity

$$-4\mathscr{A}(-q)e^{i\pi/4}\eta_1\theta_1^3=8\Sigma\frac{m^2q^2}{1-q^{2m}}+8\Sigma m^2\frac{q^{m^2+m}}{1-q^{2m}}[1+2q^{-1}+\ldots+2q^{-(m-1)^2}],$$

we obtain the class-number relation

$$(-1)^{N+1}\sum_{h\geqq 0}(-1)^h F(4N-4h-1)\Phi(4h+1)=\tfrac{1}{16}\Sigma\mu^2,$$

where $\Phi(n)$ is the sum of the divisors of n, and μ is the even minimum of an odd class of determinant $-4N$.

Similarly, from the expansion of $\mathscr{C}\eta_1^3\theta$, it is stated that

$$\sum_{h\geqq 0} I\left[\frac{4N-1-(4h+3)}{4}\right]\Phi(4h+3)=\tfrac{1}{16}\Sigma\mu^2,$$

where $I(D)=F(D)-3F_1(D)$, and $F_1(D)$ denotes the number of even classes of determinant $-D$.

Five more new class-number relations involving minima include

$$8\sum_{h\geqq 0} F[8M+3-(4h+3)]\psi(4h+3)=\Sigma\nu_2(\nu_3-\nu_1),$$

in which $\psi(n)$ denotes the sum of the divisors $<\sqrt{n}$ of n; the summation on the right extends over the even classes of $-(8M+3)$; and ν_1, ν_2, ν_3 are the three minima of a class, $\nu_1 \leqq \nu_2 \leqq \nu_3$.

To obtain class-number relations of grade 3 of the Gierster[145]-Hurwitz[167, 184] type,[298] Humbert employed the fundamental formula of Petr[299] and Humbert,[300]

$$(1)\quad \eta_1\theta_1 H_1 \Theta_1 H^2/\Theta^2 = 2H_1(x, \sqrt{q})\sum_0 q^{(8\nu+7)/8}F(8\nu+7)$$
$$-4\sum_1 q^{\frac{1}{8}(2m+1)^2}\cdot[(2m-1)q^{-\frac{1}{8}(2m-1)^2}+(2m-5)q^{-\frac{1}{8}(2m-5)^2}+\ldots]\cos(2m+1)x.$$

By setting $x=0$, and equating coefficients of q^N, we obtain

$$(2)\qquad 0=\sum_m F[8N-(2m+1)^2]-2\Sigma(\delta_1-\delta),$$

$2N=\delta\delta_1$, δ_1 even, δ odd, $\delta<\delta_1$, m arbitrary.

In (1), we put $x=\pi/3$ and use the formula for $\Theta(3x, q^3)$. In the resulting identity we equate the coefficients of q^N and use the fact that the number of solutions of

$$8N=(2x+1)^2+(2y+1)^2+3(2z+1)^2+3(2t+1)^2$$

is $16\Sigma d'$, where d' ranges over the divisors of N which have uneven conjugates and which are not multiples of 3. Whence, for $N\equiv -1 \pmod 3$, the final result is

$$-3\Sigma d'=2\Sigma F[8N-(2m+1)^2]\cos(2m+1)\pi/3-4\Sigma(\delta_1-\delta)\cos(\delta_1+\delta)\pi/3$$

in which $2N=\delta\delta_1$, so that $\cos(\delta_1+\delta)\pi/3=\frac{1}{2}$. This result combined with (2) gives, for $N\equiv -1 \pmod 3$, the relations[301] (p. 418):

$$\Sigma F[8N-9(2\mu+1)^2]=\Sigma d', \qquad \Sigma F[8N-(6\rho\pm 1)^2]=-\Sigma d'+2\Sigma(\delta'-\delta),$$

summed over all integers μ, ρ, where d' is a divisor of N which has an odd conjugate and $\delta'\delta=2N$, $\delta_1>\delta$, δ_1 is even and δ uneven. Corresponding results[301] are obtained for $N\equiv 0, 1 \pmod 3$.

Transformations of the third order yield also, for $N=6l+1$ (p. 431),

$$2\Sigma G(N-9\nu^2)=\tfrac{1}{16}\mathcal{N}_3+\tfrac{1}{2}\Sigma d,$$

summed over all integers ν, and all divisors d of N, where $G(m)$ is the number of classes of determinant $-m$, and $\mathcal{N}_3$ is the number of decompositions of N into the sum of 4 squares in which 3 of the squares are multiples of 3.

Humbert evaluated such sums[301] as $\Sigma F(N-9\nu^2)$, with N arbitrary; but it is done with less directness than by Petr.[345] New expansions lead to such relations[301] as

$$\Sigma(-1)^m F(24N-1-24m^2)=-\Sigma y(-1/y)(-1)^{\frac{1}{4}(x-2)},$$
$$48N-2=x^2-6y^2,\quad y>0,\quad x>3y;$$
$$6\Sigma(-1)^m F_1\left[\frac{24N+1-(6m+1)^2}{6}\right]=-\Sigma\left(\frac{3}{x}\right)(x-3y),$$

$24N+1=x^2-6y^2$, $y \gtreqless 0$, $x>3y$, each summed over all integers m. Terms in which $y=0$ are divided by 2.

---

[298] Cf. Klein-Fricke,[217] Elliptische Modulfunctionen, II, 1892, 231–234.
[299] Cf. Petr,[252] formula (1).
[300] Numbered (10) in Humbert's memoir.
[301] Humbert gave the results also in Comptes Rendus, Paris, 145, 1907, 5–10.

Humbert[302] gave five new class-number relations involving minima[303] of the classes.

H. Teege[304] partly by induction concluded that, when $P=8n+3$ is a product of distinct primes,

$$5\sum_{1}^{(P-3)/4}\left(\frac{a}{P}\right)a+\sum_{(P+1)/4}^{(P-1)/2}\left(\frac{a}{P}\right)a>0.16124P\sqrt{P},\quad \sum_{(P+1)/4}^{(P-1)/2}\left(\frac{a}{P}\right)a-\sum_{1}^{(P-3)/4}\left(\frac{a}{P}\right)a>0.$$

These combined confirm, in view of Dirichlet's[23] formula (6), Gauss' conjecture (Disq. Arith.,[4] Art. 303) that the number of negative determinants which have a class number h is finite for every h.

K. Petr[305] recalled that the number of representations of any number N by the representatives (Dirichlet,[93] (2)) of all the classes of positive forms $ax^2+bxy+cy^2$ of negative fundamental discriminant D is $\tau\Sigma(D,d)$, summed for the divisors d of N, where the symbol (D,d) is that of Weber[306] for the generalized quadratic character of D. Hence,[307] if $D<-4$,

$$(1)\quad \sum_{\text{class}}\sum_{x,y} q^{ax^2+bxy+cy^2}=2\sum_N q^N\left(\sum_d (D,d)\right)+h,\quad |q|<1,$$
$$x,y=0,\ \pm1,\ \pm2,\ \ldots;\ N=1,2,3,\ldots,$$

where h is the number of positive classes of D.

By methods of L. Kronecker[308] he obtained

$$(2)\qquad \sum_{m=-\infty}^{\infty}\sum_{n=-\infty}^{\infty} e^{\pi i\tau(am^2+bmn+cn^2)}=\frac{i}{\tau\sqrt{-\frac{1}{4}D}}\sum_{-\infty}^{\infty}\sum_{-\infty}^{\infty} e^{-4\pi i\tau_1(am^2+bmn+cn^2)/D},$$

where $\tau_1=-1/\tau$. Next, by the use of theta functions, he found

$$(3)\qquad \sum_{k=1}^{-D-1}(D,k)\frac{\Theta'(k\tau/D,\tau)}{\Theta(k\tau/D,\tau)}=4\pi i\sum_{N=1}^{\infty} q^{2N/|D|}\left(\sum_d (D,d)\right),$$

d is any divisor of N. Now (1), (2), (3) imply

$$(4)\quad \Sigma(D,k)\frac{\Theta'(k\tau/D,\tau)}{\Theta(k\tau/D,\tau)}=2\pi i\sum_{Cl}\sum_{x,y} q^{-2(ax^2+bxy+cy^2)/D}-2h\pi i$$
$$=-2\pi\frac{\sqrt{-D}}{\tau}\sum_{Cl}\sum_{x,y} q_1^{2(ax^2+bxy+cy^2)}-2h\pi i,$$

where $\tau_1=-1/\tau$ and $q_1=e^{\pi i\tau_1}$. For the same transformation $\tau_1=-1/\tau$,

$$(5)\qquad \sum_k (D,k)\left[2\frac{k}{D}\pi i+\frac{\Theta'(k\tau/D,\tau)}{\Theta(k\tau/D,\tau)}\right]=\frac{1}{\tau}\sum_k (D,k)\frac{\Theta'(k/D,\tau_1)}{\Theta(k/D,\tau_1)},$$

$k=1,2,\ldots,-D-1$. By use of (5), we get

$$(6)\quad 2\pi i\left[-h+\sum_k\frac{k}{D}(D,k)\right]-2\pi\frac{\sqrt{-D}}{\tau}\sum_{Cl}\sum_{x,y} q_1^{2(ax^2+bxy+cy^2)}$$
$$=\frac{1}{\tau}\Sigma(D,k)\,\Theta'\left(\frac{k}{D},\tau_1\right)\Big/\Theta\left(\frac{k}{D},\tau_1\right).$$

302 Comptes Rendus, Paris, 145, 1907, 654–658.
303 Jour. de Math., (6), 3, 1907, 393–410.
304 Mitt. Math. Gesell. Hamburg, 4, 1907, 304–314.
305 Sitzungsber. Böhmische Gesells. Wiss. (Math.-Natur.), Prague, 1907, No. 18, 8 pp.
306 Algebra, III, 1908, § 85.
307 Cf. H. Poincaré, Jour. für Math., 129, 1905, 126.
308 Sitzungsber. Akad. Wiss. Berlin, 1885, II, 761.

Now the right member of (6) is the product of $1/\tau$ by a power series in q_1. Hence the quantity in brackets in (6) is zero (Dirichlet,[19] (1)). For, otherwise (6) would imply that $\tau = -\pi i/\log q_1$ could be expressed as a power series in q_1 which converges for all q_1 such that $|q_1|<1$. Moreover, the comparison now of the two members of (6) in the light of (1) gives Lebesgue's[36] formula (1):

$$h = \frac{-1}{2\sqrt{-D}} \Sigma (D, k) \cot \frac{\pi k}{D}.$$

An alternative form of (6) is the following:

$$2\pi i \left\{ -h + \Sigma \frac{k}{D} (D, k) \right\} = \frac{1}{\tau} \{ 2\pi h \sqrt{-D} + \pi \Sigma (D, k) \cot \pi k/D \}.$$

The last two class-number formulas above follow now elegantly when τ is regarded as a variable occurring in an identity.

H. Holden[309] applied the method of his first paper[280] to a product $p=4n+3$ of distinct primes, and stated the four possible results including

$$\left(\frac{d^2 Z}{dx^2}\right)_{x=1} = -(-1)^{\frac{1}{2}h} \{ \tfrac{1}{2}\phi(p) - 1 \} H.$$

He generalized the method of his fourth paper[288] from primes $p=4n+3$ and $4n+1$ to products of primes, and gave the four possible formulas including

$$p=4n+1, \qquad \left(\frac{dT}{dx}\right)_{x=-1} = -(-1)^m \cdot \tfrac{1}{2} h,$$

where m is the number of integers between $\frac{1}{4}p$ and $\frac{1}{2}p$, and prime to p.

H. Weber[310] in a revised edition of his book on Elliptic Functions modified his earlier discussion[214] of class-number to apply to Kronecker forms,[171] in which the middle coefficient is indifferently even or uneven. He also (§ 85) replaced the Legendre-Jacobi-Kronecker symbol[220] (D/n) by (D, n) which he redefined and gave details (§§96-100) of Dedekind's[127a] solution of the Gauss[4] Problem.

M. Plancherel[311] extended certain researches of A. Hurwitz[312] and M. Lerch[313] by finding the residue of $Cl(D)$ modulo 2^m, where $D=D_1D_2\ldots D_m$ and $D, D_1, D_2, \ldots, D_m$ are fundamental discriminants. He deduced Lerch's formula[314]

$$(1) \quad \tfrac{1}{2}\phi(\Delta) - \frac{2^m}{\Delta} \sum_1^{\Delta}{}^* s = \sum_{a=1}^{m} \sum_{r, \ldots, r_a} (D_{r_1} D_{r_2} \ldots D_{r_a}, \Delta_{r_{a+1}} \ldots \Delta_{r_m}) P(D_{r_1} D_{r_2} \ldots D_{r_a}),$$

where $\Delta = |D|$, $\Delta_i = |D_i|$; $(D, Q) = \Pi(1-(D/q)$, q ranging over the different prime factors of Q, and $(D, 1)=1$; $P(D) = \frac{2}{\tau} Cl(D)$ if D is <0, $P(D)=0$ if D is >0; and Σ^* denotes that those values s only are taken which satisfy

$$(D_1/s) = (D_2/s) = \ldots = (D_m/s) = 1.$$

---

[309] Messenger Math., 37, 1908, 13–16.
[310] Lehrbuch der Algebra, Braunschweig, III, 1908, 413–427.
[311] Thesis, Pavia, 1908, 94 pp. Revista di fisica, matematica, Pavia, 17, 1908, 265–280, 505–515, 585–596; 18, 1908, 77–93, 179–196, 243–257.
[312] Acta Math.,[235] 19, 1895, 378–379.
[313] Acta Math., 30, 1906, 260–279; Mém. présentés par divers savants à l'Académie des sc., 33, 1906, Chapter III of the Prize Essay.[278]
[314] Acta Math.,[278] 30, 1906, 261. Lerch,[78] (2).

Hereafter Δ_i are assumed to be primes. Then $\frac{1}{2}\phi(\Delta)\equiv 0 \pmod{2^{m-a}}$. But

$$(D_{r_1}D_{r_2}\ldots D_{r_a},\ \Delta_{r_{a+1}}\ldots\Delta_{r_m})\equiv 0 \pmod{2^{m-a}}.$$

It follows that

$$P(D_{r_1}D_{r_2}\ldots D_{r_a})\equiv 0 \pmod{2^{a-1}},\quad P(D_1D_2\ldots D_m)\equiv 0 \pmod{2^{m-1}}.$$

The latter for $D<0$ is the rule derived from genera (cf. C. F. Gauss, Disq. Arith., Arts. 252, 231; L. Kronecker, Monatsber. Akad. Wiss. Berlin, 1864, 297; reports of both in Ch. IV). Thus (1) implies

$$P(D_1\ldots D_m)\equiv\tfrac{1}{2}\phi(\Delta_1\ldots\Delta_m) + \sum_{a=1}^{m-1}\sum_{r_1,\ldots,r_a}(D_{r_1}\ldots D_{r_a},\ \Delta_{r_{a+1}}\ldots\Delta_{r_m})P(D_{r_1}\ldots D_{r_a}) \pmod{2^m}.$$

For a negative determinant $D=-p_1p_2\ldots p_{2m+1}q_1q_2\ldots q_n$, where p, q are primes >0 and $-p\equiv q\equiv 1 \pmod 4$, this leads to

$$(2)\quad Cl(-p_1p_2\ldots p_{2m+1}q_1q_2\ldots q_n) \equiv\tfrac{1}{2}\sum_{\mu=1}^{2m+1}\sum_{r}((-1)^{\mu}p_{r_1}p_{r_2}\ldots p_{r_\mu}|-p_{r_{\mu+1}}\ldots -p_{2m+1}q_1q_2\ldots q_n)\ \phi(p_{r_1}\ldots p_{r_\mu}) \pmod{2^{2m+n+1}},$$

where the symbol $(\,|\,)$ is defined by the recurrence relation

$$(D_1D_2\ldots D_a|D_{a+1}\ldots D_m) = \sum_{\mu=0}^{m-a-1}\sum_{\rho}(D_1\ldots D_aD_{\rho_{a+1}}\ldots D_{\rho_{a+\mu}},\ \Delta_{\rho_{a+\mu+1}}\ldots\Delta_{\rho_m})\cdot(D_1\ldots D_a|D_{\rho_{a+1}}\ldots D_{\rho_{a+\mu}}),$$

and by the formula

$$(D_a|D_\beta D_\gamma)=(D_aD_\beta,\ \Delta_\gamma)(D_a,\ \Delta_\beta)+(D_aD_\gamma,\ \Delta_\beta)(D_a,\ D_\gamma)+(D_a,\ \Delta_\beta\Delta_\gamma).$$

He disposed completely of the new special case $m=5$ by (2) as in the following particular example:

$$m=5,\ D_1=-p_1,\ D_2=-p_2,\ D_3=-p_3,\ D_4=q_1,\ D_5=q_2,$$
$$\left(\frac{p_1}{q_1}\right)=\left(\frac{p_2}{q_1}\right)=-\left(\frac{p_3}{q_1}\right)=\epsilon_1,\quad \left(\frac{p_1}{q_2}\right)=\left(\frac{p_2}{q_2}\right)=-\left(\frac{p_3}{q_2}\right)=\epsilon_2.$$

The result in this case is

$$Cl(D)\equiv 2(1-\epsilon_1)(1-\epsilon_2)[1-(-4/\sigma)] +2(1-\epsilon_1\epsilon_2)[2(1-\eta_1\eta_2)+(1-\mu)(1-(-4/\sigma))] \pmod{32},$$

where

$$\eta_1=(p_2/p_3),\quad \eta_2=(p_3/p_1),\quad \eta_3=(p_1/p_2),\quad \mu=(q_1/q_2),\quad \sigma=\eta_1+\eta_2+\eta_3.$$

For $D=D_0D_1D_2\ldots D_m$, $|D_0|=8$ or 4, he obtained analogues of (1) and finally congruences (mod 2^{m+1}) for $Cl(D)$. He noted that $Cl(-4q_1q_2\ldots q_m)\equiv 0 \pmod{2^{m+1}}$ if each $q_i\equiv 1 \pmod 8$.

G. Humbert[315] obtained formulas which express new relations between the minima of odd classes of a negative determinant $-n$ and those functions of the type $\psi(n)$,

[315] Jour. de Math., (6), 4, 1908, 379–393. Abstract, Comptes Rendus, Paris, 146, 1908, 905–908.

$\chi(n)$ of the divisors of n which occur in the right member of Kronecker's[54] class-number relations. Thus were obtained alternate forms for the right members of old class-number relations.

E. Chatelain[316] obtained the ratio (see the Gauss[4] Problem) between the number of properly primitive classes of forms of determinant p^2D, p a prime, and the number of determinant D. As the representatives of the first, he chose the type (ap^2, bp, c) with c prime to p; as representatives of the second, he chose the type (a', b', c') with c' prime to p. Then between the $h(p^2\cdot D)$ forms (a, b, c) and the $h(D)$ forms $(a', b' c')$ he set up a $(k, 1)$ correspondence by means of a relative equivalence given by the unit substitution $\left(\begin{smallmatrix}\alpha & \beta\\ \gamma & \delta\end{smallmatrix}\right)$, $\beta\equiv 0 \pmod p$. Similarly he found the ratio of the number of classes of the two primitive orders of a given determinant. His proofs are similar to those of Lipschitz.[41]

M. Lerch[317] gave two deductions of

$$\sum_{k=2}^{\Delta-1}\left(\frac{-\Delta}{k}\right)\sum_{a=1}^{k-1}(-1)^{a\Delta/k}=4(1-2\epsilon)K^2-K,$$

where $-\Delta$ is a negative uneven fundamental discriminant, $K=2\tau^{-1}Cl(-\Delta)$, $\epsilon=(2/\Delta)$. Here, if $\Delta=3$, $K=\frac{1}{3}$. The second and more elementary deduction rests on Lerch's[240] formula (3).

He deduced several formulas which he had published earlier,[318] including

$$\sum_{h=1}^{\Delta-1}\left(\frac{-\Delta}{h}\right)\tan\frac{h\pi}{h}=(-1)^{\frac{1}{8}\Delta-1}\frac{4\sqrt{\Delta}}{\tau}K,\quad \Delta\equiv 0 \pmod 8).$$

K. Petr[319] reproduced his[305] discussion of 1907; and, by equating coefficients of q^n in the expansion of doubly periodic functions of the third kind, obtained Schemmel's[95] formula (4); also the number h of primitive classes of the negative fundamental discriminant $-D=D_1D_2$ for $D_2>0$ and $=4n+1$:

$$h=-\sum_{k_1,\,k_2}(D_1, k_1)(D_2, k_2)\epsilon_{k_1}\epsilon_{k_2},$$

where $k_i=0, 1, 2, \ldots, |D_i|-1$; $i=1, 2$, and where $\epsilon_{k_1}\epsilon_{k_2}=1$ or 0 according as $k_1/D_1+k_2/D_2>0$ or <0, and (D, k) is the Weber symbol.[220]

Similarly, for $-D=-D_1D_2D_3$, a negative fundamental discriminant,

$$h=-\Sigma(D_1, k_1)(D_2, k_2)(D_3, k_3)\cdot E\left(\frac{k_1}{D_1}+\frac{k_2}{D_2}+\frac{k_3}{D_3}\right),$$

where $k_i=1, 2, \ldots, |D_i|-1$; $i=1, 2, 3$; and $E(-a)=-E(a)-1$ if $a>0$. These two formulas are special cases of a formula of M. Lerch on p. 41 of his prize essay.[277] See Acta Math., 29, 1905, p. 372, formula (16). Cf. Lerch,[277] (1).

J. V. Pexider[320] for the case of a prime $p=8\mu+3$, wrote r and ρ respectively for a quadratic residue and non-residue of p, and combined the obvious identity

(1) $$\Sigma r+\Sigma\rho=\tfrac{1}{2}p(p-1)$$

with Dirichlet's[14] formula (3), viz.,

(2) $$\Sigma\rho-\Sigma r=\lambda p,$$

---

[316] Thesis, University of Zürich, 1908. Published at Paris, 1908, 79 pp.
[317] Rozpravy české Akad., Prague, 17, 1908, No. 6, 20 pp. (Bohemian).
[318] Lerch, Acta Math., 30, 1906, 237, formulas (36)-(39). Chapter III of Prize Essay.[278]
[319] Casopis, Prague, 37, 1908, 24-41 (Bohemian).
[320] Archiv Math. Phys. (3), 14, 1908-9, 84-88.

where 3λ is the number of properly primitive classes of determinant $-p$. The result is

$$\lambda=\frac{p-1}{2}-\frac{2}{p}\Sigma r=\frac{2}{p}\Sigma\rho-\frac{p-1}{2}.$$

According to M. A. Stern, if p is a prime $4\mu+3$, there exists an integer σ such that

$$(3) \qquad 2\Sigma r-\Sigma\rho=\sigma p.$$

From (1), (2), (3), we get $3\lambda=\frac{1}{2}(p-1)-2\sigma$. This result compared with Dirichlet's[14] class-number formula (3) shows that σ is the number of quadratic non-residues of p which are $<\frac{1}{2}p$.

For a prime $p=8\mu+7$, (2) holds provided now $\lambda=h(-p)$. Hence by (3), $h(-p)=\frac{1}{4}(p-3)-2\kappa$, where κ denotes the number of the quadratic non-residues of p between 0 and $\frac{1}{4}p$. Dirichlet's[14] formula (3) combined with the last result shows that

$$B=\kappa+\tfrac{1}{8}(p+1), \qquad A=R+\tfrac{1}{8}(p+1),$$

where A and R are respectively the number of positive quadratic residues of p less than $\frac{1}{2}p$ and $\frac{1}{4}p$, and B is the number of quadratic non-residues $<\frac{1}{2}p$.

A. Friedmann and J. Tamarkine,[321] in a study of quadratic residues and Bermoullian numbers, replaced $\Sigma b-\Sigma a$ in Dirichlet's[14] formula (3) so that for p a prime $\equiv 3 \pmod 4$, the latter becomes Cauchy's[28] class-number congruence (1) in the form[322]

$$h(-p)\equiv\left[2-\left(\frac{2}{p}\right)\right](-1)^{\frac{1}{4}(p+1)}\cdot 2B_{(p+1)/4} \pmod p.$$

M. Lerch[323] found that, for P a prime,

$$\prod_a \cot\frac{2\pi a}{P}=(-1)^{\frac{1}{2}G(-P)+\frac{1}{2}}\frac{1}{\sqrt{P}},$$

where a ranges over all positive integers $<P$ prime to P such that $(a/P)=1$. Cf. Stern.[31])

G. Humbert[324] introduced a parameter a in the θ-function, and considered $H(x+a)$ and $\Theta(a)$. Then, by Hermite's[69] method, he found that

$$\sum_{k=-\infty}^{+\infty}(-1)^k\cos 2ka \sum_{Cl(4N+3-4k^2)}\cos\tfrac{1}{2}(m_2-m_1)a=(-1)^N\sum_d d\cos da,$$

where m_1 and m_2 are odd minima ($m_1\leqq m_2$) of a reduced form of negative determinant $-(4N+3-4k^2)$, and d is a divisor of $4N+3$ not exceeding its square root. For $a=0$, this becomes Hermite's[69] relation (5). For $a=\frac{1}{4}\pi$, it becomes

$$\sum_{k'=-\infty}^{+\infty}(-1)^{k'}\sum_{Cl(4N+3-16k'^2)}\left(\frac{2}{\frac{1}{2}(m_2-m_1)}\right)=(-1)^N\Sigma d\left(\frac{2}{d}\right),$$

where $(2/d)$ is the Jacobi-Legendre symbol. If N is uneven, this is Kronecker's[54] relation (VII).

321 Jour. für Math., 135, 1908–9, 146–156.
322 Mém. Institut de France, 17, 1840, 445; Oeuvres (1), III, 172.
323 Encyclopédie des sc. math., 1910, tome 1, vol. 3, p. 300.
324 Comptes Rendus, Paris, 150, 1910, 431–433.

P. Bachmann[325] supplied the details of Liouville's[90] arithmetical deduction of a class-number relation of the Kronecker type (cf. Meissner[292]).

M. Lerch,[326] by a study of Kronecker's[171] generalized symbol (D/n), transformed the left member of Lerch's[240] formula (3), for a negative fundamental discriminant $-\Delta$, and m not divisible by Δ, and found that

$$(1)\quad \sum_{a=1}^{\frac{1}{2}(\Delta-1)}\left(\frac{-\Delta}{a}\right)E\left(\frac{am}{\Delta}\right)=\tfrac{1}{2}\left[(m-1)\left\{1-\left(\frac{2}{\Delta}\right)\right\}\right.$$
$$\left.-\left\{1-\left(\frac{-\Delta}{m}\right)\right\}\right]K,\quad K=\frac{2}{\tau}Cl(-\Delta).$$

Put

$$i_a=\sum_{h=1}^{n}\left\{E\left(\frac{2ha}{\Delta}\right)-2E\left(\frac{ha}{\Delta}\right)\right\}.$$

Then by formula[240] (4), we have

$$\sum_{a=1}^{\frac{1}{2}(\Delta-1)}\left(\frac{-\Delta}{a}\right)i_a=\left[n\frac{1-\epsilon}{2}+\frac{n}{2}+\frac{\epsilon-2}{2}(2-\epsilon)K\right]K,\quad \epsilon=\left(\frac{2}{\Delta}\right),\quad n=\left[\frac{\Delta-1}{2}\right],$$

Since $h(-\Delta)=(2-\epsilon)K$, we have, for $\Delta=2n+1$,

$$\sum_{a=1}^{\frac{1}{2}(\Delta-1)}\left(\frac{-\Delta}{a}\right)i_a=\left(\frac{\Delta-1}{2}-h\right)\frac{h}{2}.$$

Similarly for $\Delta=4P$,

$$\sum_{a=1}^{2P-1}(-1)^{\frac{1}{2}(a-1)}\left(\frac{a}{P}\right)i'_a=K(2P-1-2K),\quad i'_a=\sum_{m=1}^{2P-1}\left\{E\left(\frac{ma}{2P}\right)-2E\left(\frac{ma}{4P}\right)\right\}.$$

For a negative prime discriminant $-\Delta$, $\Delta=4k+3$, (1) implies:

$$h(-\Delta)=\sum_{1}^{\frac{1}{2}(\Delta-1)}(-1)^{i_a}.$$

L. E. Dickson,[327] by a method similar to the method of Landau[260] in the case of Gauss forms, showed that for $P>28$ no negative discriminant $-P\equiv 0 \pmod 4$ could have a single primitive class.

For $P\equiv 3 \pmod 4$, P with distinct factors, there are obviously two or more reduced forms. Hence, if there is only the one reduced form $[1, 1, \frac{1}{4}(1+P)]$, then $P=p^\epsilon$, where p is a prime $\equiv 3 \pmod 4$ and ϵ is odd. But for $p>3$ and $e \gtreqless 3$, a second primitive reduced form is $[\frac{1}{4}(p+1), 1, (p^\epsilon+1)/(p+1)]$. For $P=3^\epsilon$, $\epsilon \gtreqless 5$, a second primitive reduced form is $(7, 3, 9)$ or $[9, 3, \frac{1}{4}(3^{\epsilon-2}+1)]$. Hence beyond 27 we need consider only primes P. We set

$$T_j=\tfrac{1}{4}[(2j+1)^2+P]=T_0+j(j+1).$$

For any integer m and any T_j, there is some T_r, $0\leqq r\leqq \frac{1}{2}(m-1)$, such that $T_j\equiv T_r \pmod m$. From this lemma and by indirect proof it is found[328] that there is a single reduced form of discriminant $-P$ if and only if $T_0, T_1, T_2, \ldots, T_g$ are all prime numbers, where $2g+1$ denotes the greatest odd integers $\leqq \sqrt{P/3}$.

---

[325] Niedere Zahlentheorie, Leipzig, II, 1910, 423–433.
[326] Annaes scient. da Acad. Polyt., Porto, 6, 1911, 72–76.
[327] Bull. Amer. Math. Soc., (2), 17, 1911, 534–537.
[328] Cf. M. Lerch,[262] Math. Annalen, 57, 1903, 570.

When $P\equiv 7 \pmod 8$ and >7, T_0 is even and >2, and hence composite. A detailed study of $P\equiv 3 \pmod 8 = 8k-5$ shows that for all $P>163$ some T_i is composite except perhaps for $k=3t$ and $t=5l+12$ or $5l+13$. With this result and by a stencil device Dickson showed that no P under 1,500,000 except $P=3, 4, 7, 8, 11, 12, 16, 19, 27, 28, 43, 67, 163$ could have a single primitive class.

M. Lerch[329] obtained the chief results of Dirichlet by simple arithmetical methods and reproduced the deduction of several of his[240, 277] own labor-saving formulas.

E. Landau[330] established Pfeiffer's[195] asymptotic expression for $K(x)=\Sigma_{n=1}^{n=x} H_n$ where H_n denotes the number of classes of forms $ax^2+2\beta xy+\gamma y^2$ of negative determinant $-n$. Let $H_{n\nu}$ be the number of non-equivalent reduced forms of determinant $-n$ and with $|\beta|=\nu$. Then for a given n, in each reduced form $\gamma \geqq a \geqq 2\nu$, and $\nu \leqq \sqrt{n/3}$. Thus

$$K(x)=\sum_{n=1}^{x}\sum_{\nu=0}^{\sqrt{n/3}} H_{n\nu}=\sum_{n=1}^{x}(H_{n0}+\sum_{\nu=1}^{\sqrt{n/3}} H_{n\nu})$$
$$=\sum_{n=1}^{x} H_{n0}+\sum_{\nu=1}^{\sqrt{x/3}}\sum_{n=3\nu^2}^{x} H_{n\nu}=\sum_{n=1}^{x} H_{n0}+\sum_{\nu=1}^{\sqrt{x/3}} R(x,\nu).$$

But H_{n0} is the number of solutions of $a\gamma=n$, $\gamma \geqq a$. That is, if $T(n)$ is the number of divisors of n, $H_{n0}=\frac{1}{2}T(n)$, if n is not a square; but $H_{n0}=\frac{1}{2}\{T(n)+1\}$, if n is a square. Hence

$$(1) \qquad \sum_{n=1}^{x} H_{n0}=\frac{1}{2}\sum_{n=1}^{x} T(n)+\frac{1}{2}[\sqrt{x}]=\frac{1}{2}x\log x+(C-\frac{1}{2})x+O(\sqrt{x}),$$

where C is Euler's constant ($=0.57721\ldots$) and $O(k)$ is of the order[117] of k.

For a given $\nu>0$, Landau evaluated

$$R(x,\nu)=\sum_{n=3\nu^2}^{x} H_{n\nu},$$

by noting that $R(x,\nu)$ is the number of solutions of

$$a\gamma \leqq \nu^2+x, \qquad \gamma \geqq a \geqq 2\nu,$$

each solution being counted twice when $\gamma>a>2\nu$. Hence $R(x,\nu)$ is the number of lattice points in the finite area defined by these inequalities in the $a\gamma$-plane, lattice points in the interior and on the hyperbolic arc exclusive of its extremities being counted twice. The resulting value of

$$\sum_{\nu=1}^{\sqrt{x/3}} R(x,\nu)$$

combined with (1) now gives

$$K(x)=\frac{2}{9}\pi x^{\frac{3}{2}}-\frac{x}{2}+O(x^{\frac{5}{6}}\log x).$$

If $\mathcal{K}$ corresponds to K, but refers to classes having a and γ both even, the result obtained is

$$\mathcal{K}(x)=\frac{\pi}{18}x^{\frac{3}{2}}-\frac{x}{4}+O(x^{\frac{5}{6}}\log x).$$

---

329 Casopis, Prague, 40, 1911, 425–446 (Bohemian).
330 Sitzungsber. Akad. Wiss. Wien (Math.-Phys.), 121, II, a, 1912, 2246–2283.

Landau,[331] by a study of the number of lattice points in a sphere, found that if C_n is the number of solutions of $u^2+v^2+w^2=n$,

$$\sum_{n=1}^{x} C_n = \tfrac{4}{3}\pi x^{\frac{3}{2}} + O(x^{\frac{3}{4}+\epsilon}),$$

where[117] only the order of the last term is indicated and ϵ is a small arbitrary positive quantity. But by Kronecker[54] (XI) above, if $F(n)$ denotes the number of uneven classes of forms $ax^2+2bxy+cy^2$ of determinant $-n$, then $C_n=8F(n)$, if $n\equiv 3$ (mod 8); $C_n=12F(n)$, in all other cases except $n\equiv 7$ (mod 8).

In $u^2+v^2+w^2\equiv 1$ (mod 4) evidently

$$u:v:w\equiv 1:1:0,\ 1:0:1,\ 0:1:1 \pmod 2.$$

Hence

$$\sum_{n=1}^{x} F(n) = \frac{\pi}{24}x^{3} + O(x^{\frac{3}{4}+\epsilon}), \quad n\equiv 1 \pmod 4.$$

This holds also for $n\equiv 2$ (mod 4); but

$$\sum_{n=1}^{x} F(n) = \frac{\pi}{48}x^{\frac{3}{2}} + O(x^{\frac{3}{4}+\epsilon}), \quad n\equiv 3 \pmod 8.$$

J. V. Uspensky,[332] by means of lemmas of the types of Liouville's,[90] gave a complete arithmetical demonstration of each of Kronecker's[54] classic eight class-number relations. See Cresse.[374]

J. Chapelon[333] obtained a new identity derived from transformation of the 5th order of elliptic functions and with it followed the procedure of Humbert.[334] He added to Gierster's[135] list of class-number relations of the 5th grade two new ones and gave relations also for

$$\Sigma F(4N-x^2),\ x\equiv 5 \pmod{10}; \qquad \Sigma F(4N-x^2),\ x\equiv \pm 1 \pmod{10};$$

and for $\Sigma F(N-25x^2)$ summed over all integers x, where $N=5^{\mu}N'\equiv 0$ (mod 10), and N' is not divisible by 5. He gave[335] 24 class-number relations for $\Sigma F(N-x^2)$ and $H(N-x^2)$ which are characterized by various combinations of the congruences $N\equiv \pm 2, \pm 4$ (mod 10) with $x\equiv 0, \pm 1, \pm 2$ (mod 5). These 24 relations include Gierster's relations of the 5th grade.[134] The right hand members of Chapelon's class-number relations in these two memoirs are all illustrated by the following example for $N\equiv 2$ (mod 10):

$$\sum_{x} F(N-x^2) = \tfrac{3}{8}\Sigma d' - \tfrac{5}{8}\Sigma(-1)^{d'}d' + \tfrac{1}{2}\Sigma(-1)^{d_1}(d_1-d),$$

$x\equiv \pm 1$ (mod 5), where d' is any divisor of N and $N=d_1d$ with $d_1 \gtreqless d$ (see Chapelon's thesis[340]).

G. Humbert,[336] after giving an account (Humbert[185] of Ch. I) of his principal reduced forms of positive determinant D, proved that for $D=8M+3$

$$\Sigma(-1/\beta)f(a-|b|) = 2\Sigma f(2k+1), \quad \beta\equiv |b| - \tfrac{1}{2}(a+c),$$

[331] Göttingen Nachr., 1912, 764–769.
[332] Math. Sbornik, Moscow, 29, 1913, 26–52 (Russian).
[333] Comptes Rendus, Paris, 156, 1913, 675–677.
[334] Jour. de Math., (6), 8, 1907, 431.
[335] *Ibid.*, 1661–1663.
[336] Comptes Rendus, Paris, 157, 1913, 1361–1362.

where $f(x)$ is an arbitrary even function; the summation on the left extends over all principal reduced forms (a, b, c) of determinant D; and the summation on the right extends over all decompositions,

$$8M+3=(2k+1)^2+(2k'+1)^2+(2k''+1)^2, \quad k, k', k'' \text{ each} \gtreqless 0,$$

of $8M+3$ into the sum of three squares. When $f(x)=1$ and we employ the known value (cf. Kronecker's[54] formula (XI)) for the number of decompositions, we have

$$F(8M+3)=\tfrac{1}{2}\Sigma(-1/\beta).$$

If $f(x)=x^2$, we have

$$\tfrac{8}{3}(8M+3)F(8M+3)=\Sigma(-1/\beta)2\beta(a+c).$$

G. Rabinovitch[336a] proved that the class-number of the field defined by $\sqrt{-d}$, where $d=4m-1$, is unity if and only if $x^2-x+m\,(x=1, \ldots, m-1)$ are all primes. Fewer conditions are given by T. Nagel.[336b]

G. Humbert,[337] by Hermite's method of equating coefficients in theta-function expansions, found that, for all the negative determinants $-(8M+4-4k^2)$, in which M is fixed, the number of odd classes for which the even minimum is not a multiple of 8 is the sum of the divisors of $2M+1$. Similarly for determinants $-(8M-4k^2)$, the number of these classes is $\Sigma(\delta+\delta_1)$ the summation being extended over all the decompositions $2M=\delta\delta_1$, δ odd, δ_1 even, $\delta<\delta_1$. Also, by Hermite's method combined with the use of an even function (cf. Humbert[336]), he[338] obtained the following formula for the number F of odd classes having the minimum and the sum of the two odd minima $\equiv 0 \pmod p$, p arbitrary:

$$\Sigma(-1)^r F(4N+3-4r^2)=(-1)^N\Sigma' d,$$

$r<, =, >0$, and for h arbitrary, r is $\equiv h \pmod p$, $4N+3=pdd_1$, with $pd<d_1$, and $d_1\equiv \pm 4h \pmod p$.

*F. Lévy[339] discussed the determination of the number of classes of a negative determinant by means of elliptic functions.

J. Chapelon[340] gave an outline of the history of class-number relations of the general Kronecker[54] type and listed Gierster's[135] relations of the 5th grade. Examples will be given here merely to characterize each of the six exhaustive chapters of the thesis.

Chapter I contains theorems on the divisors of a number. Let $N=2^\mu N'=5^\nu N''=2^\mu 5^\nu N'''$, N' and N'' prime to 2 and 5 respectively; $N=d_1 d$, $d_1 \gtreqless d$; d' any divisor of N; and, let $\mathscr{B}=\Sigma d'(d'/5)$, $\mathscr{B}_1=\Sigma(-1)^{d'}d'(d'/5)$, $\mathscr{Q}=\Sigma(d_1-d)(d_1+d/5)$. Also let $N=da$, $d>\sqrt{N}$, $a<\sqrt{N}$. Then

$$\Sigma\left[\left(\frac{a}{5}\right)+\left(\frac{d}{5}\right)\right]a=\tfrac{1}{2}\left[1+5^\nu\left(\frac{N''}{5}\right)\right]\mathscr{B}-\tfrac{1}{2}\mathscr{Q}.$$

336a Jour. für Math., 142, 1913, 153–164; abstr. in Proc. Fifth Internat. Congress Math., Cambridge, I, 1913, 418–421.

336b Abh. Math. Seminar Hamburgischen Universität, 1, 1922, 140–150.

337 Comptes Rendus, Paris, 158, 1914, 297.

338 *Ibid.*, 1841–1845.

339 Thesis, Zürich, published A. Kündig, Geneva, 1914, 48 pp.

340 Thèse, Sur les relations entre les nombres des classes de formes quadratiques binaires, Paris, 1914, 197 pp.; Jour. de l'Ecole Polytechnique, Paris, 19, 1915.

Chapter II gives, in Hermite[80]-Humbert[293] notation, lists of standard transformation formulas for the Θ-function and expansions of Θ-functions and θ-functions.

Chapter III presents fundamental formulas for the transformation of the fifth order of Θ-functions. In

$$\Theta(5u, 5\tau)=C_1\Pi\,\Theta(u\pm\pi i/5), \quad i=0, 1, 2,$$

C_1 is found to be $\eta^5/\bar{\eta}$ where $\bar{\eta}=\eta(q^5)$ and $\eta=\eta(q)=\Sigma_{-\infty}^{+\infty}(-1)^m q^{\frac{1}{12}(6m+1)^2}\eta^5$.

Chapter IV deals with the representation of a number by certain quaternary quadratic forms. In (p. 90)

$$\eta_1^2\theta_1^2\frac{H^2}{\Theta^2}=8\Sigma_1\frac{mq^m}{1-q^{2m}}-8\Sigma\frac{mq^m}{1-q^{2m}}\cos 2mx,$$

put $x=\pi/5$ and $x=2\pi/5$ and subtract. Equating coefficients of q^M, we get

$$(1)\qquad [5\mathcal{M}_3(2M)-\mathcal{M}_1(2M)]=10\mathcal{B}+2\mathcal{B}_1,$$

where $\mathcal{M}_1(N)$ and $\mathcal{M}_3(N)$ are respectively the number of decompositions (in which the order is regarded)

$$4N=(2x+1)^2+(2y+1)^2+(2z+1)^2+5(2t+1)^2,$$
$$4N=(2x+1)^2+5(2y+1)^2+5(2z+1)^2+5(2t+1)^2.$$

From another expansion it is similarly found that

$$(2)\qquad -\tfrac{1}{2}\mathcal{M}_3+\tfrac{1}{2}\mathcal{M}_1=5^\nu(N'''/5)2^{\mu+2}\Sigma\theta'(\theta'/5),$$

summed over all divisors θ' of N'''. Then by (1) and (2),

$$\mathcal{M}_1(N)=\tfrac{1}{2}[1+5^{\nu+1}(N''/5)][5\mathcal{B}+\mathcal{B}_1].$$

Suppose that N is even. Since for a fixed value of x, $F[4N-5(2x+1)^2]$ is the number of positive solutions t, u, v of $t^2+u^2+v^2=4N-5(2x+1)^2$ (p. 118),

$$(3)\qquad \sum_{x=0}^{\infty}F(4N-5(2x+1)^2)=\tfrac{1}{16}\mathcal{M}_1(N)=\tfrac{1}{32}[1+5^{\nu+1}(N''/5)](5\mathcal{B}+\mathcal{B}_1).$$

This is a special case of Liouville's[107] (4).

In terms of functions like $\mathcal{M}_1$ and $\mathcal{M}_3$ above, Chapelon found in Ch. V expressions for

$$\Sigma F(8M-4-x^2),\ \Sigma F\left(\frac{8M\pm4-x^2}{25}\right),\ \Sigma F\left(\frac{8M-x^2}{25}\right),\ \Sigma F(N-x^2),$$
$$\Sigma E\left(\frac{N-x^2}{25}\right),\ J(N-x^2),\ J\left(\frac{N-x^2}{25}\right);$$

where $x=5\sigma\pm k$ or $10\sigma\pm k$, k constant; $E(N)=F(N)-H(N)$, $J(N)=F(N)+3H(N)$.

In Ch. VI, Chapelon found sets of relations equivalent to each one of Gierster's[135] relations of grade 5; and added large sets of new relations, the sets being distinguished by the residue of N modulo 10. He (p. 171) proved Liouville[114] (1).

H. N. Wright[341] tabulated the reduced forms $ax^2+2bxy+cy^2$ of negative determinant $-\Delta=D$ for $\Delta=1$ to 150 and 800 to 848. The values of b, c occur at the inter-

[341] University of California Publications, Berkeley, 1, 1914, No. 5, 97–114.

section of the columns giving a and the row giving Δ. For a given a, the reduced form occurs in periods, each period covering a values of Δ; and each period having the same sequence of b's. For a given D, the a's are found among those for which there is a solution of $x^2 \equiv D \pmod{a}$. For the case of Δ without square divisors, he wrote

$$\Delta = \prod_0^r k_i \qquad a = \prod_0^u h_i^{\beta_i} \prod_1^v k_i'^{\delta_i},$$

where h and k are primes; $h_0 = 2$, $\delta_i > 1$ and the k''s are those odd k_i's which in a have exponents >1. Let ν be the number of distinct factors k'^2 of a; let λ be the greatest value of ν for any a. Then for the given D, the number of reduced forms with $a \leqq \sqrt{\Delta}$ is found to be

$$\sum_{\nu=0}^{\lambda} (-1)^\nu \sum_i \left\{ \sum_{P_0} \left[\frac{[\sqrt{\Delta}]}{l_i^{(\nu)} P_0} \right] \left(\frac{D}{P_0} \right) + \sum_{P_\epsilon} \left[\frac{[\sqrt{\Delta}] + l_i^{(\nu)} P_\epsilon}{2 l_i^{(\nu)} P_\epsilon} \right] \left(\frac{D}{P_\epsilon} \right) \right\},$$

where $l_i^{(\nu)}$ is the ith product formed by taking ν factors k'^2; P_0 is a positive odd integer, P_ϵ a positive even integer, both $\leqq \sqrt{\Delta}$; (D/P_0) is a modified Jacobi symbol and if $P_\epsilon = P_0' 2^r$, P_0' odd, then $(D/P_\epsilon) = (D/P_0')(D/2^r)$, where $(D/2^r)$ is defined so that $1 + (D/2^r)$ is the number of solutions of $x^2 \equiv D \pmod{2^r}$.

The few remaining possible values of a which are $> \sqrt{\Delta}$ and $\leqq \sqrt{4\Delta/3}$ or $\leqq \frac{1}{3}[-1 + 2\sqrt{1 + 3\Delta}]$, according as a is even or odd, are to be tested by the most elementary methods. Examples show the advantage of this whole process over the classic one of Dirichlet,[23] (5).

E. Landau[342] investigated the asymptotic sum of Dirichlet's series[19]

$$\Sigma (ax^2 + 2bxy + cy^2)^{-s},$$

in the neighborhood of $s = 1$ for a form of positive determinant D. (For $D < 0$, see Ch. de la Vallée Poussin, Annales Soc. Sc. Brussells, 20, 1895-6, 372-4).

L. J. Mordell[343] announced the equivalent of two serviceable identities of Petr[344] in theta-functions. For, Mordell's Q and R are respectively Petr's C and $4 \cdot D$. By specializing the arguments in the identities and equating coefficients of like powers of q, Mordell found new representatives of five types of class-number relations such as Petr[252, 258] and Humbert[293] had deduced.

K. Petr[345] combined C. Biehler's[346] generalized Hermetian theta-function expansions, which Petr had used twice[252, 258] before, now with W. Göring's[347] formulas given by the transformation of the third order of the theta-functions. He obtained six expansions similar to the following[348]:

$$B_3 \equiv \Theta_3 \Theta_1^2 \frac{\Theta_2^2(\frac{1}{3}) \Theta_3(\frac{1}{3})}{\Theta^2(\frac{1}{3})} = \frac{1}{\sin^2 \pi/3} + B\dot{\Theta}_3(\tfrac{1}{3}) - 4 \sum_{m=2}^{\infty} \sum_{k=1}^{m-1} 2kq^{m^2-k^2} \cos 2m\pi/3,$$

in which $q = e^{\pi i \tau}$, and B is found[69] to be $8\Sigma_1^\infty q^N F(N)$, where as usual $F(n)$ is the

[342] Jahresbericht d. Deutschen Math.-Vereinigung, 24, 1915, 250–278.
[343] Messenger Math., 45, 1915, 76–80.
[344] Rozpravy české Akad., Prague, 9, 1900, No. 38 (Petr [252]).
[345] Memorial Volume for the 70th birthday of Court Councilor Dr. K. Vrby, 1915; Rozpravy české Acad., Prague, 24, 1915, No. 22, 10 pp.
[346] Thesis, Paris, 1879.
[347] Math. Annalen, 7, 1874, 311–386.
[348] Cf. G. Humbert,[293] Jour. de Math., (6), 3, 1907, 348.

number of odd Gauss forms of determinant $-n$. On expanding $\Theta_3(\frac{1}{3})$, it is found that the coefficient of q^N in B_3 is

$$-4(\Sigma d_\lambda - \Sigma d_1) + 2\Sigma(d_2 - d_1) - 6\Sigma(d_2^{(0)} - d_1^{(0)}) \\ + 12[F(N) + 2F(N - 9\cdot 1^2) + 2F(N - 9\cdot 2^2) + \dots],$$

summed for the divisors d of N; the subscript λ on d denotes that the conjugate divisor is odd; the subscript 1 denotes that the divisor agrees in parity with the conjugate and is $\leqq \sqrt{N}$; but, if it $= \sqrt{N}$, it is replaced in the sum by $\frac{1}{2}\sqrt{N}$. Also, $N = d_1 d_2$; $N = d_1^0\ d_2^0$, $d_1^{(0)} + d_2^{(0)} \equiv 0 \pmod 3$. This includes the case N odd and $\equiv 1 \pmod 3$ which G. Humbert[349] had failed to provide for in a direct way.

Similarly in

$$B_3' = q^{\frac{1}{3}} \Theta_3 \Theta_1^2 \Theta_2^2 \left(\frac{\tau}{3}\right) \Theta_3 \left(\frac{\tau}{3}\right) \Big/ \Theta^2 \left(\frac{\tau}{3}\right),$$

the coefficient of $q^{N/9}$ is

$$8\Sigma F\left(\frac{N - (9k \pm i)^2}{9}\right) - \tfrac{2}{3}\Sigma(d_2^{(\delta)} - d_1^{(\delta)}),$$

where the subscript 1 has the same meaning as before, $d_2^{(\delta)} - d_1^{(\delta)} \equiv 0 \pmod 3$ and where $i = 1, 2, 4$, according as $N \equiv 1, 4, 7 \pmod 9$. Alternative expansions of B_3 and B_3' were obtained by Petr with indication of a method of determining in them the coefficients of q^N and $q^{N/9}$ respectively in terms of divisors of N and the number of integer solutions of $x^2 + y^2 + z^2 + 9u^2 = N$ and $x^2 + 9y^2 + 9z^2 + 9u^2 = N$, respectively. The class-number relations thus resulting were given by Petr in the next paper.

K. Petr[350] completed[345] the deduction of the following class-number relations. For N arbitrary,

$$F(N) + 2F(N - 9\cdot 1^2) + 2F(N - 9\cdot 2^2) + \dots \\ = \tfrac{1}{4}[\Sigma d_\lambda + \Sigma d_\lambda^{(0)} - 4\Sigma d_1^{(0)} + \Sigma(-1)^{\frac{1}{2}(x-1) + \frac{1}{6}(y \mp 1)} x],$$

summed over the positive odd numbers x, y satisfying $3x^2 + y^2 = 4N$, such that y is not divisible by 3. [Petr in this and all the following formulas of the paper erroneously imposed the latter condition also on x.] The upper index $^{(0)}$ indicates that the sum of the corresponding divisor and its conjugate is $\equiv 0 \pmod 3$.

Again for N arbitrary,

$$F(N) - 3H(N) + 2[F(N - 9\cdot 1^2) - 3H(N - 9\cdot 1^2)] \\ + 2[F(N - 9\cdot 2^2) - 3H(N - 9\cdot 2^2)] \dots = -\tfrac{1}{2}(\Sigma \bar{d} - \Sigma d_e) \\ - \tfrac{1}{2}(\Sigma \bar{d}^{(0)} - \Sigma d_e^{(0)}) + 2\Sigma d_1^{(0)} + \tfrac{1}{2}(-1)^{\frac{1}{2}(x-1) + \frac{1}{6}(x+1)} x,$$

where $\bar{d}$ agrees in parity with its conjugate divisor of N, and d_e is odd.

For $N \equiv 1$, 4 or 7 $\pmod 9$, the two following relations are given:

$$\sum_k F\left(\frac{N - (9k \pm a)^2}{9}\right) = \tfrac{1}{24}\Sigma d_\lambda - \tfrac{1}{6}\Sigma d_1 + \tfrac{1}{24}(-1)^{\frac{1}{2}(x-1) + \frac{1}{6}(y \mp 1)} x,$$

$$\sum_k \left[F\left(\frac{N - (9k \pm a)^2}{9}\right) - 3H\left(\frac{N - (9k \pm a)}{9}\right)^2\right] \\ = -\tfrac{1}{12}(\Sigma \bar{d} - \Sigma d_c) + \tfrac{1}{3}\Sigma d_1 + \tfrac{1}{12}\Sigma(-1)^{\frac{1}{2}(x-1) + \frac{1}{6}(y \mp 1)} x,$$

in which $a = 1$, 2 or 4 according as $N \equiv 1$, 4 or 7 $\pmod 9$.

---

[349] Jour. de Math., (6), 3, 1907, 431.
[350] Rozpravy české Acad., Prague, 25, 1916, No. 23, 7 pp.

In equating coefficients of q^N in the identity[345] B_3, Petr on his page 2 of the present paper employed the identity

$$2\Theta^4(\tfrac{1}{3})=18\Sigma x q^{\frac{1}{4}(3x^2+y^2)}(-1)^{\frac{1}{2}(x-1)+\frac{1}{6}(y\pm 1)},$$

and failed to observe that x may be $\equiv 0 \pmod 3$. So he introduced an error in the denotation of all the resulting class-number relations of the paper.

L. J. Mordell[351] deduced arithmetically the first class-number relation of his preceding paper[343] in the form

$$(1) \qquad F(m)-2F(m-1^2)+2F(m-2^2)-\ldots=\Sigma(-1)^{\frac{1}{2}(a+d)+1}d,$$

where d is a divisor $\leqq\sqrt{m}$ of m and of the same parity as its conjugate divisor a; but when $d=\sqrt{m}$, the coefficient d is replaced by $\frac{1}{2}d$. Mordell considered the number of representations of an arbitrary positive integer m by the two forms

$$(2) \qquad s^2+n^2+n(2t+1)-r^2=m,$$

$$(3) \qquad d(d+2\delta)=m,$$

$$n>0, \quad -(n-1)\leqq r\leqq n, \quad t\gtreqless 0, \quad d>0, \quad \delta\gtreqless 0.$$

Then, if $f(x)$ is an arbitrary even function of x,

$$(4) \qquad \Sigma(-1)^r f(r+s)=-2\Sigma(-1)^\delta df(d)$$

where the summation on the left extends over all solutions of (2), and the summation on the right extends over all solutions of (3); but, when $\delta=0$, the coefficient 2 is replaced by unity. Now take $f(x)=(-1)^x$. Then (4) becomes $\Sigma(-1)^s=-2\Sigma(-1)^{\delta+d}d$. But for a given s, Mordell[352] found that the number of solutions of (2) is $2F(m-s^2)$. Hence we get at once the above class-number relation (1).

Mordell[352] illustrated his[343] method by writing

$$f(x)=\sum_{n=-\infty}^{\infty}\frac{q^{n^2}e^{2n\pi ix}}{1+q^{2n}}$$

and proving that

$$(1) \quad -\frac{f'(0)}{2\pi i\theta_{00}}=\sum_{n=1}^{\infty}q^{n^2+n}\frac{(1-2q^{-1}+\ldots\pm 2q^{-(n-1)^2}\mp q^{-n^2})}{1-q^{2n}}=\sum_{n=1}^{\infty}(-1)^r q^{n^2-r^2+n(2t+1)},$$

where $r=0, \pm 1, \pm 2, \ldots, \pm(n-1), n$; $t=0, 1, 2, \ldots$. But corresponding to each set of values n, r, t, there is a reduced quadratic form[353]:

$$nx^2+2rxy+(n+2t+1)y^2$$

of determinant, say, $-M$. Conversely to each reduced form $(a, 0, a)$ of determinant $-M$, there corresponds one solution, and to every other reduced form of determinant $-M$, there correspond two solutions, of the equation $M=n^2-r^2+n(2t+1)$. Hence the right member of (1) is $2\Sigma_1^\infty(-1)^M F(M)q_M$. When $f'(0)$ is given its true value, and q is replaced by $-q$, and θ_{01} by $1-2q+2q^4-2q^9+\ldots$, the

[351] Messenger Math., 45, 1916, 177–180. See a similar arithmetical deduction by Liouville.[90]
[352] Messenger Math., 46, 1916, 113–128.
[353] And so this expansion (1) suggested to Mordell his[351] arithmetical deduction.

equating of coefficients of q^n yields his[351] relation (1); which is equivalent to Kronecker's[54] (III), (VI), and is identically Petr's[252] relation (II).

Replacing $f(x)$ by $\phi(x)=f(x)\theta_{gh}(x+\xi)/\theta_{00}(x)$, where ξ is an arbitrary constant, Mordell obtained the equivalent of Kronecker's (I), (II), (V). By the use of $\chi(x)=f(x)\theta_{01}(2x, 2\omega)/\theta_{00}(x)$, he obtained a class-number relation involving an indefinite form[354] in the equation $x^2-2y^2=m$. By the use of $f(x)\theta_{00}(3x, 3\omega)/\theta_{00}(x)$, he found (cf. Petr's[258] formula (3) above) that

$$F(2m)-2F(2m-3\cdot 1^2)+2F(2m-3\cdot 2^2)-\ldots=(-1)^{m+1}\Sigma x,$$
$$x^2-3y^2=m,\quad x>0,\quad -\tfrac{1}{3}(x-1)\leqq y\leqq \tfrac{1}{3}x.$$

Replacing $f(x)$, as initially used, by

$$F(x)=\sum_{n=-\infty}^{\infty} q^{\frac{1}{4}n^2}e^{n\pi ix}/(1-q^n),\quad n \text{ odd},$$

he obtained

$$3[G(m)+2G(m-1^2)+2G(m-2^2)+\ldots]=-6\Sigma a+4\Sigma b+2\Sigma(-1)^c c,$$

where a denotes a divisor of m which is $\leqq \sqrt{m}$ and agrees with its conjugate in parity, but if $a=\sqrt{m}$ it is replaced by $a/2$; b denotes a divisor of m whose conjugate is odd, and c a divisor of m whose conjugate is even. Kronecker's[54] (IV) is the special case of this formula for m odd.

G. Humbert,[355] in a principal reduced form (Humbert[185, 186] of Ch. I), (a, b, c) of positive determinant with $b>0$, put $\beta=b-\frac{1}{2}|a+c|$, and, by Hermite's method of equating coefficients in θ-function expansions,[69] found that

$$\sum_n{}^1\left(\frac{-1}{\beta}\right)=2F(4n+2),\quad \sum_n{}^2\left(\frac{-1}{\beta}\right)=2F(8n+5),\quad \sum_n\left(\frac{-1}{\beta}\right)=2F(8n+1),$$

where $\sum_n^1$ extends over all the principal reduced forms of determinant $4n+2$ with a and c odd; $\sum_n^2$ extends over all the principal reduced forms of determinant $8n+5$ with a and c even; $\sum_n$ extends over all the principal reduced forms of determinant $8n+1$ with $\frac{1}{2}(a+c)$ even.

From the first of the three formulas is deduced the following: Among the principal reduced forms (a, b, c) of positive determinant $4n+2$, the number of those in which $b-\frac{1}{2}|a+c|$ is of the form $4k+1$ diminished by the number of those in which it is of the form $4k-1$ is double the number of positive classes of determinant $-(4n+2)$.

By denoting by $H_1(n)$ the left member of the first of these three formulas, for example, and summing as to the argument $4M+2-(2s)^2$, Kronecker's classic formulas[54] give

$$H_1(4M+2)+2H_1(4M+2-2^2)+2H_1(4M+2-4^2)+\ldots=2\phi_i(4M+2),$$

where $\phi_i(n)$ is the sum of the odd divisors of n.

---

[354] Cf. K. Petr, Rozpravy české Akad., Prag, 10, 1901, No. 40, formula (1) of the report [258]; also G. Humbert,[293] Jour. de Math., (6), 3, 1907, 381, formula (57).

[355] Comptes Rendus, Paris, 165, 1917, 321–327.

L. J. Mordell[356] recalled Dirichlet's[20] formula (2). Whence[357] if $|r|<1$,

$$(1) \qquad \sum_{a,b,c}\sum_{x,y} r^{ax^2+bxy+cy^2}=A+\tau\sum_{k=1}^{\infty}\left(\frac{D}{k}\right)\frac{r^k}{1-r^k},$$

summed for all pairs of integers $x, y>0, =0$ or <0, and for representative forms of negative discriminant D; while (D/k) is the generalized symbol of Kronecker[171] and A is the number of classes of discriminant D. We set $r=e^{2\pi i\omega}$ and write (1) as

$$(2) \qquad \phi(\omega)=A+\chi(\omega).$$

When $\chi(\omega)$ is evaluated in terms of θ-functions, (2) becomes:

$$(3) \quad \phi(\omega)=A+\frac{\tau}{4\pi\sqrt{n}}\left\{f(\omega)+\frac{2\pi}{\sqrt{n}}\sum_{v=1}^{n-1}v\left(\frac{D}{v}\right)\right\},\quad f(\omega)=\sum_{v=1}^{n-1}\left(\frac{D}{v}\right)\cdot\frac{\theta'\left(\frac{v}{n}\right)}{\theta\left(\frac{v}{n}\right)},$$

where $\theta(v)=\theta_{11}(v)$. Now[358]

$$\phi\left(-\frac{1}{\omega}\right)=\frac{\omega}{\sqrt{-n}}\phi\left(\frac{\omega}{n}\right),\quad f\left(-\frac{1}{\omega}\right)=\frac{\omega}{\sqrt{-n}}f\left(\frac{\omega}{n}\right).$$

Hence when ω is replaced by $-1/\omega$, (2) gives Kronecker's[171] formula (5_1) for the class-number.

E. Landau[358a] wrote ϵ for the fundamental unit $\frac{1}{2}(T+\sqrt{D}U)$ and by means of Kronecker's[171] class-number formula (3), obtained an upper bound of $\log\epsilon/\sqrt{D}\log D$ for very great D by noting that $K(D)\gtreqless 1$ and finding an upper bound of the sum of the Dirichlet series in that formula.

E. Landau[359] wrote $h(k)$ for the number of classes of ideals of the imaginary field defined by $\sqrt{-k}$. Let δ be any positive number. If there are infinitely many negative values $-k^{(\nu)}$ of $-k(k^{(1)}<k^{(2)}<\ldots)$ such that

$$h(k)<k^{\frac{1}{2}-\delta},$$

then, for every real $\omega>1$, $k^{(\nu+1)}>k^{(\nu)^{\omega}}$ for every ν exceeding a value depending on δ and ω. Given any $\omega>1$, if we can assign c, depending on ω, such that,

$$h(k)<c\sqrt{k}/\log k$$

holds for an infinitude of negative values $-k^{(\nu)}$ of $-k$, then $k^{(\nu+1)}>k^{(\nu)^{\omega}}$ for every $\nu\gtreqless 1$. Known facts are proved about limits to $h(k)$. He[360] derived inequalities relating to $h(k)$.

G. Humbert[361] let m_1 and m_2 be the odd minima of an odd Gaussian form (a, b, c), and $H(M)$ be the number of odd reduced forms of determinant $-M$ for which m_1

356 Messenger Math., 47, 1918, 138–142.
357 Obtained independently by Petr,[305] (1).
358 Cf. Mordell, Quar. Jour. Math., 46, 1915, 105.
358a Göttingen Nachr., 1918, 86–7.
359 Göttingen Nachr., 1918, 277–284, 285–295 (95–97).
360 Math. Annalen, 79, 1919, 388–401.
361 Unpublished letter to E. T. Bell, October 15, 1919.

or m_2 is $\equiv 0 \pmod{p}$, p being a given odd prime; and if simultaneously $m_1 \equiv m_2 \equiv 0 \pmod{p}$, he let the class count 2 units in $H(M)$; then, when $N \equiv 0 \pmod{p}$,

$$\Sigma(-1)^n H(N-n^2) + \Sigma(-1)^n H(N-n^2) = -2\Sigma(d/p)(-1)^{\frac{1}{2}(d'+d)},$$

where the first summation extends over all integers $n \equiv 0 \pmod{p}$, the second over the positive integers n not $\equiv 0 \pmod{p}$, and the third over all decompositions $N = dd'$, with $d \equiv 0 \pmod{p}$, $d < d'$ and d, d' of the same parity. The class $a(x^2+y^2)$, when $a \equiv 0 \pmod{p}$, counts here as one unit in $H(a^2)$.

Let $\phi_h(N)$ be the number of classes of positive odd Gaussian forms of determinant $-N$, for which the minimum μ is $\gtreqless 2h$; if $\mu = 2h$, the class counts for $\frac{1}{2}$ in $\phi_h(N)$. Then for N odd, positive, and prime to 3,

$$\phi_0(N) + 2\phi_1(N+3\cdot 1^2) + \ldots + \phi_h(N+3\cdot h^2) + \ldots$$
$$= \frac{3}{4}\left(\frac{-3}{N}\right)\left[1+\left(\frac{-N}{3}\right)\frac{1}{3}\right]\Sigma d\left(\frac{-3}{d}\right) + \frac{1}{2}\Sigma\left(\frac{3}{d}\right),$$

where in the second member, the summations extend over all divisors d of N. In the first member, ϕ_h certainly equals zero when h is $> \frac{1}{4}(N+1)$.

Similarly, N being odd, let $\phi'_h(N)$ be the number of classes of positive even forms for which the minimum μ is $\gtreqless 2h$; if $\mu = 2h$, the class counts $\frac{1}{2}$ in $\phi'_h(N)$. Then we have

$$\phi'_0(N) + 2\phi'_1(N+2\cdot 1^2) + 2\phi'_2(N+2\cdot 2^2) + \ldots + 2\phi'_h(N+2h^2) + \ldots$$
$$= \frac{1}{6}\left[\left(\frac{-2}{N}\right) - \frac{1}{2}\right]\Sigma d\left(\frac{-2}{d}\right) + \frac{1}{4}\Sigma\left(\frac{2}{d}\right) + \frac{1}{3}\left[1+\left(\frac{N'}{3}\right)\right]\Sigma\left(\frac{6}{d}\right),$$

where, in the second member, the summations extend over all divisors d of N; $(6/d) = 0$ if $d \equiv 0 \pmod{3}$, and N' is the quotient of N by the highest power of 3 that divides N.

And similarly,[362] let $\psi_h(M)$ be the number of reduced odd Gaussian forms (a, b, c) of determinant $-M$ for which simultaneously $a \gtreqless 2h$, $a+c-|b| \gtreqless 5h$; if in these relations, there is a single equality sign, the form counts $\frac{1}{2}$ in ψ_h; if there are two equality signs, the form counts $\frac{1}{4}$. Then, if $N \equiv 7, 17, 23$, or $33 \pmod{40}$,

$$\psi_0(N) + 2\psi_1(N+5\cdot 1^2) + \ldots + 2\psi_h(N+5\cdot h^2) + \ldots = \tfrac{1}{2}(-5/N)\Sigma d(-5/d),$$

the summation extending over all divisors d of N.

Class-number relations occur incidentally in Humbert's papers 18, 23, 24 of Ch. XV.

L. L. Mordell[363] deduced his[364] formula (1) from the identity

$$\omega\theta_{11}(x, \omega)\int_{-\infty}^{\infty} \frac{e^{\pi i\omega t^2 - 2\pi t x}}{e^{2\pi t}-1}\,dt = f\left(\frac{x}{\omega}, -\frac{1}{\omega}\right) + i\omega f(x, \omega),$$

where the path of integration may be a straight line parallel to the real axis and below it a distance less than unity, and where

$$if(x) = \sum_{n \text{ odd}}^{\pm\infty} \frac{(-1)^{\frac{1}{2}(n-1)} q^{\frac{1}{4}n^2} e^{n\pi i x}}{1+q^n}.$$

[362] Deduced by Humbert from his own formula (7), Comptes Rendus, Paris, 169, 1919, 410.
[363] Messenger Math., 49, 1919, 65–72.

By applying Kronecker's[54] formula (XI) to the right member of formula[364] (1) and integrating the left member, Mordell obtained the relation[364] (3). But by applying the identity

$$\theta_{00}(0, -1/\omega) = \sqrt{-i\omega}\,\theta_{00}(0, \omega)$$

to the right member, he found

$$\frac{1}{2a^2} + \sum_1^\infty \frac{1}{(n+a)^2} = 2\pi^2 \sum_{M=0}^\infty [4F(M) - 3G(M)]e^{-2a\pi\sqrt{M}} - 4a \sum_{M=1}^\infty \frac{F(M)}{(a^2+M^2)^2}.$$

L. J. Mordell[364] announced without proof the formulas:

$$(1)\qquad \int_{-\infty}^{\infty} \frac{te^{\pi i\omega t^2}}{e^{2\pi t}-1}\,dt = -2\sum_1^\infty F(n)q^n + \frac{2}{\omega^2}\sqrt{-i\omega}\sum_1^\infty F(n)q_1^n + \tfrac{1}{4}\theta_{00}^3(0, \omega),$$

$$(2)\qquad \int_{-\infty}^{\infty} \frac{te^{\pi i\omega t^2}}{e^{2\pi t}+1}\,dt = \sum_1^\infty (-1)^n F(4n-1)q^{\frac{1}{4}(4n-1)} + \frac{2}{\omega^2}\sqrt{-i\omega}\sum_1^\infty (-1)^{n-1}F(n)q_1^n,$$

where $R(i\omega)<0$, $q=e^{\pi i\omega}$, $q_1=e^{-\pi i/\omega}$. Proofs were given elsewhere.[363] By integrating, he deduced from (1) the relation,

$$(3)\qquad \frac{1}{2a^2} + \sum_1^\infty \frac{1}{(n+a)^2} = -4\pi^2\sum_1^\infty F(M)e^{-2a\pi\sqrt{M}} + 2a\sum_0^\infty \frac{4F(M)-3G(M)}{(a^2+M)^2},$$

where $R(a)>0$, a arbitrary.

E. T. Bell[365] proved that

$$\left.\begin{array}{lll}(1) & m=4k+1, & N_3(m)=6[\epsilon(m)+4\Sigma\xi\{\tfrac{1}{4}(m-\mu^2)\}]\\ (2) & m=4k+3, & N_3(m)=8\Sigma\xi\{\tfrac{1}{2}(m-\mu^2)\}\end{array}\right\}\ \mu \text{ odd } <\sqrt{m},$$

where $N_3(m)$ is the number of representations of n as the sum of 3 squares; $\epsilon(n)=1$ or 0, according as n is or not a square; and $\xi(n)$ is the excess of the number of divisors $4k+1$ of n over the number of divisors $4k+3$. He[366] then stated that elementary considerations yield

$$\begin{array}{lll}(3) & m=4k+1, & N_3(m)=6[\xi(m)+2\Sigma\xi(m-4a^2)],\\ (4) & m \text{ odd}, & N_3(2m)=12[\xi(m)+2\Sigma\xi(m-2a^2)],\\ (5) & m \text{ odd}, & N_3(2m)=12\Sigma\xi(2m-\mu^2),\\ (6) & n \text{ arbitrary}, & N_3(n)=2[\epsilon(n)+2\xi(n)+4\Sigma\xi(n-a^2)],\end{array}$$

where m, n, a are positive integers, μ is any positive odd integer, and where x is as always >0 in $\Sigma\xi(x)$. A comparison of (1), (2), (3), (4), (5), (6), with the well-known relations (Kronecker,[54] (XI); Hermite,[69] (7))

$$(7)\qquad \begin{cases} m=4k+1, & N_3(m)=12F(m);\quad N_3(n)=12[2F(n)-G(n)],\\ m=8k+3, & N_3(m)=\ 8F(m);\ N_3(2m)=12F(2m),\end{cases}$$

[364] Quar. Jour. Math., 48, 1920, 329–334.
[365] Quar. Jour. Math., 49, 1920, 45–51.
[366] Quar. Jour. Math., 49, 1920, 46–49.

where $G(n)$ denotes the total number of classes and $F(n)$ the number of uneven classes of determinant $-n$, gives immediately

$$\begin{array}{ll} m=4k+1, & 2G(m)=\xi(m)+2\Sigma\xi(m-4a^2), \\ m=4k+1, & 2G(m)=\epsilon(m)+4\Sigma\xi\{\tfrac{1}{2}(m-\mu^2)\}, \\ m \text{ odd}, & G(2m)=\xi(m)+2\Sigma\xi(m-2a^2), \\ m \text{ odd}, & G(2m)=\Sigma\xi(2m-\mu^2), \\ m=8k+3, & F(m)=\Sigma\xi\{\tfrac{1}{2}(m-\mu^2)\}, \\ & 12F(n)-6G(n)=\epsilon(n)+2\xi(n)+4\Sigma\xi(n-a^2). \end{array}$$

Similarly by comparing (7) with seven recursion formulas[367] such as

$$m=4k+1, \quad N_3(m)=6\zeta_1\{\tfrac{1}{2}(m+1)\}-\Sigma N_3(m-8t),$$

in which $\zeta_1(n)$ denotes the sum of all the divisors of n, and $t>0$ an arbitrary triangular number, he obtained the seven following recursion formulas for class-number:

$$\begin{array}{ll} m=4k+1, & 2G(m)+2\Sigma G(m-8t)=\zeta_1\{\tfrac{1}{2}(m+1)\}, \\ m=4k+1, & 2G(m)+4\Sigma G(m-4a^2)=\zeta_1(m), \\ m \text{ odd}, & 4G(2m)+4\Sigma G(2m-8t)=\zeta_1(2m+1), \\ m \text{ odd}, & G(2m)+2\Sigma G(2m-4a^2)=\zeta_1(m), \\ m=8k+3, & F(m)+\Sigma F(m-8t)=\zeta_1\{\tfrac{1}{4}(m+1)\}, \\ m=8k+3, & 4F(m)+8\Sigma F(m-4a^2)=\zeta_1(m), \\ & 6E(n)+12\Sigma E(n-4a^2)=4(-1)^n\lambda_1(n)-\epsilon(n), \end{array}$$

in which $\lambda_1(n)=[2(-1)^n+1]\zeta_1'(n)$, where $\zeta_1'(n)$ is the sum of the odd divisors of n; and in which $E(n)=2F(n)-G(n)$. The last of these relations is equivalent to Kronecker's[54] formula (X).

L. J. Mordell,[368] starting from Dirichlet's[20] formula (1)

$$\frac{\pi}{2}h(-n)=\sqrt{n}\sum_{r \text{ odd}}\frac{1}{r}\left(\frac{-n}{r}\right),$$

and allowing for the improper classes, proved that

$$8\sum_1^{\infty}F(n)q^n=\Sigma\frac{i^{-\frac{1}{2}(b-1)}(a/b)}{[\sqrt{-i(a+bi)}]^3}=\Sigma\left[\frac{\theta_{00}(\omega)}{\theta_{00}\left(\dfrac{c+d\omega}{a+b\omega}\right)}\right]^3, \quad q=e^{\pi i\omega},$$

where the real part of $i\omega$ is <0; the radical is taken with positive real part; the summation is carried out first for $a=0, \pm 2, \pm 4, \ldots$, and then for $b=1, 3, 5, \ldots$, in this order; (a/b) is the Legendre symbol; but if $a=0$, $b=1$, we replace (a/b) by 1. Also d is any even integer, c any odd integer, satisfying $ad-bc=1$. He also proved that $F(M)/\sqrt{M}=f(1)-\frac{1}{3}f(3)+\frac{1}{5}f(5)-\ldots$, where $f(n)$ denotes the number of solutions of $\xi^2\equiv M \pmod{n}$. Formulas of the same type are also given in which $F(n)$ is replaced by $G(n)$.

E. T. Bell,[369] by equating like powers q in the expansions of functions of elliptic theta constants, showed that the class-number relations of Kronecker, Hermite and

---

[367] Bell, Amer. Jour. Math., 42, 1820, 185–187.
[368] Messenger Math., 50, 1920, 113–128.
[369] Annals of Math., 23, 1921, 56–67; abstract in Bull. Amer. Math. Soc., 27, 1921, 151.

others may be reversed so as to give the class-number of a negative determinant explicitly in terms of the total number of representations of certain integers each as a sum of squares or triangular numbers.

Bell,[370] by paraphrasing identities between doubly periodic functions of the first and third kinds, obtained three class-number relations involving a wholly arbitrary even function $f(u)=f(-u)$. Let $\epsilon(n)=1$ or 0 according as n is or is not the square of an integer; let $F(n)$ and $F_1(n)$ denote the number of odd and even classes respectively for the determinant $-n$, $n \gtreqless 0$. The first and simplest of the three similar relations is

$$\Sigma\epsilon(\alpha')[f(\sqrt{\alpha'}-d'')-f(\sqrt{\alpha'}+d'')]+2\Sigma'\left[f\left(\frac{d'+\delta'}{2}-d''\right)-f\left(\frac{d'+\delta'}{2}+d''\right)\right]$$
$$=\Sigma F(\beta-4r^2)f(2r)-\Sigma'\left(\frac{\delta-d}{2}\right)\ f\left(\frac{\delta+d}{2}\right),$$

the Σ, Σ' extending over all indicated positive integers α', ..., d, δ such that, for β fixed,

$$\beta\equiv 3 \pmod 4),\quad \beta=\alpha'+2m''\equiv d'\delta'+2d''\delta''$$
$$(\alpha'=d'\delta',\ m''=d''\delta''),\ \text{and}\ \beta=d\delta,\ d<\sqrt{\beta};$$
$$\alpha'\equiv 1 \pmod 4),\quad d'<\sqrt{\alpha'};\quad \beta-4r^2>0.$$

Interpreting results obtained by putting $f(x)=0$, $|x|>0$, $f(0)=1$ in the three relations, it follows that the total number of representations of any prime p by $xy+yz+zx$, with x, y, z all >0, is $3[G(p)-1]$ where $G(n)=F(n)+F_1(n)$; that the like is true only when p is prime; that there are more quadratic residues than non-residues of the prime $p\equiv 3 \pmod 4$ in the series 1, 2, ..., $\frac{1}{2}(p-1)$; and so for $p\equiv 1 \pmod 4$ in the series 1, 2, ..., $\frac{1}{4}(p-1)$.

If $f(x)=1$ for all values of x, the first relation gives Hermite's[69] (3): $\Sigma F(\beta-4r^2) =\frac{1}{2}\Psi_1(\beta)$, where $\Psi_s(n)$ is the sum of the sth powers of all the divisors $>\sqrt{n}$ of n diminished by the sum of the sth powers of all the divisors $<\sqrt{n}$ of n. For $f(x)=x^2$, the first relation gives:

$$32\Sigma r^2F(\beta-4r^2)=\Psi_3(\beta)+\beta\Psi_1(\beta)-32N(4\beta),$$

the Σ extending over all integers r such that $\beta-4r^2>0$, and $N(4\beta)$ is the number of representations of 4β in the form

$$m_1^2+m_2^2+m_3^2+m_4^2+2m_5^2+2m_6^2+2m_7^2+2m_8^2$$

for which the $m_i(i=1, 2, \ldots, 8)$ are odd and $\neq 0$, and precisely 0, 2 or 4 of m_1, m_2, m_3, m_4 in each representation are included among the forms $8k\pm 1$. The paper contains a table of the value of $F(n)$, $n=1, \ldots, 100$.

E. T. Bell[371] obtained 18 class-number relations which are similar to his[370] three above and which form a complete set in the sense that no more results of the same general sort are explicit in the analysis. By specializing the arbitrary even functions which occur in these formulas, he stated that all the class-number relations of

[370] Tôhoku Math. Jour., 19, 1921, 105–116.
[371] Quar Jour. Math., 1923(?); abstract in Bull. Amer. Math. Soc., 27, 1921, 152.

Kronecker and Hermite and certain of those of Liouville and Humbert are obtained as special cases.

L. J. Mordell[372] showed that the number of solutions in positive integers of $yz+zx+xy=u$ is $3G(n)$. It is shown essentially by Hermite's[69] classical method that $x+y\equiv 1 \pmod 2$ for $2F(n)$ of the solutions; $x+y\equiv 2 \pmod 4$ for $F(n)$ of the solutions; and $x+y\equiv 0 \pmod 4$ for $3G(n)-3F(n)$ of the solutions, where always a solution is counted $\frac{1}{2}$ if one of the unknowns is 0. In particular, if n is not a perfect square, $x+y\equiv 1 \pmod 4$ for $F(n)$ of the solutions, $x+y\equiv 3 \pmod 4$ for $F(n)$ of the solutions. Particular cases had been given by Liouville[88] and Bell.[370]

G. H. Cresse[374] reproduced J. V. Uspensky's[332] arithmetical deduction of Kronecker's[54] class-number relations I, II, V and supplied some details of the proof.

R. Fricke[375] (p. 134) obtained and (p. 148) translated[373] a result of Dedekind[127a] in ideals into a solution of the Gauss Problem[4] (Cf. Weber[310]). He reproduced and amplified (pp. 269-541) Klein's theory of the modular function.[134] He denoted (p. 360) by W the substitution $\omega'=-\omega/n$ and by $\Gamma\psi(n)$ that sub-group of the modular group $\omega'=(\alpha\omega+\beta)/(\gamma\omega+\delta)$ for which $\gamma\equiv 0 \pmod n$. The fundamental polygon[134] for the group $\Gamma\psi(n)$ is called the transformation polygon T_n. Fricke found (p. 363) that in T_n, the number of fixed points for elliptic substitutions of period 2 among the substitutions of $\Gamma\psi(n)\cdot W$ is $Cl(-4n)$ if $n\equiv 0, 1, 2 \pmod 4$ and is $Cl(-4n)+Cl(-n)$ if $n\equiv 3 \pmod 4$.

Finally it should be noted that the class-number may be deduced[373] from the number of classes of ideals in an algebraic field since there is a (1, 1) correspondence between the classes of binary quadratic forms of discriminant D and the narrow classes of ideals in a quadratic field of discriminant D (Dedekind[29] of Ch. III). For the class-number of forms with complex integral coefficients, see Ch. VIII.

---

[372] Amer. Jour. Math., Jan., 1923. Abstract in Records of Proceedings of London Math. Soc., Nov. 17, 1921.
[373] Dedekind in Dirichlet's Zahlentheorie, ed. 4, 1894, 639.
[374] Annals of Math., 23, March, 1922.
[375] Die Elliptischen Functionen und ihre Anwendungen, II, 1922.

CHAPTER VII.

BINARY QUADRATIC FORMS WHOSE COEFFICIENTS ARE COMPLEX INTEGERS OR INTEGERS OF A FIELD.

G. L. Dirichlet[1] considered a form $(a, b, c) = ax^2 + 2bxy + cy^2$ in which a, b, c are given complex integers $d + ei$, where d, e are ordinary integers, and x, y are indeterminate complex integers. The determinant $D = b^2 - ac$ is assumed to be not the square of a complex integer. If to (a, b, c) we apply a linear substitution with complex integral coefficients of determinant ϵ, we obtain a form (a', b', c'), of determinant $D' = D\epsilon^2$, said to be contained in (a, b, c). If also the latter is contained in (a', b', c'), then $D' = \pm D$ and ϵ is one of the four *units* $\pm 1, \pm i$. If furthermore $D' = +D$, so that $\epsilon = \pm 1$, we call the two forms *equivalent* (properly or improperly, according as $\epsilon = +1$ or $\epsilon = -1$). Henceforth equivalence shall mean proper equivalence. It is assumed that a, b, c have no common divisor other than a unit. The g.c.d. of $a, 2b, c$ is designated by ω, which has one of the values $1, 1+i, 2$, the form (a, b, c) being of the *first, second,* or *third species* in the respective cases. Any form equivalent to it is of the same species.

Given (§ 11) two equivalent forms f, f' and one substitution S which replaces f by f', we can find all such substitutions. For, if A ranges over the transformations of f into itself, the products AS, and no other substitutions, transform f into f'. The A's are

$$\begin{pmatrix} (t-bu)/\omega & -cu/\omega \\ au/\omega & (t+bu)/\omega \end{pmatrix}, \quad t^2 - Du^2 = \omega^2,$$

the fractions being in fact equal to complex integers for every set of complex integral solutions t, u of the equation written.

The theory (§ 12) of the proper representation of a complex integer M by (a, b, c) proceeds as by Gauss (Arts. 154–6, 168–9).

For the solution (§§ 13–14) of $t^2 - Du^2 = 1$ in complex integers, see the report in this History, Vol. II, pp. 373–4.

Every form (§ 16) is equivalent to a reduced form (a, b, c) for which

$$\tfrac{1}{2}N(a) \gtreqless N(b), \quad N(a) \leqq N(c),$$

where $N(g + hi)$ denotes the norm $g^2 + h^2$ of $g + hi$. The number of reduced forms of a given determinant is finite.

H. E. Heine[2] treated binary quadratic forms whose coefficients are polynomials in a variable. Cf. König.[15]

---

[1] Jour. für Math., 24, 1842, 320–350; Werke, I, 565–596.
[2] Jour. für Math., 48, 1854, 254–266.

H. J. S. Smith[3] defined characters and genera of forms (a, b, c) with complex integral coefficients. The theory of composition (Ch. III) as given by Gauss is immediately applicable with minor alterations to complex forms; also, the congruences of Arndt[16] of Ch. III hold unchanged.

A primitive form (a, b, c) is called *uneven, semi-even,* or *even,* according as the g.c.d. of $a, 2b, c$ is $1, 1+i$, or $2i$ (i. e., is of the first, second, or third species of Dirichlet). An ambiguous form is an uneven form for which either $b=0$ or $a=kb$, $k=1+i$, 2, or $2i$. It is proved that the number of uneven ambiguous classes is half of the total number of assignable generic characters. When $D\equiv\pm1 \pmod 4$, there are as many even as semi-even ambiguous classes of determinant D. When $D\equiv1 \pmod 2$, there are as many semi-even as uneven ambiguous classes, or only half as many, according as there are altogether as many semi-even as uneven classes, or only half as many. Gauss' proof (Ch. IV) that the number of genera of uneven forms of any determinant cannot exceed the number of uneven ambiguous classes of the same determinant applies unchanged for complex coefficients. Hence half of the assignable generic characters are impossible. This leads to a proof of the law of quadratic reciprocity for complex primes and the supplementary laws.

There is extended to the complex case Gauss' theory of the representation of a binary by a ternary quadratic form, while his reduction is applicable to complex ternary forms (Ch. IX). Any binary form f of the principal genus arises from the duplication of a determinable form. Hence half of the assignable generic characters correspond to existing genera.

B. Minnigerode[4] extended to forms $f=(a, b, c)$ with complex coefficients the definition and properties of characters and the distribution of classes into genera, as given by Dirichlet[3] of Ch. IV. Let the form f be primitive and of the first species (g.c.d. of $a, 2b, c$ unity), and have a determinant D not a square. If n and n' are prime to D and representable by f, nn' is representable by x^2-Dy^2 (Gauss, Art. 229; see Ch. IV). Thus nn' is a quadratic residue of any complex prime factor l of D. With Dirichlet,[1] write $[n/l]=\pm1\equiv n^{\frac{1}{2}(p-1)} \pmod l$, where p is the norm of l, l being odd (i. e., not divisible by $1+i$). Thus $[n/l]$ has the same value for all numbers representable by f and not divisible by l. Next, if D is divisible by 4 or $4(1+i)$,

$$(-1)^{\frac{1}{4}(\lambda^2+\nu^2-1)} \quad\text{or}\quad (-1)^{\frac{1}{8}[(\lambda+\nu)^2-1]}$$

has the same value for all odd numbers $\lambda+\nu i$ representable by f. We now have the characters of f. Exactly half of the possible combinations of the characters correspond to existing genera. There are equally many classes in the various genera.

L. Bianchi[5] noted that the geometrical method of Klein[182] of Ch. I for ordinary forms can be applied to Dirichlet forms with coefficients $a+bt$, where a, b are ordinary integers and t is i or an imaginary cube root ϵ of unity. Consider the group G or G' of all linear fractional substitutions on z with coefficients $a+bt$ of determinant unity, where $t=i$ or ϵ. Apply Poincaré's geometrical interpretation * (Acta Mathematica, 3, 1883, 49) of such a substitution on $z=\xi+i\eta$ as a transformation on

* Exposition by A. R. Forsyth, Theory of Functions of a Complex Variable, ed. 3, 1918, 749.

[3] Proc. Roy. Soc. London, 13, 1864, 278–298; Coll. Math. Papers, I, 418–442.

[4] Göttinger Nachrichten, 1873, 160–180.

[5] Math. Annalen, 38, 1891, 313–333.

the points in space with the coordinates ξ, η, ζ to obtain a fundamental polyhedron P for G or P' for G'. For example, P is that part of space above the $\xi\eta$-plane which is outside the sphere $\xi^2+\eta^2+\zeta^2=1$ and between the four planes $\xi=0$, $\xi=\frac{1}{2}$, $\eta=-\frac{1}{2}$, $\eta=\frac{1}{2}$.

A Dirichlet form $ax^2+2bxy+cy^2$ is called *reduced* if P is cut by the half-circle representing it (which lies above the $\xi\eta$-plane, cuts it orthogonally and passes through the points of that plane which represent the roots of $az^2+2bz+c=0$). Every Dirichlet form is equivalent to a reduced form. The number of reduced forms is finite. Two equivalent reduced forms belong to the same period.

Bianchi[6] extended this theory to various imaginary quadratic fields.

R. Fricke[7] discussed the reduction and equivalence of forms

$$(\kappa+\lambda\sqrt{q})x^2+2\mu\sqrt{r}\,xy-(\kappa-\lambda\sqrt{q})y^2,$$

where κ, λ, μ are ordinary integers, while $q=4p-1$ and $r=4l+1$ are primes.

G. B. Mathews[8] proceeded as had Bianchi.[5] The novelty lies in the criterion (§ 13) for a principal reduced form.

A. E. Western[9] employed the representatives x^2+5y^2 and $2x^2+2xy+3y^2$ of the two classes of forms of discriminant -20. The factors $x\pm y\sqrt{-5}$ of the first are called *principal* numbers. The factors $x\sqrt{2}+y(1\pm\sqrt{-5})/\sqrt{2}$ of the second are called *secondary* numbers. There are 18 discriminants -15, -20, -24, ... for which the primary and secondary numbers together obey the ordinary laws of arithmetic as regards primality and divisibility. There are investigated quadratic forms whose coefficients are such primary or secondary numbers, including their separation into classes and their generic characters both as regards narrow and wide classes.

R. Fricke and F. Klein[10] gave an exposition largely following Bianchi.[5] Also they represented (p. 498) a Dirichlet form by the straight line secant of the Neumann sphere which joins the points on the sphere representing the zeros of the form.

J. Hurwitz[11] employed a special type of development into a continued fraction of any complex number x_0:

$$x_0=a_0-\frac{1}{x_1},\quad x_1=a_1-\frac{1}{x_2},\ \ldots,\ x_n=a_n-\frac{1}{x_{n+1}},\ \ldots,$$

where a_n is x_n itself if x_n is a complex integer, while if x_n is not a complex integer, a_n is the complex number represented by the middle point of the square which contains the point representing x_n, the squares being determined by the lines $x+y=\pm1$, ±3, ±5, ..., $x-y=\pm1$, ±3, ±5, ... This theory is used to solve Dirichlet's[1] problems: (I) To decide if two given forms with complex integral coefficients of the same determinant are equivalent under a linear substitution with complex integral

[6] Math. Annalen, 40, 1892, 384–9, 403 (43, 1893, 101-135). In preliminary form in Atti R. Accad. Lincei, Rendiconti, (4), 7, II, 1891, 3–11.

[7] Math. Annalen, 39, 1891, 62–106 (p. 73).

[8] Quar. Jour. Math., 25, 1891, 289–300.

[9] Trans. Cambridge Phil. Soc., 17, 1899, 109–148.

[10] Automorphe Functionen, Leipzig, 1, 1897, 91–93, 450–467. Brief outline by Fricke, Jahresb. Deut. Math.-Vereinigung, 6, 1899, 94–95.

[11] Acta Math., 25, 1902, 231–290 (263). Cf. Mathews.[16]

coefficients of determinant unity. (II) To find all substitutions which replace a given form by another given equivalent form.

J. V. Uspenskij[12] applied an algorithm closely related to ordinary continued fractions to the reduction of binary quadratic forms whose coefficients and variables are integral numbers of a given field. His method is more complicated than Hurwitz's[11] (whose paper he had not seen), but leads far more rapidly to the theorem that two forms are not equivalent if the periods of their reduced forms are distinct. The method of Bianchi[5] is said to be theoretically complete but requires complicated computations when applied to numerical examples.

O. Bohler[13] employed the Fricke-Klein[10] secant representation of a Dirichlet form and called the form reduced if the secant has a point in common with the fundamental octahedron or dodecahedron, according as the coefficients of the substitutions of the group are of the form $u+\rho v$, where ρ is an imaginary fourth or cube root of unity, while u and v are integers.

A. Speiser[14] considered forms $f=\alpha x^2+\beta xy+\gamma z^2$ in which the coefficients and x, y are integral numbers of an arbitrary algebraic field k. The *divisor* of f is the ideal g.c.d. $\mathfrak{a}$ of α, β, γ. If δ is the discriminant $\beta^2-4\alpha\gamma$ of f, $\delta/\mathfrak{a}^2$ is called the *primitive discriminant* $\mathfrak{d}$ of f. The representation of numbers by f proceeds as in the ordinary theory; likewise for the formula for the transformations of f into itself, except for the presence of $\frac{1}{2}(\beta\pm\sqrt{\delta})$, arising from the factorization of αf. The system of characters of f is defined by means of the symbol of Hilbert[16] of Ch. IV under certain restrictions; half of the possible systems of characters are shown to correspond to existing genera. There is determined the ratio of the numbers of classes of forms of the primitive discriminants $\mathfrak{d}$ and $\mathfrak{f}^2\mathfrak{d}$.

R. König[15] considered binary quadratic forms whose coefficients are polynomials in an independent variable z and whose discriminant is $4d$, where

$$d=(z-e_1)(z-e_2)(z-e_3).$$

There is a (1, 1) correspondence between the classes of these forms and the classes of integral "divisors" in the theory of algebraic functions when applied to the case of the function field defined by $\sqrt{d}$.

G. B. Mathews[16] modified the methods of Hurwitz[11] by introducing at the outset a geometrical definition of reduced forms, which shows that their number is finite without consideration of the possibility of points of condensation. The fact that the roots of a reduced form are expressible as pure recurrent chain-fractions is now a corollary instead of a definition. Again, the ordinary definition of proper equivalence is used without the further congruencial condition.

---

12 Applications of continuous parameters in the theory of numbers, St. Petersburg, 1910, 214 pp. Jahrbuch Fortschritte der Math., 1910, 252–3.

13 Über die Picard'schen Gruppen aus dem Zahlkörper der dritten und der vierten Einheitswurzel, Diss., Zürich, 1905, 49–74, 99–102.

14 Die Theorie der binären quad. Formen mit Koefficienten und Unbestimmten in einem beliebigen Zahlkörper, Diss. Göttingen, 1909.

15 Monatshefte Math. Phys., 23, 1912, 321–346. Generalized in Jour. für Math., 142, 1913, 191–210.

16 Proc. London Math. Soc., (2), 11, 1912–13, 329–350.

G. Cotty[17] considered forms $\phi = ax^2 + 2bxy + cy^2$ whose coefficients are integers of a real quadratic field $R(\sqrt{\Delta})$. The conjugate of such an integer a is denoted by a'. A definite form ϕ is called *perfectly* or *imperfectly definite* according as $\phi' = a'x'^2 + \dots$ is definite or indefinite. Similarly there are perfectly and imperfectly indefinite forms. The terms equivalence and class relate to the group of linear substitutions whose coefficients are integers of R of determinant unity. Reduced forms are represented by points of a fundamental domain in space of four dimensions. The number of classes of perfectly definite forms of given negative determinant is finite; each class has one and only one reduced form. The number of classes of imperfectly definite forms of given determinant is finite. Likewise for indefinite forms, there being several reduced forms in a class.

K. Hensel[18] investigated binary and ternary quadratic forms with p-adic coefficients.

---

[17] Comptes Rendus Paris, 156, 1913. 1448–51.
[18] Zahlentheorie, 1913, 292–352.

CHAPTER VIII.

NUMBER OF CLASSES OF BINARY QUADRATIC FORMS WITH COMPLEX INTEGRAL COEFFICIENTS.

G. L. Dirichlet[1] found the number H of classes of binary quadratic forms of the first species (Dirichlet[1] of Ch. VII) with integral complex coefficients of determinant D, not a square. We may write $D=\chi Q V^2$, where χ has one of the four values

$$(1) \qquad \chi=1, \quad i, \quad 1+i, \quad i(1+i),$$

while $\pm Q$ is a product of distinct odd primary complex primes, numbers of the form $4g+1+2hi$ and $1+i$ alone being called *primary*. Let $m=a+bi$ be an *odd* prime (i. e., one of a, b is odd and the other is even). According as a complex integer k, not divisible by m, is or is not the residue of the square of a complex integer modulo m, we write $[k/m]=+1$ or -1. If $M=mm'm''\ldots$ is a product of odd complex primes, no one dividing k, we write

$$[k/M]=[k/m][k/m'][k/m'']\ldots.$$

Let $n=\lambda+\nu i$ be odd, primary and relatively prime to D. Then

$$[D/n]=[\chi/n][Q/n]=[\chi/n][n/Q],$$

by the reciprocity law. In the respective cases (1),

$$[\chi/n]=1, \quad (-1)^{\frac{1}{4}(\lambda^2+\nu^2-1)}, \quad (-1)^{\frac{1}{8}[(\lambda+\nu)^2-1]}, \quad (-1)^{\frac{1}{4}(\lambda^2+\nu^2-1)+\frac{1}{8}[(\lambda+\nu)^2-1]}.$$

[These four cases can be combined by an artifice.] We have

$$H=\lim_{\rho=0} \frac{8N(\sqrt{D})}{\pi \log \sigma} \Sigma \left[\frac{\chi}{n}\right] \left[\frac{n}{Q}\right] \frac{1}{(\lambda^2+\nu^2)^{1+\rho}},$$

summed for all odd primary numbers $n=\lambda+\nu i$ relatively prime to D, where ρ is a positive variable, N denotes norm, and σ is the norm of $T+U\sqrt{D}$, T, U denoting the fundamental solution of $t^2-Du^2=1$, all of whose solutions in complex integers are given without repetition by

$$t+u\sqrt{D}=\pm(T+U\sqrt{D})^e, \quad e=0, \pm 1, \pm 2, \ldots.$$

Brief suggestions are made as to how the difficult problem of performing the summation might be accomplished.

For the case of a real positive determinant D, it is proved that $H=2h_1h_2$ or $H=h_1h_2$, according as $t^2-Du^2=-1$ is or is not solvable in real integers, where h_1

[1] Jour. für Math., 24, 1842, 350–371; Werke, I, 596–618.

and h_2 are the class numbers of properly primitive forms of divisor unity, with real integral coefficients of determinants D and $-D$, respectively. For $D=di$, where d is a positive integer whose double is not a square, $H=h_1h_2$ or $H=\frac{1}{2}h_1h_2$, according as $t^2-2du^2=-1$ is or is not solvable in real integers, where h_1 and h_2 are the class numbers of real properly primitive forms of determinants $2d$ and $-2d$ respectively.

G. Eisenstein[2] stated that the number of classes of quadratic forms with coefficients $a+b\rho$, where a and b are real integers and ρ is an imaginary cube root of unity, and with a real [positive] determinant D, not a square, is always half the product of the numbers of classes of real forms of determinants D and $-3D$.

He[3] proved this and similar theorems for such forms of negative determinants, and for forms whose coefficients involve 8th or 12th or 16th roots of unity, or numbers $a+b\sqrt{\pm 2}$.

R. Lipschitz[4] found by elementary methods the ratio of the class numbers of forms of the first species of determinants D and p^2D, where p is a complex prime different from $1+i$, and not dividing D; also the ratios of the class numbers of forms of the first, second, and third species of the same determinant.

P. Bachmann[5] simplified Dirichlet's[1] expression for the class number H of forms of determinant $D=A+Bi$ for the case in which D is a product of distinct [odd] primary complex primes, i. e., norm $N(D)$ is a product of distinct primes $\equiv 1 \pmod 4$. If $N(D)\equiv 1 \pmod 8$, then

$$N(T+U\sqrt{D})^H=N\left(\frac{\xi-\epsilon\zeta\sqrt{D}}{2}\right)^2,\qquad \epsilon=\left[\frac{1+i}{D}\right],\qquad \xi^2-D\zeta^2=4,$$

the exact values of ξ, ζ being determined. For $N(D)\equiv 5 \pmod 8$,

$$N(T+U\sqrt{D})^H=N\left(\frac{\xi-\epsilon\zeta\sqrt{D}}{\xi+\epsilon\zeta\sqrt{D}}\right)^2.$$

P. S. Nasimoff (Nazimow)[6] noted that, by the method of Dirichlet, Zahlentheorie, § 100, we may express the class number H of complex forms of determinant χQV^2 in terms of that for determinant χQ. For this case, Dirichlet's[1] formula for H is expressed as a sum of as many elliptic functions as there are terms in a complete set of residues of complex integers modulo Q.

L. Bianchi[7] noted that we can use Dirichlet's[1] method to find the numbers of classes h_2 and h_3 of forms of the second and third species of the same determinant D and find their ratios to the number h of classes of the first species. We may however employ elementary methods, either that of Lipschitz or one based on the theory of composition. That theory as presented by Dedekind in Supplement X to Dirichlet's Zahlentheorie applies unchanged to forms (a, b, c), (a', b', c') with complex coefficients of the same determinant and relatively prime divisors σ, σ', provided a, a' are relatively prime. The same is true also of his § 150, whence, if h' be the

[2] Jour. für Math., 27, 1844, 80.
[3] *Ibid.*, 311-6.
[4] Jour. für Math., 54, 1857, 193-6.
[5] Math. Annalen, 16, 1880, 537-549.
[6] Applications of Elliptic Functions to the Theory of Numbers, Moscow, 1885, Ch. 6 (in Russian). French resumé, Ann. école norm. sup., (3), 5, 1888, 164-176.
[7] Atti R. Accad. Lincei, Rendiconti, (4), 5, I, 1889, 589-599.

number of classes of divisor σ of the same determinant D, h is a multiple of h' and the quotient is the number of non-equivalent primitive forms of the first species whose first coefficient is a square dividing σ^2. There are proved by composition expressions for h_2/h and h_3/h equivalent to those by Lipschitz.

G. B. Mathews[8], for the case in which the determinant D is not divisible by a square, expressed Dirichlet's[1] formula for H as a sum of $N(D)$ terms involving elliptic functions.

D. Hilbert[9] obtained by means of algebraic numbers a theorem equivalent to the final one of Dirichlet's.[1]

For forms in an arbitrary field, see Speiser[14] of Ch. VII.

---

[8] Proc. London Math. Soc., 23, 1891–2, 159–162.
[9] Math. Annalen, 45, 1894, 309–340.

CHAPTER IX.

TERNARY QUADRATIC FORMS.

Gauss made a preliminary study of quadratic forms in three variables as a mere digression from his investigation of forms in two variables, for the purpose of determining the exact number of genera of the latter forms. Accordingly he studied especially the problem of representing binary forms by ternary forms. Seeber was the first to obtain inequalities involving the coefficients of a positive ternary form P, which segregate a single form of each class of a given determinant; but his methods and proofs were excessively complicated. In his review of Seeber's book, Gauss gave a simple geometrical representation of forms P. Dirichlet went further and defined a reduced fundamental parallelopiped corresponding uniquely to each reduced P, thereby replacing Seeber's computations by geometric intuition. In the same and succeeding years, Hermite gave arithmetical theories of reduction of quadratic forms in n variables both definite and indefinite, and in particular his theory of continual reduction. In the meantime, Eisenstein began his important studies of genera, the weight of an order or genus, and the number of classes. These studies were continued by Smith, Meyer, Mordell, and Humbert. A new method of reduction was given by Selling and simplified by Charve, Poincaré, and Got. The most complete exposition of the arithmetical theory of quadratic forms in three or more variables is Bachmann's Arithmetik der Quadratischen Formen, 1898.

Fermat[1] asserted that the double of any prime $8n-1$ is a sum of three squares.

In 1748, L. Euler[2] expressed belief that the double of any odd number is a sum of three squares and proved that this would imply that every odd number is of the form $2x^2+y^2+z^2$.

C. F. Gauss[3] employed for the ternary quadratic form the notations

$$f=\begin{pmatrix} a, & a', & a'' \\ b, & b', & b'' \end{pmatrix}=ax^2+a'x'^2+a''x''^2+2bx'x''+2b'xx''+2b''xx', \tag{1}$$

defined its *determinant* to be

$$D=ab^2+a'b'^2+a''b''^2-aa'a''-2bb'b'', \tag{2}$$

and its *adjoint* to be

$$F=\begin{pmatrix} A, & A', & A'' \\ B, & B', & B'' \end{pmatrix}\equiv\begin{pmatrix} b^2-a'a'', & b'^2-aa'', & b''^2-aa' \\ ab-b'b'', & a'b'-bb'', & a''b''-bb' \end{pmatrix}. \tag{3}$$

---

[1] Oeuvres, II, 405; III, 316; letter to K. Digby, June, 1658.

[2] This History, Vol. II, bottom of p. 260. Cf. p. 261, Legendre [19]; p. 264, Lebesque.

[3] Disquisitiones Arithmeticae, 1801, Arts. 266–285; Werke, I, 1863, pp. 299–335; German transl. by H. Maser, 1889, pp. 288–321. Smith[3] of Ch. VII noted that Gauss' theory is readily extended to forms with complex integral coefficients.

The determinant of F is equal to D^2. The adjoint of F is

$$\begin{pmatrix} aD, & a'D, & a''D \\ bD, & b'D, & b''D \end{pmatrix}.$$

It is assumed that the coefficients of f are integers and that $D \neq 0$.

If f is transformed into g by the substitution

$$(4) \qquad S=\begin{pmatrix} \alpha & \beta & \gamma \\ \alpha' & \beta' & \gamma' \\ \alpha'' & \beta'' & \gamma'' \end{pmatrix}: \quad \begin{aligned} x &= \alpha\, y+\beta\, y'+\gamma\, y'', \\ x' &= \alpha' y+\beta' y'+\gamma' y'', \\ x'' &= \alpha'' y+\beta'' y'+\gamma'' y'', \end{aligned}$$

with integral coefficients of determinant k, we say that f *contains* g and g *is contained in* f. The determinant of g is equal to k^2D.

By interchanging the rows and columns of S, we obtain a substitution said to arise by *transposition*; it replaces G by k^2F, if F and G denote the adjoints of f and g. The substitution

$$(5) \qquad S'=\begin{pmatrix} \beta'\gamma''-\beta''\gamma' & \gamma'\alpha''-\gamma''\alpha' & \alpha'\beta''-\alpha''\beta' \\ \beta''\gamma-\beta\gamma'' & \gamma''\alpha-\gamma\alpha'' & \alpha''\beta-\alpha\beta'' \\ \beta\gamma'-\beta'\gamma & \gamma\alpha'-\gamma'\alpha & \alpha\beta'-\alpha'\beta \end{pmatrix},$$

which transforms F into G, is called the *adjoint* of S. If S'' arises from S' by transposition, evidently S'' replaces g by k^2f.

If f and g contain each other, they are called *equivalent*; their determinants are equal. Their adjoints are equivalent, and conversely. If one form contains another of the same determinant, they are equivalent.

If two forms are equivalent to a third form, they are equivalent to each other. All equivalent forms are said to constitute a *class*.

An *indefinite* form is one, like $x^2+y^2-z^2$, which represents both positive and negative numbers. *Definite* forms are of two kinds: *positive* forms, like $x^2+y^2+z^2$, which represent only numbers $\geqq 0$; and *negative* forms, like $-x^2-y^2-z^2$, which represent only numbers $\leqq 0$. The same terms are applied to classes. It is proved (Art. 271) that in a definite form f the six numbers A, A', A'', aD, $a'D$, $a''D$ are negative, D itself being negative or positive according as f is a positive or negative form.

Every form (Art. 272) of determinant D is equivalent to a form whose first coefficient (a) does not exceed $\frac{4}{3}\sqrt[3]{|D|}$ numerically and the third coefficient (A'') of whose adjoint form does not exceed $\frac{4}{3}\sqrt[3]{D^2}$ numerically. A "first reduction" of f is made by means of a substitution which leaves x'' unaltered and replaces x, x' by linear functions of themselves of determinant ± 1. Thus the binary form (a, b'', a'), of determinant A'', goes into an equivalent binary form whose first coefficient may (by Art. 171, see Ch. I) be made numerically less than or equal to $\sqrt{\frac{4}{3}|A''|}$. A "second reduction" of f to g is made by means of a substitution S which leaves x unaltered and replaces x', x'' by linear functions of themselves of determinant ± 1. The adjoint substitution to S is

$$x=\pm y, \qquad x'=\gamma''y'-\beta''y'', \qquad x''=-\gamma'y'+\beta'y'',$$

and replaces F by G, and hence the binary form (A', B, A'') of determinant Da by an equivalent binary form whose last coefficient is numerically less than or equal to $\sqrt{\frac{4}{3}|Da|}$. Since we may apply these two reductions alternately, we finally reach a form f for which $a^2 \leqq \frac{4}{3}|A''|$, $A''^2 \leqq \frac{4}{3}|Da|$, whence $a^4 \leqq \frac{16}{9}\cdot\frac{4}{3}|Da|$, $|a| \leqq \frac{4}{3}\sqrt[3]{|D|}$. This proves the above theorem, whose converse need not hold.

A further reduction (Art. 274) of f is made by means of

$$x=y+\beta y'+\gamma y'', \qquad x'=y'+\gamma' y'', \qquad x''=y'';$$

let it replace f by f_1. Then $a_1=a$, $A_1''=A''$. If $A''=0$ we make $a=0$ by the first reduction. If $a=0$ we make $A''=0$ by the second reduction. First, let $aA''\neq 0$. Then by choice of β, γ', γ we can make

$$b_1''=b''+a\beta, \qquad B_1=B-A''\gamma', \qquad B_1'=B'-N\beta-A''\gamma$$

numerically $\leqq \frac{1}{2}|a|$, $\frac{1}{2}|A''|$, $\frac{1}{2}|A''|$, respectively. Second, if $a=A''=0$, then $b''=0$ and

$$a_1=0, \quad a_1'=a', a_1''=a''+2b\gamma'+2b'\gamma+a'\gamma'^2, \quad b_1=b+a'\gamma'+b'\beta, \quad b_1'=b', \quad b_1''=0,$$

whence $D=a'b'^2=a_1'b_1'^2$. We can choose β, γ' so that $|b_1|$ is less than or equal to the g.c.d. of a', b'. Then γ can be chosen so that $|a_1''|\leqq b'$.

The number of classes (Art. 276) of ternary forms of a given determinant is finite. This follows at once from the inequalities in Art. 274 with the subscripts 1 suppressed and those in Art. 272. However, the number of classes is usually smaller than the number of forms which satisfy all of these inequalities.

If (Arts. 278, 280) in a ternary form f we write

(6) $$x=mt+nu, \qquad x'=m't+n'u, \qquad x''=m''t+n''u,$$

we obtain a binary form $\phi(t, u)$ which is said to be *represented* by f. Let $F(X, X', X'')$ denote the adjoint of f, and D the determinant of ϕ. Then D is represented by F when

(7) $$X=m'n''-m''n', \qquad X'=m''n-mn'', \qquad X''=mn'-m'n.$$

This representation of D by F is said to be adjoint to the representation of ϕ by f, and is called proper or improper according as the g.c.d. of (7) is 1 or >1.

Every representation of D by F can be derived from a representation by f of a chosen binary form of determinant D.

If $\phi(t, u)$ is transformed into the equivalent form $\chi(p, q)$ by the substitution

$$t=\alpha p+\beta q, \qquad u=\gamma p+\delta q, \qquad \alpha\delta-\beta\gamma=1,$$

and if we write

$$\begin{aligned} &\alpha m+\gamma n=m_1, \qquad &&\alpha m'+\gamma n'=m_1', \qquad &&\alpha m''+\gamma n''=m_1'',\\ &\beta m+\delta n=n_1, &&\beta m'+\delta n'=n_1' &&\beta m''+\delta n''=n_1'',\end{aligned}$$

we find that χ is represented by f when

(8) $$x=m_1p+n_1q, \qquad x'=m_1'p+n_1'q, \qquad x''=m_1''p+n_1''q.$$

Further the numbers (7) are equal to the corresponding functions of the letters with the subscript 1. Thus (6) and (8) yield the same representation of D by F.

The last two facts show that we obtain all proper representations of D by F if we select arbitrarily a form from each class of binary forms of determinant D and find all proper representations of each such form by f and from every such representation deduce the representation of D by F. Forms from different classes yield distinct representations. The improper representations are readily derived from the proper.

Let (Art. 282) $\phi=pt^2+2qtu+ru^2$, of determinant D, be represented properly by the ternary form f, of determinant Δ, when

$$(9)\qquad x=\alpha t+\beta u,\qquad x'=\alpha' t+\beta' u,\qquad x''=\alpha'' t+\beta'' u.$$

Select integers $\gamma, \gamma', \gamma''$ such that the determinant of (4) is $k=\pm 1$. Let S replace f by g whose adjoint is G:

$$(10)\qquad g=\begin{pmatrix} a, & a', & a'' \\ b, & b', & b'' \end{pmatrix},\qquad G=\begin{pmatrix} A, & A', & A'' \\ B, & B', & B'' \end{pmatrix}.$$

Hence $a+p$, $b''=q$, $a'=r$, $A''=D$, and the determinant of g is Δ. Thus

$$(11)\qquad B^2=\Delta p+A'D,\qquad BB'=-\Delta q+B''D,\qquad B'^2=\Delta r+AD,$$

so that B, B' are integral solutions of the congruences

$$(12)\qquad B^2\equiv\Delta p,\qquad BB'\equiv -\Delta q,\qquad B'^2\equiv \Delta r \pmod{D}.$$

This representation of ϕ by f is said to *belong* to the pair (B, B') of solutions.

If we replace $\gamma, \gamma', \gamma''$ by $\gamma_1, \gamma_1', \gamma_1''$ such that the determinant of S_1 is $k_1=1$ or -1, and if S_1 replaces f by g_1 whose adjoint is G_1, then G_1 is equivalent to G and its B_1, B_1' are such that

$$B_1\equiv B,\quad B_1'\equiv B' \text{ or } B_1\equiv -B,\quad B_1'\equiv -B' \pmod{D}.$$

The pair (B_1, B_1') is said to be equivalent or opposite to (B, B') in the respective cases. Conversely, if (B_1, B_1') is any pair which is equivalent or opposite to (B, B'), we can find integers $\gamma_1, \gamma_1', \gamma_1''$ such that S_1 has the determinant ± 1 and replaces f by a form such that B_1 and B_1' are the fourth and fifth coefficients of its adjoint form.

Hence (Art. 283) we have a method of finding all proper representations of a given binary form ϕ of determinant $D\neq 0$ by a given ternary form f of determinant Δ. We find all non-equivalent pairs of solutions (B, B') of congruences (12), and retain only one of two opposite pairs. For each resulting pair (B, B'), we seek a ternary form g, denoted by (10), having determinant Δ and $a=p$, $b''=q$, $a'=r$, $ab-b'b''=B$, $a'b'-bb''=B'$. The last two equations uniquely determine b and b' since the determinant of their coefficients is equal to $pr-q^2=-D\neq 0$. By (11), we know $A'=b'^2-pa''$ and $B''=qa''=bb'$, one of which determines a''. Since the products of b, b', a'' by either D or Δ are integers, they are integers at least when D is prime to Δ. If b, b', a'' are integers and if f and g are equivalent, then every substitution S which replaces f by g, whose terms free of x'' are $\phi=(p, q, r)$, yields a representation (9) of ϕ by f, and all representations are found by this method. Distinct substitutions which replace f by g yield distinct representations except when (B, B') is opposite to itself.

The problems to find all representations of a given number or a given binary form by a given ternary form have therefore been both reduced to the problem to decide whether or not two given ternary forms of the same determinant are equivalent and if equivalent to find all transformations of the one into the other. The last problem is said to present serious difficulties and is discussed (Art. 285) only for the cases of determinants 1, -1, 2. Given one transformation of f into an equivalent ternary form f', we can find all if we know all the transformations of f into itself.

In case $f=ax^2+a'x'^2+a''x''^2$ and a, a', a'' all have the same sign, it is easily proved that the only transformations of f into itself are those which change the signs of the variables or permute those whose coefficients are equal.

Gauss[4] noted that all transformations of $x^2+y^2-z^2$ into itself are given by

$$\begin{pmatrix} \alpha\delta+\beta\gamma & \alpha\beta-\gamma\delta & \alpha\beta+\gamma\delta \\ \alpha\gamma-\beta\delta & \frac{1}{2}(\alpha^2+\delta^2-\beta^2-\gamma^2) & \frac{1}{2}(\alpha^2+\gamma^2-\beta^2-\delta^2) \\ \alpha\gamma+\beta\delta & \frac{1}{2}(\alpha^2+\beta^2-\gamma^2-\delta^2) & \frac{1}{2}(\alpha^2+\beta^2+\gamma^2+\delta^2) \end{pmatrix}, \quad \alpha\delta-\beta\gamma=1.$$

The editor, E. Schering noted that we obtain all transformations in which the nine coefficients are integers if we assign to $\alpha, \ldots, \delta$ all integers satisfying $\alpha\delta=\beta\gamma=1$ (two even and two odd), as well as all odd multiples of $\sqrt{\frac{1}{2}}$ satisfying the same equation.

Gauss[4a] showed how to transform any ternary quadratic form of determinant zero into a binary quadratic form.

L. A. Seeber[5] found complicated inequalities satisfied by one and only one reduced positive form of a class. While he permitted odd values of the coefficients of the product terms, we shall employ the notation (1) of Gauss[3] and Eisenstein[10] in order to facilitate comparison with the simplified conditions obtained by the latter. For a positive form (1), a, a', a'' are all positive. Seeber's chief conditions for a positive reduced form are*: (I) When b, b', b'' are all positive,

$$(13) \qquad a\leqq a'\leqq a'', \qquad 2b\leqq a', \qquad 2b'\leqq a, \qquad 2b''\leqq a.$$

(II) When b, b', b'' are all negative, $-2(b+b'+b'')\leqq a+a'$. In certain special cases there are further conditions (including quadratic inequalities), which were simplified by Eisenstein.[10] The chief content of the book is the solution (in 41 pages) of the problem to find a reduced form equivalent to any given positive form ϕ, and the proof (in 91 pages) of the theorem that no two reduced forms are equivalent. Then there is solved the problem to decide whether or not one given form can be transformed into another given form by a substitution with integral coefficients, and if so to find all such substitutions. Finally, there is discussed the determination of all reduced forms of a given negative determinant (2), which we now denote by $-D$. For this determination he used the theorem that $aa'a''\leqq 3D$ and remarked that an examination of 600 cases indicated that $aa'a''\leqq 2D$. The latter empirical theorem was later proved by Gauss,[6] Dirichlet,[8] Hermite,[9] and Lebesgue.[19] At the end of the book is a table of reduced forms which extends only to determinant -25 if we restrict attention to those Gauss forms which are properly primitive.

---

* These two cases include all, since xy, xz, $-yz$ all become negative if we put $x=X$, $y=-Y$, $z=-Z$.

[4] Posth. MS., Werke, II, 1876, 1863, 311.

[4a] Posth. MS. of 1800, Werke, X_1, 1917, 87, 88.

[5] Untersuchungen über die Eigenschaften der positiven ternären quadratischen Formen, Freiburg, 1831, 248 pp. (and in Math. Abhandlungen).

For applications to the solution of $f=0$ and to sums of three squares, see this History, Vol. II, pp. 422–3, p. 17. For applications to binary forms, see Ch. IV above.

C. F. Gauss[6] proved the final result of Seeber in a review of the latter's book and interpreted the result geometrically. He extended his own geometrical representation (Ch. I[38]) of positive binary quadratic forms to positive ternary forms. The form

$$\phi=ax^2+by^2+cz^2+2a'yz+2b'xz+2c'xy$$

gives the square of the distance between two arbitrary points in space whose coordinates with respect to three axes X, Y, Z have the differences $x\sqrt{a}, y\sqrt{b}, z\sqrt{c}$, while the cosines of the angles between the axes Y and Z, X and Z, X and Y are respectively $a'/\sqrt{bc}$, $b'/\sqrt{ac}$, $c'/\sqrt{ab}$. The points for which x, y, z are integers are the vertices of a system of parallelopipeds determined by three systems of equidistant parallel planes. The square of the volume of one of the parallelopipeds is equal to the absolute value of the determinant of the form ϕ. Equivalent forms represent the same system of points referred to different axes.

G. Eisenstein[7] defined the *mass* (weight, density) of a class K of positive ternary quadratic forms to be $1/\delta$ if one of its forms has (only a finite number) δ of automorphs of determinant $+1$ (transformations into itself). The mass of a set of classes is defined to be the sum of the masses of the classes. Let the determinant of the form f be odd (to avoid the distinction between proper and improper forms), and let the coefficients have no common divisor.

The separation of classes into orders is essentially different from that for binary forms. The positive g.c.d. Ω of the coefficients of the form F, adjoint to f, has the same value for all forms f of a class; it is called the *adjoint* (*zugeordnete*) *factor* of the class. All classes with the same determinant D and same adjoint factor Ω constitute an *order*. Since f is positive, F is negative. Write $F=-\Omega\mathfrak{F}$ where $\mathfrak{F}$ is positive and primitive. Its adjoint is equal to both Df and the product of Ω^2 by the adjoint of $\mathfrak{F}$. Since f is primitive, $D=-\Omega^2\Delta$. The following theorem is stated: Let Ω^2 be any square factor of a negative odd determinant $D=-\Omega^2\Delta$; let R be the g.c.d. of Ω and Δ; $r, r', \ldots$ the distinct prime factors of R, and $\rho=\Pi(1-1/r^2)$; then the mass of the order of ternary forms of determinant D and adjoint factor Ω is $\frac{1}{12}\Omega\Delta\rho$ if R is not a square, but is $\frac{1}{24}(2\Omega\Delta-Q)\rho$ if R is a square and Q is the largest square dividing $\Omega\Delta$.

The subdivision of orders into genera does not depend (as in the case of binary forms) solely upon the quadratic characters of the numbers represented by the forms with respect to the various prime factors of D and to 4 and 8, but also upon the characters of the adjoint forms. Let $D=-\Omega^2\Delta$ be an odd determinant; $\omega, \omega', \ldots$ the distinct prime factors of Ω which do not divide Δ; $\partial, \partial', \ldots$ those of Δ which do not divide Ω; $r, r', \ldots$ those dividing both Ω and Δ. Then f, whose adjoint is $F=-\Omega\mathfrak{F}$, has the complete character

$$\left(\frac{f}{\omega}\right), \left(\frac{f}{\omega'}\right), \ldots, \left(\frac{f}{r}\right), \left(\frac{f}{r'}\right), \ldots, \left(\frac{\mathfrak{F}}{\partial}\right), \left(\frac{\mathfrak{F}}{\partial'}\right), \ldots, \left(\frac{\mathfrak{F}}{r}\right), \left(\frac{\mathfrak{F}}{r'}\right), \ldots,$$

[6] Göttingische gelehrte Anzeigen, 1831, No. 108; reprinted, Jour. für Math., 20, 1840, 312-20; Werke, II, 1863, 188–196.

[7] Jour. für Math., 35, 1847, 117–136; Math. Abhandlungen, 1847, 177–196.

where each symbol is a Legendre sign ± 1. If we give a definite sign to each symbol, we obtain a definite genus whose mass is said to be

$$\frac{\Omega\Delta}{24}(2-\epsilon)\prod_{\omega}\left\{\frac{\omega+(-\Delta f/\omega)}{2\omega}\right\}\prod_{\partial}\left\{\frac{\partial+(-\Omega\mathfrak{F}/\partial)}{2\partial}\right\}\prod_{r}\left\{\frac{r^2-1}{4r^2}\right\},$$

where

$$\epsilon=(-1)^{\frac{1}{2}(\Omega-1)\cdot\frac{1}{2}(\Delta-1)}\left(\frac{-f}{\Omega}\right)\left(\frac{-\mathfrak{F}}{\Delta}\right),$$

the last two symbols being Jacobi's generalization of Legendre's symbol. The paper closes with a table of the characters and classes in each genus of positive ternary forms of the odd determinants $-1, -3, \ldots, -25$. A special case of one of these theorems was proved later by Eisenstein.[12] Cf. Smith.[20]

G. L. Dirichlet[8] gave a theory of reduction of positive ternary quadratic forms which is far simpler than that of Seeber.[5] He employed the notations and geometrical interpretation due to Gauss[6] and his own concept of a reduced parallelogram (Ch. I[51]). Given a lattice formed by the intersections of three systems of equidistant parallel planes, we can select a reduced fundamental parallelopiped whose faces are reduced parallelograms and none of whose edges exceeds any diagonal. For, we may take as one vertex (0) any point of the lattice. As a second vertex (1) select a lattice point at a minimum distance from (0). As (2) select a lattice point not on the line (01) but as near to (0) as possible. Then (0), (1), (2) are vertices of a reduced parallelogram. In one of the two adjacent parallel planes of the lattice choose a point (3) as near to (0) as possible. Then (0), (1), (2), (3) are vertices of a reduced parallelopiped. There is a single one if all diagonals exceed every side.

Consider a ternary form ϕ in Gauss'[6] notation. After permuting the variables or changing their signs, we may assume that $0<a\leqq b\leqq c$, that a', b', c' are either all negative or none negative, and, finally, if $b=c$, then $|c'|\leqq|b'|$, if $a=b$, then $|b'|\leqq|a'|$, if $a=b=c$, then $|c'|\leqq|b'|\leqq|a'|$. We then call ϕ reduced if it corresponds to a reduced parallelopiped with $(01)=\sqrt{a}$, $(02)=\sqrt{b}$, $(03)=\sqrt{c}$. Since the diagonals of the faces are not less than the sides, we obtain inequalities equivalent to

$$a \gtreqqless 2c'\sigma, \qquad a \gtreqqless 2b'\sigma, \qquad b \gtreqqless 2a'\sigma,$$

where $\sigma=-1$ if a', b', c' are all negative, otherwise $\sigma=+1$. The conditions on the diagonals of the parallelopiped give only

$$a+b+2a'+2b'+2c' \gtreqqless 0 \quad (a', b', c' \text{ negative}).$$

Unless an equality sign occurs, no two equivalent forms are reduced. There is given a short proof of the theorem of Seeber[5] and Gauss[6] that $abc\leqq 2|\Delta|$ for a reduced form.

Ch. Hermite[9] gave a more elementary proof than had Gauss[6] of Seeber's conjecture that in a reduced definite ternary form the product of the coefficients of the three squares is less than double the determinant.

[8] Jour. für Math., 40, 1850, 209–227; abstract in Monatsber. Akad. Wiss. Berlin, 1848, 285–8; Werke, II, 21–48. French transl. in Jour. de Math., (2), 4, 1859, 209–232. Cf. * H. Klein, Ausführung und Erläuterung von Dirichlets Abh. . . . , Hermannstadt, 1908, 58 pp.

[9] Jour. für Math., 40 1850, 173–7; 79, 1875, 17–20; Oeuvres, I, 94–99; III, 190, 190–3.

G. Eisenstein[10] tabulated the primitive reduced positive forms $\phi = ax^2 + \ldots + 2b''xy$ of determinants $-1, \ldots, -100$ and -385 (reduced according to Seeber's[5] definition). For a primitive form $a, \ldots, b''$ have no common factor. There are first given the properly primitive forms for which $a, \ldots, 2b''$ have no common factor, and then the improperly primitive forms for which a, a', a'' are all even. The table gives also the number δ of transformations of the form into itself.

He found that Seeber's quadratic inequalities for a reduced form may be replaced by linear inequalities, whence we need not employ the adjoint form. The following simplified conditions are equivalent to Seeber's:

(I) ϕ with b, b', b'' all positive; conditions (13) and

(14) if $a=a'$, then $b \leqq b'$; if $a'=a''$, then $b' \leqq b''$;
if $2b=a'$, then $b'' \leqq 2b'$; if $2b'=a$, then $b'' \leqq 2b$;
if $2b''=a$, then $b' \leqq 2b$.

(II) $ax^2 + a'y^2 + a''z^2 - 2byz - 2b'xz - 2b''xy$, b, b', b'' all $\gtreqless 0$;

conditions (13), (14), and $2(b+b'+b'') \leqq a+a'$;

if $2b=a'$, then $b''=0$; if $2b'=a$, then $b''=0$; if $2b''=a$, then $b'=0$;
if $2(b+b'+b'')=a+a'$, then $a \leqq 2b'+b''$.

Let $H(D)$ and $H'(D)$ denote the number of classes of properly and improperly primitive positive ternary forms of determinant $-D$, and $h(D)$ and $h'(D)$ the corresponding numbers for binary forms. Then if P is a product of distinct odd primes,

$$H(P) = \tfrac{1}{4}\{\Sigma h(d) + \Sigma h'(d) + \Sigma h(2d)\} + \tfrac{1}{12}(P+\lambda),$$
$$H(2P) = \tfrac{1}{2}\Sigma h(d) + \tfrac{1}{4}\Sigma h(2d) + \tfrac{1}{8}(P+\nu), \quad H'(2P) = \tfrac{1}{4}\Sigma h(d) + \tfrac{1}{4}\Sigma h'(d) + \tfrac{1}{24}(P+\rho),$$

the summations extending over all divisors $d>1$ of P, while

$$\lambda = 9 + 2\left(\frac{P}{3}\right), \quad \nu = 6 + \left(\frac{-1}{P}\right), \quad \rho = 6 - 3\left(\frac{-1}{P}\right) - 4\left(\frac{P}{3}\right),$$

where $(P/3)=0$ if P is divisible by 3. When P is a prime, $H(P)$ is expressed in other forms. Proofs by Markoff,[41] Mordell.[53]

Eisenstein[11] tabulated all the transformations leaving unaltered a reduced positive ternary form, also the non-equivalent indefinite (unbestimmten) forms of determinants <20 without a square factor. There is a single class, represented by $x^2 - y^2 + \Delta z^2$, of indefinite forms whose determinant Δ is a given odd number, without a square factor, the numbers represented by whose adjoint forms are quadratic residues of every prime factor of Δ (generalized by Meyer[24]).

Eisenstein,[12] employing the notations of Gauss,[3] noted that every form f with integral coefficients and determinant D a product of distinct odd primes $p, p', \ldots$, which is derived from $G = (x^2 + x'^2 + x''^2)/D$ by a substitution of determinant D^2, is such that every number prime to p and represented by the adjoint F of f is a quadratic residue of p, and similarly for p', etc. The latter necessary conditions are

[10] Jour. für Math., 41, 1851, 141–190.
[11] *Ibid.*, 227–242.
[12] Jour. de Math., 17, 1852, 473–7.

also sufficient. Two substitutions S and T are called right-hand equivalent if S is the product $T \cdot E$ of T by a substitution E of determinant unity. He evaluated the number of substitutions not right-hand equivalent which replace $x^2+y^2+z^2$ by a form divisible by D. Further details were given in his next paper.

Eisenstein[13] called two substitutions S and $S \cdot T$ with integral coefficients equivalent if T is a substitution with integral coefficients of determinant unity, while T and $S \cdot T$ are similar substitutions if S has integral coefficients of determinant unity. The object is to find all non-equivalent substitutions with integral coefficients of determinant D which replace $\phi=x^2+y^2+z^2$ by Df where f is some ternary form of determinant $-D$, and to find all non-similar substitutions which replace ϕ by Df where f is a given positive form.

The number of classes of ternary forms f of determinant $-\partial$ which belong to the principal genus (whence ∂f can be obtained from $x^2+y^2+z^2$ by transformation) is proved to be $\frac{1}{4}\mathfrak{H}+\frac{1}{48}(\partial+\theta)$, where $\mathfrak{H}$ is the number of classes of primitive positive binary forms ψ of determinant $-\partial$ for which $(-\psi/\partial)=1$, plus the number of determinant -2∂ for which $(-2\psi/\partial)=1$, while $\theta=47$, 19, 17, 13, 35, 31, 29, 1 when $\partial\equiv 1$, 5, 7, 11, 13, 17, 19, 23 (mod 24), respectively.

He stated that every positive definite ternary quadratic form of determinant $-D$ which has an automorph not the identity I is equivalent to a form $ax^2+\psi$ or $\frac{1}{2}[a(2x+y)^2+\psi]$, where a is a divisor of D or $2D$, respectively, while $\psi= by^2+2fyz+cz^2$ has the determinant $-D/a$ or $-2D/a$, respectively. Conversely, these ternary forms have automorphs $\neq I$ (proof by Mordell[53]).

Finally he tabulated the automorphs of a positive reduced form.

Ch. Hermite[14] considered indefinite forms reducible to $X^2+Y^2-Z^2$ by real substitution. Let Δ be the determinant and g the adjoint of f. Consider the infinitude of substitutions which transform the definite form $\phi=f+2(\lambda x+\mu y+\nu z)^2$ into a reduced form when λ, μ, ν take all real values for which $g(\lambda, \mu, \nu)=-\Delta$. These substitutions transform f into an aggregate of forms which is finite if f has integral coefficients, whence the number of classes of forms f with a given determinant is finite. He omitted the long details of this continual reduction of ϕ.

The characteristic equation of an automorph is a reciprocal equation with a root 1. The problem to find all automorphs with integral coefficients of f was made to depend on the totality of algebraic automorphs. But not all the latter were obtained, owing to a gap in the proof. P. Bachmann[15] pointed out this gap and showed how to find all algebraic automorphs. Hermite[16] later obtained all by a modification of his former method. In the meantime, G. Cantor[17] obtained all algebraic automorphs of f from those of $\Phi=Y^2-XZ$, essentially as quoted under Poincaré.[35] Cf. Meyer.[48]

---

13 Berichte Akad. Wiss. Berlin, 1852, 350–389.
14 Jour. für Math., 47, 1854, 307–312; Oeuvres, I, 1905, 193–9.
15 Jour. für Math., 76, 1873, 331–41.
16 *Ibid.*, 78, 1874, 325–8; Oeuvres, III, 185–9. Cf. J. Tannery, Bull. Sc. Math. Astr., 11, II, 1876, 221–233.
17 De Transformatione Formarum Ternariarum Quadraticarum, Halle, 1869, 12 pp. Report in Bachmann, Die Arithmetik der Quadr. Formen, 1898, 19–25. The automorphs of Φ in a general field were found by L. E. Dickson, University of Chicago Decennial Publications, 9, 1902, 29–30.

Hermite[18] proved that f has the automorph

$$S\begin{cases}(1-\gamma)x=(1+\gamma)X+\left(\mu\dfrac{\partial f}{\partial Z}-\nu\dfrac{\partial f}{\partial Y}\right)-\dfrac{\partial\gamma}{\partial\lambda}\cdot\pi,\\ (1-\gamma)y=(1+\gamma)Y+\left(\nu\dfrac{\partial f}{\partial X}-\lambda\dfrac{\partial f}{\partial Z}\right)-\dfrac{\partial\gamma}{\partial\mu}\cdot\pi,\\ (1-\gamma)z=(1+\gamma)Z+\left(\lambda\dfrac{\partial f}{\partial Y}-\mu\dfrac{\partial f}{\partial X}\right)-\dfrac{\partial\gamma}{\partial\nu}\cdot\pi,\end{cases}$$

where $\gamma=g(\lambda,\mu,\nu)$, $\pi=\lambda X+\mu Y+\nu Z$, and λ,μ,ν are arbitrary parameters. Also, $\lambda x+\mu y+\nu z=\pi$ under this substitution, as shown by multiplication by λ,μ,ν and addition. By changing the signs of λ,μ,ν, we obtain the inverse S^{-1}. The parameters of the product of any two substitutions S are found rationally.

In S replace λ,μ,ν by $\lambda/\rho,\mu/\rho,\nu/\rho$; we obtain

$$(15)\qquad (\rho^2-\gamma)x=(\rho^2+\gamma)X+\rho\left(\mu\frac{\partial f}{\partial Z}-\nu\frac{\partial f}{\partial Y}\right)-\frac{\partial\gamma}{\partial\lambda}\cdot\pi,\ \ldots,$$

in which the parameters λ,μ,ν,ρ enter homogeneously (cf. Bachmann[23]). Hermite suppressed the coefficient $\rho^2-\gamma$ of x, y, z and obtained a substitution for which

$$f(x,y,z)=f(X,Y,Z)\{\rho^2-g(\lambda,\mu,\nu)\}^2.$$

Hence any ternary form is compounded of itself and the square of a quaternary form ρ^2-g. He gave several such results. He gave the important identity (verified by Bachmann[23]):

$$\{\rho^2-g(\lambda,\mu,\nu)\}\{\rho'^2-g(\lambda',\mu',\nu')\}=\mathfrak{R}^2-g(\mathfrak{L},\mathfrak{M},\mathfrak{N}),$$

$$\mathfrak{R}=\rho\rho'+\tfrac12\left(\lambda'\frac{\partial g}{\partial\lambda}+\mu'\frac{\partial g}{\partial\mu}+\nu'\frac{\partial g}{\partial\nu}\right),\quad \mathfrak{L}=\lambda\rho'+\lambda'\rho+\tfrac12\frac{\partial f(l,m,n)}{\partial l},$$

$$\mathfrak{M}=\mu\rho'+\mu'\rho+\tfrac12\frac{\partial f}{\partial m},\qquad \mathfrak{N}=\nu\rho'+\nu'\rho+\tfrac12\frac{\partial f}{\partial n},$$

$$l=\mu\nu'-\nu\mu',\quad m=\nu\lambda'-\lambda\nu',\quad n=\lambda\mu'-\mu\lambda'.$$

Conversely, from the latter composition we can deduce the automorphs of f. He showed that $\Delta\rho^2-f(\lambda,\mu,\nu)$ may be compounded with $\rho'^2-g(\lambda',\mu',\nu')$ to give $\Delta R^2-f(L,M,N)$.

V. A. Lebesgue[19] gave a modification of Gauss'[6] proof of Seeber's[5] theorem that $abc\leqq 2D$ for a reduced positive ternary form.

H. J. S. Smith[20] proved Eisenstein's[7] theorems concerning positive forms of odd determinants and extended them to general primitive ternary forms

$$f=ax^2+a'y^2+a''z^2+2byz+2b'xz+2b''xy,$$

for which the six integers $a,\ldots,b''$ have no common divisor >1. Its *discriminant* D is the negative of Gauss'[3] determinant. Its *contravariant*

$$(a'a''-b^2)x^2+\ldots+2(bb'-a''b'')xy$$

is the negative of Gauss' adjoint form and is denoted by ΩF, so that Ω is the g.c.d.

[18] Jour. für Math., 47, 1854, 313–330; Oeuvres, 1, 1905, 200–220.
[19] Jour. de Math., (2), 1, 1856, 406–10.
[20] Trans. Phil. Soc. London, 157, 1867, 255–298; abstract in Proc. Roy. Soc. London, 15, 1867, 387–9. Coll. Math. Papers, I, 455–506, 507–9.

of its six coefficients, and $F=Ax^2+\ldots+2B''xy$ is the *primitive contravariant* of f. If f is definite, whence $\Delta>0$, we take $\Omega>0$. If f is indefinite, we choose Ω to be of sign opposite to Δ. The identity

$$f(x_1, y_1, z_1)\cdot f(x_2, y_2, z_2)-\tfrac{1}{4}\left(x_1\frac{\partial f}{\partial x_2}+y_1\frac{\partial f}{\partial y_2}+z_1\frac{\partial f}{\partial z_2}\right)^2$$
$$=\Omega F(y_1z_2-z_1y_2,\ z_1x_2-x_1z_2,\ x_1y_2-y_1x_2),$$

and that obtained by interchanging f with F and Ω with Δ, lead to the subdivision of the orders into genera. The first identity shows that the numbers, which are relatively prime to any odd prime factor ω of Ω and which are represented by f, are either all quadratic residues of ω or all non-residues of ω, whence f has the particular generic character (f/ω). The second identity shows that F has the character (F/δ), where δ is any odd prime factor of Δ. Also, as by Eisenstein, f and F have particular characters with respect to any odd prime dividing both Ω and Δ. The same identities led Smith to particular *supplementary characters* of each f and F with respect to 4 and 8, analogous to the case of binary forms. When f and F are both properly primitive and neither Ω nor Δ are multiples of 4, f and F taken separately have no particular characters with respect to 4 or 8, but have jointly a simultaneously character with respect to 4 or 8, defined by means of representations $m=f(x, y, z)$, $M=F(X, Y, Z)$ for which $xX+yY+zZ\equiv 0 \pmod 2$.

The aggregate of the particular characters of f and F gives the complete character. Two forms (or classes) with the same complete character (and same Ω and same Δ) are said to belong to the same genus. A two-page table serves to distinguish those complete characters which are possible (i. e., to which existing genera correspond) from those which are impossible, the distinction being expressed by a specified relation between the characters.

In regard to the proposal of Eisenstein[8] of Ch. XI to define a genus of forms as consisting of all the forms which can be transformed into one another by substitutions with rational coefficients of determinant unity, Smith (§ 12) remarked that, in the case of quadratic forms, it is desirable to add the limitation that the denominators of the fractional coefficients are prime to $2\Omega\Delta$, and proved that two ternary quadratic forms are transformable into each other by such substitutions if and only if their complete generic characters coincide.

Finally (§§ 13–22), Smith proved Eisenstein's[7] formulas relating to mass or weight of positive forms or genera and the corresponding formulas for the new case of an even discriminant.

J. Liouville[21] stated that $m=6\mu\pm1$ has $F(6m)$ representations by $x^2+2y^2+3z^2$ if $F(k)$ is the number of classes of binary quadratic forms of determinant $-k$.

P. Bachmann[22] used Smith's[20] identities to prove Gauss' theorems that the determinant D of every binary quadratic form ϕ representable by a ternary form f is representable by the adjoint of f, and that every proper representation of ϕ by f belongs to a pair of solutions of Gauss' congruences (12). From one representation of ϕ by f and the known automorphs of ϕ (involving the integral solutions of $\tau^2-Dv^2=1$), we obtain all the representations which belong to the same or opposite

[21] Jour. de Math., (2), 14, 1869, 359, 360.
[22] Jour. für Math., 70, 1869, 365–371. It is stated in his next paper that we should here add the condition that ϕ is a properly primitive form whose determinant D is prime to Δ.

pair of solutions and which yield the same value of (7). An application of these results leads to an arithmetical derivation of Hermite's[18] automorphs of f.

Bachmann[23] found the conditions under which Hermite's[18] formula (15) for the automorphs of f shall have integral coefficients, i. e., the coefficients of X, Y, Z shall be divisible by $\rho^2-g(\lambda, \mu, \nu)$. It is assumed that the coefficients a, a', a'', A, A', A'' of the squares of the variables in f and its adjoint g are odd and the remaining coefficients even, that the determinant Δ of f is a product of distinct odd primes, and that only one of A, A', A'' is of the form $4n+3$ [without these restrictions, Meyer[42, 44]]. Necessary and sufficient conditions for integral coefficients are that ρ, λ, μ, ν be integral solutions of $\rho^2-g(\lambda, \mu, \nu)=2^h\delta$, where δ is a divisor of both Δ and ρ, and $h=0$ or 1. He verified Hermite's[18] formula expressing the product of two quaternary forms ρ^2-g as a third such form. From this formula we obtain all solutions of $\rho^2-g=2^h\delta$, where h and δ are given and ρ is divisible by δ, when we know one solution and all solutions of $\rho^2-g=1$. As an example, there are obtained all automorphs with integral coefficients of $x^2+y^2-z^2$ (Gauss[4]).

A. Meyer[24] proved that two indefinite ternary quadratic forms of the same genus are equivalent if they have relatively prime odd values of their invariants D (the determinant of the form f) and Ω (the g.c.d. with proper sign of the coefficients of the adjoint of f). In other words, each genus contains a single class (proved by Eisenstein[11] for the case in which Δ is odd and without a square factor). For an improvement of the proof and extension to even invariants, see Meyer.[39]

E. Selling[25] employed in connection with a form, whose coefficients need not be integers,

$$F=F(x, y, z)=Ax^2+By^2+Cz^2+2Gyz+2Hzx+2Kxy,$$

four auxiliary numbers D, L, M, N defined by

$$A+L+H+K=0, \quad B+M+K+G=0, \quad C+N+G+H=0, \quad D+L+M+N=0.$$

When x, y, z are replaced by $x-t, y-t, z-t$ respectively, F becomes

$$\phi=-G(y-z)^2-H(z-x)^2-K(x-y)^2-L(x-t)^2-M(y-t)^2-N(z-t)^2.$$

A positive form F is called reduced when G, H, K, L, M, N are all negative or zero. To justify this definition, which is essentially different from Seeber's, it is proved that any positive form is equivalent to a reduced form. The proof rests on the fact that

$$(16) \qquad -(G+H+K+L+M+N)=\tfrac{1}{2}(A+B+C+D)$$

is positive, being half the sum of the values of ϕ for the four sets of values, 1, 0, 0, 0; 0, 1, 0, 0; 0, 0, 1, 0; 0, 0, 0, 1 of x, y, z, t. But the substitution

$$x=-x'+t', \quad y=-y'+t', \quad z=-x'+z', \quad t=0,$$

replaces ϕ by ϕ' in which

$$G'=-G, \quad H'=G+N, \quad K'=G+K, \quad L'=-G+L, \quad M'=G+M, \quad N'=G+H,$$
$$-(G'+\dots+N')=-(2G+H+K+L+M+N).$$

[23] Jour. für Math., 71, 1870, 296–304. Notations changed to agree with Hermite's.

[24] Zur Theorie der unbestimmten ternären quadratischen Formen, Diss., Zürich, 1871. Cf. P. Bachmann, Die Arithmetik der Quad. Formen, 1898, Ch. 9.

[25] Jour. für Math., 77, 1874, 164–229; revision (in French) in Jour. de Math., (3), 3, 1877, 43–60, 153–206. See Charve's[28] exposition of the case of positive forms, Poincaré,[9, 35] Borissow,[36] Got.[50]

Hence if G is positive we can find an equivalent form for which the sum (16) is smaller than for F. By symmetry, the same conclusion holds if $H, \ldots,$ or N is positive. Hence F is equivalent to a reduced form. Furthermore, the sum (16) has for a reduced form a value less than for any equivalent form not having the same coefficients $G, \ldots, N$ merely permuted, whence the latter form is not also reduced.

In case $G, \ldots, N$ are negative (not zero), there are 24 reduced forms derived from the given one F by the 24 permutations of x, y, z, t, and they are all equivalent to F. But if one of the coefficients is zero, the number of reduced forms exceeds 24. This furnishes a method of finding the number of automorphs.

Following Gauss,[26] let (A), (B), (C) denote three vectors whose projections on three rectangular axes are ξ, ξ_1, ξ_2; η, η_1, η_2; and ζ, ζ_1, ζ_2, respectively. When x, y, z range over all sets of three integers, the extremities of the vectors $V=x(A)+y(B)+z(C)$, whose initial point is the origin, form a lattice. Define the product of two such vectors to be the product of their lengths by the cosine of the angle between them. Then the square of V is F whose coefficients A, B, C are the squares of (A), (B), (C), while G, H, K are the products by twos of these vectors, as follows from

$$(17)\qquad \begin{cases} \xi^2+\xi_1^2+\xi_2^2=A, & \eta^2+\eta_1^2+\eta_2^2=B, \quad \zeta^2+\zeta_1^2+\zeta_2^2=C, \\ \eta\zeta+\eta_1\zeta_1+\eta_2\zeta_2=G, & \zeta\xi+\zeta_1\xi_1+\zeta_2\xi_2=H, \quad \xi\eta+\xi_1\eta_1+\xi_2\eta_2=K. \end{cases}$$

The conditions for a reduced form F are interpreted geometrically (p. 55).

Selling next considered an indefinite ternary form $f=ax^2+\ldots+2kxy$, whose discriminant is not negative, so that f can be transformed into $x^2-y^2-z^2$ by a real substitution. Take any set of real numbers $\xi, \ldots, \zeta_2$ satisfying the six equations

$$(18)\qquad \begin{cases} \xi^2-\xi_1^2-\xi_2^2=a, & \eta^2-\eta_1^2-\eta_2^2=b, \quad \zeta^2-\zeta_1^2-\zeta_2^2=c, \\ \eta\zeta-\eta_1\zeta_1-\eta_2\zeta_2=g, & \zeta\xi-\zeta_1\xi_1-\zeta_2\xi_2=h, \quad \xi\eta-\xi_1\eta_1-\xi_2\eta_2=k, \end{cases}$$

and insert them in (17); we obtain a positive form

$$F=\begin{pmatrix} A & B & C \\ G & H & K \end{pmatrix}=(\xi x+\eta y+\zeta z)^2+(\xi_1 x+\eta_1 y+\zeta_1 z)^2+(\xi_2 x+\eta_2 y+\zeta_2 z)^2,$$

called a positive form corresponding to f. Then f is called reduced if F is reduced whatever set of real solutions of (18) is employed, but with a restriction imposed later (p. 177) in the course of the long geometrical discussion. All the (infinitude of) automorphs of f are products of powers of a finite number of automorphs.

S. Réalis[27] noted the identity

$$z_1^2+z_2^2+2z_3^2=(A^2+B^2+2C^2)(\alpha^2+\beta^2+2\gamma^2)^2,$$
$$z_1=A(\beta^2-\alpha^2+2\gamma^2)-2\alpha(B\beta+2C\gamma), \quad z_2=B(\alpha^2-\beta^2+2\gamma^2)-2\beta(A\alpha+2C\gamma),$$
$$z_3=C(\alpha^2+\beta^2-2\gamma^2)-2\gamma(A\alpha+B\beta).$$

L. Charve[28] gave a clear exposition of the arithmetical part of Selling's[25] theory of reduced positive ternary quadratic forms, with additions and application to the forms considered by Hermite while seeking periodic properties in the approximation

[26] Geometrische Seite der Ternären Formen, posth. MS., Werke, II, 305.
[27] Nouv. Corresp. Math., 4, 1878, 327.
[28] Ann. sc. école norm. supér., (2), 9, 1880, suppl., 156 pp. (Thèse).

to a root of a cubic equation, thereby extending the periodic continued fraction for a root of a quadratic equation.

H. Poincaré[29] expressed in a different form the results of Hermite[14] and Selling.[25] After a linear transformation of variables, an indefinite ternary quadratic form becomes $F=\xi^2+\eta^2-\zeta^2$ [or $-F$]. Consider in a plane a point m_1 inside the circle C of radius unity and center at the origin. Given the coordinates X_1, Y_1 of m_1, the relations

$$X_1=\xi_1/(\zeta_1+1),\quad Y_1=\eta_1/(\zeta_1+1),\quad \xi_1^2+\eta_1^2-\zeta_1^2=-1$$

determine ξ_1, η_1, ζ_1 and hence determine a reduced form of F obtained by applying to F the substitution which reduces the definite form

$$\xi^2+\eta^2-\zeta^2+2(\xi_1\xi+\eta_1\eta-\zeta_1\zeta)^2.$$

Hence to each point m_1 inside C corresponds a single reduced form of F. If m_1 varies, but does not pass out of a certain region R_0, the same reduced form is obtained. Let $R_0, R_1, \ldots, R_{n-1}$ be a system of regions, each contiguous to the next, which correspond to the n distinct reduced forms. Let P be the totality of these regions. Let P' be the totality of regions $R_0', R_1', \ldots, R_{n-1}'$ mutually related as were the R's and corresponding to the same reduced forms. Joining the summits of P by circles orthogonal to C, we obtain a curvilinear polygon Q. Similarly, from $P', P'', \ldots$, we obtain polygons $Q', Q'', \ldots$. Each property of an automorph of F gives a property of Q, which is expressed in the language of non-euclidean geometry.

E. Picard[30] proved that a form $a_1x^2+\ldots+2b_3xy$ with complex coefficients satisfying the 3 conditions

$$a_1b_1'+b_2a_3'+b_3b_2'=0,\quad a_2b_3'+b_3a_2'+b_1b_1'=0,\quad a_3b_1'+b_2a_2'+b_1b_3'=0,$$

in which a' denotes the conjugate to a, is transformed into another form satisfying the same 3 conditions by all substitutions satisfying certain 5 conditions.

A. Meyer[31] defined a Null form to be one which vanishes for rational values not all zero of the variables. Let D denote the determinant of the primitive indefinite form f, and ΩF its adjoint, where F is primitive and indefinite. Then $D=\Omega^2\Delta$. Let Θ denote the positive g.c.d. of Ω and Δ, Ω' that of Θ and Ω/Θ, and Δ' that of Θ and Δ/Θ. Write $\Theta'=\Theta/(\Omega'\Delta')$, $\Omega''=\Omega/(\Theta\Omega')$, $\Delta''=\Delta/(\Theta\Delta')$, so that $\Omega'\Omega''$ is prime to $\Delta'\Delta''$, and Θ' prime to $\Omega''\Delta''$. In parts I and II it is assumed that Θ', Ω', Δ', Ω'', Δ'' are relative prime, odd, and without square factors [restrictions removed in Meyer[42]]. Such a Null form is equivalent to a reduced form

$$f(r)=\Theta'\Omega''^2\Delta''x'^2+r\Delta'\Omega''x''^2+2\Theta'\Omega'\Delta'^2\Omega''xx'',$$

where $0<r\leqq\Theta$ and r is prime to Θ. Two forms $f(r)$ and $f(r')$ belong to the same genus if r and r' have the same quadratic characters with respect to the ν prime factors of Θ. Thus there are $2^{-\nu}\phi(\Theta)$ reduced forms in each genus. Two reduced forms $f(r)$ and $f(r')$ of the same genus are equivalent if and only if there exist

[29] Assoc. franç. av. sc., 1881, 132–8.
[30] Comptes Rendus Paris, 94, 1882, 1241–3.
[31] Jour. für Math., 98, 1885, 177–230. He first proved Smith's result (this History, Vol. II, 431) on the solvability of $f=0$ when D is odd.

factorizations $\Theta=\Theta_1\Theta_2$ and $\Omega''\Delta''=D_1D_2$, a root of $rz^2\equiv r'$ (mod Θ), and a value ± 1 or ± 2 of σ', such that $\sigma'\Theta_2D_2z$ and $\sigma'\Theta_1D_1rz$ are quadratic residues of Θ_1 and Θ_2 respectively. In other words, there must exist a binary form $(a, 0, rc)$ of determinant $-r\Theta\Omega''\Delta''$ whose characters with respect to all prime factors of Θ coincide with the characters of one of the values of z or $2z$. We readily deduce the number of classes in each genus; it is independent of the prime factors $4u+3$ of Θ, and is unity if Θ contains only primes $4u+3$.

In part II, which is subject to the same assumptions, he expressed by means of Legendre-Jacobi symbols the necessary and sufficient conditions that a properly primitive binary quadratic form ϕ of determinant $M\Omega$ be representable by a ternary Null form f with the invariants Ω and Δ, where M is prime to $\Omega\Delta$. The genus of ϕ completely determines the genus of f, and conversely.

In part III is found the number of classes in a ternary Null genus whose invariants Ω and Δ are positive and odd; this number is a power of 2. The complicated rule[32] to define this number was later corrected and simplified by him.[33]

H. Poincaré[34] discussed the simultaneous reduction of a ternary quadratic form $\phi(x, y, z)$ and linear form $f(x, y, z)$. We may write $\phi=\alpha f^2+gh$, where α is a constant and g, h are linear forms. Let δ be the determinant of f, g, h. For another system ϕ_1, f_1, write $\phi_1=\alpha_1f_1^2+g_1h_1$. This system is algebraically equivalent to the system ϕ, f if and only if $\alpha=\alpha_1$, $\delta=\delta_1$. Hence a system has two independent invariants, viz., the discriminant of ϕ and the invariant S of the cubic form $f\phi$.

Write $f=\lambda x+\mu y+\nu z$ and let Θ be the adjoint of ϕ. Let a, b, c be the rational numbers determined by

$$\frac{\partial\phi}{\partial a}\equiv\phi_a(a, b, c)=2\lambda, \qquad \phi_b(a, b, c)=2\mu, \qquad \phi_c(a, b, c)=2\nu.$$

Then

$$a\phi_x(x, y, z)+b\phi_y+c\phi_z=x\phi_a(a, b, c)+y\phi_b+z\phi_c=2f.$$

Since

$$\tfrac{1}{4}[a\phi_x(x, y, z)+b\phi_y+c\phi_z]^2-\phi(a, b, c,)\phi(x, y, z) \tag{19}$$

has discriminant zero, it is a product of two linear functions, whence

$$\phi=\alpha f^2+gh, \qquad \alpha=1/\phi(a, b, c).$$

But (19) is identical with $\Theta(x_1, y_1, z_1)$, where

$$x_1=bz-cy, \qquad y_1=cx-az, \qquad z_1=ay-bx,$$

whence $ax_1+by_1+cz_1\equiv 0$. Thus

$$\phi=\alpha f^2+\alpha\Theta(x_1, y_1, -(ax_1+by_1)/c).$$

This binary form Θ can be expressed in one of the forms h^2, $\pm(h^2+k^2)$, or h^2-k^2, where h and k are real linear functions. If $\Theta=h^2+k^2$, then ϕ or $-\phi$ is a positive definite form, according as α is positive or negative, and has a single reduced form in general; the substitution which reduces $\pm\phi$ reduces the system f, ϕ. If $\Theta=-h^2-h^2$,

[32] Also stated by Meyer, Vierteljahrsschrift Naturf. Gesell. Zürich, 28, 1883, 272–4.
[33] Jour. für Math., 112, 1893, 87, 88.
[34] Jour. école polytechnique, 56, 1886, 79–142; Comptes Rendus Paris, 91, 1880, 844–6.

the substitution which reduces the positive form $af^2+ak^2+al^2$ will reduce f and ϕ. In these cases there is a single reduced system. But if $\Theta=h^2-k^2$, the reduced systems constitute a chain which is either limited at each extremity or infinite and periodic, according as the ratios of the coefficients of h and k are rational or irrational (or $4S$ an exact fourth power or not). An extended investigation is made of these two cases and the transformations of f and ϕ into themselves are found.

H. Poincaré[35] employed the automorphs (Cantor[17]) of $\Phi=Y^2-XZ$:

$$S=\begin{pmatrix} \delta^2 & -\delta\gamma & \gamma^2 \\ -2\delta\beta & a\delta+\beta\gamma & -2a\gamma \\ \beta^2 & -a\beta & a^2 \end{pmatrix}, \qquad a\delta-\beta\gamma=1.$$

To S corresponds the Fuchsian substitution

$$s:\quad z'=\frac{az+\beta}{\gamma z+\delta}.$$

If to S' corresponds s', then to SS' corresponds ss'. Let F be the form obtained from Φ by applying any linear substitution T; then $T^{-1}ST$ is an automorph of F. Let F have integral coefficients and consider the discontinuous group of all the automorphs with integral coefficients of F. To it corresponds a Fuchsian group and hence a system of arithmetical Fuchsian functions. He defined and studied the reduction of ternary quadratic forms with respect to a given group, not necessarily the "arithmetical group" of all linear substitutions with integral coefficients of determinant unity. He investigated the continuous group of automorphs with rational coefficients of a quadratic form F with integral coefficients. Such forms F constitute four categories according as they do or do not admit elliptic or parabolic substitutions.

E. Borissow[36] made an extended study of Selling's[25] method of reduction of positive forms and their automorphs, and tabulated the reduced forms of determinants $\leqq 200$.

R. Fricke[37] illustrated Poincaré's[35] ideas by finding all automorphs with integral coefficients of $qx^2-y^2-z^2$, where q is a prime, and the fundamental region of various groups of corresponding linear fractional substitutions on one variable. He[38] later treated $px^2-qy^2-rz^2$, also when the coefficients are in a quadratic field.

A. Meyer[39] proved that two properly primitive, indefinite, ternary quadratic forms f, whose adjoint forms F are properly primitive and which belong to the same genus with the invariants Ω and Δ, are equivalent if they satisfy certain conditions involving Legendre-Jacobi symbols (f/θ) and (F/η). There is a similar theorem for improperly primitive forms. In particular, two primitive indefinite ternary forms are equivalent if they belong to the same genus and their invariants are neither

[35] Jour. de Math., (4), 3, 1887, 405–464; Comptes Rendus Paris, 102, 1886, 735–7 (94, 1882, 840–3); Oeuvres, II, 463–511, 64–66 (38–40). For a further property of Poincaré's invariant $a+\delta$ of the substitution s, see G. Bagnera, Atti R. Accad. Lincei, Rendiconti, (5), 7, I, 1898, 340–6.

[36] Reduction of positive ternary quadratic forms by Selling's method, with a table of the reduced forms for all determinants from 1 to 200, St. Pétersbourg, 1890, 1–108; tables 1–116 (Russian). Cf. Fortschritte der Math., 1891, 209.

[37] Math. Annalen, 38, 1891, 50–81, 461–476.

[38] Göttingen Nachr., 1893, 705–21; 1894, 106–16; 1895, 11–18. Fricke and Klein, Automorphe Functionen, Leipzig, 1, 1897, 533–584 (502, 519).

[39] Jour. für Math., 108, 1891, 125–139.

divisible by 4 nor have a common odd divisor (Got[50]; for case of odd invariants, Meyer[24]).

S. Kempínski[40] noted that the automorphs of $f = q\xi^2 - s\eta^2 - r\zeta^2$ are $S = T^{-1}\Sigma T$, where Σ are those of $F = XY - Z^2$, and T replaces F by f. When q, r, s are primes, the conditions are found that the coefficients of S be integers, 16 cases being distinguished. There is a study of the group $\Gamma_{q, r, s}$ of corresponding linear fractional substitutions, its fundamental region, etc.

* W. A. Markoff[41] proved Eisenstein's[10] formulas for the number of classes of positive ternary quadratic forms of given determinant.

A. Meyer[42] extended his[31] results on Null forms to an arbitrary indefinite ternary quadratic form f of odd determinant. He found (§§ 1, 3) necessary and sufficient conditions for the equivalence of two forms f. He investigated (§ 2) the automorphs of f without Bachmann's[23] restriction that $\Omega^2\Delta$ is odd and has no square factor.

Meyer[43] considered the proper representation of a primitive binary quadratic form of determinant $\Omega M''$ by a primitive indefinite ternary form f with the invariants Ω and Δ, where M'' is prime to Δ. He found (p. 179) the conditions under which a number prime to $\Omega\Delta$ (assumed odd) is representable properly by f.

Meyer[44] continued his study of equivalence, obtained the number of classes in a genus with any odd invariants, removed (p. 318) some of his earlier restrictions; and discussed the solvability of $p^2 - \Omega F(q, q', q'') = \epsilon$.

A. Markoff[45] proved that the exact superior limit of the minima of all indefinite ternary quadratic forms of determinant D, for integral values not all zero of the three variables, is equal to the minimum $\sqrt[3]{\frac{4}{3}|D|}$ of the forms equivalent to

$$\phi_0 = -\sqrt[3]{\tfrac{4}{3}D}(x^2 + xy + y^2 - 2z^2).$$

For forms not equivalent to ϕ_0, this limit is equal to the minimum $\sqrt[3]{\frac{2}{5}|D|}$ of forms equivalent to

$$\phi_1 = \sqrt[3]{\tfrac{2}{5}D}(x^2 + xy - y^2 - 2z^2).$$

For forms not equivalent to ϕ_0 or ϕ_1, this limit is equal to the minimum $\sqrt[3]{\frac{1}{3}|D|}$ of forms equivalent to

$$\phi_2 = -\sqrt[3]{\tfrac{1}{3}D}(x^2 + y^2 - 3z^2).$$

Excluding forms equivalent to ϕ_0, ϕ_1 or ϕ_2, the absolute value of each further form can be made less than $\sqrt[3]{\frac{1}{3}|D|}$ for integers x, y, z not all zero.

H. Minkowski[46] applied his results (in § 7) on the thickest packing of spheres and ellipsoids to deduce most of the facts in Gauss-Dirichlet's[8] theory of the arithmetical reduction of positive ternary quadratic forms.

W. A. Markoff[47] tabulated indefinite ternary quadratic forms not representing zero for all positive determinants ≤ 50.

---

[40] Pamietnik Acad. Umiej. Krakowie, 26, 1893, 37–66 (Polish). Summary in Bull. Intern. Acad. Sc. Cracovie, 1892, 219; Fortschritte der Math., 25, 1893–4, 207–8.

[41] Proc. Math. Soc. Univ. Khrakov, (2), 4, 1894, 1–59 (Russian).

[42] Jour. für Math., 113, 1894, 186–206; 114, 1895, 233–254.

[43] Ibid., 115, 1895, 150–182.

[44] Ibid., 116, 1896, 307–325 (conclusion of preceding series).

[45] Math. Annalen, 56, 1903, 233–251; French transl. of Bull. Acad. Sc. St. Pétersbourg, (5), 14, 1901, 509.

[46] Göttingen Nachr., 1904, 330–8; Diophantische Approximationen, Leipzig, 1907, 111–7.

[47] Mém. Acad. Sc. St. Pétersbourg, (8), 23, 1909, No. 7, 22 pp.

W. F. Meyer[48] proved that any ternary substitutions (c_{jk}) of determinant D which leaves Y^2-XZ absolutely invariant can be given the form S of Poincaré[35] in one and but one way, and that $D=\pm 1$. If the c_{jk} are real, α, β, γ, δ are all real or all pure imaginaries, according as one of c_{11}, c_{13}, c_{31}, c_{33} is positive or negative. If the c_{jk} are integers (or complex integers), α, β, γ, δ are integers (or complex integers) or products of such by i (or $\sqrt{i}$).

L. Bianchi[49] discussed the linear automorphs with integral coefficients of any ternary quadratic form with integral coefficients capable of representing zero, in particular, those of $2x_1x_3-\Delta x_2^2$.

Th. Got[50] continued the investigation of Poincaré,[35] employing in particular the indefinite form $f=x^2-\phi(y, z)$, where ϕ is a positive binary quadratic form. He gave a simplification of Selling's[25] method of reduction, and (in the appendix) of Meyer's[39] proof of his final theorem.

M. Weill[51] noted that the product of two forms of type

$$f=a^2+b^2+c^2-ab-ac-bc$$

can be represented in the same form in infinitely many ways. [This is trivial since we obtain f by replacing a by $a-c$ and b by $b-c$ in a^2-ab+b^2.] Similarly for his second form $2(a+b)^2+2(c+d)^2$.

G. Julia[52] simplified Minkowski's[46] (1907) geometrical process of reducing a positive ternary quadratic form f, by proving that the determinant of the coordinates of the points which furnish the first, second, and third proper minima of f is unity.

L. J. Mordell[53] proved Eisenstein's[10] two expressions for the number of classes of primitive positive ternary quadratic forms.

G. Humbert[54] defined the Poincaré domain for an indefinite ternary quadratic form f. From each class of a given genus with odd invariants $\Omega<0$ and $\Delta>0$ select a form f_i; let F_i be its properly primitive adjoint. Let

$$(20) \qquad E=\left(\frac{f}{\bar{\Omega}}\right)\left(\frac{F}{\Delta}\right)(-1)^{\frac{1}{2}(\bar{\Omega}+1)\cdot\frac{1}{2}(\Delta+1)}, \qquad \bar{\Omega}=|\Omega|.$$

Let M be a positive integer prime to $\Omega\Delta$ such that $(-M/\delta)=(F/\delta)$ for every prime factor δ of Δ. Then the number of sets of proper solutions of $-M=F_i(x, y, z)$ $(i=1, 2, \ldots)$, such that the point (x, y, z) belongs to the Poincaré domain for F_i, is

$$\rho 2^{-\nu}[H(\bar{\Omega}M)+(2E+1)H'(\bar{\Omega}M)],$$

where ν is the number of distinct prime factors of $\bar{\Omega}$, while $H(\bar{A})$ and $H'(\bar{A})$ are the numbers of classes of positive binary forms, properly and improperly primitive, of discriminant $\bar{A}$, while $\rho=\frac{1}{2}$ if $M\not\equiv 0$, $\rho=\frac{1}{2}(E+1)$ if $M\equiv 0 \pmod 4$. Next, let Ω be odd, $\Delta=2\Delta'$, Δ' odd. Define E' by (20) with Δ replaced by Δ'. Let M be not

[48] Jahresber. d. D. Math.-Vereinigung, 20, 1911, 153–161.
[49] Atti R. Accad. Lincei, Rendiconti (classe fis. mat.), (5), 21, I, 1912, 305–315.
[50] Annales fac. sc. Toulouse, (3), 5, 1913, 1–116 (Thèse). Comptes Rendus Paris, 156, 1913, 1596–8, 1741–3; 157, 1913, 34–36; Soc. Math. France, Comptes Rendus, 1913, 47–48.
[51] Nouv. Ann. Math., (4), 16, 1916, 266–8.
[52] Comptes Rendus Paris, 162, 1916, 320–2.
[53] Messenger of Math., 47, 1918, 65–78.
[54] Comptes Rendus Paris, 166, 1918, 925–30; 167, 1918, 49–55.

divisible by 4 and such that $(-M/\delta)=(F/\delta)$. Then the number of sets of proper solutions of $-M=F_i (i=1, 2, \ldots)$ is $\rho' 2^{-\nu} H(\overline{\Omega M})$, where

$$\rho'=\tfrac{1}{2} \text{ if } |\Omega M| \equiv 1 \text{ or } 2 \pmod 4, \quad \rho'=\tfrac{1}{2}\{E'+(2/M)\} \text{ if } |\Omega M| \equiv -1 \pmod 4.$$

G. Humbert[55] found the sum of the areas of the Poincaré domains in a plane for the adjoint forms $F_1, F_2, \ldots$ (Humbert[54]).

Humbert[56] investigated the measure (weight) of the totality of classes of positive ternary forms of given determinant $\Omega^2\Delta$, whereas Smith regarded Ω and Δ as given separately.

Humbert[57] found the area of the fundamental domain of the principal subgroup of the group of automorphs of $Dx^2-y^2-Pz^2$.

E. T. Bell[58] proved that a prime n has $3\{G(n)-1\}$ representations by $f=xy+yz+zx$, x, y, z each >0, where $G(n)$ is the number of classes of binary quadratic forms of determinant $-n$. He studied also the representation of composite numbers by f.

L. J. Mordell[59] proved that the number of solutions $\gtreqless 0$ of $n=xy+yz+zx$, counting as $\frac{1}{2}$ a solution in which an unknown is zero, is triple the number of classes of binary forms of determinant $-n$, provided the classes $(k, 0, k)$ and $(2k, k, 2k)$ be reckoned as $\frac{1}{2}$ and $\frac{1}{3}$, respectively. The method differs from Bell's,[58] and n may here be composite.

C. L. Siegel[60] investigated ternary quadratic forms whose coefficients are integral algebraic numbers of any field.

A. Hurwitz[61] evaluated the number of classes of positive ternary quadratic forms of a given determinant by an extension of his method for binary forms (see Ch. VI).

The problem[62] of finding all definite ternary quadratic forms with special automorphs having integral coefficients is a chief subject of geometrical crystallography. The problem of finding a form, given the lowest numbers represented by it, is useful in the determination of the structure of a body [note due to Speiser].

Reports are given in Vol. II of this History of Libri's result (p. 429, 1820) that every integer can be expressed in the form $x^2+41y^2-113z^2$, and in certain similar forms; Dirichlet (pp. 263–4, 1850), Landau (p. 272, 1909), and Pocklington (p. 273, 1911) employed the equivalence of any positive form of determinant unity to $x^2+y^2+z^2$; Liouville[37] (p. 265, 1870; pp. 332–6, 1858–1860) evaluated sums extended over the sets of solutions of $m=i_1^2+\omega_1^2+8s_1^2$ or of $m=m_1^2+d\delta$; Torelli (pp. 294–5, 1878) discussed the number of sets of solutions of $2x^2+y^2+z^2=g$.

---

55 Comptes Rendus Paris, 167, 1918, 181–6.
56 *Ibid.*, 168, 1919, 917–23, 969–75.
57 *Ibid.*, 171, 1920, 445–450.
58 Tôhoku Math. Jour., 19, 1921, 105–116. For more details on this and the related paper by Mordell,[59] see the end of Ch. VI.
59 London Math. Soc., Records of Proceedings, Nov. 17, 1921. In full in Amer. Jour. Math., Jan., 1923.
60 Math. Zeitschrift, 11, 1921, 248–257.
61 Post. MS., current vol. of Math. Annalen.
62 Sommerfeld, Atombau und Spectrallinien, note at end, with a geometrical interpretation of the adjoint form.

CHAPTER X.

QUATERNARY QUADRATIC FORMS.

The majority of the papers relate to the representation of integers by forms of the type $ax^2+by^2+cz^2+dw^2$, with special attention to the types which represent all positive integers. While the general theory of composition of two quadratic forms in four variables is due to Brandt,[44] special results had been found by Bazin[5] and Stouff.[34] For the subject of reduced forms, see the references 28, 30, 43, 47 and 48.

L. Euler[1] stated that $4mn-m-n+x^2+y^2+y$ represents every integer (cf. Genocchi[7]); that $3a^2+3b^2+7c^2$ and $2a^2+6b^2+21c^2$ are never squares, but that he could find no similar theorem for four unknowns.

E. Waring[2] stated that if p, q, r, s are relatively prime, $pa^2+qb^2+rc^2+sd^2$ represents every integer exceeding an assignable one, and that $a^2\pm ab+b^2+c^2\pm cd+d^2$ represents every integer.

In 1840, Jacobi proved that every positive integer N can be represented by $x^2+2y^2+3z^2+6t^2$ (this History, Vol. II, p. 263). J. Liouville[3] showed that this theorem is equivalent to the fact that N is a sum of four squares. For, then $N=x^2+(y+z+t)^2+(y-z-t)^2+(2t-z)^2$. Conversely, let $N=x^2+y^2+z^2+t^2$. By changing the signs of roots, if necessary, we may assume that $x, \ldots, t$ are each of the form $3n+1$ or $3n$. Hence we can select three, say x, y, z, whose sum is a multiple of 3. Then the identity

$$3(x^2+y^2+z^2)\equiv(x+y+z)^2+2(s-z)^2+6u^2,\qquad s=\tfrac{1}{2}(x+y),\qquad u=\tfrac{1}{2}(x-y),$$

shows that $s-z$ is divisible by 3. We have only to add t^2 to each member of

$$x^2+y^2+z^2=3v^2+6w^2+2u^2.$$

G. Eisenstein[4] noted that, if m is of the form $12n+k$ ($k=1$, 5, 7, or 11), the number of proper representations of $m=\Pi p^\pi$ by $x^2+y^2+z^2+3u^2$ is $6N$, $12N$, $2N$ or $4N$, respectively, where

$$N=\Pi p^{\pi-1}\left\{p+(-1)^{\frac{1}{2}(p-1)}\left(\frac{p}{3}\right)\right\},$$

whence N is equal to the difference between the sum of those factors of m which are of the forms $12n\pm1$ and the sum of those of the forms $12n\pm5$. The number of proper representations of m by $x^2+y^2+2z^2-2uz+2u^2$ is $4N$, $2N$, $12N$, $6N$, respec-

[1] Correspondance Mathématique et Physique (ed., P. H. Fuss), St. Pétersbourg, 1, 1843, 123–4; letter to Goldbach, May 8, 1742.
[2] Meditationes algebraicae, Cambridge, ed. 3, 1782, 349.
[3] Jour. de Math., 10, 1845, 169–170. Cf. Bachmann, Niedere Zahlentheorie, 2, 1910, 320–3.
[4] Jour. für Math., 35, 1847, 134.

tively. The number of proper representations of $m=\Pi p^\pi$, where m is divisible by neither 2 nor 5, by $x^2+y^2+z^2+5u^2$ is

$$s\Pi p^{\pi-1}\left\{p+\left(\frac{p}{5}\right)\right\},$$

where $s=6$ or 4 according as m is a quadratic residue or non-residue of 5.

In 1854, Hermite gave two important formulas for the composition of quaternary forms, quoted in Ch. IX.[18] See Vol. II, pp. 277, 281.

M. Bazin[5] extended Gauss' definition of composition of binary quadratic forms to quadratic forms $f(x, y, z, v)$, $f'(x', \ldots)$, $F(X, \ldots)$ in 4 variables with integral coefficients. We say that F is *compounded* of f and f' if F is transformed into the product ff' by a bilinear substitution which expresses $X, \ldots, V$ as linear functions of xx', xy', $\ldots$, vv' with integral coefficients such that there is no common divisor >1 of the 4-rowed determinants of the coefficients of the partial derivatives of $X, \ldots, V$ with respect to any one of the 8 variables $x, \ldots, v'$. It is shown that f is not composable with another form if the determinant of f is not the negative of a square and that a definite form is not composable with an indefinite form. Two composable forms can be transformed rationally into $\phi(x, y, z)+\Delta v^2$ and $\phi'+\Delta' v'^2$, where ϕ and ϕ' are of determinants Δ and Δ', and the adjoint forms of ϕ, ϕ' are transformable into each other by a rational transformation T. All the unknowns in the problem are expressible rationally in the coefficients of T. If f, f' are composable, they and their compound F can be transformed linearly and rationally into

$$f_1=ax^2+\frac{1}{a}\psi(y, z, v), \quad f'_1=a'x'^2+\frac{1}{a'}\psi(y', z', v'), \quad F'=aa'X^2+\frac{1}{aa'}\psi(Y, Z, V),$$

where $\psi(y, z, v)=cdy^2+bdz^2+bcv^2$, and

$$X=xx'+\frac{cd}{aa'}yy'+\frac{bd}{aa'}zz'+\frac{bc}{aa'}vv', \quad Y=a'yx'-axy'+b(vz'-zv'),$$
$$Z=a'zx'-axz'+c(yv'-vy'), \quad V=a'vx'-axv'+d(zy'-yz').$$

The study of compositions $ff'=F$ with integral coefficients is not complete. Since a form f cannot always be compounded with itself the theory differs from the binary case. See Brandt.[44]

V. A. Lebesgue[6] proved Euler's[1] theorem that every integer N can be expressed in the form $4mn-m-n+x^2+y^2+y$ since the condition is

$$4\{N+(m-n)^2\}+2=\{2(m+n)-1\}^2+4x^2+(2y+1)^2.$$

But every integer of the form $4p+2$ is a sum of three squares one of which is even.

A. Genocchi[7] noted that the preceding argument does not prove the equation solvable in positive integers, and the same is true of the argument from

$$4N+2-4x^2-(2y+1)^2=(4m-1)(4n-1).$$

From $4N+2$ subtract an even and odd square whose sum is $<4N+2$; the difference

[5] Jour. de Math., 19, 1854, 215–252 (lemma, 209–214).
[6] Nouv. Ann. Math., 13, 1854, 412–3.
[7] Annali di Sc., Mat., Fisiche, 5, 1854, 503–4.

is of the form $4p+1$; let its factors be $4m-1$, $4n-1$, where m and n may be negative. But the theorem was proved elsewhere by Genocchi.[8]

J. Liouville[9] stated that, if a and b are positive integers and $a \leqq b$, $x^2+ay^2+bz^2+abt^2$ represents all positive integers only in the seven cases $a=1$, $b=1, 2, 3$; $a=2$, $b=2, 3, 4, 5$. Four of these cases are equivalent to the theorem on sums of four squares (cf. Liouville[3]). For other values of a, b, the form does not represent 2 and 3. In Vol. II of this History, pp. 330, 336–7, are quoted his theorem on the number of decompositions of m or $2m$ into $y^2+z^2+2^\alpha(u^2+v^2)$, and formulas involving summations over the solutions of $m_1^2+4m_2^2+2^\alpha d\delta=m$.

Liouville[10] proved that every positive integer except 3 can be represented by $x^2+y^2+5z^2+5t^2$, since, for an integer m of the form $8\mu\pm1$ or $8\mu+5$, $5m$ is a sum of three squares, and the same is true of $m-20$ if $m=8\mu+3$.

He[11] stated that the number of solutions of $n=x^2+y^2+3z^2+3t^2$ is the quadruple of the sum S of the divisors prime to 3 of m if $n=m$ is odd, but is $4(2^{\alpha+1}-3)S$ if $n=2^\alpha m$, m odd, $\alpha>0$. He gave the number of proper representations (g.c.d. of $x, \ldots, t$ unity), also in the cases of the next two papers.

He[12] stated that the number of all solutions of $2^\alpha m=x^2+y^2+2z^2+2t^2$ is $4\sigma(m)$, $8\sigma(m)$ or $24\sigma(m)$, according as $\alpha=0$, $\alpha=1$, or $\alpha \geqq 2$, where $\sigma(m)$ is the sum of the divisors of m (m odd). He[13] stated that the number of all representations by $x^2+y^2+4z^2+4t^2$ of $4k+3$, $4k+1$, $2m$ (m odd), $4m$, $2^\alpha m$ ($\alpha \geqq 3$) is 0, $4\sigma(m)$, $4\sigma(m)$, $8\sigma(m)$, $24\sigma(m)$, respectively.

Liouville[14] proved that each odd integer and each multiple of 4 can be represented by $x^2+3y^2+4z^2+12t^2$. Writing

$$\omega_1(m)=\sum_{d\delta=m}(-1)^{(\delta^2-1)/8}d,$$

and $A(n)$ for the number of representations of n by $x^2+y^2+z^2+2t^2$, and $B(n)$ for the number by $x^2+2y^2+2z^2+2t^2$, and taking m odd, he[15] noted that

$$A(2^\alpha m)=2\{2^{\alpha+2}-(-1)^{(m^2-1)/8}\}\omega_1(m),\qquad B(2^\alpha m)=A(2^{\alpha-1}m),$$

as follows by taking $f(x)=\cos \pi x/4$ in his formulas (F) and (I) on pp. 330–1 of Vol. II of this History. For m odd,[16] the number of representations of $2^\alpha m$ by $x^2+y^2+z^2+8t^2$ is $2\{2^\alpha-(-1)^{(m^2-1)/8}\}\omega_1(m)$ if $\alpha>1$, $12\omega_1$ if $\alpha=1$; while, if $\alpha=0$, it is $6\omega_1$ if $m=8k+1$ or $8k-3$; $4\omega_1$ if $m=8k+3$; 0 if $m=8k-1$. Treating various subcases, he found also the number of proper representations.

If m is odd,[17] the number of representations of $2m$, $4m$, $8m$, $2^\alpha m$ ($\alpha \geqq 4$), and m, by $x^2+2y^2+4z^2+8t^2$ is $2\sigma(m)$, $4\sigma(m)$, $8\sigma(m)$, $24\sigma(m)$, and

$$\sigma(m)+(-1)^{(m^2-1)/8}\Sigma(-1)^{\frac{1}{2}(i-1)}i,$$

---

[8] Nouv. Ann. Math., 12, 1853, 235–6.
[9] Jour. de Math., (2), 1, 1856, 230.
[10] *Ibid.*, (2), 4, 1859, 47, 48.
[11] *Ibid.*, (2), 5, 1860, 147–152.
[12] *Ibid.*, pp. 269–272.
[13] *Ibid.*, pp. 305–8. Proofs by P. Bachmann, Niedere Zahlentheorie, 2, 1910, 409–423.
[14] Jour. de Math., (2), 6, 1861, 135–6.
[15] *Ibid.*, pp. 225–230.
[16] *Ibid.*, pp. 324–8.
[17] *Ibid.*, pp. 409–416.

where the summation extends over the positive odd values of i in all the decompositions $m=i^2+4s^2$, s positive, negative, or zero.

Liouville[18] expressed in terms of the preceding sum, the analogous sum $\Sigma(-1)^{(r-1)/2}r(m=r^2+2u^2$, m odd, $r>0$, u positive, negative, or zero), $\sigma(m)$ and $\omega_1(m)$ the number of all representations and the number of the proper representations of $2^\alpha m$ by

$x^2+2y^2+2z^2+4t^2$ (1–4), $x^2+8y^2+8z^2+8t^2$ (5–8),
$x^2+4y^2+4z^2+8t^2$ (9–12), $x^2+8y^2+8z^2+16t^2$ (13–16),
$x^2+2y^2+4z^2+4t^2$ (62–4), $x^2+2y^2+8z^2+8t^2$ (65–8),
$x^2+8y^2+16z^2+16t^2$ (69–72), $x^2+4y^2+4z^2+16t^2$ (73–6),
$x^2+16(y^2+z^2+t^2)$ (77–80), $x^2+y^2+2z^2+4t^2$ (99–102),
$x^2+y^2+4z^2+8t^2$ (103–4), $x^2+4y^2+16z^2+16t^2$ (105–8),
$x^2+y^2+8z^2+8t^2$ (109–12), $x^2+4y^2+8z^2+8t^2$ (113–6),
$x^2+y^2+16(z^2+t^2)$ (117–20), $x^2+4y^2+8z^2+16t^2$ (143–4),
$x^2+2y^2+16(z^2+t^2)$ (145–7), $x^2+2y^2+2z^2+8t^2$ (148–9),
$x^2+2y^2+4z^2+16t^2$ (150–2), $x^2+2y^2+8z^2+16t^2$ (153–4),
$x^2+y^2+2z^2+8t^2$ (155–6), $x^2+y^2+4z^2+16t^2$ (157–160),
$x^2+2y^2+2z^2+16t^2$ (161–4), $x^2+y^2+z^2+16t^2$ (165–8),
$x^2+y^2+8z^2+16t^2$ (201–4), $x^2+y^2+2z^2+16t^2$ (205–8),
$x^2+8y^2+8z^2+64t^2$ (246–8), $x^2+8y^2+16z^2+64t^2$ (249–252),
$x^2+8y^2+64(z^2+t^2)$ (421–4).

Liouville[19] noted that the number $N(n)$ of representations of $n=2^\alpha 3^\beta m$ (m prime to 6) by $x^2+y^2+z^2+3t^2$ is

$$-\left\{3^{\beta+1}-(-1)^{\alpha+\beta}\left(\frac{m}{3}\right)\right\}\left\{2^{\alpha+1}+(-1)^{\alpha+\beta+(m-1)/2}\right\}\Sigma,\quad \Sigma\equiv\sum_{d\delta=m}(-1)^{\frac{1}{2}(\delta-1)}\left(\frac{\delta}{3}\right)d,$$

the case $\alpha=\beta=0$ being due to Eisenstein.[4] He enumerated the representations by $x^2+y^2+z^2+zt+t^2$. The number of representations of $2^\alpha 3^\beta m$ by $x^2+y^2+2z^2+6t^2$ is $N(2^{\alpha-1}3^\beta m)$ if $\alpha>0$, and $\{3^{\beta+1}+(-1)^\beta(m/3)\}\Sigma$ if $\alpha=0$.

In the same volume he enumerated, in terms of $N(n)$ or Σ, the representations by

$x^2+2y^2+2z^2+3t^2$ (129–133), $x^2+2y^2+4z^2+6t^2$ (134–6),
$x^2+y^2+z^2+12t^2$ (161–8), $x^2+2y^2+2z^2+12t^2$ (169–172),
$x^2+y^2+4z^2+12t^2$ (173–6), $x^2+4y^2+4z^2+12t^2$ (177–8),
$3x^2+4y^2+4z^2+4t^2$ (179–181), $x^2+y^2+3z^2+4t^2$ (182–4),
$x^2+3y^2+4z^2+4t^2$ (185–8), $2x^2+2y^2+3z^2+4t^2$ (189–192),
$x^2+y^2+z^2+3t^2$ (193–204), $x^2+4y^2+12z^2+16t^2$ (205–8),
$x^2+3y^2+6z^2+6t^2$ (209–213), $2x^2+3y^2+3z^2+6t^2$ (214–8),
$x^2+3y^2+3z^2+3t^2$ (219–24), $3x^2+3y^2+3z^2+4t^2$ (229–38),
$3x^2+3y^2+4z^2+12t^2$ (239–240), $3x^2+4y^2+12z^2+12t^2$ (241–2),
$x^2+3y^2+3z^2+12t^2$ (243–8), $x^2+3y^2+12y^2+12t^2$ (249–52),
$x^2+12y^2+12z^2+12t^2$ (253–4), $3x^2+4y^2+12z^2+48t^2$ (255–6),
$2x^2+2xy+2y^2+3z^2+3t^2$ (225–6), $x^2+xy+y^2+3z^2+3t^2$ (227–8),
$x^2+y^2+3z^2+3t^2$, x, y, z, t odd (296), $x^2+xy+y^2+2z^2+2zt+2t^2$ (308–10).

[18] Jour. de Math., (2), 7, 1862, pages cited in parenthesis after the forms.
[19] Jour. de Math., (2), 8, 1863, 105-128.

If $m=4\mu+1=r^2+4s^2$, $2m=i^2+i_1^2$, where r, i, i_1 are positive odd integers, then (p. 311) his former sums[17] are connected by the relation

$$\Sigma(-1)^{\frac{i-1}{2}}i=(-1)^{\mu}\Sigma(-1)^{\frac{r-1}{2}}r.$$

Liouville[20] noted that, if m is prime to 10, the number of representations of $2^{\alpha}5^{\beta}m$ by $x^2+y^2+z^2+5t^2$ is

$$\tfrac{1}{3}\left\{5^{\beta+1}+(-1)^{\alpha}\left(\frac{m}{5}\right)\right\}\{2^{\alpha+1}-(-1)^{\alpha}\cdot 5\}\Sigma,\quad \left\{5^{\beta+1}+\left(\frac{m}{5}\right)\right\}\Sigma,\quad \Sigma\equiv\sum_{d\delta=m}\left(\frac{\delta}{5}\right)d,$$

according as $\alpha>0$, $\alpha=0$; and enumerated also the proper representations. Eisenstein[4] had treated only the case $\alpha=\beta=0$.

He[21] enumerated the representations by $x^2+5(y^2+z^2+t^2)$, $F+6G$, $2F+3G$, $F+3G$, where $F=x^2+xy+y^2$, $G=z^2+zt+t^2$. The number of representations of $2^{\alpha}3^{\beta}m$ (m prime to 6) by $x^2+2y^2+2yz+2z^2+3t^2$ is $N=2(3^{\beta+1}-2)\sigma(m)$ if $\alpha=0$ and $3N$ if $\alpha>0$ (p. 160). The number of representations of $2^{\alpha}3^{\beta}m$ by $x^2+2y^2+3z^2+6t^2$ is N if $\alpha=1$, $3N$ if $\alpha>1$, but was not found in general if $\alpha=0$ (pp. 299–312).

Liouville[22] noted that, if m is prime to 10, the number of representations of $2^{\alpha}5^{\beta}m$ by $x^2+y^2+5z^2+5t^2$ is $2(5^{\beta+1}-3)\sigma(m)$ if $\alpha>0$, but was unable to treat the case $\alpha=0$.

He[23] showed how to deduce the number of representations of $p^{\alpha}q$ (where q is not divisible by the prime p) by $x^2+y^2+p(z^2+t^2)$ from the number for q, and likewise for $x^2+2y^2+pz^2+2pt^2$ and $x^2+3y^2+pz^2+3pt^2$.

He[24] gave theorems on sums of the numbers of representatives by two forms.

A. Korkine and G. Zolotareff[25] employed the known limit of the minima of ternary forms to obtain the precise limit of the minima of positive quaternary quadratic forms f: we can assign to the variables in any f of determinant $-D$ integral values such that $f\leqq\sqrt[4]{4D}$, while there exist forms f whose minimum is $\sqrt[4]{4D}$.

R. Gent[26] used the method of Dirichlet (this History, Vol II, p. 290) to prove that the number of solutions of $8m=t^2+u^2+3v^2+3w^2$, where $m, t, \ldots, w$ are odd and >0, is the sum of the divisors prime to 3 of m. This is also obtained by applying his theorems (Ch. I[94]) on binary quadratic forms to $t^2+3v^2=4p$, $u^2+3w^2=4q$, $p+q=2m$, p and q odd. From his conjecture on $8n=x^2+7y^2$, quoted there, it follows that the number of solutions of $16m=t^2+u^2+7v^2+7w^2$ is the sum of the divisors prime to 7 of m. There is a similar conjecture for 15 instead of 3 and 7.

H. J. S. Smith[27] proved that every positive integer can be represented by either of the forms $x^2+y^2+3u^2+3v^2$, $x^2+2y^2+3u^2+6v^2$.

---

[20] Jour. de Math., (2), 9, 1864, 1–16.
[21] *Ibid.*, 17–24, 181–4, 223–4.
[22] Jour. de Math., (2), 10, 1865, 1-13.
[23] *Ibid.*, 43–54; 11, 1866, 211–6.
[24] *Ibid.*, (2), 10, 1865, 359–360; 11, 1866, 39–40, 103–4, 131–2, 280–2. Cf. Humbert.[51]
[25] Math. Annalen, 5, 1872, 581–3; Korkine's Coll. Papers, 1, 1911, 283–8.
[26] Zur Zerlegung der Zahlen in Quadrate, Progr. Liegnitz, 1877.
[27] Coll. Math. in Memoriam D. Chelini, Milan, 1881, 117–43; Coll. Math. Papers, II, 309–11.

L. Charve[28] applied to positive quaternary quadratic forms $f(x, y, z, t)$ the method of reduction given for ternary forms by Selling[25] of Ch. IX. In f replace x, y, z, t by $x-u, y-u, z-u, t-u$; we get

$$\phi=a(x-y)^2+b(x-z)^2+c(x-t)^2+d(x-u)^2+e(y-z)^2 \\ +f(y-t)^2+g(y-u)^2+h(z-t)^2+k(z-u)^2+l(t-u)^2.$$

It is called *reduced* if it satisfies one of the three sets of conditions:

(i) All of the coefficients $a, \ldots, l$ are positive.

(ii) a alone is negative and is less in absolute value than b, c, d, e, f, g.

(iii) a and h alone are negative; $|a|<b, c, d, e, f, g$; $|h|<b, c, e, f, k, l$; and $|a+h|<b, c, e, f$.

It is proved that every form is equivalent to one reduced form and to only one if we do not distinguish a form from that obtained by permuting the variables. The reduction can be effected by repetitions of the two substitutions

$$(x, y, z, t, u); \quad x=T-Z, \quad y=X-U, \quad z=Y-U, \quad t=T-U, \quad u=0.$$

While the method is applicable to forms in n variables, the computations would increase very rapidly with n.

Charve[29] tabulated the positive quaternary quadratic reduced forms of determinant $\leqq 20$.

On pp. 310–1 of Vol. II of this History is quoted Pepin's evaluation of the excess of the number of solutions of $m=x^2+y^2+z^2+2t^2$ with x even over the number of solutions with x odd. On p. 313 is quoted Gegenbauer's theorem on the number of all solutions.

E. Picard[30] recalled that an indefinite quaternary quadratic form with integral coefficients can be reduced to $\pm(u_1^2+u_2^2+u_3^2-u_4^2)$ or to $f=u_1^2+u_2^2-u_3^2-u_4^2$. With the last associate the definite form

$$\phi=-(\eta-\eta_0)(\xi-\xi_0)f+2\,\text{norm}[(\eta-\xi)u_1-(1+\xi\eta)u_2+(\xi+\eta)u_3+(1-\xi\eta)u_4],$$

where ξ, η are any complex parameters in which the coefficients of $\sqrt{-1}$ are both positive, while ξ_0 is the conjugate to ξ. When applied to ϕ, a substitution of determinant unity with integral coefficients, which leaves f invariant, induces a substitution on ξ, η of the type

$$\xi'=\frac{a\xi+b}{c\xi+d}, \qquad \eta'=\frac{l\eta+m}{n\eta+p},$$

or with the fractions interchanged. We obtain a discontinuous group.* The existence of this hyperabelian group is proved simply by a method not yielding its properties. Application is made to functions invariant under a given hyperabelian group.

*Cf. Bourget,[40] Cotty,[43] E. Hecke, Math. Annalen, 71, 1912, 1–37; 74, 1913, 465–510; O. Blumenthal, *ibid.*, 56, 1903, 509–548; 58, 1904, 497.

[28] Annales sc. école norm. sup., (2), 11, 1882, 119–34; Comptes Rendus Paris, 92, 1881, 782–3.

[29] Comptes Rendus Paris, 96, 1883, 773.

[30] Jour. de Math., (4), 1, 1885, 87–128. Summary in Comptes Rendus Paris, 98, 1884, 904–6.

T. Pepin[31] noted that, if we take $f(x)=\cos xt$, $t=\pi/2$ or $\pi/4$, in Liouville's formulas for an even function $f(x)$ [this History, Vol. II, Ch. XI], we obtain the number of solutions of various equations of the type

$$(1) \qquad 2^a m = x^2+2^b y^2+2^c z^2+2^d w^2.$$

There is a direct proof (p. 188) that the number of representations of a positive odd integer m as a sum of four squares is double the number of representations of m by $x^2+y^2+2z^2+2t^2$, and that the number of solutions of $m=x^2+2y^2+2z^2+2t^2$ is equal to that of $m=x^2+p^2+q^2+2t^2$ (x odd).

J. W. L. Glaisher[32] proved by the use of products of infinite series that $R_1-R_2=6\chi(4n+1)$, if R_1 is the number of compositions of $24n+6$ in the form $x^2+y^2+2z^2+2w^2$, where $x^2, \ldots, w^2$ are all of the form $(12m\pm 1)^2$ or all of the form $(12m\pm 5)^2$, or two are of one form and two of the other, while R_2 is the number of compositions in which three are of one form and one of the other, composition and the function χ being defined on p. 296 of Vol. II of this History.

Pepin[33] gave many theorems on the representations of numbers by $x^2+y^2+3z^2+3t^2$, $x^2+3y^2+4z^2+12t^2$ and (1). He noted that many of Liouville's theorems of this type can be deduced from his series of eighteen articles [this History, Vol. II, Ch. XI], while others have been obtained only by the use of elliptic functions.

X. Stouff[34] verified that the form

$$\Phi(x,y,z,u)=A(x^2+u^2)+A'y^2+A''z^2+(By+Cz)(x-u)+Dxu+Eyz,$$

subject to the conditions $A/A'=A''/A=-C/B$, $A(D+E)=BC$, may be compounded with itself:

$$\Phi(x_1,y_1,z_1,u_1)=(D+2A)\Phi(X,Y,Z,U)\Phi(x,y,z,u),$$
$$x_1=X\{-(A+D)x+Cz-Au\}+Y\{Bx+A'y+(E-A)z-Bu\}$$
$$+Z(Ay+A''z)+U(-Ax-Cz+Au),$$

y_1, z_1, u_1 being similar long bilinear functions. As an application, let

$$\frac{c'\zeta+c_2'}{c_3'\zeta+c_4'}=\frac{C_1\left(\dfrac{c_1\zeta+c_2}{c_3\zeta+c_4}\right)+C_2}{C_3\left(\dfrac{c_1\zeta+c_2}{c_3\zeta+c_4}\right)+C_4},$$

identically in x, y, z, u, X, Y, Z, U, when the values of $x_1, \ldots, u_1$ are inserted and

$$c_i=\eta_{i1}x+\ldots+\eta_{i4}u, \quad c_i'=\eta_{i1}x_1+\ldots+\eta_{i4}u_1, \quad C_i=\eta_{i1}X+\ldots+\eta_{i4}U.$$

This application is to Fuchsian groups of linear fractional substitutions.

F. Klein and R. Fricke[35] proved analytically that the number of representations of $4m$ by $x^2+y^2+7z^2+7w^2$ with $x+z$ even is the quadruple of the sum of the divisors prime to 7 of m. Cf. Humbert.[49]

[31] Atti Accad. Pont. Nuovi Lincei, 38, 1884–5, 171–196.
[32] Quar. Jour. Math., 20, 1885, 93.
[33] Jour. de Math., (4), 6, 1890, 5–67.
[34] Annales fac. sc. Toulouse, 6, 1892, G, 19 pp.
[35] Elliptischen Modulfunctionen, 2, 1892, 400.

R. Fricke[36] studied the linear substitutions with integral coefficients of determinant unity which leave $pz_1^2+qz_2^2+rz_3^2-sz_4^2$ invariant, where $p, \ldots, s$ are positive odd integers no two with a common factor.

L. Bianchi[37] gave a new derivation of Fricke's results when $p, \ldots, s$ are primes. By means of linear fractional substitutions on one variable with coefficients in an imaginary quadratic field, he[38] investigated the groups leaving invariant $x_1^2+x_2^2+\mu(x_3^2-\nu x_4^2)$, $x_1^2+Dx_2^2-x_1x_4$, and $\mu(x_1^2+x_2^2)+Dx_3^2-x_4^2$, etc., where μ, ν, D are integers.

*J. P. Bauer[39] obtained limits on the coefficients of reduced forms.

H. Bourget[40] considered the group of automorphs of $u_1^2-Du_2^2+u_3u_4$ and its relation to the hyperabelian group of Picard.[30]

K. Petr[41] evaluated in terms of the class-number of binary quadratic forms the number of solutions of equations like

$$x^2+y^2+z^2+2u^2=n, \qquad x^2+3y^2+3z^2+3u^2=n, \qquad x^2+y^2+z^2+5u^2=8n.$$

For example, if $n=2^\lambda 3^\mu N$ (N odd and prime to 3), the number of solutions of $x^2+y^2+z^2+3u^2=n^2$ is

$$\{2^{\lambda+1}(-1)^{\frac{1}{2}(N-1)}+(-1)^{\lambda+\mu}\}\{3^{\mu+1}(N/3)-(-1)^{\lambda+\mu}\}\Sigma(d_{12m\pm1}-d_{12m\pm5}).$$

Petr[42] enumerated by use of θ-functions the solutions of $x^2+y^2+z^2+5u^2=n$.

G. Cotty[43] considered quaternary quadratic forms f whose coefficients satisfy certain quadratic equations of the theory of abelian functions. Each f has an adjoint binary quadratic form ϕ. If two f's are arithmetically equivalent, their ϕ's are equivalent, but not conversely. f is definite if ϕ is definite and negative. The number of classes of definite forms f of a given discriminant is finite (since true for those whose ϕ's belong to a particular class). To each class corresponds one and only one reduced form. The same questions are treated also for indefinite forms f of the same type.

H. Brandt[44] proved that if a bilinear substitution transforms a quaternary quadratic form A into the product BC of two such forms, there exist two further bilinear substitutions, derived rationally from the given substitution, one of which transforms B into AC, and the other transforms C into AB, so that the three substitutions form a symmetric triple. [For the corresponding theorem for binary quadratic forms, see Dedekind[39] and Speiser[46] of Ch. III.] A form A with rational coefficients is transformable into a product of forms with rational coefficients by means of a bilinear substitution with rational coefficients if and only if the determinant of A is the square of a rational number [cf. Bazin[5]]. For a bilinear substitution with

[36] Göttinger Nachr., 1893, 705–21. Fricke and Klein, Automorphen Functionen, Leipzig, 1, 1897, 577–582.
[37] Atti R. Acc. Lincei, Rendiconti, classe fis. mat., (5), 3, I, 1894, 3–12.
[38] Annali di Mat., (2), 21, 1893, 237–288; 23, 1895, 1–44; Math. Annalen, 42, 1893, 30–57; 43, 1893, 101–135.
[39] Bestimmung des Grenzwertes für das Product der Hauptcoeffizienten in reduzierten quad. quaternären Formen, Diss. Bonn, Leipzig, 1894.
[40] Annales fac. sc. Toulouse, 12, 1898, No. 4, 90 pp.
[41] Rospravy České Akad. Prague, 10, 1901, No. 40 (Bohemian).
[42] Bull. Intern Acad. Sc. Prague, 7, 1903, 180–7 (Abstract of Petr[41]). See pp. 163, 187 above.
[43] Annales fac. sc. Toulouse, (3), 3, 1911, 316–374. Summary in Comptes Rendus Paris, 154, 1912, 266–8, 337–9.
[44] Jour. für Math., 143, 1913, 106–127.

integral coefficients, and for properly primitive forms A, B, C with integral coefficients and with equal determinants, necessary and sufficient conditions for a composition $A=BC$ are that the determinant D of A be a square and that the form adjoint to A be divisible by $\sqrt{D}$.

S. Ramanujan[45] proved that $ax^2+by^2+cz^2+du^2$ represents all positive integers for only 55 sets of positive integers a, b, c, d, including the 12 considered by Liouville[9] and Pepin.[33]

H. S. Vandiver[46] proved that the number of representations of a prime p in the form $xy+zw$, where $x, \ldots, w$ are all positive integers $<\sqrt{p}$, is $-p-1+4\{\phi(1)+\phi(2)+\ldots+\phi(P)\}$, where P is the largest integer $\leqq \sqrt{p}$, and ϕ is Euler's ϕ-function.

G. Giraud[47] discussed the automorphs of $u_1^2+u_2^2-u_3^2-u_4^2$, and the reduction of quaternary quadratic forms, supplementing the conditions of Korkine and Zolotareff[18] of Ch. XI to obtain a unique reduced form.

G. Julia[48] employed a four-dimensional lattice of points with integral coordinates to reduce a quaternary form $f=\Sigma a_{ij}x_ix_j$ to one in which $a_{11}, \ldots, a_{44}$ are the first four proper minima of f and proved that the determinant of the coordinates of the points which furnish these minima is unity.

G. Humbert[49] proved that, if m is a positive integer prime to 10, the number of representations of $4m$ by $x^2+y^2+10z^2+10t^2$ is the quadruple of the sum of the divisors of m (Liouville[22]). He deduced arithmetically the theorem of Klein and Fricke[35] for m odd. He[50] later proved that, if m is a positive odd number prime to 11, the number of representations of $4m$ by $x^2+11y^2+2z^2+22t^2$ is double the sum of the divisors of m. Again,[51] the number of representations of an odd integer m by the totality of the two forms $x^2+6y^2+z^2+6t^2$ and $2x^2+3y^2+2z^2+3t^2$ is the quadruple of the sum of the divisors prime to 3 of m.

For further results see the reports on papers 11, 15, and 25 of Ch. XV.

---

45 Proc. Cambridge Phil. Soc., 19, I, 1916, 11–21.
46 Bull. Amer. Math. Soc., 23, 1916–7, 114.
47 Annales école norm. sup., (3), 33, 1916, 303–330; Comptes Rendus Paris, 163, 1916, 193.
48 Comptes Rendus Paris, 162, 1916, 498–501.
49 Comptes Rendus Paris, 169, 1919, 407–414.
50 *Ibid.*, 170, 1920, 354.
51 *Ibid.*, 170, 1920, 547. Cf. Liouville.[24]

CHAPTER XI.

QUADRATIC FORMS IN n VARIABLES.

This chapter deals with various methods of reduction of quadratic forms, their equivalence, number of classes, transformations into themselves, characters, genera, rational transformation, upper limits of the minima of forms for integral values of the variables, and the representation of integers or quadratic forms by other quadratic forms.

G. Eisenstein[1] stated that the quadratic forms in n variables with a given determinant fall into a finite number of classes.

C. G. J. Jacobi[2] noted that a quadratic form in 3, 4, 5, ... variables can be transformed into one having 1, 3, 6, ... coefficients zero, since a quadratic form $V(x_1, \ldots, x_n)$ is equivalent to one which contains besides the squares z_i^2 only $n-1$ products $z_i z_{i+1}$. Let the terms involving x_n be $x_n(a_1x_1+\ldots+a_{n-1}x_{n-1})+a_nx_n^2$. We can choose linear homogeneous functions $x_1', \ldots, x_{n-1}'$ of $x_1, \ldots, x_{n-1}$ with integral coefficients of determinant unity such that $a_1x_1+\ldots+a_{n-1}x_{n-1}=f_1x_{n-1}'$, where f_1 is the g.c.d. of $a_1, \ldots, a_{n-1}$. Then $V=a_nx_n^2+f_1x_nx_{n-1}'+V_1$, where V_1 is a quadratic form in $x_1', \ldots, x_{n-1}'$. Repeating for V_1, etc., we establish the theorem.

Ch. Hermite[3] proved that, if D is the absolute value of the determinant of a quadratic form $f(x_0, x_1, \ldots, x_n)$ whose coefficients may be irrational, we can assign integral values to the x's such that

$$f<(\tfrac{4}{3})^{n/2}\sqrt[n+1]{D}\equiv M.$$

If D is the absolute value of the determinant of $F(X_0, \ldots, X_n)$ and

$$\tfrac{1}{2}\frac{\partial F}{\partial X_0}=AX_0+BX_1+CX_2+\ldots+LX_n,$$

then F is called *reduced* if

$$|A|<M,\quad |B|<\tfrac{1}{2}|A|,\quad |C|<\tfrac{1}{2}|A|,\ \ldots,\ |L|<\tfrac{1}{2}|A|,$$

and if the adjoint form $G(Y_0, \ldots, Y_n)$ of F becomes for $Y_0=0$ a reduced form. Hence if we can find the reduced quadratic forms in n variables, we can obtain those in $n+1$ variables. The number of reduced forms with integral coefficients of a given determinant is finite.

---

[1] Jour. für Math., 35, 1847, 118–9.

[2] Monatsber. Akad. Wiss. Berlin, 1848, 414–7; Jour. für Math., 39, 1850, 290–2; Werke, 6, 1891, 318–21.

[3] Jour. für Math., 40, 1850, 261–278; Oeuvres, 1, 1905, 100–121. First letter to Jacobi.

If $a_0, a_1, \ldots, a_n$ are the coefficients of the squares in the adjoint of a reduced form F, so that $a_n<(\frac{4}{3})^{n/2}\sqrt[n+1]{D^n}$, it is stated that

$$a_i a_{n-i}<\mu\sqrt[n+1]{D^{n(i+1)}} \qquad (i=1, \ldots, n),$$

where μ depends only on n and i (proof by Stouff[34]).

Hermite[4] gave a simpler method of reducing quadratic forms $f=\Sigma a_{ij}x_ix_j$ in $n+1$ variables which is analogous to Lagrange's method for the binary case. He employed the derived forms in n variables $g_\mu=\Sigma(a_{\mu\mu}a_{ij}-a_{\mu i}a_{\mu j})y_iy_j$, summed for $i, j=0, 1, \ldots, \mu-1, \mu+1, \ldots, n$. Any definite form can be transformed into a form f for which g_μ is reduced and $a_{\mu i}<\frac{1}{2}a_{\mu\mu}(i\neq\mu)$ numerically, while $a_{\mu\mu}$ is the least of the a_{ii}; then f is called reduced. In a definite reduced form the product of the coefficients of the squares is $<(\frac{4}{3})^{n(n+1)/2}D$. Although there is only a finite number of reduced (definite or indefinite) forms for a given determinant, two or more such forms may be equivalent, i. e., a reduced form is not unique in its class. For the case of determinant unity and fewer than 8 variables, there is a single class [the error of including the case of 8 variables was corrected by Minkowski[23]].

Hermite[5] conjectured that the upper limit of the minima of all definite quadratic forms in n variables of determinant D for integral values of the variables is

$$2\sqrt[n]{D/(n+1)},$$

and gave details only for $n\leqq 3$. But for $n=4$ this conjectured limit is less than the limit $\sqrt{2}\sqrt[4]{D}$ obtained by Korkine and Zolotareff.[18] He again (p. 302) defined a reduced form.

Hermite[6] applied to definite ternary quadratic forms f the method of Jacobi[2] to obtain a transformed form whose coefficients are limited in terms of the determinant.

Hermite[7] outlined the parallelism between the equivalence of indefinite quadratic forms in n variables and that of forms decomposable into n linear factors, both as regards algebraic equivalence under real linear transformation and arithmetical equivalence under linear transformation with integral coefficients of determinant unity. An indefinite quadratic form $F(x_1, \ldots, x_n)$ is said to be of the type of *index* j if real linear functions $U_1, \ldots, U_n$ can be chosen so that

$$F\equiv -U_1^2-\ldots-U_j^2+U_{j+1}^2+\ldots+U_n^2.$$

Since there are $\frac{1}{2}n(n-1)$ parameters in the general transformation of F into itself, there are that many "arguments" in the U's. For all values of these arguments suppose we have found all substitutions with integral coefficients of determinant 1 which transform

$$\phi=U_1^2+\ldots+U_n^2$$

into a reduced form (Hermite[4]). Applying these substitutions to F, we obtain an infinitude (F) of forms. Let F have integral coefficients. Then the coefficients of the forms (F) are integers limited in terms of the determinant Δ of F. He stated

[4] Jour. für Math., 279–90; Oeuvres, I, 122–135. Second letter to Jacobi.
[5] *Ibid.*, 291–307; Oeuvres, I, 136–155. Third letter to Jacobi.
[6] *Ibid.*, 308–315; Oeuvres, I, 155–163. Fourth letter to Jacobi.
[7] Jour. für Math., 47, 1854, 330–42; Oeuvres, I, 220–33.

the following theorem whose proof he found only after many attempts: All quadratic forms with integral coefficients which belong to the same type and have the same determinant are reducible to a finite number of classes. All automorphs of an indefinite quadratic form or decomposable form, each with integral coefficients, are products of powers of a finite number of automorphs. Cf. Stouff[9] of Ch. XIV.

G. Eisenstein[8] stated that two forms of any degree in any number of variables can be transformed into each other by a linear substitution with rational coefficients of determinant unity if and only if they belong to the same genus (cf. Smith,[15] Ch. IV,[21, 22], Ch. IX[20]). Given any primitive representation of D by any form Φ, we obtain from Φ an equivalent form Ψ whose first coefficient is D by applying a linear substitution of determinant unity the elements of the first column of whose matrix are the integers defining the representation, and conversely. Hence we obtain all primitive representations of D by Φ if we set up all forms Ψ equivalent to Φ and having D as first coefficient, find all substitutions replacing Φ by Ψ and select the first columns of the matrices.

A quadratic form with the first coefficient D may be written

$$\Psi=\{(Du+\xi)^2-F\}/D, \tag{1}$$

where ξ is linear and F quadratic in the variables $x, y, \ldots$ other than u. The condition that Ψ shall have integral coefficients is evidently that

$$\xi^2\equiv F \pmod{D}, \tag{2}$$

identically in $x, y, \ldots$. Let Φ be a positive n-ary quadratic form of determinant $-\Delta$ ($+\Delta$ in modern notation), so that the determinant of F is $-D^{n-2}\Delta$. Let $F_1, F_2, \ldots$ denote the non-equivalent $(n-1)$-ary forms F for which congruence (2) is solvable. Let ω_k denote the number of incongruent solutions of $\xi^2\equiv F_k \pmod{D}$ such that (1) with $F=F_k$ gives a form equivalent to Φ, whence $\omega_1+\omega_2+\ldots$ is the number of forms Ψ with first coefficient D and equivalent to Φ. Let δ_k be the number of linear transformations of F_k into itself, and ϵ the number for Φ. Then the number of representations of D by Φ is $\Sigma\epsilon\omega_k/\delta_k$. In particular, let Φ be a sum of 4 squares, whence $\Delta=-1$. Let D be a product of μ distinct primes p_k, so that (2) has 2^μ incongruent roots. Hence each ω_k is 2^μ. Also $\epsilon=8\cdot 24$. Using his value for $\Sigma 1/\delta_k$, we find that the number of proper representations of $D=p_1\ldots p_\mu$ as a sum of 4 squares is $8\Pi(p_k+1)$. The corresponding number is found when D has multiple factors or factors 2. The number of improper and proper representations is deduced.

V. A. Lebesgue[9] wrote $f=f(x_1, \ldots, x_n)$ for a quadratic form with real coefficients (not necessarily rational), $f(x_1, x_2)$ for the form obtained from f by taking $x_3=0, \ldots, x_n=0$, and designated by $D_1, D_2, \ldots, D_n$ the determinants of $f(x_1), f(x_1, x_2), \ldots, f$. As known,

$$f=D_1X_1^2+\frac{D_2}{D_1}X_2^2+\ldots+\frac{D_n}{D_{n-1}}X_n^2,$$

[8] Berichte Akad. Wiss. Berlin, 1852, 352, 374–84.
[9] Jour. de Math., (2), 1, 1856, 401–6.

whence f is definite and positive if and only if $D_1>0, \ldots, D_n>0$. If such an f does not exceed a certain limit L for integral values of $x_1, \ldots, x_n$, then, since $X_n=x_n$,

$$x_n<\sqrt{LD_{n-1}/D_n}.$$

Similarly, $x_{n-1}, \ldots, x_1$ are limited. Thus f has a minimum for integral values. This theorem was assumed by Hermite,[3] who deduced from it the existence of reduced forms; Lebesgue found the proofs insufficient since the values of f for integral x's need not be integers and might approach a limit without reaching it. Every definite positive quadratic form with real coefficients is equivalent to a form $f(u_1, \ldots, u_n)$ for which each of the binary forms $f(u_i, u_j)$ is a reduced form.

On pp. 331–5 of Vol. II of this History are quoted J. Liouville's theorems on the number of representations of $4m$ or $8m$ by $s_4+2^a s_4'$ $s_4+2^a s_2$, $s_8+2^a s_4$, where s_n and s_n' are sums of n odd squares, and of m by $2x^2+\sigma$, where σ is a sum of 4 or 6 squares.

Liouville[10] published a series of papers on quadratic forms in six variables. He noted that the number of representations of $2^\alpha 3^\beta m$ (m prime to 6) by σ_5+3v^2, where σ_n (and σ_n' below) is a sum of n squares, is

$$\tfrac{1}{5}(-1)^{2\alpha}\left\{9^{\beta+1}+(-1)^\alpha\left(\frac{m}{3}\right)\right\}\{4^{\alpha+1}-(-1)^\alpha\cdot 9\}\sum_{d\delta=m}\left(\frac{\delta}{3}\right)d^2,$$

and found the number of proper representations. To this problem he reduced that of the number of representations by $x^2+3\sigma_5$, σ_4+q, $3\sigma_4+q$, $\sigma_2+3\sigma_2'+q$, where $q=2u^2+2uv+2v^2$.

He[11] noted that the number of representations of $2^\alpha m$ (m odd) by σ_5+2v^2 is

$$\tfrac{2}{3}\left\{4^{\alpha+2}-\left(\frac{-2}{m}\right)\right\}\sum_{d\delta=m}\left(\frac{-2}{\delta}\right)d,$$

and reduced to this problem that for $2\sigma_5+v^2$.

Liouville[12] noted that the number of representations of an odd integer m by $\sigma_4+2\sigma_2$ is the product of 8 by

$$\rho_2(m)=\sum_{d\delta=m}(-1)^{\frac{1}{2}(\delta-1)}d^2,$$

and that of $2^\alpha m$ ($\alpha>0$) is

$$4\left\{2^{2\alpha+1}-\left(\frac{-1}{m}\right)\right\}\rho_2(m).$$

In terms of $\rho_2(m)$ he expressed the number of representations by $\sigma_2+2\sigma_4$, $x^2+4y^2+2\sigma_4$, $x^2+2\sigma_2+4\sigma_3$, $\sigma_2+2\sigma_2'+4\sigma_2''$, $\sigma_3+2\sigma_2+4v^2$. He[13] treated also $x_1x_2+x_2x_3+x_3x_4+x_4x_5$.

H. J. S. Smith[14] considered an n-ary quadratic f_1 with the symmetrical matrix A_1, its ith derived matrix A_i of $\binom{n}{i}=I$ rows and I columns whose elements are the i-rowed minors of A_1. Let f_i be a quadratic form whose matrix is A_i. It is a concomitant of the ith *species* of f_1. For, if a linear transformation of matrix α_1 replaces f_1 by f_1' and if the transformation whose matrix is the ith derived matrix α_i of α_1

10 Jour. de Math., (2), 9, 1864, 89–128.
11 *Ibid.*, pp. 161–180.
12 *Ibid.*, 257–280, 421–4; (2), 10, 1865, 73–6, 145–150, 155–160.
13 Comptes Rendus Paris, 62, 1866, 714.
14 Proc. Roy. Soc. London, 13, 1864, 199–203; Coll. Math. Papers, I, 412–7.

replaces f_i by f'_i, the matrix of f'_i is the product of the ith derived matrix of f'_1 by a power of the determinant of a_1. Then $f_1, \ldots, f_{n-1}$ are called the fundamental concomitants of f_1. Let the coefficients of f_1 be integers. The *primary divisor* of $\Sigma a_i x_i^2 + 2\Sigma b_{ij} x_i x_j$ $(i<j)$ is the positive g.c.d. of the a_i, b_{ij}; while the *secondary divisor* is the g.c.d. of the a_i, $2b_{ij}$. Thus the g.c.d. ∇_i of the i-rowed minors of A_1 is the primary divisor of f_i.

Consider forms of the same determinant and same *index k of inertia* (number of plus signs in the canonical form $\Sigma \pm x_j^2$). Two such forms are said to belong to the same *order* if the primary and secondary divisors of their corresponding concomitants are identical. Those forms of an order whose particular characters coincide are said to constitute a *genus*. To define these characters, note that the primary divisor of the concomitant ψ_i of the second species of $\theta_i \equiv f_i/\nabla_i$ is the integer

$$I_i = \frac{\nabla_{i+1}}{\nabla_i} \div \frac{\nabla_i}{\nabla_{i-1}}.$$

Let δ_i be any odd prime divisor of I_i. Then the numbers prime to δ_i which are representable by θ_i are either all quadratic residues of δ_i or are all quadratic non-residues of δ_i, and in the respective cases we attribute to f_i the particular character $(\theta_i/\delta_i)=+1$, $(\theta_i/\delta_i)=-1$. This follows from the identity

$$\theta_i(x_1, \ldots, x_I)\cdot\theta_i(y_1, \ldots, y_I) - \tfrac{1}{4}\left\{\sum_{k=1}^{I} y_k \frac{\partial \theta_i}{\partial x_k}\right\}^2 = \psi_i,$$

the variables of ψ_i being $x_j y_k - x_k y_j$ $(j, k=1, \ldots, I; j<k)$. In case the determinant of f_1 is even, there are supplementary characters with regard to 4 or 8, not enumerated here.

Smith[15] perfected his preceding investigation. Let $\nabla_1 = 1$ so that f_1 is primitive. If its index of inertia is k, attribute to the invariant I_k the sign $-$, and to the invariants $I_1, \ldots, I_{k-1}, I_{k+1}, \ldots, I_{n-1}$ the signs $+$. Write $I_0 = I_n = 0$. If the series $I_0, I_1, \ldots, I_n$ present ω different sequences each consisting of an odd number of odd invariants, preceded and followed by even invariants, there are 2^ω assignable orders, which all exist except in specified cases. Rules are given to find the existing supplementary characters with regard to 4 or 8. Only those total characters which satisfy a specified equation correspond to existing forms. Every genus whose character satisfies this equation actually exists. Two forms, having the same invariants, of the same order and of the same genus are transformable into each other by linear substitutions with rational coefficients of determinants unity such that the denominators of the coefficients are prime to any given number.

The second half of the paper relates to the determination of the weight of a given genus of definite n-ary quadratic forms (cf., for $n=3$, Eisenstein[7] of Ch. IX). The concluding applications to sums of 5 and 7 squares are quoted in full on pp. 308–9 of Vol. II of this History.

Smith[16] elaborated and gave proofs of various theorems in his two preceding papers. In particular he enumerated the supplementary and simultaneous characters and the number of solutions of $f \equiv \mu$ (mod p, or 2^δ), investigated the weight (density)

---

[15] Proc. Roy. Soc. London, 16, 1867, 197–208, Coll. Math. Papers, I, 510–23.
[16] Mém. divers savants Institut de France, (2), 29, 1884; Coll. Math. Papers, II, 623–680.

of a class of forms. This memoir was awarded the Grand Prix of the French Academy (along with Minkowski[23]).

R. Lipschitz[17] discussed the asymptotic value of the number $\phi(m)$ of sets of integral solutions without a common factor of $f=m$, $C_1>0, \ldots, C_\rho>0$, where $f, C_1, \ldots, C_\rho$ are any forms in $x_1, \ldots, x_\nu$ with integral coefficients such that $\phi(m)$ is finite for each positive integer m. When f is a quadratic form which takes only positive values and if Δ is its determinant, the median value of the number $\phi(m)$ of primitive representations of m by f is

$$\frac{2^k\pi^{[\nu/2]}m^{\frac{1}{2}\nu-1}}{(\nu-2)\ldots(\nu-2k)\sqrt{\Delta}\sum_{s=1}^{\infty} s^{-\nu}},$$

where $k=[\frac{1}{2}(\nu-1)]$, $[t]$ denoting the greatest integer $\leqq t$.

A. Korkine and G. Zolotareff[18] considered positive real n-ary quadratic forms f of determinant $-D$. For integral values, not all zero, of the variables, a given f has an unique minimum. Let the coefficients vary continuously such that the determinant remains $-D$. Then the minimum varies continuously and takes one or more maxima:

$$2\sqrt[n]{\frac{D}{n+1}}\ (n\gtreqqless 2),\quad \sqrt[n]{2^{n-2}D}\ (n\gtreqqless 3),\quad 2\sqrt[n]{D}\ (n\gtreqqless 8),\quad 2\sqrt[6]{\frac{D}{3}}\ (n=6),\quad \sqrt[7]{64D}\ (n=7).$$

The first limit was conjectured by Hermite.[5] Another limit due to Hermite[3] is proved, as well as the more precise limit:

$$\frac{3^{\frac{1}{2}(m-2)}}{2^{\frac{1}{2}(m-3)}}\sqrt[n]{D}(n=2m),\qquad \sqrt[n]{\frac{3^{m(m-1)}D}{2^{m(m-2)}}}\ (n=2m+1),$$

actually reached when $n=2, 3, 4$ only. Use is made of several particular forms, called *extreme*, whose minima decrease under all infinitesimal variations of the coefficients not altering the determinant.

It is shown that any positive form f can be reduced to

$$AX_1^2+A'X_2^2+\ldots+A^{(n-1)}X_n^2,$$

where

$$X_1=x_1+\alpha x_2+\beta x_3+\ldots+\gamma x_n,\quad X_2=x_2+\delta x_3+\ldots+\zeta x_n,\ \ldots,\quad X_n=x_n,$$

where $\alpha, \beta, \ldots$ are numerically $\leqq \frac{1}{2}$, A being the minimum of f, A' the minimum of $A'X_2^2+\ldots+A^{(n-1)}X_n^2$, A'' the minimum of $A''X_3^2+\ldots$, etc. For details see the report of the extension by Jordan[2] of Ch. XVI to Hermitian forms.

They[19] gave all extreme positive quadratic forms in 2, 3, 4, 5 variables, and proved that the precise limits of the minima of forms of determinant $-D$ are $\sqrt{\frac{4}{3}D}$, $\sqrt[3]{2D}$, $\sqrt[4]{4D}$, $\sqrt[5]{8D}$, respectively. They obtained several theorems on extreme n-ary forms.

---

[17] Berichte Akad. Wiss. Berlin, Jahre 1865, 1866, 174–185. Report by P. Bachmann, Die Analytische Zahlentheorie, 1894, 438–447.
[18] Math. Annalen, 6, 1873, 366–389; Korkine's Coll. Papers, 1, 1911, 289–327.
[19] *Ibid.*, 11, 1877, 242–292; Korkine's Coll. Papers, I, 351–425.

C. Jordan[20] treated first the following question. Given two n-ary quadratic forms F and G with complex integral coefficients $a+bi$ of determinants D and Δ, to decide if F can be transformed into G by a substitution S with complex integral coefficients of determinant δ, and if so to find all such substitutions. It is evidently necessary that Δ/D be the square of a complex integer δ. Any S can be expressed uniquely as a product TU, where U has complex integral coefficients of determinant unity and

$$T=\begin{pmatrix} a_{11} & 0 & 0 \dots 0 \\ a_{21} & a_{22} & 0 \dots 0 \\ \dots & \dots & \dots \end{pmatrix}, \quad a_{11}\dots a_{nn}=\delta, \quad a_{kl}=a_{ll}(p_{kl}+iq_{kl}) \text{ if } k>l,$$

where the p and q are $>-\frac{1}{2}$ and $\leqq \frac{1}{2}$, the a's being complex integers. Thus T is one of a limited set $T, T', \dots$. It remains to find the substitutions U of determinant unity which transform one of $FT, FT', \dots$ into G. Hence the problem is reduced to substitutions of determinant unity. A substitution

$$x'_j=a_{j1}x_1+\dots+a_{jn}x_n \ (j=1, \dots, n)$$

of determinant δ is called *reduced* if

$$\phi=\sum_{j=1}^{n} N(x'_j)=\mu_1 N(x_1+\epsilon_{12}x_2+\dots+\epsilon_{1n}x_n)+ \\ \mu_2 N(x_2+\epsilon_{23}x_3+\dots+\epsilon_{2n}x_n)+\dots+\mu_n N(x_n),$$

identically, where $N(x)$ is the norm of x, while the μ's are positive and

$$\mu_1\dots\mu_n=N(\delta), \quad \mu_{k+1} \geqq \tfrac{1}{2}\mu_k, \quad N(\epsilon_{jk}) \leqq \tfrac{1}{2}.$$

Then also the bilinear form ϕ is called *reduced*. He had proved (*ibid.*, 48, 1880; see Ch. XVI[2]) that every ϕ is properly equivalent to a reduced form and that the coefficients of a substitution which transforms a reduced form into itself are limited in terms of n. Let some reduced substitution transform F into a reduced form $G=\Sigma b_{jk}x_jx_k$. In the first of two cases, the modulus of every b_{jk} is limited, and G is an ordinary reduced form. In the second case, only certain $|b_{ik}|$ are limited and G is a singular reduced form which is equivalent to a simple (unreduced) form whose coefficients are limited. Hence there is a limited number of classes of forms of determinant Δ.

Let F and G be two n-ary quadratic forms of the same determinant whose coefficients are complex integers and limited. It is proved by induction on n that every substitution with integral coefficients of determinant unity which transforms F into G is a product of substitutions of determinant unity, whose coefficients are integers and limited, such that the first substitution transforms F into G, while the others transform G into itself.

The question of the representation of an m-ary quadratic form by an n-ary form is reduced to the above problem of the equivalence of forms and their automorphs. For $m=1$. it is a question of the representation of numbers.

Finally forms of determinant zero are discussed.

[20] Jour. école polyt., 51, 1882, 1–43; extracts in Comptes Rendus Paris, 93, 1881, 113–7, 181–5, 234–7.

H. Poincaré[21] put into the same *order* two algebraically equivalent forms of any degree when the g.c.d. of their coefficients is the same, likewise that of these coefficients affected with multinomial coefficients, as well as the g.c.d. of the coefficients (affected or not with multinomial coefficients) of their concomitants, including covariants and contravariants. He called two n-ary forms *equivalent modulo m* if there exists a linear substitution with integral coefficients whose determinant is $\equiv 1 \pmod{m}$ which transforms the one form into the other modulo m. He put into the same *genus* two algebraically equivalent forms which are equivalent with respect to every modulus (which follows if they are equivalent with respect to every prime power modulus). Forms of the same genus belong to the same order.

The obvious extension of Eisenstein's classification of ternary quadratic forms to n-ary quadratic forms f does not give some of the true characters. But let α_p denote the g.c.d. of all the p-rowed minors of the matrix of f, and $\alpha_p\beta_p$ that of the principal p-rowed minors and the doubles of the non-principal minors. Write

$$\alpha_1=\gamma_1,\quad \alpha_2=\gamma_1^2\gamma_2,\quad \alpha_3=\gamma_1^3\gamma_2^2\gamma_3,\ \ldots,\quad \alpha_{n-1}=\gamma_1^{n-1}\gamma^{n-2}\ldots\gamma_{n-2}^2\gamma_{n-1},\quad \Delta=\gamma_1^n\gamma_2^{n-1}\ldots\gamma_n,$$

where Δ is the determinant of f. Then the three sets of integers α_i, β_i, γ_i ($i=1, \ldots, n-1$) are ordinal *characters* of the first, second, and third kinds respectively. Two forms belong to the same order if and only if they have the same ordinal character of the first and second kind, or hence of the second and third kind.

Since two n-ary quadratic forms belonging to the same order and having the same determinant Δ are always equivalent with respect to any odd modulus prime to Δ, they belong to the same genus if equivalent with respect to any power of 2 and the odd prime factors p of Δ. Let $\lambda_1, \ldots, \lambda_n$ be integers for which γ_i is divisible by p^{λ_i}, but not by p^{λ_i+1}. The *chief* i-rowed minor (whose elements lie in the first i rows and first i columns of the matrix of f) is divisible by p^μ, where

$$\mu=i\lambda_1+(i-1)\lambda_2+\ldots+2\lambda_{i+1}+\lambda_i,$$

and hence is Ap^μ. The integer A may be assumed prime to p after applying a transformation to f. Similarly, the chief i-rowed minor of ϕ is $Bp^{\mu'}$, where B is prime to p. Then f and ϕ are equivalent with respect to an arbitrary power of p as modulus if and only if, for $i=1, \ldots, n-1$, $\mu'=\mu$ and, when $\lambda_{i+1}>0$, A and B are both quadratic residues of p or both non-residues.

As to characters with respect to a power of 2, only the following example is given. Let the first coefficient of f, all its chief minors, and Δ itself be $\equiv 1 \pmod 4$. Then the α's are odd and each $\beta_i=1$. If ϕ is of the same genus as f, we may assume after applying a suitable transformation to ϕ that its chief minors (including its first coefficients and determinant) are all odd. Then f and ϕ are equivalent with respect to an arbitrary power of 2 as modulus if the number of the chief minors of ϕ which are $\equiv 3 \pmod 4$ is divisible by 4. His illustrative results for binary cubic forms are quoted under that topic (Ch. XII[12]).

H. Minkowski[22] proved that an n-ary positive quadratic form $\Sigma a_{ik}x_ix_k$ takes, only for a finite number of sets of integral values of the x's, a value not exceeding a given positive number. Hence among all the forms in the class of a given f occur certain

[21] Comptes Rendus Paris, 94, 1882, 67–69, 124–7.
[22] Comptes Rendus Paris, 96, 1883, 1205–10; revised in Gesammelte Abh., I, 145–8.

forms $\phi=\Sigma a_{ik}\xi_i\xi_k$ for which the first a_{ii} not equal to a_{ii} is $<a_{ii}$. These ϕ's, whose number is usually 1, are called reduced by Hermite.[4] If $n \leqq 4$, ϕ is proved to be reduced if and only if $a_{11} \leqq a_{22} \leqq \dots \leqq a_{nn}$ and

$$\phi(\epsilon_1, \dots, \epsilon_n) \gtreqless a_{ii} \qquad (\epsilon_i=\pm 1, \epsilon_k=0 \text{ or } 1 \text{ or } -1; k\neq i; i=1, \dots, n).$$

Minkowski[23] employed Poincaré's[21] definition of order, genus, and equivalence modulo N, but wrote d_{p-1}, σ_p, o_{p-1} for a_p, β_p, γ_p, and took $d_0=a_1=\gamma_1=1$, thus restricting to primitive forms. Define the *index* I to be the number of negative terms in the canonical form $\Sigma\pm x_i^2$. If ϕ is equivalent to f modulo N, write $\sigma_h d_{h-1}\phi_h$ for the minor determinant and (ϕ_h) for the matrix whose elements lie in the first h rows and first h columns of the matrix of ϕ, and write $\phi_0=1$. There exists a *characteristic* form ϕ of the class of forms equivalent to f modulo N for which ϕ_h is relatively prime to $2o_1 \dots o_{n-1}\phi_{h-1}\phi_{h+1}$ for every h. Let I_h denote the index of the h-ary form whose matrix is (ϕ_h). Write $\epsilon=(-1)^{I_h}$. For all characteristic forms ϕ of the various classes of forms f of a genus, the following units (± 1), called the *characters* of the genus, possess the same values:

$$\left(\frac{\phi_h}{p}\right) \text{ if } \pi=\sigma_{h-1}o_h\sigma_{h+1}\equiv 0 \pmod{p=\text{odd prime}};$$

$$(-1)^{\frac{1}{2}(\phi_h-1)},\ \left(\frac{\phi_{h-1}}{\epsilon_h\phi_h}\right)\cdot(-1)^{I_h(I_h-1)/2},\ \left(\frac{\phi_{h+1}}{\epsilon_h\phi_h}\right)\cdot(-1)^{I_h(I_h+1)/2} \text{ if } \pi\equiv 0 \pmod 4;$$

$$\left(\frac{2}{\phi_h}\right) \text{ if } \pi\equiv 0 \pmod 8,$$

where the symbols are those of Legendre-Jacobi for quadratic residue character. Conversely, two forms f belong to the same genus if, for the characteristic forms ϕ of their classes of forms, these units possess the same values. If the characters of a genus satisfy the conditions implied by a specified congruence, the genus exists.

As a generalization of Gauss' theory of the representation of numbers and binary forms by ternary, there is developed at length a theory of the representation of numbers and m-ary quadratic forms by n-ary forms, especially for $m=n-1$. His determination of the mass (weight) of positive genera was later simplified and generalized by him.[25]

The number of classes of forms in n variables of determinant unity is $\gtreqless [n/8]+1$, where $[x]$ denotes the greatest integer $\leqq x$. Thus there are at least 2 classes if $n=8$, contrary to Hermite.[4]

This memoir, which was written in his seventeenth year, won (along with Smith[16]) the Grand Prix of the French Academy for the problem of the representation of numbers by a sum of five squares. His results on this special problem are quoted on p. 312 of Vol. II of this History, while on p. 327 there is an account of the auxiliary problem of the number of solutions of $f\equiv m \pmod N$.

Minkowski[24] determined the number of classes of n-ary quadratic forms f in a genus by means of Dirichlet's transcendental method for $n=2$, and a further[23] study of the number of solutions of $f\equiv m \pmod N$.

---

[23] Mém. divers savants Institut de France, (2), 29, 1884, No. 2, 180 pp. Original German MS with additions to correspond to the French text, Gesamm. Abh., I, 1-144.

[24] Diss., Königsberg; Acta Math., 7, 1885, 201-258; Gesamm. Abh., I, 157-202.

Minkowski[25] found that the expression for the mass of a genus, obtained by him[23] for special genera and for any genus by Smith,[15] becomes far simpler if we employ the definition given by Poincaré[21] and himself.[23] Write $o_0, o_1, \ldots, o_{n-1}$ for $\gamma_1, \gamma_2, \ldots, \gamma_n$ of Poincaré and restrict attention to positive forms f of matrix (a_{ik}). Let $f(N)$ denote the number of substitutions incongruent modulo N whose determinant is $\equiv 1 \pmod{N}$ which when applied to f leave unaltered all the residues $a_{ik} \pmod{N}$. If t exceeds the exponent of the highest power of the prime q which divides $2\, o_0 o_1 \ldots o_{n-1}$ and if q^U is the highest power of q which divides

$$\prod_{h=0}^{n-1} o_h^{\frac{1}{2}(n-h)(n-h+1)-1},$$

define $f\{q\}$ by

$$q^U f\{q\} = f(q^t)/q^{\frac{1}{2}n(n-1)t}.$$

The mass of the genus of f is the integer

$$\frac{2\prod_{j=1}^{n}\Gamma\left(\frac{j}{2}\right)}{\{\Gamma(\frac{1}{2})\}^{n(n+1)/2}} \cdot \sqrt{\prod_{h=1}^{n-1} o_h^{h(n-h)}} \cdot \frac{1}{f\{2\}\cdot f\{3\}\cdot f\{5\}\ldots},$$

where the final product extends over all primes.

For $n \leqq 5$ he characterized a Hermite[5] reduced form by a finite number of simple linear inequalities. When $\frac{1}{2}n(n+1)-1$ of them become equalities, the reduced form is called a limit form. The latter, when positive, are of the same class as the extreme forms of Korkine and Zolotareff.[18, 19] For $n=6$ see Minkowski.[28]

L. Gegenbauer[26] obtained at once the recursion formula for the number $F_k(n)$ of representations of n by $f = a_1x_1^2 + \ldots + a_kx_k^2$ (a_i positive integer, x_i integer $\gtreqless 0$):

$$\sum_{x=1}^{n} F_{k+1}(x) = \sum_{x=1}^{n-a_{k+1}} F_k(x)\,[1+\sqrt{(n-x)/a_{k+1}}],$$

from which he proved by induction that

$$\sum_{x=1}^{n} F_k(x) = (\pi n)^{k/2}/\{2^k(\tfrac{1}{2}k)\,!\sqrt{a_1\ldots a_k}\} + \epsilon n^{(k-1)/2},$$

where ϵ is finite for all values of n. Hence follows the known mean value of the number of representations of an integer, in the neighborhood of n, by f:

$$\pi^{\frac{1}{2}k} n^{\frac{1}{2}k-1}/\{2^k(\tfrac{1}{2}k-1)\,!\sqrt{a_1\ldots a_k}\}.$$

Minkowski[27] investigated forms of any degree which are transformed into themselves by only a finite number of linear homogeneous transformations S with integral coefficients. The identity I is the only S of finite order which is $\equiv I \pmod{4}$. An S of finite order which is $\equiv I \pmod{2}$ can be transformed by a substitution with integral coefficients of determinant ± 1 into a transformation which multiplies each variable by 1 or -1. It follows that the order of any finite group of n-ary transformations S divides $2^n(2^n-1)(2^n-2)\ldots(2^n-2^{n-1})$.

Minkowski[28] proved that the identity I is the only linear homogeneous transformation with integral coefficients and finite order which is $\equiv I \pmod{p}$, where p is

[25] Jour. für Math., 99, 1886, 1-9; Gesamm. Abh., I, 149-156.
[26] Sitzungsber. Akad. Wiss. Wien (Math.,), 93, II, 1886, 215-221.
[27] Jour. für Math., 100, 1887, 449-58; Gesamm. Abh., I, 203-211.
[28] *Ibid.*, 101, 1887, 196-202; Gesamm. Abh., I, 212-8.

an odd prime [simpler proof, Minkowski[32]]. Let f denote a real positive n-ary quadratic form of determinant $D\neq 0$. Since f can be transformed into a sum of n squares, it is evidently transformed into itself by only a finite number $t(f)$ of linear transformations (*automorphs*) with integral coefficients. By a small variation of the coefficients of f we obtain a positive form the ratios of whose coefficients are all rational and admitting the same automorphs as f. Hence let f have integral coefficients without a common divisor. The group of the $t(f)$ automorphs of f is simply isomorphic with the group of their residues with respect to any odd prime p. Hence $t(f)$ divides the known order of the group of all transformation-residues modulo p whose determinant is $\equiv \pm 1 \pmod{p}$, as well as the order (Minkowski,[24] p. 218) of the group of all transformation-residues modulo p which leave f unaltered modulo p. It follows that $t(f)$ divides the product Πq^k extended over all primes $q=2, 3, 5, 7, \ldots$, where

$$k=\left[\frac{n}{q-1}\right]+\left[\frac{n}{q(q-1)}\right]+\left[\frac{n}{q^2(q-1)}\right]+\ldots,$$

$[a]$ denoting the largest integer $\leqq a$. Furthermore, Πq^k is the least common multiple of all possible numbers $t(f)$. Finally, he extended to $n=6$ his[25] characterization of reduced forms.

Minkowski[29] proved that two n-ary quadratic forms f and f' with rational coefficients of determinants $\neq 0$ can be transformed rationally into each other if and only if the invariants J, A, B have the same value for each. Here J denotes the number of negative terms in the canonical form $\Sigma \pm x_i^2$ of f. Under a rational transformation the determinant Δ of f is multiplied by the square of the (rational) determinant of the transformation, whence the totality of the primes occurring in Δ to odd powers is invariant. Write A for the product of these primes prefixed with the sign of $(-1)^J$ or, when such primes are absent, write $A=(-1)^J$. The invariant B is the product of those odd primes p for which a certain unit $C_p=\pm 1$ has the value -1, the definition and expression for C_p being rather complicated.

Two n-ary quadratic forms can be transformed rationally into rational multiples of each other if their determinants are $\neq 0$, and $n-2J$ and D have the same absolute values for each. The invariant D is, for n odd, the value of the invariant B of the form Af, while its definition is more complicated for n even.

Special cases and corollaries to these two theorems are noted. Zero is represented rationally by every indefinite quadratic form in 5 or more variables, by every one in 4 variables if D has no square factor, by one in 3 variables if $D=1$, and by one in 2 variables if $D=-1$.

Minkowski[30] considered an essentially positive quadratic form f in $x_1, \ldots, x_n$. It becomes $\xi_1^2+\ldots+\xi_n^2$ under a real transformation

$$\xi_a=\pi_{a1}x_1+\ldots+\pi_{an}x_n, \qquad |\pi_{ab}|\neq 0.$$

Interpret $\xi_1, \ldots, \xi_n$ as coordinates of a point P, of an n-fold space, such that the square of the linear element from P is the sum of the squares of the differentials $d\xi_1, \ldots, d\xi_n$. Let $P_1, \ldots, P_n$ be the points for which a single one of $x_1, \ldots, x_n$ is 1

[29] Jour. für Math., 106, 1890, 5–26; Gesamm. Abh., I, 219–239.
[30] Jour. für Math., 107, 1891, 278–297; Gesamm. Abh., I, 243–260.

and the others are 0. Let $\mathfrak{p}_1, \ldots, \mathfrak{p}_n$ denote the vectors from the origin O of coordinates to $P_1, \ldots, P_n$. Then $\Sigma x_i \mathfrak{p}_i$ denotes the vector from O to the point determined by $x_1, \ldots, x_n$. The vectors $\mathfrak{p}_1, \ldots, \mathfrak{p}_n$ determine an n-dimensional parallelopiped. It with similar parallelopipeds fill the entire space. Their vertices give all points for which $x_1, \ldots, x_n$ are integers, and form a regular lattice L. To the fundamental parallelopiped F of a lattice therefore corresponds the quadratic form f. To all possible arrangements of the points of the lattice L into parallelopipeds of the same volume as F correspond a class of equivalent forms. We are led geometrically also to the existence of certain limits to the minimum M of f, including $M < n\sqrt[n]{D}$, which imply important results on algebraic numbers.

A. Meyer[31] proved by induction on n that two properly primitive indefinite n-ary quadratic forms of odd invariants $o_1, \ldots, o_{n-1}$ (and $\sigma_1 = \ldots \sigma_{n-1} = 1$) are properly or improperly equivalent if they belong to the same genus and if two successive terms of $o_1, \ldots, o_{n-1}$ are relatively prime, the theorem being known and presupposed for $n=3$. Use is made of the notations of Minkowski.[23]

H. Minkowski[32] proved that the order of a finite group of linear substitutions on n variables with integral coefficients is always $\leqq (2^{n+1}-2)^n$. This is a limit to the number of such automorphs of a positive n-ary quadratic form. Integral values, not all zero, may be assigned to the n variables of a positive definite quadratic form f of determinant D such that

$$f \leqq \frac{4}{\pi} \{\Gamma(1+\tfrac{1}{2}n)\}^{2/n} \sqrt[n]{D},$$

where Γ denotes the ordinary gamma function [this is the case $p=2$ of a theorem on sums of pth powers of linear forms quoted in this History, Vol. II, p. 95]. The number of classes of positive quadratic forms of given determinant is finite.

P. Bachmann[33] gave a systematic exposition of quadratic forms.

X. Stouff[34] proved the final statement in the report of Hermite.[3]

Stouff[35] proved Hermite's[4] theorem that in a reduced definite n-ary quadratic form of determinant D the product of the coefficients of the squares is $< \mu D$, where μ depends only on n; and Hermite's[7] statement that the coefficients of the forms (F) have limits depending only on D.

G. Humbert[36] established connections between the arithmetical theory of quadratic forms and the theory of singular abelian functions. The normal periods $(1, 0)$, $(0, 1)$, (g, h), (h, g') of an abelian function of two variables are said to satisfy a *singular relation* if

$$F \equiv Ag + Bh + Cg' + D(h^2 - gg') + E = 0$$

for integers $A, \ldots, E$ which may be taken free of a common factor. If there is a single such relation any transformation of order $n=1$ of the periods changes it into

31 Vierteljahrsschrift Natur. Gesell. Zürich, 36, 1891, 241–250.
32 Geometrie der Zahlen, Leipzig, 1896, 180–7, 122–3, 196–9.
33 Die Arithmetik der Quad. Formen, Leipzig, 1898, 371–668.
34 Bull. sc. math., (2), 26, I, 1902, 302–308.
35 Annales sc. école norm. sup., (3), 19, 1902, 89–118.
36 Jour. de Math., (5), 9, 1903, 43–137 [5, 1899, 233–7]. Cf. E. Hecke, Math. Annalen, 71, 1912, 1–37; 74, 1913, 465–510.

an analogous relation $A_1G+B_1H+\ldots+E_1=0$ in the new periods (G, H), etc., where

$$A_1=A(db)_{31}+B(ad)_{31}+C(ac)_{31}+D(cd)_{31}+E(ab)_{31},$$

with similar expressions for $B_1, \ldots, E_1$, where $(ab)_{ij}\equiv a_ib_j-a_jb_i$, while $a_i, \ldots d_i$ are integers verifying the classic relations $(ad)_{12}+(bc)_{12}=1$, etc., and hence are the 16 coefficients of an abelian substitution on 4 variables. Then $\Delta=B^2-4AC-4DE$ is invariant. Changing the notation, we conclude that $f=x^2-4yz-4tu$ is transformed into itself by

$$X=x\{2(ad)_{03}-1\}+2y(db)_{03}+2z(ac)_{03}+2t(cd)_{03}+2u(ab)_{03},$$
$$Y=x(ad)_{31}+y(db)_{31}+z(ac)_{31}+t(cd)_{31}+u(ab)_{31}, \ldots.$$

Thus, given one representation of a positive integer δ by f, we obtain all representations by these formulas. To deduce the representations of δ, when $\delta\equiv 0$ or 1 (mod 4), by $x^2+\eta^2+\zeta^2-\tau^2-v^2$, we write

$$\tau+\eta=2y, \quad \tau-\eta=2z, \quad v+\zeta=2t, \quad v-\zeta=2u;$$

and get $\delta=f$. The representations of binary quadratic forms by f is discussed (p. 134).

In Part II it is assumed there are two singular relations $F=0$ and

$$F_1\equiv A_1g+B_1h+C_1g'+D_1(h^2-gg')+E_1=0.$$

Let $\Delta_1=B_1^2-\ldots$ be the invariant of F_1. The invariant of $xF+yF_1=0$ is $Q=\Delta x^2+2\delta xy+\Delta_1y^2$, where $\delta=BB_1-2AC_1-2CA_1-2DE_1-2ED_1$. If we replace the system $F=F_1=0$ by an arithmetically equivalent system,

$$\lambda F+\mu F_1=0, \qquad \lambda' F+\mu' F_1=0, \qquad \lambda\mu'-\lambda'\mu=\pm 1,$$

where $\lambda, \ldots$ are integers, the corresponding quadratic form is obtained from Q by replacing x by $\lambda x'+\lambda' y'$ and y by $\mu x'+\mu' y'$. Hence to every system of two singular relations $F=F_1=0$ corresponds a *class* of positive binary quadratic forms Q. Conversely, if two systems of two singular relations lead to equivalent forms Q and Q', is one system reducible to the other by an ordinary abelian transformation? The answer (pp. 81, 116, 130) is not as simple as implied in his[37] preliminary note.

Take the three absolute invariants of a binary sextic Φ as the modules of abelian functions related to $\sqrt{\Phi}$ and also as the Cartesian coordinates of a point M in space (p. 91). If there are two singular relations between the periods of these abelian functions, M describes a skew hyperabelian space curve which therefore corresponds to a class of positive binary quadratic forms. Consider (pp. 111, 131) the classes of positive primitive binary quadratic forms belonging to the same genus whose determinant is the same odd or double of the same odd number; let $\phi_1, \phi_2, \ldots$ be forms selected one from each class; then the hyperabelian curves associated with the classes of forms $4\phi_1, 4\phi_2, \ldots$ are of the same genus and correspond point to point. In this connection it is shown that ϕ_1 and ϕ_2 belong to the same genus if $z^2-\phi_1$ and $z^2-\phi_2$ are equivalent.

[37] Comptes Rendus Paris, 134, 1902, 876–882; 136, 1903, 717–23.

Humbert[38] in Part III assumed that the periods satisfy three singular relations $F_0=F_1=F_2=0$. The Δ invariant for $xF_0+yF_1+zF_2=0$ is a positive ternary quadratic form T which becomes an equivalent form when we replace the initial system $F_0=F_1=F_2=0$ by an arithmetically equivalent system $\lambda_i F_0+\mu_i F_1+\nu_i F_2=0$ $(i=1, 2, 3)$, where $\lambda_1, \ldots, \nu_3$ are integers of determinant ± 1, or when we apply an ordinary abelian transformation of degree 1. Hence to each system corresponds a class of forms T. Such a form is not an arbitrary positive ternary form since it is representable properly by $x^2-4yz-4tu$. The ternary forms so representable are studied at length, also in connection with hyperabelian curves and surfaces.

H. Minkowski[39] simplified Hermite's method of reduction. The form

$$f(x_1, \ldots, x_n)=\Sigma a_{hk}x_h x_k \qquad (a_{kh}=a_{kh} \text{ if } h\neq k)$$

is called *reduced* if $a_{12} \geqq 0$, $a_{23} \geqq 0, \ldots, a_{n-1n} \geqq 0$ and if

$$f(s_1^{(l)}, \ldots, s_n^{(l)}) \geqq a_{ll} \qquad (l=1, \ldots, n)$$

for every set of integers $s_1^{(l)}, \ldots, s_n^{(l)}$ such that the g.c.d. of $s_l^{(l)}, s_{l+1}^{(l)}, \ldots, s_n^{(l)}$ is unity. This infinitude of inequalities reduces to a finite system. Every definite form f is equivalent to one and but one reduced form.

Consider the positive form *

$$f=\xi_1^2+\ldots+\xi_n^2, \qquad \xi_j=\sum_{k=1}^{n} a_{jk}x_k,$$

the coefficient of $2x_h x_k$ in f being $a_{hk}=\Sigma_{j=1}^{j=n} a_{jh}a_{jk}$. In the space A of the $\frac{1}{2}n(n+1)$ arbitrary real variables a_{hk}, every point (a_{hk}) for which f is an essentially positive form corresponds, in view of the preceding relations, to a domain $A(f)$, of $n^2-\frac{1}{2}n(n+1)$ dimensions, of points (a_{hk}). In A we seek a domain B such that every class of positive quadratic forms is represented by a point of B, and, when the point is not on the boundary of B, by no other point of B. We can choose for B a convex cone bounded by a finite number of planes and having as its vertex the origin $x_1=0, \ldots, x_n=0$. The part of B which corresponds to forms f of determinant $\leqq 1$ has a finite volume. With this volume are connected certain asymptotic expressions for the number of classes of forms f. Application is made to the finding of all extreme forms (Korkine and Zolotareff[18, 19]).

L. E. Dickson[40] spoke of a form or substitution as being in a field (domain of rationality) F if its coefficients are all numbers of F. Within any field F, not having modulus 2 (so that $2x$ does not count as zero), any n-ary quadratic form whose determinant is not zero can evidently be transformed into

$$q=\sum_{i=1}^{n} a_i x_i^2 \qquad (\text{each } a_i \neq 0).$$

An obvious necessary condition that q be equivalent to $Q=\Sigma a_i X_i^2$ under linear transformation in F is that a_1 be representable by q, viz., that there exist solutions b_i in F

* We may regard the system of linear forms $\xi_1, \ldots, \xi_n$ as reduced if f is.

38 Jour. de Math., (5), 10, 1904, 209–273; Comptes Rendus Paris, 134, 1902, 1261–6.

39 Jour. für Math., 129, 1905, 220–274; Gesamm. Abh., II, 53–100. (Application to finiteness of classes of linear groups by L. Bieberbach, Göttingen Nachr., 1912, 207–216 (Fortschritte, 1912, 197, for gap in proof).)

40 Bull. Amer. Math. Soc., 14, 1907, 108–115.

of $\alpha_1 = \Sigma a_i b_i^2$. Assume that this condition is satisfied and that F does not have a modulus. After a suitable permutation of $x_1, \ldots, x_n$, we may write

$$W_k \equiv \sum_{i=1}^{k} a_i b_i^2 \neq 0 \qquad (k=1, \ldots, n).$$

Then the substitution

$$x_i = b_i y_1 + W_{i-1} y_i - b_i \sum_{j=i+1}^{n} a_j b_j y_j \qquad (i=1, \ldots, n)$$

has a determinant $\neq 0$ and replaces q by

$$q' = \alpha_1 y_1^2 + \sum_{j=2}^{n} a_j W_j W_{j-1} y_j^2.$$

It is shown that q' and Q, with like first coefficients, are equivalent in F if and only if

(3) $$\sum_{i=2}^{n} \alpha_i X_i^2 = \sum_{j=2}^{n} a_j W_j W_{j-1} y_j$$

under a transformation in F on $n-1$ variables. Hence q and Q are equivalent in F if and only if α_1 is representable by q and the forms (3) are equivalent under $(n-1)$-ary transformation in F. The final criteria are therefore that $\alpha_1, \alpha_2, \ldots, \alpha_n$ be representable by certain forms in $n, n-1, \ldots, 1$ variables, respectively, whose coefficients are given functions of the a_i. For example, if $n=3$, the conditions are that α_1 be representable by q, and α_2 by $a_1 a_2 W_2 \xi^2 + a_3 \alpha_1 W_2 \eta^2$, and that $a_1 a_2 a_3 \alpha_1 \alpha_2 \alpha_3$ be a square in F.

If F is the field of all rational numbers, there exist rational values of $b_1, \ldots, b_4$ such that $\Sigma a_i b_i^2 = +1$ or -1, according as $a_1, \ldots, a_4$ are not all negative or are all negative. Hence any n-ary quadratic form with rational coefficients of determinant $\neq 0$ is reducible by a linear substitution with rational coefficients to one of the types

$$f_{p,a,b,c} \equiv \sum_{i=1}^{p} x^2 - \sum_{i=p+1}^{n-3} x^2 + a x_{n-2}^2 + b x_{n-1}^2 + c x_n^2,$$

in which a, b, c are all negative if $p < n-3$, while $f_{p,a,b,c}$ is reducible to $f_{p,\alpha,\beta,\gamma}$ if and only if $ax^2+by^2+cz^2$ is reducible to $\alpha X^2 + \beta Y^2 + \gamma Z^2$ by a substitution with rational coefficients.

Dickson[41] had obtained part of the preceding results less simply. He also determined all quadratic forms in a general field which are invariant under a given substitution and proved that their reduction to canonical types depends upon the above problem of the normalization of a fixed quadratic form.

G. Voronoï[42] gave new applications to quadratic forms of Hermite's[3, 7] principle of continuous parameters (Hermite[53] of Ch. I). Let $\phi = \Sigma a_{ij} x_i x_j$ be any n-ary positive quadratic form. For $k=1, \ldots, s$, let $(l_{1k}, \ldots, l_{nk})$ be the different representations of the minimum M of ϕ, taking one of two sets $(\pm l_{1k}, \ldots, \pm l_{nk})$, considered not distinct. First, let

(4) $$\sum_{i,j=1}^{n} a_{ij} l_{ik} l_{jk} = M \qquad (k=1, \ldots, s),$$

considered as equations in a_{ij}, have an infinitude of sets of solutions a_{ij}, whence

[41] Trans. Amer. Math. Soc., 7, 1906, 275–280, 285–292.
[42] Jour. für Math., 133, 1907, 97–178.

there is an infinitude of sets of values not all zero of the parameters $p_{ij}=p_{ji}$ $(i, j=1, \ldots, n)$ satisfying $\Sigma p_{ij}l_{ij}l_{jk}=0$ for $k=1, \ldots, s$. Write $\psi=\Sigma p_{ij}x_ix_j$ and

(5) $$f=\phi+\rho\psi \qquad (\rho \text{ arbitrary}).$$

This f is a positive form if and only if ρ lies in a certain interval $-R'<\rho<R$. If $R=+\infty$, then $-R'$ is finite. Replacing ψ by $-\psi$ in (5), we have the interval $-R<\rho<R'$. Hence we may assume that R is finite. It is shown that the set (f) of positive forms (5) with $0<\rho<R$ contains a form $\phi_1=\phi+\rho_1\psi$ determined by the conditions that all the representations of the minimum M of ϕ are also representations of the minimum M of ϕ_1, while ϕ_1 possesses at least one further representation of M. Hence there is a series $\phi, \phi_1, \phi_2, \ldots$ of positive quadratic forms such that, if s_k is the number of representations of the minimum of ϕ_k, then $s<s_1<s_2<\ldots$. But such a series terminates since the number of different representations of the minimum of an n-ary positive quadratic form is $\leqq 2^n-1$. If the series terminates with ϕ_k, the latter is determined by the representations of its minimum and is called a *perfect* form.

Let ϕ itself be perfect. Then (4) have a single set of solutions $a_{ij}=\alpha_{ij}M$, where the α_{ij} are rational, so that ϕ/M has rational coefficients. Perfect forms with proportional coefficients are not regarded as different.

Evidently any linear substitution with integral coefficients of determinant ± 1 transforms any perfect form into a perfect form. The number of classes of equivalent perfect forms is proved to be finite.

Given a positive integer σ, consider the domain R composed of all the points $(x_1, \ldots, x_n)$ for which

(6) $$y_k(x)\equiv p_{1k}x_1+\ldots+p_{mk}x_m \gtreqless 0 \qquad (k=1, \ldots, \sigma).$$

A point for which each $y_k(x)>0$ is said to be interior to R. If R has an interior point, R is said to have m dimensions. For $\mu=1, \ldots, m-1$, define a *face* of μ dimensions of R to be a domain $P(\mu)$ formed of the points of R for which $y_k(x)=0$ $(k=1, \ldots, \tau)$, provided these equations define a domain of μ dimensions composed of all the points which do not also satisfy one of the equations $y_j(x)=0$ $(j=\tau+1, \ldots, \sigma)$.

To the m-dimensional domain R defined by (6) corresponds an m-dimensional domain $\mathfrak{R}$ formed of all the points (x) for which

(7) $$x_i=\sum_{k=1}^{\sigma} \rho_k p_{ik}, \quad \rho_1 \gtreqless 0, \ldots, \rho_\sigma \gtreqless 0 \quad (i=1, \ldots, m).$$

Conversely, the domain $\mathfrak{R}$ determines the corresponding domain R.

Let $(l_{1k}, \ldots, l_{nk})$, for $k=1, \ldots, s$, denote all the representations of the minimum of any perfect form ϕ, and write $\lambda_k=l_{1k}x_1+\ldots+l_{nk}x_n$. Consider

$$f=\Sigma a_{ij}x_ix_j=\sum_{k=1}^{s} \rho_k\lambda_k^2, \quad \rho_k \gtreqless 0,$$

whose coefficients are

$$a_{ij}=\sum_{k=1}^{s} \rho_k l_{ik}l_{jk}, \quad \rho_1 \gtreqless 0, \ldots, \rho_s \gtreqless 0.$$

To compare these relations with (7), regard the $\frac{1}{2}n(n+1)$ distinct a_{ij} as the former variables x_i. Hence there is defined a domain $\mathfrak{R}$ and then a corresponding domain R, determined by linear inequalities (6). This $\frac{1}{2}n(n+1)$-dimensional domain R is called the domain corresponding to the perfect form ϕ.

Let ϕ be transformed into ϕ' by a substitution $x_i=\Sigma a_{ik}x_k'$ with integral coefficients of determinant ±1. The adjoint substitution $\Sigma a_{ki}x_k=x_i'$ transforms R into R', where R and R' are the domains corresponding to the perfect forms ϕ and ϕ'. The set (R) of all domains corresponding to all n-ary perfect forms can be separated into as many classes of equivalent domains as there are classes of perfect forms.

If a quadratic form f is interior to a face $P(\mu)$ of μ dimensions of R, f belongs only to those domains of the set (R) which are contiguous to the face $P(\mu)$. There is a single domain contiguous to R by a face of $\frac{1}{2}n(n+1)-1$ dimensions. Hence we can find a series $R, R_1, R_2, \ldots, R_{\tau-1}$ of domains no two of which are equivalent and such that every domain contiguous to one of the series is equivalent to one of them. This series gives a complete system of representatives of the different classes of the set (R).

A positive quadratic form is called *reduced* if it belongs to any domain of such a series giving a complete system. A reduced form can be transformed into another or the same reduced form only by a substitution which transforms into itself a domain or a face of a domain of the series $R, \ldots, R_{\tau-1}$.

All binary perfect forms constitute a single class of forms equivalent to x^2+xy+y^2. The domain R of the latter is composed of the forms

$$ax^2+2bxy+cy^2=\rho x^2+\rho'y^2+\rho''(x-y)^2, \qquad \rho\geqq 0,\ \rho'\geqq 0,\ \rho''\geqq 0.$$

Thus R is determined by the conditions $a+b\geqq 0$, $-b\geqq 0$, $c+b\geqq 0$. This agrees with Selling's definition of a reduced form (Ch. I[92]). By use of a modified domain, we obtain Lagrange's conditions for a reduced form. We are led similarly to Selling's conditions for a reduced positive ternary form. All perfect forms in 4 (5) variables fall into 2 (3) classes.

Let M be the minimum and D the determinant of a positive n-ary quadratic form $f=\Sigma a_{ij}x_ix_j$. Then $f/\sqrt[n]{D}$ has determinant unity and the minimum $\mathfrak{M}=M/\sqrt[n]{D}$. When $\mathfrak{M}$ is a maximum, f is called an *extreme* form. It is proved that f is extreme if and only if it is perfect and if its adjoint form (with the coefficients $\partial D/\partial a_{ij}$) is interior to the domain corresponding to f. There exist forms in $n\geqq 6$ variables which are perfect without being extreme.

Voronoï[43] investigated the conditions that

$$\sum_{i,j=1}^{n} a_{ij}x_ix_j+2\sum_{i=1}^{n} a_ix_i\geqq 0 \quad (a\text{'s arbitrary parameters}),$$

for all sets of integers $x_1, \ldots, x_n$, when $f=\Sigma a_{ij}x_ix_j$ is a positive form. Interpret $a_1, \ldots, a_n$ as coordinates of a point in n-space. For $n=2$, Dirichlet[51] of Ch. I noted that the conditions determine a hexagon with three pairs of parallel edges. The hexagon is here replaced by a convex polyhedron R in n-space, determined by 2τ independent inequalities

$$\Sigma a_{ij}l_{ik}l_{jk}\pm 2\Sigma a_il_{ik}\geqq 0 \qquad (k=1, \ldots, \tau).$$

The systems $\pm(l_{1k}, \ldots, l_{nk})$ of integers for which the equality signs hold include all

[43] Jour. für Math., 134, 1908, 198–287.

representations of the minimum of f, as well as all sets representing n consecutive minima of the classic literature. By means of translations we obtain from R congruent polyhedra (R) which together fill n-space uniformly. A vertex (a_i) is called *simple* if it belongs to only $n+1$ polyhedra of (R). A polyhedron is called *primitive* if all its vertices are simple. A vertex (a_i) is defined by $n+1$ equations

$$\Sigma a_{ij}l_{ik}l_{jk}+2\Sigma a_i l_{ik}=A \qquad (k=0, 1, \ldots, n).$$

Define the corresponding *simplex* L as the totality of points:

$$x_i=\sum_{k=0}^{n}\theta_k l_{ik}, \qquad \sum_{k=0}^{n}\theta_k=1,\ \theta_k \gtreqless 0\ (i=1, \ldots, n).$$

The totality (L) of the simplexes corresponding to the summits of the set (R) of all primitive polyhedra fills n-space uniformly. All quadratic forms which define primitive polyhedra of the type characterized by (L) are interior to a domain D of quadratic forms of $\frac{1}{2}n(n+1)$ dimensions defined by linear inequalities. Details are given for $n=2, 3, 4$.

Voronoï[44] investigated the domains of quadratic forms corresponding to different types of primitive polyhedra. Corresponding to the different incongruent faces of $n-1$ dimensions of the simplexes belonging to (L), there exist numbers

$$\rho_k=\sum_{i,j=1}^{n} p_{ij}^{(k)}a_{ij} \qquad (k=1, \ldots, \sigma),$$

called regulators. The fundamental theorem (p. 96) states that $\Sigma a_{ij}x_ix_j$ defines a set (R) of primitive polyhedra belonging to the type characterized by (L) if and only if each $\rho_k>0$. The final twenty pages are devoted to polyhedra of 2, 3, 4 dimensions.

L. E. Dickson[45] reduced the problem of the equivalence of two pairs of quadratic forms with coefficients in any given field to that for single quadratic forms (Dickson[40]).

K. Petr[46] gave an elementary proof by induction of Hermite's theorem on the minimum of a quadratic form, as well as the following theorem. If in the definition of a reduced positive quadratic form f of discriminant D, we assume that a_{11} is the minimum of f and that, among all forms of the same class, the product $a_{11}a_{22}\ldots a_{nn}$ has for the reduced form the least value, then, for the reduced form,

$$a_{11}a_{22}\ldots a_{nn}<\left(\tfrac{4}{3}\right)^{n(n-1)/2}|D|.$$

T. Astuti[47] gave a simple proof that

$$\lambda(a+b+c)^2+2\mu(a^2+b^2+c^2+\tfrac{1}{2}f^2+\tfrac{1}{2}g^2+\tfrac{1}{2}h^2)$$

is a definite positive form if and only if $3\lambda+\mu>0$, $\mu>0$.

H. Blichfeldt[48] proved by means of a new principle in the geometry of numbers that a positive definite quadratic form in n variables and of determinant D has a value not numerically greater than

$$\frac{2}{\pi}\left\{\Gamma\left(1+\frac{n+2}{2}\right)\right\}^{2/n}\sqrt[n]{D}$$

---

[44] Jour. für Math., 136, 1909, 67–181.
[45] Trans. Amer. Math. Soc., 10, 1909, 347–360.
[46] Casopis, 40, 1911, 485–7 (Fortschritte der Math., 1911, 240).
[47] Annaes Sc. Acad. Polyt. do Porto, 8, 1913, 119–120.
[48] Trans. Amer. Math. Soc., 15, 1914, 227–235; Bull. Amer. Math. Soc., 25, 1918–9, 449–453.

for integral values, not all zero, of the variables. The asymptotic value $n\sqrt[n]{D}/\pi e$ of this expression is half of that of Minkowski's[32] limit.

E. Landau[49] considered a definite positive real quadratic form $Q(u_1, \ldots, u_k)$ of determinant D, and the number $A(x)$ of sets of integers $u_1, \ldots, u_k$ for which $Q \leqq x$. He proved that

$$(8) \qquad A(x)=\alpha x^{k/2}+O(x^s) \quad s=\tfrac{1}{2}(k-1)k/(k+1),$$

where α is the volume

$$V=\pi^{k/2}/\{\sqrt{D}\Gamma(1+\tfrac{1}{2}k)\}$$

of the k-dimensional ellipsoid $Q=1$. Here Γ denotes the gamma function, and $Of(x)$ means a function whose quotient by $f(x)$ remains numerically less than a fixed finite value for all sufficiently large values of x. To generalize to sets of integers $u_1, \ldots, u_k$ for which not only $Q \leqq x$, but also $u_1 \equiv z_1 \pmod{M_1}, \ldots, u_k \equiv z_k \pmod{M_k}$, where the M's are positive integers and the z's are integers, we have only to replace α by $V/(M_1 \ldots M_k)$ in (8).

Landau[50] gave another proof of his preceding results.

G. Giraud[51] investigated the linear transformations of $x_1x_5+x_2x_4+x_3^2$ into itself (cf. Humbert[36]), as well as those of $u_1^2+u_2^2+u_3^2-u_4^2-u_5^2$.

G. Humbert[52] indicated a method to obtain arithmetically Liouville's[12] results on forms $\sigma_4+2\sigma_2$, $\sigma_2+2\sigma_4$, σ_5+2v^2 and $\sigma_3+2\sigma_3'$ in six variables.

E. T. Bell[53] recalled that in 1860–1864 Liouville stated many theorems on the number $T(n)$ of representations of n by quadratic forms in 4 and 6 variables and the number $P(n)$ of proper representations. Write $f(n)$ for $(-1)^{\pi(n)}$ or 0, according as n is or is not the square of an integer not divisible by a square >1, where $\pi(n)$ is the number of distinct prime factors of n. Then $P(n)=\Sigma T(d)f(n/d)$, summed for the divisors d of n. By means of a formula which generalized all of Liouville's, we may express $P(n)$ in terms of the function T for various arguments, and hence compute $P(n)$ from T.

A. Walfisz[54] expressed as an infinite series involving Bessel's functions the sum of the numbers of representations of 1, 2, ..., x by a positive definite quadratic form in κ variables with integral coefficients, generalizing the result for $\kappa=2$ by Hardy[178] of Ch. I.

---

49 Sitzungsber. Akad. Wiss. Berlin, 1915, 458–476.
50 Sitzungsber. Akad. Wiss. Wien (Math.), 124, IIa, 1915, 445–468.
51 Annales sc. école norm. sup., (3), 32, 1915, 237–403; 33, 1916, 331–362.
52 Comptes Rendus Paris, 169, 1919, 407–14; 172, 1921, 505–511; Jour. de Math., (8), 4, 1921, 11–35.
53 Annals of Math., (2), 21, 1919–20, 166–179; Jour. de Math., (8), 2, 1919, 249–271; Comptes Rendus Paris, 169, 1919, 711–2.
54 Über die summatorischen Funktionen einiger Dirichletscher Reihen, Diss. Göttingen, 1922, p. 55.

CHAPTER XII.

BINARY CUBIC FORMS.

G. Eisenstein[1] considered the cubic form

$$f=(a, b, c, d)=ax^3+3bx^2y+3cxy^2+dy^3$$

with integral coefficients. Its *corresponding* (determining) quadratic form is

$$F=Ax^2+Bxy+Cy^2, \qquad A=b^2-ac, \qquad B=bc-ad, \qquad C=c^2-bd.$$

[The Hessian of f is $-F$.] The Gaussian determinant $D=B^2-4AC$ of $2F$ is called the *determinant* of f. Let ω be the g.c.d. of a, b, c, d; ω_1 that of $a, 3b, 3c, d$; and Ω that of A, B, C. Then ω^2 divides Ω, and Ω^2 divides D.

If the substitution with integral coefficients

$$(1) \qquad \binom{\alpha\ \beta}{\gamma\ \delta}: \ x=\alpha x'+\beta y', \quad y=\gamma x'+\delta y', \quad \alpha\delta-\beta\gamma=\epsilon\neq 0,$$

replaces f by $f'=(a', b', c', d')$, f' is said to be *contained* in f, and the substitution is called *proper* or *improper*, according as ϵ is positive or negative. If $\epsilon=\pm 1$, the inverse of the substitution is $\binom{\pm\delta\ \mp\beta}{\mp\gamma\ \pm\alpha}$, which therefore transforms f' into f, and f, f' are called *equivalent*, properly or improperly according as $\epsilon=+1$ or $\epsilon=-1$. All (properly) equivalent cubic forms constitute a *class* and have the same values for both ω and ω_1.

If (1) transforms f into f', $\binom{\epsilon\alpha\ \epsilon\beta}{\epsilon\gamma\ \epsilon\delta}$ transforms F into the quadratic form F' which corresponds to f'. The determinant of f' equals $\epsilon^6 D$. In particular, equivalent cubic forms have the same determinant and equivalent corresponding quadratic forms. But several classes of forms f may correspond to the same class of forms F.

It is proved that there is a single substitution which transforms a cubic form of determinant $\neq 0$ into an equivalent form.

Since $4A^3\equiv(3abc-2b^3-a^2d)^2-Da^2$, $4A^3$ is of the form U^2-DV^2. If A' is any integer representable by F, we can transform F into a form F' whose first coefficient is A'. The same substitution replaces f by a cubic f' whose corresponding quadratic form is F' and whose determinant is D. Hence $4A'^3$ is representable by U^2-DV^2. Similarly, if A' is representable by F, A'^2 is representable by $Ax^2-Bxy+Cy^2$.

By means of the above theorems and the theory of composition of classes of quadratic forms it is proved that, when p is a prime $\equiv 3 \pmod 4$ and $-p$ is a regular determinant (Ch. V), every class of quadratic forms of determinant $-p$, which by triplication produces the principal class, corresponds to a class of cubic forms of determinant $-4p$, while the remaining quadratic classes correspond to no cubic class with the same determinant (Pepin[13]).

[1] Jour. für Math., 27, 1844, 89–106 (319 for algebraic indentities and covariants).

Eisenstein[2] had stated, with no restriction on the determinant except that it have no square factor, that for each class of the principal genus of quadratic forms which by triplication produces the principal class there corresponds a unique cubic class, while no cubic classes correspond to the remaining quadratic classes [error noted by Arndt,[5] Cayley,[6] and Pepin[13]]. If $bc-ad$ is even and if $N, N_1, \ldots$ are the integers prime to $2D$ which can be represented properly by (A, B, C), it is stated that all integers prime to $2D$ representable properly by the cubic form f are given by the values of V occurring among the relatively prime solutions U, V of

$$U^2-DV^2=N^3, \qquad U^2-DV^2=N_1^3, \ldots.$$

F. Arndt,[3] who was not acquainted with Eisenstein's[1, 2] work, used new notations and wrote

$$(2) \qquad A=2(b^2-ac), \quad B=bc-ad, \quad C=2(c^2-bd), \quad D=B^2-AC\neq 0,$$

and called $\phi=(A, B, C)=Ax^2+2Bxy+Cy^2$ the *characteristic* of the cubic form $f=(a, b, c, d)$ and D the determinant of f. Thus ϕ and D are identical with $2F$ and D of Eisenstein. f is called *contrary* to $-f$ and *opposite* to $(a, -b, c, -d)$.

To decide whether two given cubic forms f and f' with the same characteristic ϕ are equivalent or not, we determine whether or not f' is identical with a form obtained by applying to f one of the substitutions which transforms ϕ into itself:

$$\begin{pmatrix}(T-BU)/m, & -CU/m \\ AU/m, & (T+BU)/m\end{pmatrix}, \qquad T^2-DU^2=m^2,$$

where m is the g.c.d. of $A, 2B, C$. When D is a square or when D is negative and $-4D/m^2>4$ or $=3$, f and f' with the same characteristic ϕ are equivalent if and only if they are identical or contrary. For other D's the answer requires a computation employing the successive solutions of $T^2-DU^2=m^2$.

If f and f' are equivalent, but their characteristics ϕ and ϕ' are not identical, the latter are equivalent and we first seek a substitution of ϕ into ϕ'.

Given a form $\phi=(A, B, C)$ in which A and C are even and B positive, to find all cubic forms f having ϕ as characteristic, we seek the integral solutions a, b, c, d of (2). These equations are the necessary and sufficient conditions that $\binom{b\ d}{a\ c}$ shall replace $(A, -B, C)$ by $\psi=(\frac{1}{2}A^2, B^2-\frac{1}{2}AC, \frac{1}{2}C^2)$, whose determinant is DB^2. There is a discussion of this problem on quadratic forms (cf. Arndt[4]). The resulting cubic forms f fall into a finite (unspecified) number of systems, each system containing as many forms as there are sets of solutions of $X^2-DY^2=m^2$. Equivalent forms f with the same ϕ belong to the same system. The number of classes of cubics with the same ϕ is equal to the number of systems if D is a square or $-4D/m^2 \gtreqless 4$, but is the triple of the number of systems if $D>0$ or $-4D/m^2=3$. The number of classes of cubics with the same determinant is finite. Details are given for the number of classes when D is a square. The excluded case $D=0$ is finally treated at length.

Arndt[4] evaluated the number of systems by a further study of the form ψ, and obtained the number of classes of cubic forms of an arbitrary determinant.

---

[2] Jour. für Math., 27, 1844, 75–79.
[3] Archiv Math. Phys., 17, 1851, 1–53.
[4] Archiv Math. Phys., 19, 1852, 408–418.

Arndt[5] gave a simpler theory. He showed how to find all cubic forms f with the same characteristic F and investigated their separation into classes.

A. Cayley[6] gave a simplification of Eisenstein's theory and an extension to any negative determinant. Let (A, B, C) be a properly primitive reduced quadratic form of determinant $-D$. We can find a (single) pair of cubic forms (a, b, c, d) and $(-a, -b, -c, -d)$ the negative of whose Hessian is (A, B, C) if and only if $(A, B, C)^2=(A, -B, C)$, i. e., the triplication of (A, B, C) produces the principal form.

Ch. Hermite[7] employed the Hessian, $-\phi$, of f, where

$$\phi=px^2+2qxy+ry^2, \quad p=b^2-ac, \quad 2q=bc-ad, \quad r=c^2-bd,$$

and the cubic covariant

$$F=(ax^2+2bxy+cy^2)(qx+ry)-(bx^2+2cxy+dy^2)(px+qy).$$

The most general cubic form having the quadratic covariant ϕ is $tf+uF$, where $t^2-\Delta u^2=1$, $\Delta=q^2-pr$. Let f have integral coefficients $a, \ldots, d$, and let ϕ be properly primitive. Then $tf+uF$ has integral coefficients if and only if t and u are integers.

F. Arndt[8] tabulated the reduced binary cubic forms f and their classes for all negative determinants $-D$, $D\leqq 2000$. Here f is called *reduced* if its characteristic ϕ is a reduced quadratic form, i. e., $|B|\leqq\frac{1}{2}|A|$, $|C|\gtreqless|A|$.

Ch. Hermite[9] applied his method of continual reduction (Ch. XIV[1]) to cubic forms f of determinant D. For $D=-\Delta<0$, we easily find that, for a reduced form,

$$ad<\sqrt{\tfrac{16}{27}\Delta}, \qquad bc<\sqrt{\tfrac{1}{3}\Delta},$$

whereas Arndt[8] obtained the same limit for bc as for ad. For the more difficult case $D>0$, use is made of the irrational covariant

$$\psi=\tfrac{1}{9}a^2\{2(\alpha-\beta)(\alpha-\gamma)(x-\beta y)(x-\gamma y)-(\beta-\gamma)^2(x-\alpha y)^2\}$$

of f, with the real root α, and conjugate imaginary roots β, γ. Since

$$4\psi^3-3\phi^2\psi-\phi^3-Df^2=0,$$

where $-\frac{1}{2}\phi$ is the Hessian of f, we may replace the usual conditions that ψ be a reduced quadratic form by conditions rational in a, b, c, d:

$$\tfrac{1}{4}(A+2\epsilon B)^3+3D(A+2\epsilon B)+Df(1,0)f(1,2\epsilon)>0, \qquad \epsilon=\pm 1,$$
$$\tfrac{1}{4}(C-A)^3+3D(C-A)+Df(1,1)f(-1,1)>0.$$

A. Cayley[10] remodeled Arndt's table by arranging it in the manner of Cayley's tables for binary quadratic forms (Ch. IV[7]).

Th. Pepin[11] found the linear forms of the prime divisors of linear functions of $X=x(x^2-9y^2)$, $Y=y(x^2-y^2)$, where x, y are relatively prime integers. There are

[5] Jour. für Math., 53, 1857, 309–321.
[6] Quar. Jour. Math., 1, 1857, 85, 90; Coll. Math. Papers, III, 9, 11.
[7] Quar. Jour. Math., 1, 1857, 88–89 [20–22]; Oeuvres, I, 437–9 [434–6].
[8] Archiv Math. Phys., 31, 1858, 335–448.
[9] Comptes Rendus Paris, 48, 1859, 351; Oeuvres, II, 93–99.
[10] Quar. Jour. Math., 11, 1871, 246–261; Coll. Math. Papers, VIII, 51.
[11] Comptes Rendus Paris, 92, 1881, 173–5.

given eleven theorems similar to the first one following: The prime divisors other than 2, 3, 7 of $X+3Y$ and of $X+9Y$ are the primes of one of the forms $18l\pm 1$.

H. Poincaré[12] noted that, in accord with his theory of orders and genera of forms (see Ch. XI[21]), the complete ordinal characters of

$$f=ax^3+3bx^2y+3cxy^2+dy^3$$

are given by the g.c.d. of a, b, c, d, that of $a, 3b, 3c, d$, that of r, s, t, and that of $2r, s, 2t$, if the Hessian of f is

$$6rx^2+6sxy+6ty^2, \qquad r=ac-b^2, \qquad s=ad-bc, \qquad t=bd-c^2.$$

Every form f is equivalent with respect to an arbitrary power of 2 as modulus to one of the six forms *

$$2x^3+6x^2y+6xy^2+2y^3, \quad x^3, \quad x^3+y^3, \quad 3x^2y+3xy^2, \quad x^3+3xy^2, \quad x^3+3xy^2+y^3,$$

which belong to different genera. The fourth and sixth have the same discriminant and belong to the same order for the modulus 2. The others belong to different orders.

Every form f is equivalent with respect to an arbitrary power of 3 as modulus to one of the six forms

$$3x^2y+3xy^2, \quad 3x^2y, \quad 3x^3+9x^2y+9xy^2+3y^3, \quad x^3, \quad x^3+3xy^2, \quad x^3+6xy^2,$$

whose discriminants D are all different.

The forms with $D\equiv 0 \pmod 5$ fall into 3 orders each with one genus. Those with $D\equiv 1$ or 4 form a single order and single genus. The forms with $D\equiv 2$ or 3 form one order with 3 genera; then f is equivalent to one of

$$x^3+6xy^2+y^3, \quad 2x^3+12xy^2+2y^3, \quad x^3+9xy^2 \text{ if } D\equiv 2 \pmod 5,$$
$$x^3+12xy^2+y^3, \quad 2x^3+24xy^2+2y^3, \quad x^3+6xy^2 \text{ if } D\equiv 3 \pmod 5.$$

Th. Pepin[13] noted that the first theorem of Eisenstein[2] is in error for a positive determinant, since three distinct cubic classes correspond to a single quadratic class (pp. 260, 271); while for a negative determinant without a square factor cubic forms may correspond to quadratic forms belonging neither to the principal genus nor to the properly primitive order. Thus there should be deleted the final clause of the last theorem quoted from Eisenstein.[1]

Pepin employed the notations of Eisenstein, and his definitions except that, when D and hence B is odd, $2F$ (and not F) is taken as the quadratic form corresponding to the cubic f and designated by $(2A, B, 2C)$ in Gauss' notation. Whether the latter form or $(A, \frac{1}{2}B, C)$ corresponds to f, the conditions are $A=b^2-ac$, $B=bc-ad$, $C=c^2-bd$. This system of quadratic equations is equivalent to the simpler system

$$A^2=Ab^2-Bab+Ca^2, \qquad Ac-Bb+Ca=0, \qquad Ad-Bc+Cb=0.$$

First, let B be odd. Let m denote the g.c.d. of $A=mA_1$, $B=mB_1$ $C=mC_1$. After a preliminary transformation of variables, we may assume that A_1 is a prime not

* The first form $2(x+y)^3$ is equivalent to $2x^3$, and the third form $3(x+y)^3$ of the next set is equivalent to $3x^3$.

[12] Comptes Rendus Paris, 94, 1882, 67–69, 124–7.

[13] Atti Accad. Pont. Nuovi Lincei, 37, 1883–4, 227–294.

dividing $2mB_1$. The determinant D of f is then odd and the corresponding quadratic form is $2F=(2mA_1, mB_1, 2mC_1)$. This form actually corresponds to a cubic if and only if $mA_1^2=A_1b^2-B_1ab+C_1a^2$ has integral solutions b, a such that a, A_1 are relatively prime. Multiplying this condition by $4A_1$, we obtain the equivalent condition that one of the two numbers mA_1^3, $4mA_1^3$ be representable properly by $t^2-\Delta a^2$, where $\Delta=B_1^2-4A_1C_1$. There are analogous results when B is even.

It follows that the only classes of properly primitive quadratic forms which correspond to cubic forms are those which by triplication produce the principal class. There are proved many theorems serving to determine the classes of cubics which correspond to given classes of quadratic forms.

Pepin[14] gave many numerical examples of the classification of cubic forms of negative determinant.

G. B. Mathews[15] called a class K *subtriplicate* if K^3 is the principal class. If (a, b, c) is subtriplicate, there exists a bilinear substitution

$$\begin{pmatrix} a_1 & a_2 & a_2 & a_3 \\ a_0 & a_1 & a_1 & a_2 \end{pmatrix}, \quad a_1^2-a_0a_2=a, \quad a_1a_2-a_0a_3=2b, \quad a_2^2-a_1a_3=c,$$

which transforms $(a, -b, c)$ into the compound of (a, b, c) with itself. Thus a_0, $3a_1$, $3a_2$, a_3 are the coefficients of a binary cubic form whose Hessian is (a, b, c). The question of integral solutions is discussed.

A. S. Werebrusow[16] noted that if the cubic form f is of determinant D, the solutions of $f=m$ in integers depends on $x^3-y^2=-m^2D$. Each has only a finite number of sets of solutions.[16a] For, if x is not a square and if y is the greatest integer $\leqq x^{3/2}$, then x^3-y^2 is of the same order as $\sqrt{x}$ when x increases indefinitely.

G. B. Mathews[17] proved that every binary cubic with integral coefficients which has a Hessian of the form $\mu(Ax^2+Bxy+Cy^2)$, where A, B, C are given integers, is expressible in the form $m\phi+n\psi$, where ϕ, ψ are particular cubics and m, n are integers. Cf. Hermite.[7]

Mathews and W. E. H. Berwick[18] noted that, for the reduction of a binary cubic f with a single real root, the Hessian is an indefinite form and does not lead to a unique reduced form for f. Let $f=(x-\alpha y)\phi$, where α is real and irrational and ϕ is a definite form. By the geometrical theory of quadratic forms, the fundamental triangle T_0 lies outside the circle $x^2+y^2=1$, between the lines $x=\pm\frac{1}{2}$, and above the x-axis. Applying unitary substitutions, we obtain curvilinear triangles T which fill the plane. The plots of the complex roots β, γ of $\phi=0$ are within or on the boundaries of two conjugate triangles T, T'. Let the coefficient of i in β be positive and let β lie within T. There is a unique substitution S which carries T into T_0, taking β to β_0. Then S replaces f by f_0, which is reduced by definition. If β is on a boundary, two substitutions carry β to the boundary of T_0; but we select as the reducing substitution the one which carries β to the left of the y-axis. Unique reduced forms f_0

[14] Atti Accad. Pont. Nuovi Lincei, 39, 1885–6, 23–87.
[15] Messenger Math., 20, 1891, 70–74.
[16] Math. Soc. Moscow, 26, 1907, 115–129 (Fortschritte der Math., 38, 1907, 241).
[16a] L. J. Mordell gave a proof in Proc. London Math. Soc., (2), 13, 1914, 60–80; 18, 1919, v, Records of Nov. 14, 1918.
[17] Proc. London Math. Soc., (2), 9, 1911, 200–4.
[18] *Ibid.*, (2), 10, 1912, 48–53.

thus defined are not properly equivalent. There is found a finite process to calculate S. Finally there is treated the case of cubics with a rational factor.

Mathews[19] quoted from the preceding paper the result that a binary cubic with an irrational real root and two complex roots is equivalent to a unique reduced form f whose roots satisfy the inequalities

$$\beta\gamma-1>0, \quad \beta+\gamma-1<0, \quad \beta+\gamma+1>0.$$

Since β and γ are conjugate, these are equivalent to the inequalities

$$\Pi(\beta\gamma-1)>0, \quad \Pi(\beta+\gamma-1)<0, \quad \Pi(\beta+\gamma+1)>0,$$

involving symmetric functions of the roots, and hence equivalent to

$$d(d-b)+a(c-a)>0, \quad ad-(a+b)(a+b+c)<0, \quad ad+(a-b)(a-b+c)>0.$$

The latter are the necessary and sufficient conditions that $f=(a, b, c, d)$ be reduced, no binomial coefficients being attached to f, nor to $(a, b, c)=ax^2+bxy+cy^2$.

To classify cubics f with a negative discriminant* $-\Delta$ and Hessian $H=(A, B, C)$, employ the identity noted by Eisenstein[1]:

$$(3) \qquad -A^2 \equiv Ab^2-3Bba+9Ca^2, \quad A=3ac-b^2, \ldots.$$

Thus the first two coefficients a, b, of any cubic with an assigned Hessian H furnish a set of integral solutions of $L(b, a)=-A^2$, where L denotes $(A, -3B, 9C)$. Conversely, if b, a is a set of integral solutions of (3), there is a unique cubic f with the Hessian H. Thus there is a one-to-one correspondence between cubics and representations of $-A^2$ by L. By means of the automorphs of L, these cubics can be arranged in a finite number of sequences. Then we can find a complete set of representative cubics of determinant $-\Delta$. There is a table, calculated by Berwick, of the non-composite reduced cubics for $\Delta<1000$.

F. Levi[20] established a (1, 1) correspondence between cubic algebraic fields and classes of binary cubic forms $f=ax^3+bx^2y+\ldots$ for which $a, b, \ldots$ have no common divisor, and such that no integer k exists for which kf arises from a form with integral coefficients by a substitution of determinant k.

B. Delaunay[21] considered forms $f\equiv x^3+nx^2y-pxy^2+qy^3$ of negative discriminant and proved that $f=1$ has, apart from the trivial solution $x=1, y=0$, at most four solutions, and usually not more than two. To apply this result to $f(x, y)\equiv Ax^3+Bx^2y+Cxy^2+Ey^3$ of negative discriminant, employ a remark of Lagrange (this History, Vol. II, p. 673). Let σ_1 be the number of roots of $f(x, 1)\equiv 0 \pmod{\sigma}$. Then $f(x, y)=\sigma$ has in general less than $2\sigma_1$ solutions.

* The discriminant δ of $f=ax^3+\ldots$ is a^4 times the product of the squares of the differences of the roots x/y of $f=0$. Thus $\delta=-27D$, for D as in Eisenstein.[1]

[19] Proc. London Math. Soc., (2), 10, 1912, 128–138. The paper admittedly has some duplication with Pepin's.[13]

[20] Berichte Gesell. Wiss. Leipzig (Math.), 66, 1914, 26–37.

[21] Comptes Rendus Paris, 171, 1920, 336–8.

CHAPTER XIII.

CUBIC FORMS IN THREE OR MORE VARIABLES.

G. Eisenstein[1] employed the primary complex prime factors p_1 and p_2, in the domain of an imaginary cube root ρ of unity, of a prime $p\equiv 1 \pmod 3$, and

$$\Phi=u^3+pp_1y^3+pp_2z^3-3puyz=\prod_{j}^{0,1,2}\{u+\rho^j\eta y+\rho^{-j}\theta z\},$$

where $\eta=\sqrt[3]{pp_1}$, $\theta=\sqrt[3]{pp_2}$. The product of two forms Φ is of that form. Write $y=v+\rho w$, $z=v+\rho^2 w$. Then Φ represents only real integers when u, v, w are real integers. A ternary form $F=au^3+3bu^2v+3b'u^2w+\ldots$, with real integral coefficients without a common factor, is called (p. 318) *associated* with Φ if a^2F is the product of the three factors

$$au+\{b+(c+d\rho)\rho^j\eta+(c+d\rho^2)\rho^{-j}\theta\}v+\{b'+(c'+d'\rho)\rho^j\eta+(c'+d'\rho^2)\rho^{-j}\theta\}w,$$

for $j=0, 1, 2$, where a, b, b', c, d, c', d' are real integers for which $cd'-c'd=a$. Any form equivalent to F is also associated with Φ. There is an extended investigation of the representation of real integers by associated forms F, also of the forms F.

In the sequel he[2] evaluated the number of classes of associated forms.

A. Meyer[3] considered a form $f(x_1, \ldots, x_n)$ with real integral coefficients which is a product of m linear factors such that f is not a product of two forms with rational coefficients. After applying a preliminary linear transformation with integral coefficients, we may assume that $f(x, -1, 0, \ldots, 0)=0$ has no factor with integral coefficients, and that, if its roots are called $\omega_1, \ldots, \omega_m$, $f(x_1, \ldots, x_n)=u_1\ldots u_m/g$, where u_k is a linear function of $x_1, \ldots, x_n$ whose coefficients are polynomials in ω_k with integral coefficients, and g is the g.c.d. of the coefficients of f. Since the u_k are linear combinations of only m linear functions of $x_1, \ldots, x_n$, we may assume that $n\leqq m$. But forms f with $n<m$ arise from forms f having $n=m$ by equating to zero $n-m$ of the x's. Hence we may take $n=m$.

The further investigation is restricted to forms

$$f=(u_1x_1+u_2x_2+u_3x_3)(u_1'x_1+u_2'x_2+u_3'x_3)(u_1''x_1+u_2''x_2+u_3''x_3)/g,$$
$$u_1=a_1+b_1\omega+c_1\omega^2,\quad u_2=a_2+b_2\omega+c_2\omega^2,\quad u_3=a_3+b_3\omega+c_3\omega^2,$$

while u_i', u_i'' are derived from u_i by replacing ω by ω', ω'', where ω is the real and ω', ω'' are the imaginary cube roots of a positive integer D not divisible by a cube. Also, $a_1, \ldots, c_3$ are real integers without a common factor. Here g is the g.c.d. of the coefficients of $x_1^3, 3x_1^2x_2, \ldots, 3x_1x_2x_3$ and also of $x_1^3, x_1^2x_2, \ldots, x_1x_2x_3$. If Θ is the

[1] Jour. für Math., 28, 1844, 289–274; cf. this History, Vol. II, 594–5.
[2] *Ibid.*, 29, 1845, 19–53. Both papers reprinted in Eisenstein's Math. Abhand., 1847, 1–120.
[3] Vierteljahr. Naturf. Gesell. Zürich, 42, 1897, 149–201 (Habilitationsschrift, 1870).

product of those primes whose squares divide D, then $g=\theta^3\cdot n$, where θ divides Θ, and n is prime to $3D$ and is the product $n_1n_1'n_1''$ of conjugate ideals, such that n_1 is the g.c.d. ideal factor of u_1, u_2, u_3. Let M_k be an ideal not divisible by a principal ideal and with norm prime to $3D$ such that M_kn_1 is a principal ideal m_k. Then $v_j=u_jM_kM_k'M_k''/m_k$ is an integral algebraic number divisible by $M_k'M_k''$. Then

$$f=\frac{1}{\theta^3}N\left(\frac{v_1x_1+v_2x_2+v_3x_3}{M_k'M_k''}\right).$$

After applying a linear substitution, we obtain a reduced form with

$$v_1=a_1,\quad v_2=a_2+b_2\omega,\quad v_3=a_3+b_3\omega+c_3\omega^2,$$
$$0\leqq a_2<a_1,\quad 0\leqq a_3<a_1,\quad 0\leqq b_3<b_2,\quad 0<c_3,\quad a_1b_2c_3=\Delta,$$

where Δ is the determinant of the coefficients of the initial u_1, u_2, u_3. The number of reduced forms with a given Δ is therefore finite. There is given a process to find a set of non-equivalent reduced forms; also to find all the transformations of a reduced form into itself.

H. Poincaré[4] solved the problems: to decide whether or not two given ternary cubic forms are arithmetically equivalent; to distribute the forms into classes, genera, and orders; to find the substitutions with integral coefficients which transform a form into itself. The method is a generalization to forms of any degree of that employed by Hermite for quadratic forms and those decomposable into linear factors (Ch. XI[7]).

In order that two forms be arithmetically equivalent, they must be equivalent under real transformation, which can be decided by algebraic considerations which lead also to a transformation replacing one of the forms by the other. Hence let F and F' be two forms derivable by real transformations from the same canonical form H. To decide whether or not F and F' are arithmetically equivalent, we must define forms which play with respect to F and F' the same rôle that reduced forms play with respect to quadratic forms.

Call a substitution *reduced* if it replaces Σx_i^2 by a definite reduced quadratic form; the coefficients of the substitution need not be rational. Forms derivable from the canonical form H by a reduced substitution are called *reduced forms*. We desire all reduced forms arithmetically equivalent to a form F which can be derived from H by a real substitution T. If τ reproduces H, τT replaces H by F. To find all reduced forms equivalent to F, seek all transformations E with integral coefficients such that τTE is a reduced substitution, i. e., replaces Σx_i^2 by a reduced form. There is always one E and in general only one. This definition of the reduced form of F depends upon the particular canonical form H chosen among all the forms equivalent to F under real transformation, and also upon the particular definition chosen for a reduced definite quadratic form (the choice here being usually that by Korkine and Zolotareff[18] of Ch. XI).

Jordan[6] of Ch. XIV had proved that the forms with integral coefficients algebraically equivalent to a given form separate into a finite number of classes, provided the discriminant D is not zero. Here a new proof is given, also for the case in which $D=0$, but certain other invariants are not all zero.

---

[4] Jour. école polyt., cah. 51, 1882, 45–91 (algebraic part in 50, 1881, 199–253). Comptes Rendus Paris, 90, 1880, 1336; 91, 1880, 844–6.

These principles are applied to ternary cubic forms. There are seven types of canonical forms H to consider in turn. First, if $H=6axyz+\beta(x^3+y^3+z^3)$, it is reproduced only by a substitution permuting x, y, z, which therefore leaves Σx^2 unaltered. In general, there is a single substitution E which reduces $(\Sigma x^2)T$. Hence $F=HT$ has a single reduced form FE. It is shown that the various coefficients of a reduced form do not exceed specified limits depending on the invariants S and T of H, so that there is a finite number of classes. For certain other H's, the reduced forms constitute a finite number of genera each with an infinitude of classes.

Ph. Furtwängler[5] studied decomposable ternary cubic forms F with integral coefficients. Let $F=a^{-2}l_1l_2l_3$, where $l_j=ax_1+P_jx_2+Q_jx_3$. Write $r=1/\sqrt[3]{a^4}$. With F we associate the family of definite quadratic forms

$$r\lambda_1^2l_1^2+r\lambda_2 l_2 l_3\,(\lambda_2>0,\ \lambda_1\lambda_2=1),\quad r\lambda_1^2l_1^2+r\lambda_2^2l_2^2+r\lambda_3^2l_3^2\,(\lambda_1\lambda_2\lambda_3=1),$$

according as l_2 and l_3 are conjugate imaginary, or l_1, l_2, l_3 are all real. Call F *reduced* if any one of the associated definite quadratic forms is reduced. Apply the Seeber-Gauss condition (Ch. IX[5]) that in a reduced form $Ax_1^2+Bx_2^2+Cx_3^2+\ldots$, ABC does not exceed double the absolute value of the determinant ($-\frac{1}{4}D$ and D in the above cases if D is the discriminant of F). We find that all the coefficients of the l_j are limited in terms of D. Hence there is only a finite number of reduced cubic forms with a given discriminant, and therefore a finite of classes.

Applying Hermite's method of continual reduction to the associated quadratic form, we see that, in the case of imaginery l_2 and l_3, the equivalent reduced cubic forms constitute a finite period which repeats when the parameter λ_2 varies continuously; the automorphs constitute an infinite cyclic group. When the l's are all real, the reduction process is discussed geometrically; the automorphs are generated by two substitutions. The problem to represent a given integer by F is solved as in Eisenstein's[1] case.

The determination of the factors of a decomposable cubic form F depends on the solution of a single cubic equation, which is assumed to be irreducible. For F's of the same discriminant D, we saw that there is a limited number of sets of coefficients of the factors and therefore of cubic equations, and hence of cubic fields defined by them. Given a cubic field of discriminant D, we seek the forms F, whose discriminants must be of the type m^2D, where m is an integer. If $m=1$, F is called a *Stamm form*; those of its automorphs which do not permute its linear factors multiply them by three conjugate units, and every triple of conjugate units belonging to the field of F determines such an automorph. Since the units depend only on the field, all Stamm forms of a field have the same automorphs.

The discriminant of any F is not less than the discriminant of the field. If the coefficients of F are relatively prime and if H is a Stamm form which represents unity, and if F arises by composition of F with H, the discriminant of F is the same as that of the field. Similar theorems follow for the composition of the corresponding three-dimensional lattices.

---

[5] Zur Theorie der in Linearfaktoren zerlegbaren, ganzzahligen ternären cubischen Formen, Diss., Göttingen, 1896, 62 pp.

CHAPTER XIV.

FORMS OF DEGREE $n \gtreqless 4$.

Reports on important papers on this topic have already been given under Eisenstein,[8] Lipschitz,[17] Poincaré,[21] Minkowski,[27] and (for decomposable forms) Hermite,[7] all of Ch. XI. Hermite's method was extended to arbitrary forms by Poincaré[4] of Ch. XIII. Decomposable forms, chiefly ternary cubics, were treated by Meyer[3] of Ch. XIII.

Ch. Hermite[1] investigated the reduction and equivalence of forms

$$(1) \qquad f(x,y)=a_0x^n+a_1x^{n-1}y+\ldots+a_ny^n, \qquad a_0\neq 0.$$

From its μ real linear factors $x+\alpha_k y$ and ν pairs of conjugate imaginary linear factors $x+\beta_j y$, $x+\gamma_j y$, construct the definite quadratic form

$$(2) \qquad \phi=\sum_{k=1}^{\mu} t_k^2(x+\alpha_k y)^2+2\sum_{j=1}^{\nu} u_j^2(x+\beta_j y)(x+\gamma_j y),$$

where the t_k and u_j are real variables. For chosen values of the t_k, u_j, let the substitution

$$S: \qquad x=mX+m_0Y, \quad y=nX+n_0Y, \quad mn_0-m_0n=1,$$

with integral coefficients, replace ϕ by Φ, of type (2) with

$$T_k=t_k(m+\alpha_k n), \qquad U_j^2=u_j^2(m+\beta_j n)(m+\gamma_j n).$$

Let S replace f by $F=A_0X^n+\ldots$. Then A_0 is evidently equal to

$$f(m,n)=a_0\prod_{k=1}^{\mu}(m+\alpha_k n)\cdot\prod_{j=1}^{\nu}(m+\beta_j n)(m+\gamma_j n),$$

whence

$$(3) \qquad A_0^2/[TU]^2=a_0^2/[tu]^2, \qquad [tu]\equiv t_1\ldots t_\mu u_1^2\ldots u_\nu^2.$$

Write $\Phi=PX^2+2QXY+RY^2$. It is proved that

$$(4) \qquad A_iA_{n-i}=\frac{A_0^2(PR)^{\frac{1}{2}n}}{[TU]^2}\{i\},\quad \{i\}<\frac{1}{n^n}\binom{n}{i}^2 (0\leqq i\leqq n).$$

Assume that the substitution S replaces ϕ by a reduced form Φ, whence $PR<\frac{4}{3}D$, if D is the determinant of ϕ. Hence

$$(5) \qquad A_iA_{n-i}<(\tfrac{4}{3})^{\frac{1}{2}n}\{i\}\theta, \qquad \theta\equiv\frac{a_0^2D^{\frac{1}{2}n}}{[tu]^2}.$$

---

[1] Jour. für Math., 41, 1851, 197–203, 213–6, §§ V, VI, XII; Oeuvres, I, 171–8, 189–192. In a preliminary paper, Jour. für Math., 36, 1848, 357–364; Oeuvres, I, 84–93, he took, for the case in which all the roots are real, certain functions Δ_k of the roots α_i in place of the variables t_k^2. For $n=3$, the Δ_k are the three squares of the differences of the roots. For $n=4$, $\Delta_1=-(\alpha_2-\alpha_3)(\alpha_3-\alpha_4)(\alpha_4-\alpha_2)$, ... , $\Delta_4=-(\alpha_1-\alpha_2)(\alpha_2-\alpha_3)(\alpha_3-\alpha_1)$, whence $\Omega=\Delta_1\Delta_2\Delta_3\Delta_4$ is essentially the discriminant of f. The determinant of $\Sigma\Delta_k(x+\alpha_k y)^2$ is $4\Omega^{\frac{1}{2}}(\alpha_1-\alpha_2)(\alpha_3-\alpha_4)$, an irrational invariant of f.

When the variables t_k, u_j range over all real numbers, θ has an absolute minimum, which is equal to the minimum of the analogous function Θ of the T_k, U_k for any equivalent form F. This minimum is defined to be the *determinant* of f.

Seek the sets of values of the t_k, u_j which render θ an absolute minimum; insert these values into ϕ to obtain the quadratic forms corresponding to f. The substitution S which reduce the latter replace f by the *reduced forms* of f. Equivalent forms have the same reduced forms. By (5), the forms with integral coefficients of the same determinant θ fall into a finite number of classes.

For a binary cubic f with three real roots α_i, the corresponding quadratic form ϕ becomes the Hessian of f when the t's are the differences of the roots. Cf. Hermite[9] of Ch. XII.

Hermite[2] applied his preceding results to a quartic form

$$f = ax^4 + 4bx^3y + 6cx^2y^2 + 4b'xy^3 + a'y^4$$

with binomial coefficients prefixed, and with four real roots α_k. Consider $\phi = \Sigma t_k(x-\alpha_k y)^2$, thus replacing his former t_k^2 by t_k. Taking into account all these changes of notation, we see that (5) give

$$(6) \qquad AA' < \frac{T}{144}, \quad BB' < \frac{T}{144}, \quad C^2 < \frac{T}{144}, \quad T \equiv \frac{a^2D^2}{t_1t_2t_3t_4}.$$

For positive real values of the variables $t_1, \dots, t_4$, the minimum of T is proved to be $16a^2(\alpha_1-\alpha_3)^2(\alpha_2-\alpha_4)^2 = 16^2(\theta_1-\theta_2)^2$, where θ_1 and θ_2 are the largest and smallest roots of $4\theta^3 - i\theta + j = 0$, i and j being the familiar invariants of f. Call the medium root θ_3. Hence to compute the complete set of reduced forms F with the invariants i and j, we compute the preceding minimum of T, and then find the (finite number of) sets of integers $A, B, \dots$ satisfying the inequalities (6). If G denotes the Hessian of F,

$$(7) \qquad \phi = \sqrt{G+\theta_3 F}/\{2(\theta_1-\theta_3)(\theta_2-\theta_3)\}.$$

We retain only the F's for which this quadratic form ϕ is reduced.

In the introduction (p. 2), Hermite employed Eisenstein's[8] (Ch. XI) conception of a genus as the aggregate of forms equivalent under a linear substitution with rational coefficients of determinant unity, and stated without proof the theorem that, if m is odd and >3, all binary forms having given values for its invariants constitute a single genus.

Hermite[3] called $f = ax^m + mbx^{m-1}y + \dots$ *primitive* if the g.c.d. of $a, b, \dots$ is unity, and *properly* primitive if also the g.c.d. of $a, mb, \dots$ is unity. Any covariant of f is arithmetically equivalent to the same covariant of a form equivalent to f. Those forms for which the g.c.d. of the coefficients (with or without prefixed binomial coefficients) of every covariant is constant constitute an *order*.

Hermite[4] considered *decomposable* forms ϕ in n variables $x_1, \dots, x_n$ with complex integral coefficients, such that ϕ is a product of linear factors L_i, while $\phi=0$ has no

---

[2] Jour. für Math., 52, 1856, 1–17; Oeuvres, I, 350–371.
[3] Jour. für Math., 52, 1856, 18–38; Oeuvres, I, 374–396.
[4] Jour. für Math., 53, 1857, 182–192; Oeuvres, I, 415–428.

integral solutions other than $x_1=0, \ldots, x_n=0$. The substitution which reduces $\Sigma L_i L_i'$, where L_i' is the conjugate of L_i (Hermite[7] of Ch. XI) replaces ϕ by a form Φ whose coefficients are complex integers limited in terms of $\Delta\Delta'$, where Δ is the determinant of the coefficients of the linear forms L_i. This Φ is called a *reduced* decomposable form.

R. Dedekind[5] studied forms decomposable into linear factors in connection with norms of algebraic numbers and ideals. In particular, he discussed the composition $XY=Z$ of such forms.

C. Jordan[6] investigated the equivalence of forms of degree m in n variables. If one such form F is transformed into a second one G by a substitution

$$S: \qquad y_1=a_{11}x_1+\ldots+a_{1n}x_n, \ldots, \qquad y_n=a_{n1}x_1+\ldots+a_{nn}x_n,$$

of determinant unity, F and G are said to be *algebraically equivalent* and to belong to the same *family*. If each a_{ij} is a complex integer, F and G are said to be *arithmetically equivalent* and to belong to the same *class*.

Writing $N(y)$ for the product of y and the complex number conjugate to y, we make correspond to the substitution S the Hermitian form

$$g=N(a_{11}x_1+\ldots+a_{1n}x_n)+\ldots+N(a_{n1}x_1+\ldots+a_{nn}x_n),$$

and call S *reduced* when g is reduced (Jordan[2] of Ch. XVI). In that case, $N(a_{\lambda k}) \leqq m_k \equiv 2^{k-1}\mu_k$, where the μ's are such that $\mu_{k+1} \gtreqless \frac{1}{2}\mu_k$, and $\mu_1\mu_2\ldots\mu_n=\Delta$, Δ denoting the norm of the determinant of S.

Any form G belonging to the same family as F is called *reduced* with respect to F if among the substitutions which transform F into G there exists a reduced substitution. Certainly G is arithmetically equivalent to a reduced form. For, if S transforms F into G, we can find a substitution T with complex integral coefficients of determinant unity such that ST is reduced; then T replaces G by an equivalent form which is reduced since ST transforms F into it.

Let G be a reduced form derived from F by the reduced substitution S corresponding to the above Hermitian form g. For the modulus of each coefficient of G there is found a superior limit in terms of the m_k and the sum s of the moduli of the coefficients of F. Assume now that the coefficients of G are complex integers; it is shown that each m_k has a superior limit in terms of s provided the degree m of F is >2 and its discriminant is not zero. But it was shown at the outset that in any family of forms whose invariants are complex integers there occurs a form F the moduli of whose coefficients have superior limits which are integral functions of the invariants (and perhaps also integers occurring in the coefficients of identically vanishing covariants). It follows that the number of classes of forms with integral complex coefficients in a family of forms of degree >2 and discriminant $\neq 0$ is limited in terms of the invariants; each class contains a reduced form the moduli of whose coefficients

[5] Dirichlet's Zahlentheorie, ed. 2, 1871, pp. 424, 465; ed. 3, 1879, 544; ed. 4, 1894, 580. Report by D. Hilbert, Jahresbericht d. Deutschen Math.—Vereinigung, 4, 1897, 235-6; French transl., Annales fac. sc. Toulouse, (3), 1, 1909, 318; H. Weber, Algebra, III, 1908, 330-7. P. Bachmann, Arith. der Zahlenkörper, V, 1905, Ch. 10.

[6] Jour. école polyt., t. 29, cah. 48, 1880, 111-150. Summary in Comptes Rendus Paris, 88, 1879, 906; 90, 1880, 598-601, 1422-3.

are likewise limited. If two forms F and G of degree m in n variables are algebraically equivalent and have complex integral coefficients, and if l is the number of such forms algebraically equivalent to G and reduced with respect to G, then every substitution which transforms F into G is the product of one such substitution (the moduli of whose coefficients are limited in terms of l, m, n and the moduli of the coefficients of F and G) by several substitutions T_j which transform G into itself and are generated by infinitesimal substitutions leaving G unaltered, while the moduli of the coefficients of each T_j are limited in terms of l and n. In case the discriminant of G is not zero, and $m>2$, the T_j do not occur, so that only a limited number of substitutions transform F into G. Then we can decide by a limited number of trials whether or not F is equivalent to G and if equivalent find all the the substitutions with complex integral coefficients which transform F into G.

H. Poincaré[7] discussed the representation of an integer N by

$$F=B_m x^m+B_{m-1}x^{m-1}y+\ldots+B_0 y^m.$$

Then $P=B_m^{m-1}N$ is represented by the form $\Phi=x_1^m+\ldots$ obtained by writing $B_m x=x_1$ in $B_m^{m-1}F$. Write

$$\Phi=\prod_{j=1}^{m}(x+a_j y)\equiv\text{norm}(x+a_1 y).$$

Consider the norm Ψ of $x_0+a_1x_1+a_1^2x_2+\ldots+a_1^{m-1}x_{m-1}$. Consider an ideal composed of such numbers $x_0+a_1x_1+\ldots$. For each ideal I of norm P, examine whether or not norm I is equivalent to $P\Psi$ by Hermite's[4] method, and if equivalent find a substitution replacing one by the other, and hence a representation $x_0=\beta_0$, $\ldots$, $x_{m-1}=\beta_{m-1}$ of P by Ψ. In case each $\beta_j=0$ for $j\gtreqless 2$, we have $P=\Phi(\beta_0, \beta_1)$, as desired. This method is essentially the same as that of Lagrange, this History, Vol. II, p. 691 (cf. pp. 677–8, especially Dirichlet).

L. Gegenbauer[8] proved that, if p is an odd prime,

$$a_{p-1}(x^p+y^p)-\sum_{k=1}^{\pi}(-1)^k(a_{p-k}-a_{p-k-1})x^ky^k(x^{p-2k}+y^{p-2k}),\quad \pi=\tfrac{1}{2}(p-1),$$

represents no power of a prime for positive integers x, y, provided $a_\pi, \ldots, a_{p-1}$ are integers $\gtreqless 0$ satisfying the relations

$$a_k\gtreqless a_{k-1},\quad 2\sum_{k=\pi+1}^{p-1}a_k+a_\pi=q,\quad p\sum_{k=\pi+1}^{p-1}a_k+\pi a_\pi=aq,$$

where q is a prime and a a positive integer, but satisfying none of the three sets of conditions

$$a_{p-1}=1,\quad a_{p-1-k}=0\ (k>0);\quad 2\sum_{k=\pi+1}^{p-1}(-1)^ka_k+(-1)^\pi a_\pi=1;\quad a_k=1.$$

In case the first or second set of conditions hold, there exist such representations of 2^e for any p, where $e=mp+2$ or $mp+1$, respectively. In case $a_k=1$, only when $p=3$ is 3^2 or 3^{2+3a} representable by the form properly or improperly, respectively.

X. Stouff[9] proved that, if a form $F(x_1, \ldots, x_m)$ of degree n with integral coefficients is a product of linear factors, but is irreducible (i. e., is not the product of two

[7] Bull. Soc. Math. France, 13, 1885, 162–194; summary, Comptes Rendus Paris, 92, 1881, 777–9.
[8] Sitzungsber. Akad. Wiss. Wien (Math.), 97, IIa, 1889, 368–373.
[9] Annales fac. sc. Toulouse, (2), 5, 1903, 129–155.

forms with integral coefficients), its linear factors may be found by solving an irreducible equation of degree n. Henceforth, let $m=n$. The automorphs of F are easily found when its linear factors are known. There is an amplification of the proof sketched by Hermite[7] in Ch. XI that there exist only a finite number of classes of forms F with a given invariant (the square of the determinant of the coefficients of the linear factors) and his method of finding the automorphs.

G. Bisconcini[10] noted that every rational number A representable simultaneously by binary forms $f(x, y)$ and $g(x, y)$ of degrees n and $n-1$, respectively, is given by

$$A=g(\lambda, \mu)^n/f(\lambda, \mu)^{n-1},$$

where λ, μ are arbitrary rational numbers. For, if we take $x=y\lambda/\mu$ in $f=g$, we get

$$x=\lambda g(\lambda, \mu)/f(\lambda, \mu), \qquad y=\mu g(\lambda, \mu)/f(\lambda, \mu).$$

Inserting these in $A=g(x, y)$, we evidently obtain the result stated.

*A. S. Werebrusow[11] showed how to test whether or not a given quartic in x, y can be expressed as a quadratic form in quadratic functions of x, y.

G. Julia[12] gave simplifications and extensions of Hermite's[1, 2] method of continual reduction. The three parts will be considered in turn. Part I[13] treats of the reduction of a binary form (1) with real coefficients. Let $f(z, 1)=0$ have μ real roots a_k and ν pairs of conjugate imaginary roots β_j, β'_j, the coefficient of i in β_j being positive. With f associate the definite quadratic form [(2) of Hermite]

$$\phi=\sum_{k=1}^{\mu} t_k^2(x-a_ky)^2+2\sum_{j=1}^{\nu} u_j^2(x-\beta_jy)(x-\beta'_jy)=px^2-2qxy+ry^2.$$

Represent ϕ by the point ζ in the upper half plane for which $pz^2-2qz+r=0$. We reduce ϕ by applying to ζ a linear substitution S on x, y with real integral coefficients of determinant unity which replaces ζ by a point interior to the classic fundamental domain D_0 of the modular group. When the t_k, u_j take all real values, the totality of substitutions S which reduce ϕ is the totality of substitutions which bring upon D_0 each of the triangles of the modular division having at least one point in common with the locus of ζ which is the least convex polygon D containing in its interior or on its boundary all the points represented by $a_1, \ldots, a_\mu, \beta_1, \ldots, \beta_\nu$. The sides of D are arcs of circles orthogonal to the real axis, and D has as summits all the a_k.

This geometrical interpretation of Hermite's method leads to a simple presentation of the majority of the results obtained by him by computation and also results for quartic forms[2] whose roots are not all real.

In Part II, Julia[14] extended the preceding discussion to binary forms (1) of degree n with complex coefficients and variables. Let $z_1, \ldots, z_n$ be the roots of $f(z, 1)=0$, and associate with (1) the definite Hermitian form

$$\phi=\sum_{i=1}^{n} t_i^2(x-z_iy)(x'-z'_iy')=pxx'-qxy'-q'x'y+ryy',$$

[10] Periodico di Mat., 22, 1907, 119–129.
[11] Math. Soc. Moscow, 27, 1909, 170–4 (Russian).
[12] Mém. Acad. Sc. l'Institut de France, 55, 1917, 1–293. Report is made from the summary in seven notes in the Comptes Rendus.
[13] Comptes Rendus Paris, 164, 1917, 32–35.
[14] Comptes Rendus Paris, 164, 1917, 352–5.

where x' denotes the conjugate imaginary to x, etc. Mark in the plane $O\xi\eta$ of $z=\xi+i\eta$ the points $z_1, \ldots, z_n$. Represent ϕ by a point of the half-space $O\xi\eta\tau$ ($\tau>0$) defined by its projection q'/p on $O\xi\eta$ and by its distance $O\zeta$ from the origin O such that $\overline{O\zeta}^2=r/p$. Then for all values of the t's, ζ describes the interior and surface of a convex polyhedron D with summits $z_1, \ldots, z_n$, the edges being semi-circles orthogonal to the plane $O\xi\eta$, and the faces being portions of the spheres orthogonal to $O\xi\eta$ and passing through three of the points z_j. Take

$$\theta=\frac{a_0a_0\delta^{n/2}}{t_1^2\ldots t_n^2},\qquad \delta=pr-qq'.$$

If the values of the t's which render θ an absolute minimum are substituted in ϕ, we get the *correspondent* of f, and by definition the substitution which reduces it reduces also f. Details are given for $n=3$ and $n=4$, without subdivision of cases (as was necessary for real cubic and quartic forms in Part I).

The last method may be applied[15] to forms with real coefficients to deduce the same reduced forms as by Hermite's method, but without requiring a separation of cases. In the notations of Part I, take

$$\phi=\sum_{k=1}^{\mu} t_k^2N(x-\alpha_ky)+\sum_{j=1}^{\nu} u^2N(x-\beta_jy)+\sum_{j=1}^{\nu} u_j'^2N(x-\beta_j'y)$$
$$=pxx'-qxy'-q'x'y+ryy',$$

where t_k^2, u^2, $u_j'^2$ are arbitrary positive parameters. Since f has real coefficients, the polyhedron D is now symmetrical with respect to the plane $O\xi\tau$. Hermite's polygon is the section of D by the plane $O\xi\tau$ if we represent the roots of $f(z, 1)=0$ in the plane $O\xi\tau$ (and not in $O\xi\eta$). To obtain the minimum of

$$\theta=\frac{a_0^2\delta^{n/2}}{t_1^2\ldots t_\mu^2\; u_1^2u_1'^2\ldots u_\nu^2u_\nu'^2},$$

we must take $u_j^2=u_j'^2$, whence $q'=q$, and then ϕ becomes the form which Hermite associated with f. Hence the method leads to Hermite's reduced forms.

In Part III, Julia[16] discussed the reduction of higher forms with conjugate variables. Let f be a product of n binary Hermitian forms

$$f_j=a_jxx'-b_jxy'-b_j'x'y+c_jyy',$$

where a_j and c_j are real, b_j and b_j' are conjugate imaginaries. Let $f_1, \ldots, f_\mu$ be indefinite forms of determinants $-\delta_1, \ldots, -\delta_\mu$, while $f_{\mu+1}, \ldots, f_n$ are definite forms of determinants $\delta_{\mu+1}, \ldots, \delta_n$ (all δ's being positive). For $j\leqq\mu$, f_j is represented in the half-space $O\xi\eta\tau$ ($\tau>0$) by a hemisphere σ_j whose great circle in the plane $O\xi\eta$ is $f_j(z, 1)=0$. Picard[4] of Ch. XV introduced for the continual reduction of such a form f_j a definite Hermitian form ϕ_j of determinant δ_j whose representative point ζ_j is a variable point of σ_j. Associate with f the definite Hermitian form

$$\phi=t_1^2\phi_1+\ldots+t_\mu^2\phi_\mu+t_{\mu+1}^2f_{\mu+1}+\ldots+t_n^2f_n.$$

Vary the real parameters t_j and the complex parameters which determine the

[15] Comptes Rendus Paris, 164, 1917, 484–6.
[16] *Ibid.*, 571–4, 619–622, 910–913 ($n=2$).

points ζ_j, and effect the continual reduction of f. The totality of substitutions which reduce ϕ replace f by a set of forms including the reduced forms of f.

Every binary form of degree $2n$ with conjugate variables[17] which is invariant under an infinite group of linear substitutions is of the type $a_0\phi^n+a_1\phi^{n-1}\psi+\ldots+a_n\psi^n$, where ϕ and ψ are two distinct Hermitian forms (not necessarily with integral coefficients), while the a's are real. Conversely, every such form admits the infinite group leaving ϕ and ψ separately invariant (*ibid.*, 163, 1916, 599). Since such forms evidently decompose, an indecomposable form admits only a finite group.

L. E. Dickson[18] investigated triples of forms $f(x)\equiv f(x_1, \ldots, x_n)$, Φ, F of general degree for which $f(x)\Phi(\xi)\equiv F(X)$, where $X_1, \ldots, X_n$ are bilinear functions of $x_1, \ldots, x_n; \xi_1, \ldots, \xi_n$. Assume that the determinant $\Delta(\xi)$ of the coefficients of the linear functions $X_1, \ldots, X_n$ of $x_1, \ldots, x_n$ is not identically zero, and likewise for the determinant $\Delta'(x)$ of the coefficients of the linear functions $X_1, \ldots, X_n$ of $\xi_1, \ldots, \xi_n$. It is proved that $f(x)$ admits a *composition* $f(x)f(\xi)\equiv F(X)$. Without either assumption it is proved that if $f(x_1, x_2, x_3)$ and $F(X_1, X_2, X_3)$ are indecomposable cubic forms for which $f\cdot\Phi(\xi_1, \xi_2, \xi_3)=F$, then Φ is the cube of a linear form. In particular, no indecomposable ternary cubic admits composition.

Dickson[19] gave a general theory of forms f admitting composition. By interpreting $x_1, \ldots, x_n$ as the coordinates of a hypercomplex number $x=\Sigma x_i e_i$, we may express the bilinear relations in the single formula $x\xi=X$. After applying a linear transformation on the x's which leaves $f(x)$ unaltered, and one on the ξ's leaving $f(\xi)$ unaltered, we obtain a *normalized* composition $x\xi=X$ of f such that e_1 is a principal unit (modulus). Then each irreducible factor of f divides both $\Delta(x)$ and $\Delta'(x)$ of the preceding report. Conversely, if f has a covariant (of weight δ and index λ) which is not identically zero, f^δ is divisible by Δ^λ and Δ'^λ. If $n<5$, the Hessian of f is not identically zero, if f is not expressible in fewer than n variables, so that f has the same irreducible factors as Δ and Δ'. Hence, if $n=3$, f is a product of powers of linear forms; if $n=4$, f is a power of a quadratic form or a product of powers of linear forms.

On the composition of special forms of degree >2, see this History, Vol. II, pp. 470, 570, 593–5, 677–8, 691, 697, 727–8.

Report was made in Vol. II of this History on Minkowski's maximum value of the minimum of $|f_1|^p+\ldots+|f_n|^p$, where the f_i are linear forms (pp. 95, 96), on representation by x^3+y^3 (pp. 572–8) or as a sum of powers (pp. 717–729), and on definite forms representable as sums of squares or fourth powers of polynomials (pp. 720, 723, papers 21, 22, 38*a*).

---

[17] Comptes Rendus Paris, 164, 1917, 991–3.
[18] *Ibid.*, 172, 1921, 636–640; errata, 1262.
[19] Comptes Rendus du Congrès Internat. Math. (Strasbourg), 1921, 131–146.

CHAPTER XV.

BINARY HERMITIAN FORMS.

Ch. Hermite[1] wrote $v_0=x-iy$ for the conjugate of $v=x+iy$ and u_0 for the conjugate of u and considered the form

$$f(v, u; v_0, u_0)=Avv_0+Bvu_0+B_0v_0u+Cuu_0,$$

where A and C are real, while B, B_0 are conjugate imaginaries. Thus f takes only real values. Such a form f is called a binary *Hermitian* form.* The substitution

$$(1)\qquad v=aV+bU,\quad v_0=a_0V_0+b_0U_0,\quad u=cV+dU,\quad u_0=c_0V_0+d_0U_0$$

transforms f into a Hermitian form

$$F=A'VV_0+B'VU_0+B_0'V_0U+C'UU_0,$$

where

$$A'=f(a, c; a_0, c_0),\qquad C'=f(b, d; b_0, d_0),\qquad B'=a\frac{\partial C'}{\partial b}+c\frac{\partial C'}{\partial d}.$$

It follows that

$$B'B_0'-A'C'=(ad-bc)(a_0d_0-b_0c_0)(BB_0-AC),$$

so that $\Delta=BB_0-AC$ is an invariant, called the *determinant*, of f. If in the last relation we replace a, c, b, d, by[2] m, n, v, u, we obtain

$$(2)\qquad f(v, u; v_0, u_0)\cdot f(m, n; m_0, n_0)\equiv VV_0-\Delta UU_0.$$

for

$$U=nv-mu,\qquad V=m\frac{\partial f}{\partial v}+n\frac{\partial f}{\partial u}.$$

Replace V by $V\sqrt{\Delta}$ and hence V_0 by $\pm V_0\sqrt{\Delta}$, where the sign is that of Δ; we get

$$f(v, u; v_0, u_0)=\pm(VV_0\mp UU_0)\Delta/M,\qquad M=f(m, n; m_0, n_0).$$

Thus f is definite or indefinite, according as Δ is negative or positive. In the first case, all numbers represented by f have the same sign; and f is called a positive or negative (definite) form according as those numbers are all positive or all negative.

Let m, n be relatively prime complex integers. We can determine complex integers μ, ν such that $m\nu-n\mu=1$. In (2) replace v by μ, u by ν, and write $x+yi$ for $V=m\partial f/\partial\mu+n\partial f/\partial\nu$; we get

$$f(\mu, \nu; \mu_0, \nu_0)\ M=x^2+y^2-\Delta.$$

---

* Gauss barely mentioned the general f and its determinant in a posth. MS. of later than 1834, Werke, X_1, 1917, 94.

[1] Jour. für Math., 47, 1854, 345–368; Oeuvres, I, 237–263.

[2] Incorrectly v, u, m, n in the original.

Thus the substitution (1), with a, b, c, d replaced by m, μ, n, ν, replaces f by

$$(3) \qquad F=MVV_0+(x+iy)VU_0+(x-iy)V_0U+\frac{x^2+y^2-\Delta}{M}UU_0,$$

where $M=f(m, n;\ m_0, n_0)$. Hence to find all proper representations of a given integer M by a form f, employ in turn the various sets of incongruent solutions x, y of $x^2+y^2\equiv\Delta \pmod{M}$, and determine the substitutions which replace f by the form (3). From each such substitution

$$v=mV+\mu U,\quad v_0=m_0V_0+\mu_0U_0,\quad u=nV+\nu U,\quad u_0=n_0V_0+\nu_0U_0,$$

where $m\nu-n\mu=+1$, we obtain a proper representation of M by f. All such representations are distinct. This theory is parallel to that of Gauss, Disq. Arith., Arts. 154–5, 168.

Every positive definite form can be transformed, by a substitution (1) with integral complex coefficients of determinant $ad-bc=1$, into a *reduced* form f in which $A\leqq C$, and $2m$ is numerically $\leqq A$ and $2n$ is numerically $\leqq A$, where $m+ni=B$. It follows that $A\leqq\sqrt{-2\Delta}$. Hence for a reduced positive form f with integral coefficients and a given negative determinant Δ, A and hence m and n have only a limited number of values, and likewise for C since $\Delta=m^2+n^2-AC$. For example, if $\Delta=-1$, then $A=1$, $B=0$, $C=1$, $f=VV_0+UU_0$. The application to sums of four squares were quoted on pp. 288–9 of Vol. II of this History; while on p. 225 of Vol. I were quoted the auxiliary results on the number of solutions of $x^2+Ay^2\equiv\Delta \pmod{M}$ and of general congruences.

E. Picard[3] employed a complex variable x and a positive real number y. If a, b, c, d are complex integers, the substitution

$$(4) \qquad X=dX'+bY',\quad Y=cX'+aY',\quad ad-bc=1,$$

replaces the positive definite form

$$(5) \qquad XX_0+xXY_0+x_0X_0Y+(xx_0+y^2)YY_0$$

by $AX'X'_0+BX'Y'_0+\ldots$, where

$$A=cc_0(xx_0+y^2)+dc_0x+d_0cx_0+dd_0,\ \ldots.$$

Hence we are led to a discontinuous group G of substitutions

$$(S) \quad x'=\frac{B}{A},\qquad x'x'_0+y'^2=\frac{C}{A}.$$

Write $x=\xi+i\eta$, $y=\zeta>0$, and interpret ξ, η, ζ as rectangular coordinates of a point in space. Any form (5) can be transformed by some substitution (4) with complex integral coefficients into a reduced form (Hermite[1]). Then x', y' are found by (S). By the conditions for a reduced form, $x'x'_0+y'^2\geqq 1$, and the coefficient of i in x' is numerically $\leqq\frac{1}{2}$. Hence if (ξ', η', ζ') is the point into which (ξ, η, ζ) is transformed,

$$(6) \qquad -\tfrac{1}{2}\leqq\xi'\leqq\tfrac{1}{2},\quad -\tfrac{1}{2}\leqq\eta'\leqq\tfrac{1}{2},\quad \xi'^2+\eta'^2+\zeta'^2\geqq 1.$$

[3] Bull. Soc. Math. France, 12, 1883–4, 43–47; extract in Math. Annalen, 39, 1891, 142–4.

The fundamental polyhedron P of the group G is exterior to the sphere of radius unity and limited by the four planes $\xi=\frac{1}{2}$, $\xi=-\frac{1}{2}$, $\eta=\frac{1}{2}$, $\eta=-\frac{1}{2}$. There is a single point of P corresponding to any chosen point in space with $\zeta>0$.

Picard[4] considered the canonical form $F=uu_0-vv_0$ of any indefinite Hermitian form $F=axx_0+\ldots+cyy_0$ with complex integral coefficients. Let

$$U=\mu D_0u+\mu C_0v, \qquad V=Cu+Dv, \qquad \mu\mu_0=1$$

be the general automorph of determinant unity of F. The form

$$UU_0+VV_0=uu_0-vv_0+2(Cu+Dv)(C_0u_0+D_0v_0)$$

becomes $\Phi(x, y; x_0, y_0)$ when u, v are replaced by their values in x, y. This definite form Φ involves three arbitrary parameters appearing in C and D, subject only to the condition $DD_0-CC_0=1$. Apply to Φ Hermite's method of continual reduction (Ch. I[53]). Suppose we have found all the substitutions which, for all values of the three parameters, replace Φ by a reduced form in the sense of Hermite,[1] and apply each of these substitutions to F. The number of resulting forms f is limited, their coefficients being limited as functions of the determinant Δ of F. These f's are called the *reduced* forms of F. Two F's are arithmetically equivalent if and only if their f's are identical.

The preceding problem is simplified by interpreting geometrically each of the five conditions

$$a<c, \quad 2m<a, \quad -2m<a, \quad 2n<a, \quad -2n<a$$

that

$$\Phi=axx_0+(m+ni)xy_0+(m-ni)x_0y+cyy_0$$

be a positive reduced form. Writing $z=C/D$, we see that $a<c$ becomes

$$(zz_0+1)(\alpha\alpha_0+\gamma\gamma_0-\beta\beta_0-\delta\delta_0)+2z(\alpha\gamma_0-\beta\delta_0)+2z_0(\alpha_0\gamma-\beta_0\delta)<0,$$

if F becomes uu_0-vv_0 for $u=\alpha x+\beta y$, $v=\gamma x+\delta y$. Equating to zero the left member of the inequality, we obtain in the complex z-plane the equation of a circle which cuts orthogonally the circle K of radius unity and center at the origin. Hence Φ is reduced if the point representing z is inside a curvilinear polygon P whose sides (at most five in number) are circles cutting K orthogonally. Details are given for $F=xx_0-3yy_0$, for which there are eight contiguous forms.

For simplicity let the given indefinite form F be reduced. The corresponding form Φ will be reduced for certain values of the parameter z for which we may assume $|z|\leqq 1$. These values are represented by points interior to the polygon P. When the point z departs from P, Φ ceases to be reduced. Let the substitution which replaces Φ by a reduced form Φ_1 replace F by F_1. The points z for which Φ_1 is reduced are inside a contiguous polygon P_1 having in common with P only a side or a vertex. Treating each of these polygons contiguous to P as we did P, we obtain a series of polygons whose corresponding forms, F_1, etc., give all the reduced forms arithmetically equivalent to F. The polygon formed of all the P's is a fundamental

[4] Annales sc. école norm. sup., (3), 1, 1884, 9–54; extract in Comptes Rendus Paris, 96, 1883, 1567–1571, 1779–1782; 97, 1883, 745–7 (corresponding to the three paragraphs of this report).

region for the infinite discontinuous group of the linear fractional substitutions

$$z'=\frac{D_0z+\mu_0C}{C_0z+\mu_0D}$$

corresponding to the automorphs of F. Details are given for $F=xx_0-2yy_0$.

Picard[5] proved that a binary indefinite Hermitian form with complex integral coefficients is not invariant in general under an elliptic substitution.

L. Bianchi[6] recalled that Picard[4] noticed a distinction between forms F whose determinant Δ is a sum of two squares and those for which Δ is not. It is here proved directly, without using the method of continual reduction, that if Δ is a sum of two squares and has no square factor the number of classes of forms F of determinant Δ is 2 or 3, according as Δ is even or odd. In each case the group of automorphs of F is transformed into a modular group.

Bianchi[7] gave a simple complete theory of equivalence of forms $f=axx_0+bxy_0+b_0x_0y+cyy_0$, where a and c are ordinary integers, while $b=r+st$, r and s being ordinary integers, and t is i or an imaginary cube root ϵ of unity, and b_0 is the conjugate of b. Employ the space coordinates ξ, η, ζ and fundamental polyhedron P defined by Bianchi[5] of Ch. VII. If f is indefinite, $azz_0+bz+b_0z_0+c=0$ represents a real circle C in the $\xi\eta$-plane. Over it describe a hemisphere, which is said to represent f. If the hemisphere cuts through the polyhedron P, f is reduced. Two reduced forms belong to the same period. But if f is definite of determinant $D<0$, C is an imaginary circle, through which go a pencil of spheres, two reducing to the points

$$\xi=-\frac{m}{a},\quad \eta=\frac{n}{a},\quad \zeta=\pm\frac{\sqrt{-D}}{a}\qquad (b=m+in).$$

That one of these points which lies above the $\xi\eta$-plane is taken as the point representing f. If it lies in P, f is reduced. Every f is equivalent to a reduced form.

Bianchi[8] extended this theory to various imaginary quadratic fields.

R. Fricke and F. Klein[9] gave an exposition largely following Bianchi.[7] Further, they represented (pp. 497–8) $axx_0+bxy_0+b_0x_0y+cyy_0$ $(b=b_1+ib_2)$ by the point in space with the homogeneous coordinates $(c, -b_1, b_2, a)$.

O. Bohler[10] employed the Fricke-Klein[9] representation of a definite positive form $f=(a, b, b_0, c)$, where $b=b_1+ib_2$, by the point with the homogeneous coordinates a, b_1, b_2, c. Call f reduced if the representative point is on or inside the fundamental ocahedron or dodecahedron, according as the coefficients of the substitutions of the group are of the form $u+\rho v$, where ρ is an imaginary fourth or cube root of unity, while u and v are integers. In the first case, the arithmetical conditions are

$$a\gtreqless 2|b_1|,\quad a\gtreqless 2|b_2|,\quad c\gtreqless 2|b_1|,\quad c\gtreqless 2|b_2|.$$

[5] Amer. Jour. Math., II, 1889, 187–194.
[6] Atti R. Accad. Lincei, Rendiconti, (4), 6, I, 1890, 375–384.
[7] Math. Annalen, 38, 1891, 313–333.
[8] *Ibid.*, 40, 1892, 389–412 (42, 1893, 30–57; 43, 1893, 101–135). In preliminary form in Atti R. Accad. Lincei, Rendiconti, (4), 6, I, 1890, 331–9; (4), 7, II, 1891, 3–11, where he admits Picard's[3] prior determination of the fundamental polyhedron for the domain (1, i). Groups with coefficients in any field were studied by A. Viterbi, Gior. di Mat., 36, 1898, 346–361.
[9] Automorphe Funktionen, Leipzig, 1, 1897, 92, 452–8, 467–500.
[10] Über die Picard'schen Gruppen aus dem Zahlkörper der dritten und vierten Einheitswurzeln, Diss., Zürich, 1905, 36–49, 96–99.

P. Fatou[11] called $axx_0+bxy_0+\dots$ *primitive* if for $b=b_1+ib_2$ the integers a, b_1, b_2, c have no common divisor, and primtive of the first or second *species*, according as the g.c.d. of a, $2b_1$, $2b_2$, c is 1 or 2. Let $f, f', \dots$ be representatives of the different classes of positive primitive forms of the first species (i. e., properly primitive) of negative determinant $-\Delta$. Dirichlet's method is said to give the identity

$$\frac{1}{k}\Sigma\frac{1}{f^s}+\frac{1}{k'}\Sigma\frac{1}{f'^s}+\dots=\Sigma\frac{1}{n^s}\cdot\Sigma\frac{1}{n^{s-1}}\qquad(s>2),$$

where the summations on the left extend over the complex integers x, y for which the corresponding form represents a number prime to 2Δ, while the summations on the right extend over the positive integers n prime to 2Δ. Further, k denotes the number of automorphs of f with complex integral coefficients of determinant unity, k' the number of automorphs of f', etc. Thus if a representation by f' is counted as $1/k'$, the total number of proper and improper representations of an integer m prime to 2Δ by the totality of forms $f, f', \dots$, is equal to the sum of the divisors of m. For $\Delta=1$, 2 or 3, there is a single class represented by $xx_0+\Delta yy_0$, whence follows Jacobi's theorem on the number of representations of an odd integer as a sum of 4 squares and two analogous theorems by Liouville.

To deduce from the above identity the class-number

$$\frac{\Delta}{2}\Pi\left\{1+\left(\frac{-1}{p}\right)\frac{1}{p}\right\}\qquad(\Delta>1),$$

where the product extends over the different odd prime factors p of Δ, we have only to seek the limits for $s=2$ of the products of the two members by $s-2$. The class number is 1 only for $\Delta=1$, 2, 3.

G. Humbert[12] doubted that the method used by Bianchi[8] for the field R defined by $\sqrt{-D}$ for various special values of D is applicable for a general $D>0$. Hence he returned to the method of Hermite, starting with the reduction of

$$(A,\ b,\ C)=Axx_0-bxy_0-b_0x_0y+Cyy_0,\qquad AC-bb_0>0.$$

We can find a properly equivalent form whose first coefficient A is its proper minimum (the least real integer >0 properly representable by it). Then by replacing x by $x+\lambda y$ we can find a form (A, b, C) for which, if $b=B_1-iB_2\sqrt{D}$, we have

$$(7)\qquad -\tfrac{1}{2}\leqq B_1/A,\qquad B_2/A<\tfrac{1}{2},$$

and obviously also $C\gtreqless A$. But these inequalities do not now imply that the proper minimum is actually A; we must require that

$$(8)\qquad A\lambda\lambda_0-b\lambda\mu_0-b_0\lambda_0\mu+C\mu\mu_0\gtreqless A$$

for all sets of integers λ, μ of the field R for which the principal ideals (λ), (μ) are relatively prime. When (7) and (8) hold, the form is called *reduced.* The form (A, b, C) is represented by the point (ξ, η, ζ) defined by

$$\xi+\eta i=b_0/A,\qquad \xi-\eta i=b/A,\qquad \xi^2+\eta^2+\zeta^2=C/A,\qquad \zeta>0.$$

[11] Comptes Rendus Paris, 142, 1906, 505–6 (in full, and after making various corrections indicated, *ibid.*, 166, 1918, 581.

[12] Comptes Rendus Paris, 161, 1915, 189–196, 227–234.

Then (8) requires that (ξ, η, ζ) be on or without the sphere (λ, μ) whose center is the point $z=\lambda/\mu$ in the plane $\zeta=0$ and whose radius is the reciprocal of the modulus of μ. The point representing a reduced form is thus in a prismatic domain bounded by the four planes $\xi=\pm\frac{1}{2}$ and $\eta=\pm\frac{1}{2}\sqrt{D}$, and proved to be closed below by the spheres (λ, μ). Proof is made by use of the theory of ideals. There are points in $\zeta=0$ on certain spheres (λ, μ), but inside of none, and called *singular* vertices of the domain. This domain is a fundamental region for the group of linear fractional substitutions on one variable with integral coefficients in R of determinant unity. To find the reduced forms whose proper minima are the least integers represented properly or not by the forms, use is made also of the spheres (λ, μ) for which λ, μ are not relatively prime.

Humbert[13] gave a rapid method of finding a fundamental domain of the group of automorphs of xx_0-Dyy_0, where D is a positive real integer:

$$z'=\frac{\lambda z+D\nu_0}{\nu z+\lambda_0}, \qquad \lambda\lambda_0-D\nu\nu_0=1.$$

The domain is the locus of the points ζ on or within the circle $X^2+Y^2=D$, above the X-axis, and exterior to all the circles

$$\left(\zeta-\frac{\lambda}{\mu}\right)\left(\zeta_0-\frac{\lambda_0}{\nu_0}\right)=\frac{1}{\nu\nu_0}.$$

G. Julia[14] recalled Picard's[3, 4] result that every indefinite binary Hermitian form H is invariant under an infinite subgroup G of the complex modular group, which constitutes a Fuchsian group conserving the half sphere Σ representing H. On such a Σ consider the half-circle Γ representing a quadratic form f with complex integral coefficients (Ch. VII). Then f is said to be *contained* in H. There exists a hyperbolic substitution T of G which conserves f. An f is contained in an H if and only if the norm of the determinant of f is the square of a real integer, and then there exist infinitely many H's containing the f, and there exists a modular hyperbolic substitution leaving f invariant.

G. Humbert[15] extended the method of Fatou[11] and obtained the number of representations by a properly primitive indefinite form

$$f(x,y)=axx_0+bx_0y+b_0xy_0+cyy_0, \qquad D=bb_0-ac>0,$$

where a and c are not both even and a, b, b_0, c have no common factor. In each class of forms of determinant D choose one representative f with $a>0$; let $f, f', \ldots$ be the forms chosen. From one representation ξ, η of a positive odd integer m by f are deduced an infinite series of representations by applying to ξ, η the automorphs S of f; the group of the S has a fundamental domain R of points in a region exterior to the circle

$$a(\xi^2+\eta^2)+b_0(\xi+i\eta)+b(\xi-i\eta)+c=0.$$

Among the representations (of the same series) of m by f there are only two,

[13] Comptes Rendus Paris, 162, 1916, 697–702.
[14] Comptes Rendus Paris, 163, 1916, 599–600, 691–4.
[15] Comptes Rendus Paris, 166, 1918, 581–7.

$m=f(x_i, y_i)=f(-x_i, -y_i)$ for which the analytic point $z_i=x_i/y_i$ is in or on the boundary of R. Dirichlet's classic method is said to give the identity

$$(9)\qquad \Sigma\{f(x_i, y_i)\}^{-s}+\Sigma'\{f'(x_i', y_i')\}^{-s}+\ldots=2\Sigma\frac{1}{n^s}\cdot\Sigma\frac{1}{n^{s-1}}\qquad (s>2),$$

where the first summation on the left extends over all complex integers x_i, y_i for which $f(x_i, y_i)$ is positive and prime to $2D$, with $z_i=x_i/y_i$ on or within the boundary of R; the second summation Σ' relates similarly to f' and the analogous domain R'. The summations on the right extend over the positive real integers n prime to $2D$.

Hence the total number of representations of a positive integer m prime to $2D$ by the forms $f, f', \ldots$ is double the sum of the divisors of m, provided among the representations, $m=f^{(h)}(x, y)$, we count only those for which the analytic point x/y is on or within the boundary of the domain R_h which corresponds to $f^{(h)}$; if on the boundary, the corresponding representation is counted as $\frac{1}{2}$; if at one of a cycle of ν equivalent vertices of R_h, it is counted as $1/\nu$.

Application is made to representations by $\psi=z^2+t^2-D(u^2+v^2)$, where $D=1, 2, 3, 5, 6$, when there is a single class of properly primitive forms of determinant D. For example, if $D=1$, the number of representations of a positive odd integer m by ψ for which $z, \ldots, v$ are real integers such that

$$z^2+t^2+u^2+v^2>2|zu+tv|+2|zv-tu|,$$

is quadruple the sum of the divisors of m.

Humbert[16] wrote $s=2+\rho$, multiplied the two members of his identity (9) by ρ and evaluated (by the method of Dirichlet) the two limits when ρ approaches zero over positive values. The final formula evaluates the sum of the non-euclidean areas (expressed by integrals) of the fundamental spherical domains of the groups of automorphs of $f, f', \ldots$.

Humbert[17] proved that if P is positive and $\equiv 1$ or 2 (mod 4), all properly primitive indefinite Hermitian forms f of given positive determinant D which is odd or double an odd integer, of the field $R(i\sqrt{P})$, belong to a single class, when D and P have no common odd divisor. Each form represents properly every odd integer prime to DP. If $P=1$, $D\equiv 1$ (mod 4), the improperly primitive forms belong to two classes (transformable into each other by substitutions of determinant i).

In f write $x=z+it$, $y=u+iv$ to obtain a real quaternary form. Then replace z by nz, and v by $2mz$ (where m, n are integers) to obtain an indefinite ternary quadratic form Φ. To the latter apply the theorems quoted at the end of the reports on A. Meyer[39, 43] of Ch. IX to conclude that Φ represents $+1$, so that we have the first theorem above.

G. Humbert[18] considered Hermitian forms $f=(a, b, b_0, c)$ of discriminant $\Delta=ac-bb_0$, where a and c are real integers and b is an integral algebraic number of the field $\mathcal{C}$ defined by $\sqrt{-P}$, P being a positive integer without square factor such that $P\equiv 1$ or 2 (mod 4). The *measure* of the totality of the H classes of positive,

[16] Comptes Rendus Paris, 166, 1918, 753–8. Generalized to the field $R(i\sqrt{P})$, *ibid.*, 171, 1920, 377–382, 445–450.
[17] Comptes Rendus Paris, 166, 1918, 865–870.
[18] Comptes Rendus Paris, 168, 1919, 1240–6. Proofs in Humbert.[21]

properly primitive forms f (a, c not both even, a, b, b_0, c with no real integral factor in common), of given discriminant Δ, having no common odd factor with P, is defined to be $M(\Delta)=\Sigma 1/k_j$, where k_j is the number of transformations whose coefficients are integers of $\mathcal{C}$ of determinant unity of f_j into itself, if $f_1, \ldots, f_H$ are representatives of the classes. It is stated that (proof in Humbert[21])

$$M(\Delta)=\tfrac{1}{8}P\Delta\prod_{\delta}\left\{1+\left(\frac{-P}{\delta}\right)\frac{1}{\delta}\right\}\prod_{\omega}\left\{1+\left(\frac{-\Delta}{\omega}\right)\frac{1}{\omega}\right\},$$

where δ ranges over the real odd prime factors >1 of Δ, and ω over those of P. For $P=1$ this becomes Fatou's[11] final result.

The measure of the totality of classes of positive, primitive or imprimitive, but proper forms (a, c not both even), of given discriminant Δ, prime to $2P$, is[19]

$$\mathcal{M}(\Delta)=\tfrac{1}{8}P\left(\frac{-P}{\Delta}\right)\prod_{\omega}\left\{1+\left(\frac{-\Delta}{\omega}\right)\frac{1}{\omega}\right\}\Sigma d\left(\frac{-P}{d}\right),$$

summed for all positive integral divisors d (including 1) of P.

For $P=1$, the number of classes of positive, proper forms of odd discriminant Δ is

$$\tfrac{1}{4}\left(\frac{-1}{\Delta}\right)\Sigma\delta\left(\frac{-1}{\delta}\right)+\tfrac{1}{4}\left\{2+\left(\frac{-1}{\Delta}\right)\right\}T(\Delta),$$

where $T(\Delta)$ is the number of divisors δ of Δ. Corresponding results are given for $P=2$.

In a positive, proper reduced form of Hermite, let $b=b_1+i\sqrt{2}b_2$. The corresponding quadratic form (a, b_1, c) is a positive, proper reduced form of determinant $-\Delta-2b_2^2$, $0<2|b_2|<a$. Conversely, if the latter is given, we obtain the proper reduced forms $(a, b_1, \pm b_2, c)$. Hence a certain sum of class-numbers of quadratic forms of determinants $-\Delta-2b_2^2$, with b_2 variable, is equal to the number of classes of Hermitian forms of discriminant Δ. Similarly for $P=1$.

Humbert[20] discussed the determination of a fundamental domain of a linear fractional group Γ leaving invariant a circle. It is shown that the method of "rayonnement," previously regarded as purely theoretic, can be reduced to computations and made as manageable as the usual method of symmetry.

Humbert[21] considered positive Hermitian forms f, with a, c, Δ positive, and used his[18] former notations. Let I be an ideal, and I_0 its conjugate, of the field $\mathcal{C}$. If u, v, are algebraic integers of I, $f(u, v)$ is the product of II_0 by a rational integer m, and we may write $m=f(u/I, v/I)$ symbolically, and say that we have a generalized representation, *belonging to* I, of m by f. The representation is proper if I is the g.c.d. of u, v. The numbers obtained by representations belonging to I coincide with those obtained by representations belonging to any equivalent ideal.

From each of the H classes of properly primitive, positive forms select a representative form f_j. To each proper representation, belonging to I, of a positive integer m, prime to 2Δ, by f_j corresponds a solution B of

$$BB_0/I^2I_0^2\equiv-\Delta \pmod{m},$$

[19] Proof, Comptes Rendus Paris, 169, 1919, 448–454.
[20] *Ibid.*, 169, 1919, 205–211.
[21] Comptes Rendus Paris, 169, 1919, 309–315, 360–5, 407–414.

where B is in the ideal I^2, and conversely. Two solutions B and B' are called distinct if $(B'-B)/m$ is not in I^2. The number of proper representations, belonging to I, of m by the totality of the forms f_j is the number of distinct solutions of the congruence, provided a representation is counted as $1/k_j$, where k_j is the number of transformations whose coefficients are integers of $\mathcal{C}$ of determinant unity of f_j into itself. If Δ has no odd divisor in common with P, the number of solutions of the congruence is

$$N=\Pi p^{a-1}\{p-(-P/p)\}\cdot\Pi(2\rho^{\beta}),$$

if $m=\Pi p^{a}\cdot\Pi\rho^{\beta}$, the primes p not dividing P, while the ρ's are odd prime factors of P of which $-\Delta$ is a quadratic residue. Since N is independent of I, it gives the number of ordinary proper representations of m by the f_j. Hence

$$\sum_{j=1}^{H}\sum_{x,\,y}\frac{1}{k_j}f_j^{-s}\left(\frac{x}{I},\ \frac{y}{I}\right)=\sum\frac{N}{m^s}\qquad(s>2),$$

where the second summation on the left extends over all sets of algebraic integers x, y of $\mathcal{C}$ such that x/I and y/I are relatively prime ideals and f_j is prime to 2Δ, while the summation on the right extends over the above integers $m=\Pi p^{a}\cdot\Pi\rho^{\beta}$ prime to 2Δ.

From each of the h classes of ideals of $\mathcal{C}$ select an ideal I_c. After modifying the second member of the preceding formula by the classic method of Dirichlet, we get the fundamental formula

$$\sum_{j=1}^{H}\sum_{c=1}^{h}\sum_{X,\,Y}\frac{1}{k_j}f_j^{-s}\left(\frac{X}{I_c},\ \frac{Y}{I_c}\right)=h\sum\frac{1}{n^s}\cdot\sum\frac{1}{n^{s-1}}\cdot\prod_{\omega}\left\{1+\left(\frac{-\Delta}{\omega}\right)\frac{1}{\omega^{s-1}}\right\},$$

where n ranges over all positive odd integers prime to 2Δ, and ω over all the distinct odd prime factors of P, while X, Y are algebraic integers of I_c such that f_j is prime to 2Δ. Passing to the limit $s=2$, we obtain his[18] expression for $M(\Delta)$.

Humbert[22] gave an immediate extension of his[18] formula for $M(\Delta)$ to apply when P and Δ have any common divisor, also for the new case $P\equiv3 \pmod 4$, and treated completely also the case of improperly primitive forms.

Humbert[23] proved that, if Δ is not a quadratic residue of all odd prime factors of P (so that a form has only two automorphs), the number of classes of positive, proper Hermitian forms, whether primitive or not, of discriminant Δ of the field defined by $\sqrt{-P}$, is the double of the measure $\mathcal{M}(\Delta)$ of the totality of the classes (Humbert[18]). As an application there is deduced a relation between class-numbers of binary quadratic forms of discriminants $\Delta+5t^2$, $t=0, 1, \ldots$.

Humbert[24] employed an ideal I of the field R defined by $i\sqrt{P}$, where $P\equiv1$, 2 or 3 (mod 4). If λ, μ, ν, ρ are numbers of I and if in the Hermitian form $f(x, y)$ we replace x, y by the respective symbolic expressions

$$\frac{\lambda}{I}x+\frac{\mu}{I}y,\qquad \frac{\nu}{I}x+\frac{\rho}{I}y,$$

[22] Comptes Rendus Paris, 170, 1920, 349–355.
[23] *Ibid.*, 481–6. Errata, 171, 1920, 450.
[24] *Ibid.*, 170, 1920, 541–7, 625-630.

and x_0, y_0 by the expressions conjugate to them, and if we replace II_0 by its real integral value, we obtain a Hermitian form which, like f, has integral algebraic coefficients in $R(i\sqrt{P})$. This symbolic substitution is said to belong to I. If we employ for I the non-equivalent ambiguous ideals and take $\lambda\rho-\mu\nu=II_0$, we obtain a group Γ essentially that of Bianchi[8] (Math. Annalen, 42, 1893). Defining equivalence and classes with respect to this new group, we obtain the measure of its classes by an extension of Humbert's[21] method. The fundamental domain of Γ is determined. For $P=6$ there is found the number of these extended classes of positive proper Hermitian forms of discriminant Δ. There is deduced a relation between class-numbers of binary quadratic forms of discriminants $\Delta+6t^2$, $t=0, 1, \ldots$.

Humbert[25] noted that his[17] theorem holds also for the forms of the ring defined by $i\sqrt{P}$, when $P\equiv 3 \pmod 4$. He made here a further[21] study of representations, belonging to an ideal I of the field or ring, of m by $f=(a, b, b_0, c)$, and especially of restricted representations in which (X, Y) gives a point of the fundamental domain $\mathscr{P}$. There is given the analogue for indefinite forms of his[21] fundamental formula for positive forms. If $\mathscr{P}_1$ is the domain symmetrical with $\mathscr{P}$ with respect to the origin, the number of representations of a positive integer m prime to $2D$ by $x^2+y^2-D(z^2+t^2)$, such that the point $(x+iy):(z+it)$ belongs to the domain composed of both $\mathscr{P}$ and $\mathscr{P}_1$, is the quadruple of the sum of the divisors of m.

[25] Comptes Rendus Paris, 171, 1920, 287–293. Errata, 450.

CHAPTER XVI.

HERMITIAN FORMS IN n VARIABLES AND THEIR CONJUGATES.

Ch. Hermite[1] considered the form

$$f=x_0(a_{00}x+a_{01}y+\ldots+a_{0n}v)+\ldots+v_0(a_{n0}x+a_{n1}y+\ldots+a_{nn}v)$$

in the $n+1$ complex variables $x, y, \ldots, v$ and their conjugates $x_0, \ldots, v_0$, such that a_{jk} and a_{kj} are conjugate imaginaries. It is proved that there exists a linear substitution S on $x, \ldots, v$ with complex integral coefficients of determinant unity such that S and the conjugate substitution S_0 on $x_0, \ldots, v_0$ transform any given definite form into a form f having

$$a_{00}a_{11}\ldots a_{nn}<2^{\frac{1}{2}n(n+1)}d, \qquad d=|a_{ij}|,$$

and called a *reduced* form. Assume that also the coefficients of the form are complex integers. By the last relation, each real integer a_{jj} is limited in terms of the invariant d. Since f is definite, $a_{jj}a_{kk}-a_{jk}a_{kj}>0$, whence the absolute value of a_{jk} is limited. Hence there is a limited number of reduced forms (and hence of classes) of a given determinant d.

C. Jordan[2] noted that the method used by Korkine and Zolotareff[18] of Ch. XI for the reduction of n-ary quadratic forms may be extended to Hermitian forms

$$g=\sum_{j=1}^{n} N(a_{j1}x_1+\ldots+a_{jn}x_n)=\sum_{j,\,k=1}^{n} b_{jk}x_jx'_k,$$

in which x, x' are conjugate complex variables and $N(x)$ denotes xx'. Since

$$g=b_{11}N\left(x_1+\frac{b_{21}}{b_{11}}x_2+\ldots\right)+\sum_{j,\,k=2}^{n} c_{jk}x_jx'_k,$$

we ultimately obtain

$$g=\sum_{j=1}^{n} m_jN(y_j), \quad y_1=x_1+\epsilon_{12}x_2+\ldots+\epsilon_{1n}x_n,$$
$$y_2=x_2+\epsilon_{23}x_3+\ldots+\epsilon_{2n}x_n, \ \ldots, \ y_n=x_n,$$

where the m_j are positive real numbers and the ϵ_{jk} are complex numbers.

Consider the systems of values of the complex variables $x_1, \ldots, x_n$ for which $g<q$, where q is any assigned positive number. Then $m_jN(y_j)<q$, so that the moduli of $x_n, x_{n-1}, \ldots, x_1$ are limited. Hence there is a limited number of sets of complex integers $x_1, \ldots, x_n$ for which $g<q$. Let $\alpha_1, \ldots, \alpha_n$ be one of these sets for

[1] Jour. für Math., 53, 1857, 182–192; Oeuvres, I, 415–428.
[2] Jour. école polyt., t. 29, cah. 48, 1880, 111, 119–134; summary in Comptes Rendus Paris, 90, 1880, 1422–3. For application to forms of degree m, see Ch. XIV.[6]

which g has a minimum value μ_1. Evidently the a's have no common divisor not a unit. Hence there exist complex integers β_j, γ_j, ... such that the determinant of

$$a_1x_1+\beta_1x_2+\gamma_1x_3+\ldots,\quad \ldots,\quad a_nx_1+\beta_nx_2+\gamma_nx_3+\ldots$$

is unity. Substituting these for $x_1, \ldots, x_n$ and their conjugates for $x_1', \ldots, x_n'$ in g, we obtain a Hermitian form g' having μ_1 as the coefficient of x_1x' Hence, as above, we may write $g'=\mu_1N(y_1)+g'$, where g_1' involves neither x_1 nor x'. At least one system of complex integral values of $x_2, \ldots, x_n$ makes g' a minimum μ_2. Then $\mu_2 \overline{\geqq} \frac{1}{2}\mu_1$. Proceeding with g' as we did with g, etc., we see that there exists a linear substitution with complex integral coefficients of determinant unity which replaces g by

$$r=\sum_{j=1}^{n}\mu_jN(y_j),\qquad \mu_{k+1}\overline{\geqq}\tfrac{1}{2}\mu_k,$$

with y's as above. Replacing x_j by $x_j+\delta_{jj+1}x_{j+1}+\ldots+\delta_{jn}x_n$ for $j=1, \ldots, n$, we can choose complex integers δ_{jk} such that $N(\epsilon_{jk})\leqq\frac{1}{2}$. Such a form r is called *reduced.* Hence any g is arithmetically equivalent to a reduced form.

The determinants Δ and $\mu_1\mu_2\ldots\mu_n$ of g and r are equal. If $\Delta\neq 0$ it follows that $\mu_{j+1}^{n-j}\leqq c\Delta/\mu_1$, where c depends only on n, j. Hence the modulus of each coefficient c_{kl} of $r=\Sigma c_{kl}x_kx_l'$ has a superior limit depending on μ_1 and Δ.

Assume that g has complex integral coefficients. Then evidently $\mu_1\overline{\geqq}1$, so that each μ_k has a superior limit depending on Δ only, and the same is true of the modulus of each coefficient c_{kl} of r. Hence every form g of a given determinant $\Delta\neq 0$ with complex integral coefficients is arithmetically equivalent to a reduced form r the moduli of whose coefficients have superior limits depending on Δ only. Since the coefficients of r are also complex integers, they have only a finite number of values. Hence the forms g of a given determinant $\Delta\neq 0$ with complex integral coefficients fall into a finite number of classes.

Finally, for the substitutions with complex integral coefficients of determinant unity which transform any reduced form r into another reduced form, it is proved that the modulus of each coefficient has a superior limit depending only on the number n of variables.

E. Picard[3] considered a ternary Hermitian form

$$(1)\qquad f=axx_0+a'yy_0+a''zz_0+byz_0+b_0y_0z+b'zx_0+b_0'z_0x+b''xy_0+b_0''x_0y,$$

where a, a', a'' are real, b and b_0 are conjugate imaginaries, etc. Call

$$\delta=\begin{vmatrix} a & b'' & b_0' \\ b_0'' & a' & b \\ b' & b_0 & a'' \end{vmatrix}$$

the *determinant* of f. Under the linear substitution

$$(2)\qquad x=\alpha X+\beta Y+\gamma Z,\quad y=\alpha'X+\beta'Y+\gamma'Z,\quad z=\alpha''X+\beta''Y+\gamma''Z,$$

[3] Acta Math., 1, 1882–3, 297–320. Summary of first part in Comptes Rendus Paris, 95, 1882, 763–6.

of determinant D, and $x_0=\alpha_0X_0+\beta_0Y_0+\gamma_0Z_0$, etc., f becomes a form whose determinant is equal to $DD_0\delta$. If a and $l\equiv aa'-b''b''_0$ are not zero, we have

$$f=\phi/al,\quad \phi\equiv vv_0+luu_0+a\delta ww_0,$$

$$(3)\qquad u=ax+b''_0y+b'z,\quad v=ly+(ab_0-b'b'')z,\quad w=z.$$

If l and $a\delta$ are positive, ϕ is a sum of positive terms, and f is called *definite* and is reducible to $\pm(UU_0+VV_0+WW_0)$. Otherwise, f is reducible to $\pm(UU_0+VV_0-WW_0)$ and is *indefinite.*

Let the coefficients of f and the substitution (2) be integral algebraic numbers of an imaginary quadratic field. From (2), (3) and the similar relations $U=aX+b''_0Y+b'Z, \ldots, W=Z$, we at once obtain u, v, w as linear functions of U, V, W, whose coefficients are fractions whose denominators divide al. This gives an automorph of ϕ if (2) is an automorph of f. It follows that a definite form f has only a finite number of automorphs with integral algebraic coefficients.

Next, consider an indefinite form $\alpha uu_0+\beta vv_0-\gamma ww_0$, where α, β, γ are positive real integers. It has an infinitude of automorphs

$$(4)\qquad U=Mu+Pv+Rw,\quad V=M'u+P'v+R'w,\quad W=M''u+P''v+R''w,$$

with integral algebraic coefficients. The group of corresponding substitutions

$$(5)\qquad X=\frac{Mx+Py+R}{M''x+P''y+R''},\qquad Y=\frac{M'x+P'y+R'}{M''x+P''y+R''},$$

is proved to be discontinuous for all pairs of values $x=x'+ix''$, $y=y'+iy''$ of the domain D defined by

$$\alpha(x'^2+x''^2)+\beta(y'^2+y''^2)-\gamma<0.$$

There are defined uniform functions of x, y, obtained as series convergent in the domain D, which are invariant under the group of substitutions (5). These hyperfuchsian functions are the analogues of the thetafuchsian functions of one variable obtained by Poincaré.

Picard[4] investigated arithmetically forms F of type (1) in which the coefficients are complex integers, but stated that the conclusions may be readily extended to the case of integral algebraic numbers of an imaginary quadratic field. Let F be reducible to $F=uu_0+vv_0-ww_0$ by the substitution

$$u=\alpha x+\beta y+\gamma z,\qquad v=\alpha'x+\beta'y+\gamma'z,\qquad w=\alpha''x+\beta''y+\gamma''z,$$

where $\alpha, \ldots, \gamma''$ need not be complex integers. Let (4) be the general automorph of F. With F associate the definite form

$$\Phi=UU_0+VV_0+WW_0,$$

where U, V, W are given by (4). By the conditions for an automorph,

$$\Phi=F+2\cdot\text{Norm}(M''u+P''v+R''w),$$

where the three parameters are subject to the single condition

$$(6)\qquad M''M''_0+P''P''_0-R''R''_0=-1.$$

[4] Acta Math., 5, 1884-5, 121-182.

For certain values of these parameters seek a substitution with complex integral coefficients which reduces Φ, according to Jordan's[2] definition of a reduced form. When the parameters vary continuously subject to (6), the substitution will cease after a time to reduce Φ and a new substitution must be employed. This process is called the continual reduction of Φ.

We say that the indefinite form F is *reduced* when the corresponding definite form Φ is reduced for certain values of the parameters M'', P'', R''. The number of reduced forms arithmetically equivalent to a given form F of determinant $\Delta \neq 0$ is finite. For, the number of reduced forms F of determinant Δ is limited, since each coefficient of Φ is limited as a function of Δ.

To simplify the further discussion,[5] write $\xi = M''/R''$, $\eta = P''/R''$, $\phi = \Phi/(R''R''_0)$. In view of (6), we have

$$\phi = (1 - \xi\xi_0 - \eta\eta_0)F + 2 \cdot \text{Norm}(u\xi + v\eta + w), \qquad \xi\xi_0 + \eta\eta_0 < 1.$$

The points (ξ, η) for which $\xi\xi_0 + \eta\eta_0 \leqq 1$ form a domain S. The points interior to S for which ϕ is a reduced form constitute a domain D, corresponding to F. It is shown that D has at most one point in common with the boundary $\xi\xi_0 + \eta\eta_0 = 1$ of S. The domain D of a reduced form (1) has a point on the boundary of S if and only if $a = 0$, $b'' = 0$. Whatever be the given indefinite form (1) there exists an arithmetically equivalent reduced form with $a = b'' = 0$. Zero can be represented by every indefinite form (1), in contrast with the theory of ternary quadratic forms. When the point (ξ, η) departs from the domain D, we again employ a substitution to reduce Φ; all such substitutions give rise to adjacent reduced forms to which correspond domains D', D'', ... They with D determine a domain δ which is a fundamental region of the infinite discontinuous group G of linear fractional substitutions on ξ and η obtained when the point (ξ, η) departs from δ. To each point interior to S corresponds by a substitution of G one or a limited number of points within δ, and only one point of sub-domain R called a fundamental domain of G.

Finally,[6] there is an investigation of the hyperfuchsian invariants of the group G in the neighborhood of a point $\xi = 0$, $\eta = -1$ in common with R and the boundary of S.

H. Poincaré[7] started with the canonical forms of ternary linear homogeneous substitutions, found the conditions that each leaves invariant a ternary Hermitian form, and classified the resulting substitutions as elliptic, hyperbolic, parabolic and loxodromic.

Picard[8] obtained results similar to those of Poincaré.

L. Kollros[9] considered a positive definite Hermitian form

$$f = a_{11}xx_0 + a_{12}xy_0 + a_{13}xz_0 + a_{21}yx_0 + \ldots + a_{33}zz_0, \qquad a_{ki} = \text{conjugate of } a_{ik}.$$

Writing $a_{12} = b_{12} + ic_{12}$, $x = x_1 + iy_1$, etc., we obtain a real quadratic form to which is applied the reduction process of Hermite (Jour. für Math., 40, 1850, 302). From f is obtained a corresponding form by changing i to $-i$. One of such a pair of corre-

[5] Summary in Comptes Rendus Paris, 97, 1883, 845–8.
[6] Summary in Comptes Rendus Paris, 97, 1883, 1045–8.
[7] Comptes Rendus Paris, 98, 1884, 349–352.
[8] *Ibid.*, 416–7.
[9] Comptes Rendus Paris, 131, 1900, 173–5.

sponding forms f can always be transformed, by a linear substitution with complex integral coefficients of determinant ± 1 or $\pm i$, in general in a single way, into a reduced form characterized by one of two sets of 20 or 18 inequalities between the a_{ii}, b_{ij}, c_{ij}. It is conjectured that $a_{11}a_{22}a_{33} \leqq 4D$ in a reduced form.

R. Alezais[10] studied the automorphs of $uu_0+vw_0+v_0w$ in connection with hyperfuchsian functions.

L. E. Dickson[11] started with any field (or domain of rationality) F such that there exists a quadratic equation whose coefficients are numbers in F, but whose roots ω and ω' are not in F. Then if a, b range independently over F, the numbers $e=a+b\omega$ constitute a field Q. Write $e'=a+b\omega'$. Then $H=\Sigma a_{ij}\xi_i\xi_j'$ is called a *Hermitian form in* Q if each a_{ij} is in Q, while $a_{ij}'=a_{ji}$. If $|a_{ij}|\neq 0$, we can find a linear transformation $\xi_i=\Sigma\beta_{ik}\eta_k$, $\xi_i'=\Sigma\beta_{ik}'\eta_k'$, with coefficients in Q, which reduces H to $\Sigma\gamma_i\eta_i\eta_i'$, where each γ_i is in F. In case F is a finite (Galois) field, we may take each $\gamma_i=1$. Also the cases in which F is the field of all rational or all real numbers are solved completely.

E. Picard[12] proved that the subgroup of substitutions (5) with real integral coefficients defined by the automorphs (4) with real integral coefficients of $uu_0+vv_0-ww_0$ is discontinuous at every real point (u_1, v_1) for which $u_1^2+v_1^2<1$ and at every point $(u=u_1+iu_2,\ v=v_1+iv_2)$ which is not real $(u_2^2+v_2^2\neq 0)$, but is at a finite distance. He investigated briefly the functions invariant under this subgroup.

G. Giraud[13] classified the linear substitutions leaving $xx_0+yy_0-zz_0$ invariant and studied the fundamental domain of the group of Picard.

G. Humbert[14] found the measure (cf. CH. XV[18]) of the number of representations by a ternary Hermitian form in the field defined by $\sqrt{-1}$ or $\sqrt{-2}$ and the measure of their classes.

G. Giraud[15] studied the group Γ of the automorphs of

$$x_1x_1^0+\ldots+x_nx_n^0-x_{n+1}x_{n+1},$$

where x^0 is the conjugate imaginary of x, and certain integrals invariant under Γ.

---

10 Annales sc. école norm. sup., (3), 19, 1902, 261–323; (3), 21, 1904, 269–295 (Thèse).
11 Trans. Amer. Math., Soc., 7, 1906, 280–3.
12 Annales sc. école norm. sup., (3), 33, 1916, 363–371; extract in Comptes Rendus Paris, 163, 1916, 284–9.
13 Annales sc. école norm. sup., (3), 38, 1921, 43–164.
14 Jour. de Math., (8), 4, 1921, 3–35. Summary in Comptes Rendus Paris, 172, 1921, 497–511.
15 Leçons sur les fonctions automorphes, Paris, 1920, Ch. II.

CHAPTER XVII.

BILINEAR FORMS, MATRICES, LINEAR SUBSTITUTIONS.

If in a bilinear form $\Sigma a_{ij}x_iy_j$ of matrix $A=(a_{ij})$ we introduce new variables by means of linear substitutions

$$x_i=\Sigma p_{ij}x'_j, \qquad y_i=\Sigma q_{ij}y'_j,$$

the determinants of whose matrices $P=(p_{ij})$, $Q=(q_{ij})$ are $\neq 0$, we obtain a bilinear form in the x'_j, y'_j whose matrix is $M=P'AQ$, where $P'=(p_{ji})$ is obtained from P by interchanging rows and columns. Write R for P'. Accordingly, two matrices A and M with integral elements, or two bilinear forms whose matrices are A and M, are called *equivalent* if there exist matrices R and Q whose elements are integers of determinants 1 or -1 such that $M=RAQ$.

Ch. Hermite[1] showed how to find a k-rowed square matrix with integral elements of determinant ± 1, the elements of whose first row are k given integers whose g.c.d. is 1. Another method was based by K. Weihrauch[2] upon his solution of a linear Diophantine equation (this History, Vol. II, p. 75). G. Eisenstein[3] had solved the problem when $k=3$ by means of a canonical form of a substitution with integral coefficients.

H. J. S. Smith[4] wrote $\nabla_0=1$ and ∇_k for the g.c.d. of the k-rowed minors of an n-rowed square matrix A with integral elements and proved that we can always find two n-rowed unit matrices R and Q (with integral elements of determinant unity) such that RAQ is the matrix whose diagonal elements are $e_1, \ldots, e_n$, where $e_j=\nabla_{n-j+1}/\nabla_{n-j}$, so that M is the matrix of

$$e_1x_1y_1+e_2x_2y_2+\ldots+e_nx_ny_n.$$

He first proved that ∇_k is also the g.c.d. of the k-rowed minors of RAQ, when R and Q are any unit matrices. Also e_j is divisible by e_{j+1}. There are noted generalizations to the case in which A is rectangular and not a square matrix. He[5] later established related results.

L. Kronecker[6] proved that every n-ary linear substitution with integral coefficients of determinant unity is generated[7] by $x_1=x'_1+x'_2$, $x_i=x'_i$ $(i>1)$ and the $n-1$ sub-

[1] Jour. de Math., 14, 1849, 21–30; Oeuvres, I, 265–273.
[2] Zeitschrift Math. Phys., 21, 1876, 134–7.
[3] Jour. für Math., 28, 1844, 327–9.
[4] Phil. Trans. London, 151, 1861, 293–326; Coll. Math. Papers, I, 367–409. Report was made under Smith[20] of Ch. III of the part dealing with matrices with assigned minors.
[5] Proc. London Math. Soc., 4, 1873, 236–253; Coll. Math. Papers, II, 67–85.
[6] Monatsber. Akad. Berlin, 1866, 597–612; reprinted in Jour. für Math., 68, 1868, 273–285; Werke, I, 145–162.
[7] A. Krazer, Annali di Mat., (2), 12, 1884, 283–300, noted that three generators suffice, their coefficients being 0, ± 1.

stitutions $x_1=-x'_k$, $x_k=x'_1$, $x_i=x'_i$ $(i\neq 1, k)$ for $k=2, \ldots, n$. He gave $n+2$ simple generators of all substitutions with integral coefficients which leave unaltered the bilinear form

$$\sum_{r=1}^{n} (x_r y_{n+r} - x_{n+r} y_r).$$

He also investigated algebraically the substitutions which multiply the last form by a constant.

G. Frobenius[8] considered a matrix $A=(a_{ij})$ with m rows and n columns whose elements a_{ij} are integers (or polynomials in a parameter with integral coefficients). Write d_k for the g.c.d. of the k-rowed minors of A. Then d_k is evidently divisible by d_{k-1}. The quotient $e_k=d_k/d_{k-1}$ (with $e_1=d_1$) is called the kth *elementary divisor* of A. It is shown that e_k is divisible by e_{k-1}, and that two matrices (or bilinear forms) are equivalent if and only if each elementary divisor of one is equal to the corresponding one of the other. This was proved by transforming[9] the bilinear form of matrix A into the reduced form $e_1x_1y_1+\ldots+e_lx_ly_l$, where l is the rank of A and is such that $e_l\neq 0$, $e_{l+1}=0$.

In particular (p. 160), let $l=m=n$, so that A is a square matrix of determinant $\pm d\neq 0$. Then the bilinear form of matrix A can be reduced to

$$f_1x_1y_1+f_1f_2x_2y_2+\ldots+f_1f_2\ldots f_nx_ny_n,$$

where the f's are integers. Thus $d=f_1^nf_2^{n-1}\ldots f_n$. The number of ways in which d can be so decomposed is therefore the number $h(d)$ of classes of bilinear forms of determinant $\pm d\neq 0$ in two sets each of n variables. If d is the product $d'd''$ of two relatively prime factors, then $h(d)=h(d')\cdot h(d'')$. It thus remains to find $h(p^a)$, where p is a prime; this is the number h_a of sets of integral solutions $a_i \gtreqless 0$ of

$$a=na_1+(n-1)a_2+\ldots+a_n,$$

and hence (Vol. II, Ch. III, of this History) is the coefficient of x^a in the development of the reciprocal of $(1-x)(1-x^2)\ldots(1-x^n)$ into a power series in x. Hence if $\pm d=\Pi p^a$, $h(d)=\Pi h_a$, which is independent of the primes p.

The equation (p. 151) $\Sigma a_{ij}x_ix_j=f$ is solvable[10] in integers if and only if f is divisible by the g.c.d. of the a_{ij}.

An alternate bilinear form (p. 165)

$$\sum_{i,j=1}^{n} a_{ij}x_iy_j, \quad a_{ij}=-a_{ji},$$

with integral coefficients can be transformed *cogrediently* (the same substitution on the x's as on the y's and having integral coefficients of determinant unity) into

$$e_2(x_1y_2-x_2y_1)+e_4(x_3y_4-x_4y_3)+\ldots+e_{2l}(x_{2l-1}y_{2l}-x_{2l}y_{2l-1}),$$

where $2l$ is the rank, and $e_{2k}=e_{2k-1}$ $(k=1, \ldots, l)$ are the elementary divisors. Hence two equivalent alternate forms can be transformed into each other cogrediently.

[8] Jour. für Math., 86, 1879, 146–208.

[9] The possibility of this reduction when $m=n$ was later proved by induction on n by L. Kronecker, Jour. für Math., 107, 1891, 135–6.

[10] Generalized to algebraic domains of genus zero by M. Lerch, Monatsh. Math. Phys., 2, 1891, 465–8.

Two pairs (pp. 202–4) of bilinear forms with coefficients in a field F, such that the determinants of the first forms of each pair are not zero, are equivalent in F if and only if their characteristic matrices have the same elementary divisors (due to Weierstrass when F is the field of all complex numbers).

Two bilinear forms are called *congruent* modulo k if their corresponding coefficients are congruent (p. 187). Two forms are called *equivalent modulo* k if each can be transformed into a form congruent to the other by substitutions with integral coefficients whose determinants are prime to k.

Frobenius[11] continued his discussion of the final topic and proved that every bilinear form is equivalent modulo k to a *reduced* form $g_1x_1y_1+\dots+g_rx_ry_r$, where g_ρ is divisible by $g_{\rho-1}$ for every $\rho>1$, and k is divisible by and exceeds g_r. Here r is the rank modulo k. The question whether or not one bilinear form contains another in the ordinary arithmetical sense is reduced to the corresponding question with respect to the modulus which is the final elementary divisor. Given two square matrices A and B with integral elements, we can find square matrices P and Q with integral elements such that $PAQ=B$ if and only if each elementary divisor of B is a multiple of the corresponding elementary divisor of A.

Frobenius[12] gave an elegant algebraic proof of the last theorem by means of identities between determinants.

K. Hensel[13] gave another simple proof of the same theorem.

L. Kronecker[14] applied to $Ax_1y_1+Bx_1y_2-Cx_2y_1+Dx_2y_2$ of determinant $\Delta=AD+BC$ the same transformation $T=\binom{\alpha\ \beta}{\gamma\ \delta}$ on the y's as on the x's and obtained $A'x_1y_1+\dots$ of determinant $\epsilon^2\Delta$, where $\epsilon=\alpha\delta-\beta\gamma$. These bilinear forms are called *equivalent* if their coefficients and those of T are integers and $\epsilon=1$. Then $B+C$, as well as Δ, is an invariant. Between the values of A, $B-C$, D and the values of A', $B'-C'$, D' evidently hold the same relations as between the coefficients of $Q=Ax^2+(B-C)xy+Dy^2$ and the form by which T replaces Q. Write $\theta=\Delta-\frac{1}{4}(B+C)^2$ for the negative of the determinant of Q. Hence we obtain representatives of all classes of equivalent bilinear forms of determinant Δ if we take for $B+C$ all values numerically $<2\sqrt{\Delta}$ and for each such value take all sets of values of A, $B-C$, D given by representative quadratic forms Q of determinant $-\theta$. Call the bilinear form *reduced* if Q is reduced. Two bilinear forms are called *completely equivalent* if one can be transformed into the other by the same transformation T on the y's as on the x's, where T is of determinant unity and $T\equiv\binom{1\ 0}{0\ 1}$ (mod 2). For his analogous definition for quadratic forms, see Kronecker[113] of Ch. I. Applications to class number are cited under Kronecker[144] of Ch. VI.

T. J. Stieltjes[15] gave an exposition of the results by Smith[4] and proved that two systems of linear forms represent the same systems of numbers if and only if they are equivalent.

---

[11] Jour. für Math., 88, 1879, 96–116.
[12] Sitzungsber. Akad. Wiss. Berlin, 1894, 31–44. Reproduced by Bachmann.[18]
[13] Jour. für Math., 114, 1895, 109–115 (25–30 for an arithmetical proof of theorems on regular minors employed by Frobenius[12]).
[14] Abh. Akad. Wiss. Berlin, 2, 1883, No. 2; Werke, II, 425–495.
[15] Annales Fac. Sc. Toulouse, 4, 1890, final paper, 85–97.

G. Landsberg[16] called two rectangular matrices, each with s rows and t columns, *equivalent* if their corresponding integral elements are congruent modulo p, a prime. Hence there are p^{st} non-equivalent matrices. He found how many of them are of a given rank modulo p.

Landsberg[17] gave a modification, more convenient to apply, of Frobenius'[8] proof of his theorem that two pairs of bilinear forms with the same elementary divisors can be transformed into each other rationally.

P. Bachmann[18] gave an exposition of the theory of the equivalence of matrices.

P. Muth[19] gave an exposition of the theory of pairs of bilinear forms, including the generalization to elements which are polynomials or integers of an algebraic field.

L. Kronecker[20] treated square matrices with integral coefficients and (pp. 78, 90) the arithmetical equivalence of forms $axx'+a'xy'+byx'+b'yy'$.

L. E. Dickson[21] found the necessary and sufficient conditions for the existence of a bilinear form, with coefficients in any given field (domain of rationality) F, invariant under a given substitution S with coefficients in F and cogredient in the two sets of variables ξ_i, η_i. When these conditions are satisfied the existing bilinear forms are all reducible to a single one by a transformation on the ξ's commutative with S_ξ and a (possibly different) transformation on the η's commutative with S_η.

A. Ranum[22] discussed linear substitutions of finite period with rational or integral[23] coefficients. Cf. Minkowski[27, 28] of Ch. XI.

O. Nicoletti[24] and L. E. Dickson[25] each proved Frobenius'[8] theorem on the equivalence of pairs of bilinear forms in a general field by a suitable modification of Weierstrass' earlier proof for the case of the field of all complex numbers.

A. Châtelet[26] called the matrix $T=\binom{\alpha\ \alpha'}{\beta\ \beta'}$ a principal *reduced* matrix if $\alpha>0$, $\alpha/\beta>1$, $-1<\alpha'/\beta'<0$. He called T *equivalent* to T_1 if $T=MT_1$, where M is a matrix whose elements are integers of determinant ± 1.

S. Lattès[27] noted that if to each root of the characteristic equation

$$x^n=c_1+c_2x+\ldots+c_nx^{n-1}$$

of a substitution S there corresponds a single elementary divisor, S has the rational canonical form

$$Y_1=y_2,\quad Y_2=y_3,\ \ldots,\quad Y_{n-1}=y_n,\quad Y_n=\sum_{i=1}^{n} c_iy_i.$$

For a general S there are several such sets of variables, one set for each factor of the above type of the characteristic determinant of S. But this canonical form is not

---

16 Jour. für Math., 111, 1893, 87–88.
17 Jour. für. Math., 116, 1896, 331–349.
18 Die Arithmetik der Quad. Formen, Leipzig, 1898, 275–316.
19 Theorie und Anwendung der Elementartheiler, Leipzig, 1899, 43–69; Preface, xiv, xv.
20 Vorlesungen über Determinanten, 1, 1903, 64–84, 163–171, 373–390.
21 Trans. Amer. Math. Soc., 7, 1906, 283–5.
22 Trans. Amer. Math. Soc., 9, 1908, 183–202; Jahresb. d. Deutschen Math.-Vereinigung, 17, 1908, 234–6.
23 Bull. Amer. Math. Soc., 15, 1908–9, 4–6.
24 Annali di Mat., (3), 14, 1908, 265–325. B. Calò, *ibid.*, (2), 23, 1895, 159–179, had given an algebraic (but not rational) proof of the case of Weierstrass.
25 Trans. Amer. Math. Soc., 10, 1909, 347–351.
26 Comptes Rendus Paris, 148, 1909, 1746–9; 150, 1910, 1502–5. Leçons sur la Théorie des Nombres, Paris, 1913, Ch. VI.
27 Comptes Rendus Paris, 155, 1912, 1482–4.

as convenient for applications as the classic canonical form, involving conjugate irrationalities, which was extended to substitutions in an arbitrary field by L. E. Dickson.[28]

E. Cahen[29] treated bilinear forms in two sets each of n variables and square matrices with integral coefficients.

C. Cellitti[30] expressed a binary substitution as a product of powers of

$$\begin{pmatrix}1 & 1\\ 0 & 1\end{pmatrix}, \quad \begin{pmatrix}1 & 0\\ 1 & 0\end{pmatrix}, \quad \begin{pmatrix}D & 0\\ 0 & 1\end{pmatrix}.$$

W. H. Bussey[31] discussed the linear dependence of m sets of n integers modulo p.

O. Veblen and P. Franklin[32] gave an exposition of the theory of matrices whose elements are integers.

For miscellaneous theorems on minors of a matrix with integral coefficients, see Encyclopédie des sc. math., t. I, vol. 3, pp. 87–89.

For reports on the literature of systems of linear forms, equations, and congruences, and on matrices, see this History, Vol. II, pp. 82–98.

---

[28] Amer. Jour. Math., 24, 1902, 101–8.
[29] Théorie des nombres, 1, 1914, 268–284, 329–367, 376–7, 387–8.
[30] Atti R. Accad. Lincei, Rendiconti, (5), 23, II, 1914, 208–212.
[31] Amer. Math. Monthly, 21, 1914, 7–11.
[32] Annals of Math., (2), 23, 1922, 1–15.

CHAPTER XVIII.

REPRESENTATION BY POLYNOMIALS MODULO p.

ANALYTIC REPRESENTATION OF SUBSTITUTIONS, POLYNOMIALS REPRESENTING ALL INTEGERS MODULO p.

When x ranges over a complete set of residues modulo 5, x^3 ranges over the same residues rearranged and hence represents the substitution

$$\begin{pmatrix} 0 & 1 & 2 & 3 & 4 \\ 0 & 1 & 3 & 2 & 4 \end{pmatrix},$$

which replaces each number in the first row by the number below it. The problem is to find polynomials (like x^3) which represent all integers modulo p. The most important papers are those by Hermite[4] and Dickson.[14]

E. Betti[1] proved that all 120 substitutions on 5 letters are represented by $ax+b$ and $(ax+b)^3+c$ modulo 5.

Betti[2] noted that if x and each B_i are elements of the Galois field of order p^ν (see this History, Vol. I, pp. 233–252), the function

$$\psi(x) = \sum_{i=0}^{\nu-1} B_i x^{p^i} + B_\nu$$

will represent a substitution on the p^ν elements of the $GF[p^\nu]$ if and only if $\psi(x)=k_0$ has one and only one root in the field whatever value k_0 has in the field.

E. Mathieu[3] noted that the preceding $\psi(x)$ represents a substitution if and only if $\Sigma B_i x^{p^i}$ vanishes only when $x=0$. The function $\psi(x)$ is the case $\eta=1$ of

$$\sum_{i=0}^{\nu-1} B_i x^{p^{\eta i}} + B_\nu,$$

where x and each B_i are elements of the $GF[p^{\eta\nu}]$.

Ch. Hermite[4] noted that, if p is a prime, the substitution which replaces $0, 1, \ldots, p-1$ by a rearrangement $a_0, a_1, \ldots, a_{p-1}$ of them is represented analytically by Lagrange's interpolation formula

$$\theta(x) = \sum_{i=0}^{p-1} \frac{a_i \phi(x)}{(x-i)\phi'(i)}, \qquad \phi(x) = \prod_{i=0}^{p-1} (x-i) \equiv x^p - x \pmod{p},$$

whence $\phi'(x) \equiv -1$. Thus $\theta(x)$ is a polynomial in x of degree $p-2$ with integral coefficients if $p>2$. Any such polynomial represents a substitution modulo p if and

---

1 Annali di Sc. Mat. e Fis., 2, 1851, 17–19.
2 *Ibid.*, 3, 1852, 72–74 (6, 1855, 5–34).
3 Jour. de Math., (2), 6, 1861, 275 (282–7), 301.
4 Comptes Rendus Paris, 57, 1863, 750; Oeuvres, II, 280–8. Report in J. A. Serret's Algèbre Supérieure, ed. 5, 2, 1885, 383–390, 405–412.

only if the tth power of $\theta(x)$, for $t=1, \ldots, p-2$, reduces to a polynomial of degree $\leqq p-2$ on applying $x^p \equiv x \pmod p$. He applied this theorem to find all polynomials which represent substitutions on $p=7$ letters.

F. Brioschi[5] gave properties of substitutions represented by

$$\epsilon(x^{p-2}+ax^{\frac{1}{2}(p-3)}).$$

A. de Polignac[6] gave a long proof of a result which is a corollary to Hermite's expansion of the interpolation formula.

F. Brioschi[7] proved that, if p is a prime, $x^{p-2}+ax^{(p-1)/2}+bx$ cannot represent a substitution on p letters unless $p=7$, $b\equiv 3a^2 \pmod 7$.

A. Grandi[8] proved the generalization that

$$x^{p-s}+ax^{(p-s+1)/2}+br$$

cannot represent a substitution on p letters if $p>4(s-1)d+1$, where d is the g.c.d. of $p-1$ and $(p-s-1)/2$. If p lies under this limit and exceeds $2(s-1)d+1$, then must $b\equiv\frac{1}{2}(p-1)(s-1)a^2 \pmod p$. Grandi[9] gave the further generalization that

$$x^{2\mu-(2s-1)}+\sum_{i=1}^{h} a_i x^{\mu-(is-1)}+br, \qquad \mu=\tfrac{1}{2}(p-1), \quad h>1,$$

cannot represent a substitution on p letters if the g.c.d. d of μ and s is $<\frac{1}{3}\mu/(hs-1)$ and no one of $a_1, \ldots, a_h, b$ is divisible by p. But if

$$2(hs-1)<\mu/d\leqq 3(hs-1), \qquad b\not\equiv 0 \pmod p, \qquad s>1,$$

a necessary condition that it represent a substitution is

$$b\equiv\mu(2s-1)a_1^2 \pmod p.$$

G. Raussnitz[10] proved that $f(x)\equiv a_0x^{p-2}+a_1x^{p-3}+\ldots+a_{p-2}$ represents a substitution on p letters, where p is a prime, if and only if

$$\begin{vmatrix} a_0 & a_1 & a_2\ldots a_{p-3} & a_{p-2}-k \\ a_1 & a_2 & a_3\ldots a_{p-2}-k & a_0 \\ \cdots & \cdots & \cdots & \cdots \\ a_{p-2}-k & a_0 & a_1\ldots a_{p-4} & a_{p-3} \end{vmatrix} \equiv 0 \pmod p$$

for $k=0, 1, \ldots, a_{p-2}-1, a_{p-2}+1, \ldots, p-1$. For. $f\equiv 0, f-1\equiv 0, \ldots, f-(p-1)\equiv 0 \pmod p$ must each have a real root, and not the root 0 except for $f-a_{p-2}\equiv 0$. Hence the theorem follows from his result quoted in this History, Vol. I, p. 226.

*F. Rinecker[11] discussed the cases $p=5, 7, 11$.

L. J. Rogers[12] proved that $x^r\{f(x^s)\}^{(p-1)/s}$ represents a substitution on p letters (p a prime) if r is less than and prime to $p-1$ and $f(x^s)$ is a polynomial in x^s with

---

[5] Göttingen Nachr., 1869, 491; Math. Annalen, 2, 1870, 467–470; Comptes Rendus Paris, 95, 1882, 665 (816, 1254).
[6] Bull. Soc. Math. de France, 9, 1881, 59–67. Cf. Dickson.[14]
[7] Reale Istituto Lombardo di Sc. Let., Rendiconti, Milan, (2), 12, 1879, 483–5.
[8] Giornale di Mat., 19, 1881, 238–244.
[9] Reale Istituto Lombardo di Sc. Let., Rendiconti, Milan, (2), 16, 1883, 101-110.
[10] Math. und Naturw. Berichte aus Ungarn, 1, 1882–3, 275–8.
[11] Ueber Substitutionsfunktionen modulo 11 und die analytische Darstellung der Permutationen von 5, 7, 11 Elementen, Diss. Erlangen, 1886, 29 pp.
[12] Proc. London Math. Soc., 22, 1890, 37–52.

integral coefficients which is never zero modulo p. He treated at length (and especially the interpretation by a polygon of 7 sides) the representation of substitutions on 7 letters, showing how to compute the inverse of such a substitution (his proof being objectionable since he used $x^6 \equiv 1$ mod 7 and then took $x \equiv 0$).

Rogers[13] proved that if a congruence has all its roots real for a prime modulus p and if $s_k \equiv 0 \{k=1, \ldots, \frac{1}{2}(p-1)\}$, where s_k is the sum of the kth powers of its roots, then will $s_k \equiv 0 \{k=\frac{1}{2}(p+1), \ldots, p-2\}$. Hence we need employ only the first $\frac{1}{2}(p-1)$ conditions of Hermite to decide whether or not a given polynomial represents a substitution.

L. E. Dickson[14] generalized the theorems of Hermite[4] and Rogers[12] to substitutions on p^n letters by employing Galois imaginaries of the $GF[p^n]$, found all polynomials of degree <7 suitable to represent substitutions on p^n letters, and proved that, if k is an odd integer prime to $p^{2n}-1$,

$$\xi^k + ka\xi^{k-2} + k \sum_{l=2}^{\frac{1}{2}(k-1)} \frac{(k-l-1)(k-l-2)\ldots(k-2l+1)}{2\cdot 3\ldots l} a^l \xi^{k-2l}$$

represents a substitution on the p^n elements of the $GF[p^n]$, since it is the sum of the kth powers of the roots of $x^2-\xi x-a=0$. Also for $\xi(\xi^d-\nu)^{\rho/d}$, if d is a divisor of $\rho=p^r-1$ and if ν is not the dth power of an element of the $GF[p^n]$. It is shown that Mathieu's[3] function

$$\chi(X) = \sum_{i=1}^{m} A_i X^{p^{n(m-i)}},$$

with coefficients A_i in the $GF[p^{mn}]$, represents a substitution on its p^{mn} elements if and only if

$$\begin{vmatrix} A_1 & A_2 & \ldots & A_m \\ A_2^{p^n} & A_3^{p^n} & \ldots & A_1^{p^n} \\ A_3^{p^{2n}} & A_4^{p^{2n}} & \ldots & A_2^{p^{2n}} \\ \ldots & \ldots & \ldots & \ldots \\ A_m^{p^{n(m-1)}} & A_1^{p^{n(m-1)}} & \ldots & A_{m-1}^{p^{n(m-1)}} \end{vmatrix} \neq 0,$$

this being the resultant of $\chi(X)=0$ and $X^t=1$, where $t=p^{nm}-1$. The group of all such substitutions is proved (p. 178) to be identical with the group of all m-ary linear homogeneous substitutions with coefficients in the $GF[p^n]$.

POLYNOMIALS REPRESENTING ONLY NUMBERS OF PRESCRIBED NATURE.

H. W. Lloyd Tanner[15] considered a polynomial $F(x)$ with integral coefficients such that for any integer x not divisible by the odd prime p the value of $F(x)$ is 1 or -1. We may set $F(x)=f(x^2)+x^q\phi(x^2)$, where f and ϕ are of the form $a+bx^2+cx^4+\ldots+kx^{p-3}$, while q is the largest odd factor of $p-1$. Then

$$(fx^2 \pm x^q\phi x^2)^2 \equiv 1, \qquad x^q f(x^2)\phi(x^2) \equiv 0 \pmod{p}.$$

Thus for each $x \not\equiv 0$, either $f(x^2) \equiv 0$ or $\phi(x^2) \equiv 0$.

---

13 Messenger Math., 21, 1891–2, 44–47.

14 Annals of Math., 11, 1896–7, 65–120, 161–183. Partial summary in Dickson's Linear Groups, Leipzig, 1901, 54–71, covering also Amer. Jour. Math., 18, 1896, 210–8; 22, 1900, 49–54.

15 Messenger of Math., 21, 1891–2, 139–144.

The function $F_1(x)=3+2x+2x^2+\ldots+2x^{p-2}$ is congruent to -1 when $x=1$ and to $+1$ when $x=2, \ldots, p-1$. Hence the product

$$F_1(a^{-1}x) \quad F_1(b^{-1}x)\ldots \quad F_1(k^{-1}x)$$

is congruent to -1 when $x=a, b, \ldots, k$, and to $+1$ for the remaining x's $\not\equiv 0$. The product is computed by a device.

L. E. Dickson[16] proved that a ternary cubic form C vanishes for no set of values of x, y, z in the $GF[p^n]$, $p>2$, other than $x=y=z=0$ if and only if its Hessian is a constant multiple of C, and if the binary form $C(x, y, 0)$ is irreducible in the field. All such forms C are equivalent under linear transformation in the field. Another criterion for forms C is found.

Dickson[17] proved for $m=2$ and $m=3$ that every form of degree m in $m+1$ variables with coefficients in the $GF[p^n]$, $p>m$, vanishes for values, not all zero, in the field. No binary cubic form represents only cubes in the $GF[p^n]$, $p^n\equiv 1 \pmod 3$. An investigation is made of sextic forms in two or more variables which represent only cubes.

Dickson[18] investigated quartic and sextic forms in two or more variables which represent only quadratic residues. For example, a binary quartic modulo $p>2$ which represents only quadratic residues of p is identically congruent to the square of a quadratic form.

Dickson[19] treated forms $F(x_1, \ldots, x_n)$ of degree m with integral coefficients which are congruent to zero modulo 2 only when each $x_i\equiv 0 \pmod 2$. After replacing each x_i^a $(a>1)$ by x_i, F becomes $\Pi(1+x_i)-1 (\text{mod } 2.)$ For $m=4$, $n=3$, F can be transformed linearly into

$$[egkr]\equiv x_1^4+x_2^4+x_3^4+x_1^3x_2+ex_1^2x_3^2+gx_1x_2^2x_3+(g+1)x_1x_2x_3^2 \\ +(e+1)x_1x_3^3+kx_2^3x_3+rx_2^2x_3^2+(k+r+1)x_2x_3^3,$$

or $\Sigma x_1^4+\Sigma x_1^2x_2^2+x_1x_2x_3(x_1+x_2+x_3)$, which is unaltered under all linear transformations. The latter and [1100], [1001], [1111], [1000], [1101], [0010] give all the non-equivalent types. The case $m=6$, $n=3$, is treated partially.

---

[16] Bull. Amer. Math. Soc., 14, 1908, 160-9.
[17] *Ibid.*, 15, 1909, 338-347.
[18] Trans. Amer. Math. Soc., 10, 1909, 109-122.
[19] Quar. Jour. Math., 42, 1911, 162-171.

CHAPTER XIX.

CONGRUENCIAL THEORY OF FORMS.

Modular Invariants and Covariants.

Let $f_1, \ldots, f_l$ be any system of forms in the arbitrary variables $x_1, \ldots, x_m$ with undetermined integral coefficients taken modulo p, a prime. Let $c_1, c_2, \ldots$ denote the coefficients arranged in any order. Under the transformation

$$T: \quad x_i \equiv \sum_{j=1}^{m} t_{ij} x'_j \pmod p \qquad (i=1, \ldots, m),$$

with integral coefficients, let f_i become f'_i, and let $c'_1, c'_2, \ldots$ denote the coefficients of $f'_1, \ldots, f'_l$ corresponding in position to $c_1, c_2, \ldots$, respectively. A polynomial $K(c_1, c_2, \ldots; x_1, \ldots, x_m)$ with integral coefficients taken modulo p is called a *modular covariant* of the forms $f_1, \ldots, f_l$ if, for every transformation T,

$$K(c'_1, c'_2, \ldots; x'_1, \ldots, x'_m) \equiv |t_{ij}|^{\mu} \cdot K(c_1, c_2, \ldots; x_1, \ldots, x_m) \pmod p$$

holds identically in $c_1, c_2, \ldots, x'_1, \ldots, x'_m$ after $x_1, \ldots, x_m$ have been eliminated by means of the congruences T, and $c'_1, c'_2, \ldots$ have been replaced by their expressions in terms of $c_1, c_2, \ldots$, and finally the exponent of each c_i has been reduced to a value $<p$ by means of Fermat's theorem $c^p \equiv c \pmod p$. The exponent μ is called the index of K.

As an immediate generalization, we may take the coefficients c_i, t_{ij} and the coefficients of K to be Galois imaginaries of the $GF[p^n]$ (cf. Vol. I, pp. 233–252). For $n=1$, we have the above case.

The ordinary algebraic covariants of a system of algebraic forms f_i become modular covariants when the coefficients of the f_i are interpreted as arbitrary elements of any $GF[p^n]$. But we obtain in this way only a relatively small proportion of the modular covariants.

The fundamental paper is that by Dickson[4] who based a complete theory on the notion of classes; the report gives a simple example.

L. E. Dickson[1] extended to modular invariants the annihilators of algebraic invariants and computed a complete set of linearly independent invariants of the binary quadratic form in the $GF[p^n]$ and binary cubic form in the $GF[5]$ or $GF[3^n]$. The binary form $\Sigma a_i x^{m-i} y^i$ in the $GF[p^n]$ has the absolute invariant

$$\prod_{i=0}^{m} (a_i^{p^n-1} - 1).$$

Dickson[2] found for the m-ary quadratic form modulo 2, with $m<6$, a complete set (m in number) of independent invariants, as well as a complete set of linearly

[1] Trans. Amer. Math. Soc., 8, 1907, 205–232.
[2] Proc. London Math. Soc., (2), 5, 1907, 301–324.

independent invariants; also certain invariants when $m=6$. The canonical forms are characterized by the invariants.

Dickson[3] gave an invariantive reduction of quadratic forms

$$Q_m=\sum_{i<j}^{1,\ldots,m} c_{ij}x_ix_j+b_ix_i^2$$

in the $GF[2^n]$. Its algebraic discriminant Δ is a skewsymmetic determinant modulo 2 and hence is zero if m is odd. In that case we define the *semi-discriminant* of Q_m to be the expression derived algebraically by dividing by 2 each of the (even) coefficients in the expansion of Δ. According as m is even or odd, the vanishing of the discriminant Δ or semi-discriminant is a necessary and sufficient condition that an m-ary quadratic form in the $GF[2^n]$ shall be transformable linearly into a form of fewer than m variables. It is reducible to a form in r (but not fewer than r) variables if and only if every $\mu^{(m)}, \ldots, \mu^{(r+1)}$ vanishes, but not every $\mu^{(r)}$, where $\mu^{(s)}$ ranges over the minors or semi-minors of order s of Δ, according as s is even or odd.

For $n\leqq 4$ all invariants of Q_3 are expressible in terms of three.

Dickson[4] gave a simple, complete theory of modular invariants from a new standpoint. In his former papers the test for the invariance of a polynomial P was the direct verification that it remained unaltered (up to a power of the determinant of transformation) under the total linear group G in the field. Now the transformation concept is employed only to furnish a complete set of non-equivalent classes $C_0, \ldots, C_{k-1}$ of systems of s forms under the group G. Then P is an absolute invariant if it takes the same value for all systems of s forms in a class.

For example, consider a single form $f=ax^2+2bxy+cy^2$ whose coefficients are undetermined integers taken modulo p, a prime >2. The particular forms f which are congruent to squares of linear functions constitute a *class* $C_{1,1}$ with the representative form x^2. Again, there is a class $C_{1,2}$ represented by νx^2, where ν is a fixed quadratic non-residue of p. Also, for $D=1, 2, \ldots, p-1$, there are classes $C_{2,D}$ represented by x^2+Dy^2. Finally, there is the class C_0 of forms all of whose coefficients are divisible by p. A single-valued function of the coefficients of f is called a *modular invariant* of f if and only if the function has the same value modulo p for all forms in the class $C_{1,1}$, the same (usually new) value for all forms in the class $C_{1,2}$, and similarly for each class $C_{2,D}$ and C_0. One such function is the determinant $D=b^2-ac$ of f. Another modular invariant is

$$I_0=(1-a^{p-1})(1-b^{p-1})(1-c^{p-1}),$$

which has the value 1 for any form of class C_0 and the value 0 for all remaining forms f. Finally, the function

$$A=\{a^{\frac{1}{2}(p-1)}+c^{\frac{1}{2}(p-1)}(1-a^{p-1})\}(1-D^{p-1})$$

is an invariant of f. For, if $D\not\equiv 0$, $A\equiv 0$. If $D\equiv a\equiv c\equiv 0$, f is in class C_0 and $A\equiv 0$. If $D\equiv 0$, $a\not\equiv 0$, then $f\equiv a(x+yb/a)^2$ is in the class $C_{1,1}$ or $C_{1,2}$, according as a is a quadratic residue or non-residue of p, and $A\equiv +1$ or -1, respectively. If $D\equiv a\equiv 0$, $c\not\equiv 0$, then $f\equiv cy^2$, $A\equiv c^{\frac{1}{2}(p-1)}$. Hence A has the same value for all the forms in each

[3] Amer. Jour. Math., 30, 1908, 263–281. Cf. Dickson.[4]
[4] Trans. Amer. Math., Soc., 10, 1909, 123–158. Cf. Dickson,[15] 4–15.

class. Further, the values of D, I_0, A fully differentiate the various classes. Hence they form a fundamental system of modular invariants of f. A complete system of linearly independent invariants of f is furnished by I_0, A, D^j $(j=0, 1, \ldots, p-1)$.

In general, the number of linearly independent modular invariants of any system of forms is the number of classes. There are developed complete theories of reduction and invariants of an m-ary quadratic form and binary cubic form is the $GF[p^n]$, $p>2$ or $p=2$.

Dickson[5] considered combinants of a system of s forms f_i, viz., invariants which remain unaltered, apart from the factor δ^v, when the f_i are replaced by linear homogeneous functions of themselves of determinant $\delta \neq 0$. The theory of classes[4] is applicable here and leads to a general theory of combinants. There is found a fundamental system of combinants of two binary or two ternary quadratic forms in the $GF[p^n]$, $p>2$.

Dickson[6] found, by the theory of classes, a complete set of linearly independent invariants of q linear forms on m variables in the $GF[p^n]$. When $q>m$, every invariant is a polynomial in the invariants of systems of m forms in m variables.

Dickson[7] considered the classes C_i of systems of forms under any linear group in any field (finite or infinite). Let the invariants I_1, I_2, ... completely *characterize* the classes, i. e., let each I_k have the same value for two classes only when the latter are identical. Then any (single-valued) invariant is a single-valued function of I_1, I_2, ... But it does not always follow, as stated, that a polynomial invariant is a polynomial in I_1, I_2, ...

For the $GF[p^n]$, the *characteristic* invariants I_k are exhibited explicitly. There is found a complete set of linearly independent invariants of a binary quadratic and linear form and of two binary quadratic forms in the $GF[p^n]$, both for $p=2$ and $p>2$.

Dickson[8] had already treated the last problem without the theory of classes and by a very long computation found the invariants of two binary quadratic forms in the $GF[2^n]$, $n=1, 2, 3$. It was the meditation on those results that led him to conceive the idea of a theory based on classes, which so greatly simplified the whole subject.

Dickson[9] laid the foundation of the theory of modular covariants by finding the universal modular covariants of all systems of binary forms. In fact, he proved that every polynomial in x, y with coefficients in the $GF[p^n]$ which is (relatively) invariant under all binary linear transformations in that field is a polynomial, with coefficients in the field, in the two invariants

$$L=\begin{vmatrix} x^{p^n} & y^{p^n} \\ x & y \end{vmatrix}, \qquad Q=\begin{vmatrix} x^{p^{2n}} & y^{p^{2n}} \\ x & y \end{vmatrix} \div L.$$

L and Q are congruent to the products of all non-proportional linear and irreducible

[5] Quar. Jour. Math., 40, 1909, 349–366.
[6] Proc. London Math. Soc., (2), 7, 1909, 430–444.
[7] Amer. Jour. Math., 31, 1909, 337–354.
[8] *Ibid.*, 103–146.
[9] Trans. Amer. Math. Soc., 12, 1911, 1–18. Simplified by Dickson,[15] pp. 33–38, 61–64. Still simpler is the proof that certain coefficients of an invariant are zero, Quar. Jour. Math., 42, 1911, 158–161.

quadratic forms in the field, respectively. There is a remarkable application to the classification of all irreducible binary forms.

Dickson[10] extended the results last cited to m variables. By induction on m it is proved that a fundamental system of polynomial invariants of the group G_m of all linear homogeneous substitutions on $x_1, \ldots, x_m$ with coefficients in the $GF[k]$, where $k=p^n$, are furnished by

$$L_m=\begin{vmatrix} x_1^{k^{m-1}} & \cdots & x_m^{k^{m-1}} \\ x_1^{k^{m-2}} & \cdots & x_m^{k^{m-2}} \\ \cdots & \cdots & \cdots \\ x_1^{k} & \cdots & x_m^{k} \\ x_1 & \cdots & x_m \end{vmatrix}, \qquad Q_{m,s}=\begin{vmatrix} x_1^{k^m} & \cdots & x_m^{k^m} \\ \cdots & \cdots & \cdots \\ x_1^{k^{s+1}} & \cdots & x_m^{k^{s+1}} \\ x_1^{k^{s-1}} & \cdots & x_m^{k^{s-1}} \\ \cdots & \cdots & \cdots \\ x_1 & \cdots & x_m \end{vmatrix} \div L_m \quad (s=1,\ldots,m-1).$$

Here L_m is the product of all non-proportional linear forms in the field. There is found a complete solution of the *form problem* for the group G_m, viz., the determination of the sets of values of $x_1, \ldots, x_m$ for which the fundamental absolute invariants $L_m^{k-1}, Q_{m,1}, \ldots, Q_{m,m-1}$ take assigned values in the infinite field composed of all roots of all congruences modulo p. For a simple account of this theory when $m=2$, see Dickson,[15] pp. 58–61. Finally there are found simple expressions for the product of all distinct ternary cubic forms equivalent to irreducible binary forms, the product of all non-vanishing ternary cubic forms, the product of all distinct ternary quadratic forms of non-vanishing discriminant, etc. R. Le Vavasseur[11] had obtained the product $D=0$ of all congruences of the first and second degrees in x, y modulo p by replacing $x_1, \ldots, x_6$ in L_6 (with $n=1$) by $x^2, xy, y^2, x, y, 1$. Let L_3 (with $n=1$) become B when x_1, x_2, x_3 are replaced by $x, y, 1$, so that $B=0$ is the product of the linear congruences. Dividing D by B^{p^2+p+2} to remove the factorable quadratic congruences, we see that the quotient is the product of all irreducible quadratic congruences.

Dickson[12] obtained a fundamental system of invariants of each type of subgroup of the group of all binary substitutions of determinant unity in the $GF[p^n]$, $p>2$.

Dickson[13] proved that the set of all modular covariants of any system of forms in m variables is *finite*, in the sense that they are all polynomials, with coefficients in the initial finite field, of a finite number of covariants of the set. There is found a fundamental system of covariants of the binary quadratic form modulo 3.

Dickson[14] found for the binary quartic form modulo 2 a complete system of 20 linearly independent semi-invariants, one of 10 invariants, and one of 10 linear covariants.

Dickson[15] gave an exposition of known results, found (pp. 21–32) a fundamental system of semi-invariants of a binary modular form of order n, and deduced from

---

10 Trans. Amer. Math. Soc., 12, 1911, 75–98.
11 Mém. Acad. Sc. Toulouse, (10), 3, 1903, 43–44.
12 Amer. Jour. Math., 33, 1911, 175–192; Bull. Amer. Math., Soc., 20, 1913, 132–4.
13 Trans. Amer. Math. Soc., 14, 1913, 299–310.
14 Annals of Math., (2), 15, 1913–4, 114–7.
15 On Invariants and the Theory of Numbers, The Madison Colloquium of 1913, Amer. Math. Soc., 1914, 110 pp. Report in Bull. Amer. Math. Soc., 20, 1913–4, 116–9; 21, 1914–5, 464–470.

them a fundamental system of invariants of the quadratic modulo p and cubic modulo 3. He discussed (pp. 65–73) the modular geometry[16] and covariantive theory of the general quadratic form in m variables modulo 2, found (pp. 73–98) a fundamental system of covariants of the ternary quadratic form modulo 2, and (pp. 99–110) gave a theory of plane cubic curves with a real inflexion point valid in ordinary and in modular geometry, treating especially the number of real inflexion points.

Dickson[17] determined a fundamental set of invariants of the system of a binary cubic f and quadratic g, and of the system f, g and a linear form modulo 2 and also modulo 3.

W. C. Krathwohl[18] proved that a fundamental system of invariants modulo p under linear transformation acting cogrediently on x_1, y_1 and x_2, y_2 is furnished by

$$L_i=\begin{vmatrix} x_i^p & y_i^p \\ x_i & y_i \end{vmatrix}, \quad \begin{vmatrix} x_i^{p^2} & y_i^{p^2} \\ x_i & y_i \end{vmatrix} \div L_i, \quad M=\begin{vmatrix} x_2 & y_2 \\ x_1 & y_1 \end{vmatrix}, \quad M_1=\begin{vmatrix} x_2 & y_2 \\ x_1^p & y_1^p \end{vmatrix} \quad (i=1, 2),$$

$$M_2=\begin{vmatrix} x_2^p & y_2^p \\ x_1 & y_1 \end{vmatrix}, \quad \frac{M_2^{s+1}L_1^{p-s-1}+(-1)^s M_1^{p-s}L_2^s}{M^p} \quad (s=1, \ldots, p-2).$$

F. B. Wiley[19] proved the finiteness of the modular covariants of any system of binary forms and cogredient points, thus generalizing theorems of Dickson and Krathwohl.

Dickson[20] gave a new method of deriving all modular invariants from the semi-invariants which is more direct and simpler than his[15] former method.

Dickson[21] obtained a fundamental system of semi-invariants of the ternary and quarternary quadratic form modulo 2 by a method simpler than his[15] former one for the ternary case; also the linear and quadratic covariants of the quaternary form.

Dickson[22] proved that the inflexion and singular points of a plane cubic curve $u\equiv 0 \pmod 2$ are given by its intersections with $H\equiv 0$, where H is cubic form which plays a rôle analogous to the Hessian in the theory of algebraic curves. Two u's are equivalent under the group G of linear transformations with integral coefficients modulo 2 if and only if they have the same number of real points (i. e., with integral coordinates), real inflexion points, and real or imaginary singular points. The 22 canonical types under G are characterized by modular invariants. There are only 10 types under imaginary transformations.

Dickson[23] classified quartic curves modulo 2 by means of their real or imaginary bitangents and distinguished the numerous types invariantively. The process yields a fundamental set of modular invariants.

Dickson[24] noted that with a conic modulo 2 is associated covariantly its apex and covariant line. Two pairs of conics are projectively equivalent modulo 2 if and only

---

16 To the references (p. 98) on modular geometry, add G. Tarry, Assoc. franç. av. sc., 33, 1910, 22–47, on the existence of primitive angles (cf. Arnoux's book).
17 Quar. Jour. Math., 45, 1914, 373–384.
18 Amer. Jour. Math., 36, 1914, 449–460.
19 Trans. Amer. Math. Soc., 15, 1914, 431–8.
20 *Ibid.*, 502–3.
21 Bull. Amer. Math. Soc., 21, 1914–5, 174–9.
22 Amer. Jour. Math., 37, 1915, 107–116. Report in Proc. Nat. Acad. Sc., 1, 1915, 1–4.
23 Amer. Jour. Math., 37, 1915, 337–354.
24 *Ibid.*, 355–8. Abstract in Bull. Amer. Math. Soc., 19, 1912–13, 456–7.

if they have the same properties as regards existence of apices and covariant lines, distinctness of apices and lines, and incidence of apices and lines. A fundamental system of invariants of two conics modulo 2 is obtained, as well as certain formal invariants.

J. E. McAtee[25] gave a complete invariantive classification of n-ary quadratic forms modulo P^λ, where P is an odd prime, and found modular invariants which completely characterize the classes. The invariants determine the values of α, β, $\ldots, p, q, \ldots$ in Jordan's[28] canonical form. He found numerous modular invariants of a binary quadratic form modulo P^λ, P any prime $\gtreqless 2$, and a fundamental system modulo 2^2.

O. C. Hazlett[26] extended Hermite's results on covariants associated with a binary form f (Jour. für Math., 52, 1856, 21–23) to modular covariants. If the order of f is not divisible by p, the product of any modular covariant of f in the $GF[p^n]$ by a power of f is expressible as the sum of a polynomial in Q and the covariants associated with f together with the product of L by a modular covariant. Hence if the variables x, y are restricted to values in the field, the product of a power of f by any modular covariant of f is congruent to an ordinary algebraic covariant of f.

Hazlett[27] gave a new proof of Dickson's[13] theorem on the finiteness of modular covariants and of Wiley's[19] theorem.

REDUCTION OF MODULAR FORMS TO CANONICAL TYPES.

Many of the preceding papers derived the canonical types as a basis of the construction of the modular invariants. The following papers give no invariants, but stop with the determination of the canonical types.

C. Jordan[28] proved that, if P is an odd prime, every quadratic form with integral coefficients can be transformed linearly modulo P into $\theta x_1^2+x_2^2+\ldots+x_p^2$, where θ is 1 or a particular quadratic non-residue of P. It can be transformed linearly modulo P^λ into

$$P^\alpha(\theta_1x_1^2+x_2^2+\ldots+x_p^2)+P^\beta(\theta_2y_1^2+y_2^2+\ldots+y_q^2)+\ldots \qquad (\alpha>\beta>\ldots).$$

For modulus 2^λ, we obtain $2^\alpha\Sigma_\alpha+2^\beta\Sigma_\beta+\ldots$, where each Σ_ρ is of one of the four types

$$S_\rho, \qquad S_\rho+az^2, \qquad S_\rho+Az^2+A_1z_1^2, \qquad S_\rho+u^2+uv+v^2,$$

where $S_\rho=x_1y_1+\ldots+x_py_p$; $a=1$, 3, 5, or 7; A and A_1 odd, $A\leqq A_1$, $A<4$, $A_1<8$. Further restrictions on a, A, A_1 are found when the form contains two or more Σ_ρ of the same type. There is no mention of the question as to whether or not no two of the resulting canonical forms are equivalent.

L. E. Dickson[29] found independently of Jordan that his first result holds true also for the Galois field of order p^n, $p>2$ (defined in this History, Vol. I, pp. 233–252).

---

[25] Amer. Jour. Math., 41, 1919, 225–242.
[26] Amer. Jour. Math., 43, 1921, 189–198.
[27] Trans. Amer. Math. Soc., 22, 1921, 144–157.
[28] Jour. de Math., (2), 17, 1872, 368–402; (7), 2, 1916, 253–80 (for modulus p, $p>2$). Cf. McAtee.[25]
[29] Amer. Jour. Math., 21, 1899, 194, 222–5. Linear Groups, Leipzig, 1901, 157–8, 197–200 (126, 218, for automorphs of higher forms). Also, Dickson.[15]

Any k-ary quadratic form f in the Galois field of order 2^n, such that f cannot be expressed as a quadratic form in fewer than k variables, can be transformed linearly into

$$\xi_0^2+\sum_{i=1}^{m}\xi_i\eta_i, \qquad \lambda\xi_1^2+\lambda\eta_1^2+\sum_{i=1}^{m}\xi_i\eta_i,$$

according as k is odd or even, where λ is 0 or a particular element such that $\lambda\xi_1^2+\xi_1\eta_1+\lambda\eta_1^2$ is irreducible in the field. The groups leaving these forms invariant were investigated at length.

C. Jordan[30] investigated the quadratic forms f invariant modulo p under a given linear substitution S, where p is a prime. By use of the canonical form of S, there are obtained necessary and sufficient conditions for the existence of forms f of determinant prime to p. When these conditions are satisfied, the general f is found and reduced to canonical types by linear transformation of the variables not altering S.

J. A. de Séguier[31] made the analogous investigation for bilinear forms.

Dickson[32] obtained the canonical forms and linear automorphs of all ternary cubic forms in the Galois field of order 3^n.

Dickson[33] found all the canonical types of families $\lambda q_1+\mu q_2$ of ternary quadratic forms for any finite field, and for the fields of all real or all complex numbers.

A. H. Wilson[34] obtained all canonical types of nets of modular conics

$$xC_1+yC_2+zC_3, \qquad C_i=\sum_{j,\,k=1}^{3} a_{ijk}t_it_j,$$

in the $GF[\,\gamma^n]$ under linear transformation on t_1, t_2, t_3 and also on x, y, z.

Dickson[35] found the canonical types of all cubic forms in four variables with integral coefficients under linear transformation modulo 2. For each type without a singular point, all the real straight lines on the cubic surface are given. He[36] later examined in detail the configuration of real and imaginary lines on typical cubic surfaces modulo 2.

Dickson[37] proved that a quartic curve modulo 2 has at most 7 bitangents except a special type which has an infinitude. There are found all non-equivalent quartic curves with 0, 5, 6 or 7 real points. He[38] elsewhere gave another classification.

FORMAL MODULAR INVARIANTS AND COVARIANTS.

The definition of a formal modular invariant differs from that on page 293 in two respects. First, the coefficients c_1, c_2, ... of the ground forms are now arbitrary

[30] Jour. de Math., (6), 1, 1905, 217–284. Abstract in Comptes Rendus Paris, 138, 1904, 537–541, 725–8.
[31] Jour. de Math., (6), 5, 1909, 1–63 (44, 52). Abstract in Comptes Rendus Paris, 146, 1908, 1247–8.
[32] Amer. Jour. Math., 30, 1908, 117–128.
[33] Quar. Jour. Math., 39, 1908, 316–333.
[34] Amer. Jour. Math., 36, 1914, 187–210.
[35] Annals of Math., (2), 16, 1914–5, 139–157.
[36] Proc. Nat. Acad. Sc., 1, 1915, 248–253.
[37] Trans. Amer. Math. Soc., 16, 1915, 111–120.
[38] Messenger Math., 44, 1914–5, 189–192.

variables and not undetermined integers taken modulo p. Second, the final congruence is to hold identically in $c_1, c_2, \ldots, x'_1, \ldots, x'_m$, without reducing exponents of the c_i by Fermat's theorem.

A. Hurwitz[39] gave the first example of a formal modular invariant of a binary form f, with its interpretation as to the number of solutions of $f \equiv 0 \pmod{p}$. For a report, also of the generalizations by H. Kühne and Dickson, see this History, Vol. I, pp. 231–3. Hurwitz raised the question of the finiteness of a fundamental system of formal invariants of a given group L of linear substitutions with integral coefficients taken modulo p, and answered it affirmatively for the special case in which the order of L is prime to p. He regarded the general case as offering an essential difficulty, not removable by known methods.

M. Sanderson[40] proved the existence of a formal modular invariant I of any system of forms under any modular group G such that $I \equiv i \pmod{p}$ for all integral values of the coefficients of the forms, where i is any given modular invariant of the forms under G. This theorem enables us, as in the algebraic theory, to construct covariants of binary forms from invariants of this system and an additional linear form whose coefficients are y and $-x$, provided the invariants have been made formally invariant as regards x, y. Certain modular covariants of a binary quadratic form are expressed in symbolic notation.

L. E. Dickson[15] (pp. 40–58) was the first to construct complete systems of formal invariants and semi-invariants. This was done for the binary quadratic form for $p=2$ and for any $p>2$. For a binary cubic, all formal semi-invariants are found when $p=2$ and $p=5$, while certain formal invariants are found for any $p \neq 3$. Illustrations of Miss Sanderson's theorem are given (pp. 54–55).

Dickson[41] gave a simple, effective method of finding formal modular invariants. For example, the only points with integral coordinates modulo 2 are (1, 0), (0, 1) and (1, 1). The values of $Q=ax^2+bxy+cy^2$ at these points are $a, c, s=a+b+c$. Their elementary symmetric functions furnish a fundamental set of formal invariants of Q modulo 2. Similarly, the values of $l=\eta x+\xi y$ at the same points are $\eta, \xi, \eta+\xi$, which undergo the same permutations as the points when l is transformed linearly. Hence any symmetric function of $\phi(a, \eta), \phi(c, \xi), \phi(s, \eta+\xi)$ is a formal invariant of Q and l modulo 2, where ϕ is any polynomial. For moduli >2, we first raise to suitable powers the values of the ground forms at the points with integral coordinates. The method applies also to semi-invariants. It leads simply to criteria for the equivalence of forms.

O. E. Glenn[42] gave simple differential operators which convert one formal modular covariant into another. He also employed modular transvectants.

Glenn[43] employed an annihilator of formal modular invariants as had Dickson for modular invariants. He noted that Dickson's universal covariant Q is a covariant of L; take the Jacobian J_1 of the Hessian of L with L, the Jacobian J_2 of J_1 with $L, \ldots$; after $p-2$ such operations we obtain Q. He expressed at trans-

[39] Archiv Math. Phys., (3), 5, 1903, 17–27.
[40] Trans. Amer. Math. Soc., 14, 1913, 489–500.
[41] Trans. Amer. Math. Soc., 15, 1914, 497–503.
[42] Bull. Amer. Math. Soc., 21, 1914–5, 167–173.
[43] Amer. Jour. Math., 37, 1915, 73–78.

vectants many formal covariants of the binary cubic modulo 2 and the binary quadratic modulo 3.

Glenn[44] considered the construction of covariants with given formal semi-invariant leaders. He reduced the question of the finiteness of the formal covariants of a binary form f_m of degree m modulo 2 to the question for systems of forms f_1, f_2, f_3. The product of any formal covariant of order >3 of f_3 modulo 2 by a power of $K=a_1+a_2$ is expressible in terms of certain 14 covariants.

Glenn[45] conjectured that p^2-1 is the maximum order of an irreducible covariant of any system of binary forms modulo p.

Glenn[46] discussed the determination of formal modular covariants ϕ_1, ϕ_2 of the binary form f_m of order m modulo p for which

$$f_m \equiv Q\phi_1 + L\phi_2 \pmod{p},$$

identically in the coefficients and variables of f_m.

Glenn[47] obtained a fundamental system of 20 formal covariants of the binary cubic modulo 2 and one of 18 for the binary quadratic form modulo 3, making use of Dickson's[15] fundamental system of formal semi-invariants.

O. C. Hazlett[48] proved the conjecture of Miss Sanderson[40] that if S be any system of binary forms in ξ, η and if S' be the system consisting of S and $x\eta-y\xi$, every modular covariant of S is polynomial in L and a specified set of modular invariants of S'. The theorem is extended to binary forms in any number of pairs of cogredient variables.

Glenn[49] gave processes to construct formal modular semi-invariants and covariants of binary forms f_m of order m, found a complete system of 6 semi-invariants and 19 covariants of the pair f_1, f_2 modulo 2, a complete system of 9 semi-invariants (and many covariants) of f_4 modulo 3, and a complete system of 19 covariants of f_4 modulo 2.

Hazlett gave a theorem on formal covariants analogous to that quoted[26] for modular covariants.

W. L. G. Williams[50] gave general theorems on the formal semi-invariants of the binary cubic modulo p and obtained a fundamental system for $p=5$ and $p=7$.

The following Chicago dissertations are in course of publication: B. F. Yanney, modular invariants of a binary quartic; J. S. Turner, invariants of the binary group modulo p^2; M. M. Felstein, invariants of the n-ary group modulo p^k; Constance R. Ballantine, invariants of the binary group with a composite modulus.

---

44 Trans. Amer. Math. Soc., 17, 1916, 545–556.
45 *Ibid.*, 18, 1917, 460–2.
46 Annals of Math., (2), 19, 1917–8, 201–6.
47 Trans. Amer. Math. Soc., 19, 1918, 109–118; 20, 1919, 154–168. Lists reproduced in Proc. Nat. Acad. Sc., 5, 1919, 107–110.
48 Trans. Amer. Math. Soc., 21, 1920, 247–254; 22, 1921, 148–157 for related results.
49 *Ibid.*, 21, 1920, 285–312.
50 Trans. Amer. Math. Soc., 22, 1921, 56–79.

AUTHOR INDEX.

The numbers refer to pages. Those in parenthesis relate to cross-references.

CH. I. REDUCTION AND EQUIVALENCE OF BINARY QUADRATIC FORMS, REPRESENTATION OF INTEGERS.

CH. II. EXPLICIT VALUES OF x, y IN $x^2+\Delta y^2=g$.

CH. III. COMPOSITION OF BINARY QUADRATIC FORMS.

CH. IV. ORDERS AND GENERA; THEIR COMPOSITION.

CH. V. IRREGULAR DETERMINANTS.

CH. VI. NUMBER OF CLASSES OF BINARY QUADRATIC FORMS WITH INTEGRAL COEFFICIENTS.

CHS. VII, VIII. BINARY QUADRATIC FORMS WITH COMPLEX COEFFICIENTS.

CH. IX. TERNARY QUADRATIC FORMS.

CH. X. QUATERNARY QUADRATIC FORMS.

CH. XI. QUADRATIC FORMS IN n VARIABLES.

CHS. XII, XIII. CUBIC FORMS.

CH. XIV. FORMS OF DEGREE $n \geqq 4$.

CHS. XV, XVI. HERMITIAN FORMS.

CH. XVII. BILINEAR FORMS, MATRICES, LINEAR SUBSTITUTIONS.

CH. XVIII. REPRESENTATION BY POLYNOMIALS MODULO p.

CH. XIX. CONGRUENCIAL THEORY OF FORMS.

SUBJECT INDEX.

Binary quadratic forms—Cont.
reciprocal, 11
reduced (see equivalence, principal), 6, 9, 10, 14–16, 22, 24, 25, 28, 30, 31, 35, 37, 39, 41, 43, 44, 48–53, 85, 94, 123, 157, 163, 175, 184, 188, 198, 200–2
— parallelogram, 21
representation (see mean, number), 2, 6, 11–23, 25, 26, 28, 29, 31, 32, 34, 35, 38, 39, 41, 44, 49–51, 54–59, 66, 67, 81, 86, 99, 198, 201
— belongs to root, 12, 14, 66
— by ternary (see ternary)
—, proper, improper, 2, 3, 5, 9, 77
representatives of classes, 97, 99–100, 106, 123–4, 138, 147–8, 153, 155, 169, 178, 181, 192
restricted, 74, 77
roots of, 24, 27, 30, 43, 51
schlicht, 74, 77, 155
simplest, 74, 82, 95
species, 198–9
subaltern class, 126
symbols, 93, 108–9, 123, 127
tables, 5, 8, 10, 11, 20, 22, 31, 40, 49, 50, 54, 56, 57, 59, 85, 89–91, 93–5, 97, 109, 111, 150, 172, 187–8, 196
transformations (all) into another form, 13, 15–17, 21, 25, 47–48, 105, 198, 201
trinary divisor, 11
triplication, 103
x^2+y^2, 5, 7, 10, 15, 17, 51, 55–57, 59, 60, 111, 122, 134–5, 142–5, 162, 175–7, 191
x^2+2y^2, 3–5, 7–10, 15, 17, 19, 20, 23–25, 29, 35, 37–39, 42, 46, 47, 49, 51, 54, 56–59, 134, 145
x^2+3y^2, 3–5, 15, 17, 20, 23, 25, 29–31, 35, 38, 42, 46, 47, 54–59, 134, 162
$x^2\pm 5y^2$, 3, 7, 29, 53, 57, 59
x^2-y^2, 8
x^2-2y^2, 3, 7, 9, 10, 19, 25, 29, 35, 47, 50, 54, 59, 135, 161, 175, 191
$x^2+xy\pm y^2$, 45, 47, 48, 59
x^2+27y^2, 55, 57–59

Biquadratic character, 17, 55, 57
— reciprocity, 18

Class number of binary quadratic forms
approximation, 153, 164
arithmetical methods, 134, 157, 184
as coefficient, 124–5, 134–5, 143–5, 160, 188, 190, 195
as exponent of -1, 103, 182
asymptotic mean, 133–4, 143, 146, 184–5
— median, 95, 123, 143–4, 151
by correspondence with ideals, 125, 159, 192, 197
complex coefficients, 203–5
computation formulas, 157, 170, 188
congruencial conditions on (see parity), 94, 117, 151, 164–5, 182
— — modulo 2^m, 155, 171, 179–80
— — modulo 3, 103, 159
— — modulo 8, 164
convention, 109

Class number of binary quadratic forms—Cont.
determinant negative, 92–136, 140–50, 154–5, 157–67, 171–9, 181–97
— — fundamental, 93, 97–8, 100–3, 118–9, 124–5, 132–3, 154–5, 158, 166–7, 171–3, 182
— positive, 94–7, 99–105, 107, 111, 115, 117–20, 150, 185–6
— — fundamental, 100–2, 107, 118–9
— prime, 97–8, 102–3, 119, 122, 125, 132–3, 154, 158, 166–7, 171–3, 182
discriminant negative, 137–40, 146–7, 151–8, 162, 166–71, 178–9, 181, 183–4
— — fundamental, 93, 137, 146–7, 151–8, 162, 169, 178–9, 181, 183
— positive, 137–40, 151, 153, 169–70
— — fundamental, 137, 151, 164
— prime, 155
— product of fundamental discriminants, 169, 170, 179–81
divisible by three, 103, 159
even classes, 127, 130–2, 137, 142–3, 150, 161–2, 176–7, 184, 189, 196
— — restricted by Humbert, 193
formula analogous to Gauss sums, 170
— of Cauchy, 103, 124, 169
— — Lebesgue, 103, 118, 151, 155, 169, 171, 179
fundamental equation of Dirichlet, 100, 138, 152–3, 166, 169
— formulas of Kronecker, 139, 148, 151
Gauss problem (see ratio), 95, 114
improperly primitive classes, 99, 100, 103, 105, 107, 109–10, 124, 126–7, 135, 148–50, 175, 181
indefinite forms, 120
involving minima, 120–1, 175, 186
limit ratio of Kronecker, 148, 165
linear function of divisors, 186, 195
lower bound of, 178
odd classes (see properly), 111–16, 121–5, 130, 133–6, 143–4, 147, 160–2, 175–7, 185–97
— —, Humbert's restricted, 192–3
— —, Joubert's restricted, 111
of D in terms of $h(-D)$, 140, 153
parity (see congruencial), 94, 107, 117, 151, 155, 164–5, 179
primitive, 138–40, 146–7, 152–9, 163–4, 168–71, 178–81, 183–4
properly primitive, 97–104, 107, 111, 115, 117–20, 122–7, 132–3, 135, 148, 150, 154–5, 159–60, 163–8, 172–4, 178–9, 181–2
product of class numbers, 171
ratio of class numbers for different orders (see Gauss problem), 109, 114, 124, 135, 148, 175, 203–5
recurrence formula (see relations), 191, 195
relations (see sum), 92, 105–6, 108–17, 121–2, 124–5, 128–38, 140–5, 149–50, 160–2, 174–7, 185–91, 195–7

A CATALOG OF SELECTED
DOVER BOOKS
IN SCIENCE AND MATHEMATICS

Astronomy

BURNHAM'S CELESTIAL HANDBOOK, Robert Burnham, Jr. Thorough guide to the stars beyond our solar system. Exhaustive treatment. Alphabetical by constellation: Andromeda to Cetus in Vol. 1; Chamaeleon to Orion in Vol. 2; and Pavo to Vulpecula in Vol. 3. Hundreds of illustrations. Index in Vol. 3. 2,000pp. 6⅛ x 9¼.
Vol. I: 23567-X
Vol. II: 23568-8
Vol. III: 23673-0

EXPLORING THE MOON THROUGH BINOCULARS AND SMALL TELESCOPES, Ernest H. Cherrington, Jr. Informative, profusely illustrated guide to locating and identifying craters, rills, seas, mountains, other lunar features. Newly revised and updated with special section of new photos. Over 100 photos and diagrams. 240pp. 8¼ x 11. 24491-1

THE EXTRATERRESTRIAL LIFE DEBATE, 1750–1900, Michael J. Crowe. First detailed, scholarly study in English of the many ideas that developed from 1750 to 1900 regarding the existence of intelligent extraterrestrial life. Examines ideas of Kant, Herschel, Voltaire, Percival Lowell, many other scientists and thinkers. 16 illustrations. 704pp. 5⅜ x 8½. 40675-X

THEORIES OF THE WORLD FROM ANTIQUITY TO THE COPERNICAN REVOLUTION, Michael J. Crowe. Newly revised edition of an accessible, enlightening book recreates the change from an earth-centered to a sun-centered conception of the solar system. 242pp. 5⅜ x 8½. 41444-2

A HISTORY OF ASTRONOMY, A. Pannekoek. Well-balanced, carefully reasoned study covers such topics as Ptolemaic theory, work of Copernicus, Kepler, Newton, Eddington's work on stars, much more. Illustrated. References. 521pp. 5⅜ x 8½.
65994-1

A COMPLETE MANUAL OF AMATEUR ASTRONOMY: Tools and Techniques for Astronomical Observations, P. Clay Sherrod with Thomas L. Koed. Concise, highly readable book discusses: selecting, setting up and maintaining a telescope; amateur studies of the sun; lunar topography and occultations; observations of Mars, Jupiter, Saturn, the minor planets and the stars; an introduction to photoelectric photometry; more. 1981 ed. 124 figures. 26 halftones. 37 tables. 335pp. 6½ x 9¼.
42820-6

AMATEUR ASTRONOMER'S HANDBOOK, J. B. Sidgwick. Timeless, comprehensive coverage of telescopes, mirrors, lenses, mountings, telescope drives, micrometers, spectroscopes, more. 189 illustrations. 576pp. 5⅝ x 8¼. (Available in U.S. only.)
24034-7

STARS AND RELATIVITY, Ya. B. Zel'dovich and I. D. Novikov. Vol. 1 of *Relativistic Astrophysics* by famed Russian scientists. General relativity, properties of matter under astrophysical conditions, stars, and stellar systems. Deep physical insights, clear presentation. 1971 edition. References. 544pp. 5⅝ x 8¼. 69424-0

Chemistry

THE SCEPTICAL CHYMIST: The Classic 1661 Text, Robert Boyle. Boyle defines the term "element," asserting that all natural phenomena can be explained by the motion and organization of primary particles. 1911 ed. viii+232pp. 5⅜ x 8½. 42825-7

RADIOACTIVE SUBSTANCES, Marie Curie. Here is the celebrated scientist's doctoral thesis, the prelude to her receipt of the 1903 Nobel Prize. Curie discusses establishing atomic character of radioactivity found in compounds of uranium and thorium; extraction from pitchblende of polonium and radium; isolation of pure radium chloride; determination of atomic weight of radium; plus electric, photographic, luminous, heat, color effects of radioactivity. ii+94pp. 5⅜ x 8½. 42550-9

CHEMICAL MAGIC, Leonard A. Ford. Second Edition, Revised by E. Winston Grundmeier. Over 100 unusual stunts demonstrating cold fire, dust explosions, much more. Text explains scientific principles and stresses safety precautions. 128pp. 5⅜ x 8½. 67628-5

THE DEVELOPMENT OF MODERN CHEMISTRY, Aaron J. Ihde. Authoritative history of chemistry from ancient Greek theory to 20th-century innovation. Covers major chemists and their discoveries. 209 illustrations. 14 tables. Bibliographies. Indices. Appendices. 851pp. 5⅜ x 8½. 64235-6

CATALYSIS IN CHEMISTRY AND ENZYMOLOGY, William P. Jencks. Exceptionally clear coverage of mechanisms for catalysis, forces in aqueous solution, carbonyl- and acyl-group reactions, practical kinetics, more. 864pp. 5⅜ x 8½. 65460-5

ELEMENTS OF CHEMISTRY, Antoine Lavoisier. Monumental classic by founder of modern chemistry in remarkable reprint of rare 1790 Kerr translation. A must for every student of chemistry or the history of science. 539pp. 5⅜ x 8½. 64624-6

THE HISTORICAL BACKGROUND OF CHEMISTRY, Henry M. Leicester. Evolution of ideas, not individual biography. Concentrates on formulation of a coherent set of chemical laws. 260pp. 5⅜ x 8½. 61053-5

A SHORT HISTORY OF CHEMISTRY, J. R. Partington. Classic exposition explores origins of chemistry, alchemy, early medical chemistry, nature of atmosphere, theory of valency, laws and structure of atomic theory, much more. 428pp. 5⅜ x 8½. (Available in U.S. only.) 65977-1

GENERAL CHEMISTRY, Linus Pauling. Revised 3rd edition of classic first-year text by Nobel laureate. Atomic and molecular structure, quantum mechanics, statistical mechanics, thermodynamics correlated with descriptive chemistry. Problems. 992pp. 5⅜ x 8½. 65622-5

FROM ALCHEMY TO CHEMISTRY, John Read. Broad, humanistic treatment focuses on great figures of chemistry and ideas that revolutionized the science. 50 illustrations. 240pp. 5⅜ x 8½. 28690-8

Engineering

DE RE METALLICA, Georgius Agricola. The famous Hoover translation of greatest treatise on technological chemistry, engineering, geology, mining of early modern times (1556). All 289 original woodcuts. 638pp. 6¾ x 11. 60006-8

FUNDAMENTALS OF ASTRODYNAMICS, Roger Bate et al. Modern approach developed by U.S. Air Force Academy. Designed as a first course. Problems, exercises. Numerous illustrations. 455pp. 5⅜ x 8½. 60061-0

DYNAMICS OF FLUIDS IN POROUS MEDIA, Jacob Bear. For advanced students of ground water hydrology, soil mechanics and physics, drainage and irrigation engineering, and more. 335 illustrations. Exercises, with answers. 784pp. 6⅛ x 9¼. 65675-6

THEORY OF VISCOELASTICITY (Second Edition), Richard M. Christensen. Complete, consistent description of the linear theory of the viscoelastic behavior of materials. Problem-solving techniques discussed. 1982 edition. 29 figures. xiv+364pp. 6⅛ x 9¼. 42880-X

MECHANICS, J. P. Den Hartog. A classic introductory text or refresher. Hundreds of applications and design problems illuminate fundamentals of trusses, loaded beams and cables, etc. 334 answered problems. 462pp. 5⅜ x 8½. 60754-2

MECHANICAL VIBRATIONS, J. P. Den Hartog. Classic textbook offers lucid explanations and illustrative models, applying theories of vibrations to a variety of practical industrial engineering problems. Numerous figures. 233 problems, solutions. Appendix. Index. Preface. 436pp. 5⅜ x 8½. 64785-4

STRENGTH OF MATERIALS, J. P. Den Hartog. Full, clear treatment of basic material (tension, torsion, bending, etc.) plus advanced material on engineering methods, applications. 350 answered problems. 323pp. 5⅜ x 8½. 60755-0

A HISTORY OF MECHANICS, René Dugas. Monumental study of mechanical principles from antiquity to quantum mechanics. Contributions of ancient Greeks, Galileo, Leonardo, Kepler, Lagrange, many others. 671pp. 5⅜ x 8½. 65632-2

STABILITY THEORY AND ITS APPLICATIONS TO STRUCTURAL MECHANICS, Clive L. Dym. Self-contained text focuses on Koiter postbuckling analyses, with mathematical notions of stability of motion. Basing minimum energy principles for static stability upon dynamic concepts of stability of motion, it develops asymptotic buckling and postbuckling analyses from potential energy considerations, with applications to columns, plates, and arches. 1974 ed. 208pp. 5⅜ x 8½. 42541-X

METAL FATIGUE, N. E. Frost, K. J. Marsh, and L. P. Pook. Definitive, clearly written, and well-illustrated volume addresses all aspects of the subject, from the historical development of understanding metal fatigue to vital concepts of the cyclic stress that causes a crack to grow. Includes 7 appendixes. 544pp. 5⅜ x 8½. 40927-9

ROCKETS, Robert Goddard. Two of the most significant publications in the history of rocketry and jet propulsion: "A Method of Reaching Extreme Altitudes" (1919) and "Liquid Propellant Rocket Development" (1936). 128pp. 5⅜ x 8½. 42537-1

STATISTICAL MECHANICS: Principles and Applications, Terrell L. Hill. Standard text covers fundamentals of statistical mechanics, applications to fluctuation theory, imperfect gases, distribution functions, more. 448pp. 5⅜ x 8½. 65390-0

ENGINEERING AND TECHNOLOGY 1650–1750: Illustrations and Texts from Original Sources, Martin Jensen. Highly readable text with more than 200 contemporary drawings and detailed engravings of engineering projects dealing with surveying, leveling, materials, hand tools, lifting equipment, transport and erection, piling, bailing, water supply, hydraulic engineering, and more. Among the specific projects outlined–transporting a 50-ton stone to the Louvre, erecting an obelisk, building timber locks, and dredging canals. 207pp. 8⅜ x 11¼. 42232-1

THE VARIATIONAL PRINCIPLES OF MECHANICS, Cornelius Lanczos. Graduate level coverage of calculus of variations, equations of motion, relativistic mechanics, more. First inexpensive paperbound edition of classic treatise. Index. Bibliography. 418pp. 5⅜ x 8½. 65067-7

PROTECTION OF ELECTRONIC CIRCUITS FROM OVERVOLTAGES, Ronald B. Standler. Five-part treatment presents practical rules and strategies for circuits designed to protect electronic systems from damage by transient overvoltages. 1989 ed. xxiv+434pp. 6⅛ x 9¼. 42552-5

ROTARY WING AERODYNAMICS, W. Z. Stepniewski. Clear, concise text covers aerodynamic phenomena of the rotor and offers guidelines for helicopter performance evaluation. Originally prepared for NASA. 537 figures. 640pp. 6⅛ x 9¼. 64647-5

INTRODUCTION TO SPACE DYNAMICS, William Tyrrell Thomson. Comprehensive, classic introduction to space-flight engineering for advanced undergraduate and graduate students. Includes vector algebra, kinematics, transformation of coordinates. Bibliography. Index. 352pp. 5⅜ x 8½. 65113-4

HISTORY OF STRENGTH OF MATERIALS, Stephen P. Timoshenko. Excellent historical survey of the strength of materials with many references to the theories of elasticity and structure. 245 figures. 452pp. 5⅜ x 8½. 61187-6

ANALYTICAL FRACTURE MECHANICS, David J. Unger. Self-contained text supplements standard fracture mechanics texts by focusing on analytical methods for determining crack-tip stress and strain fields. 336pp. 6⅛ x 9¼. 41737-9

STATISTICAL MECHANICS OF ELASTICITY, J. H. Weiner. Advanced, self-contained treatment illustrates general principles and elastic behavior of solids. Part 1, based on classical mechanics, studies thermoelastic behavior of crystalline and polymeric solids. Part 2, based on quantum mechanics, focuses on interatomic force laws, behavior of solids, and thermally activated processes. For students of physics and chemistry and for polymer physicists. 1983 ed. 96 figures. 496pp. 5⅜ x 8½. 42260-7

Mathematics

FUNCTIONAL ANALYSIS (Second Corrected Edition), George Bachman and Lawrence Narici. Excellent treatment of subject geared toward students with background in linear algebra, advanced calculus, physics, and engineering. Text covers introduction to inner-product spaces, normed, metric spaces, and topological spaces; complete orthonormal sets, the Hahn-Banach Theorem and its consequences, and many other related subjects. 1966 ed. 544pp. 6⅛ x 9¼. 40251-7

ASYMPTOTIC EXPANSIONS OF INTEGRALS, Norman Bleistein & Richard A. Handelsman. Best introduction to important field with applications in a variety of scientific disciplines. New preface. Problems. Diagrams. Tables. Bibliography. Index. 448pp. 5⅜ x 8½. 65082-0

VECTOR AND TENSOR ANALYSIS WITH APPLICATIONS, A. I. Borisenko and I. E. Tarapov. Concise introduction. Worked-out problems, solutions, exercises. 257pp. 5⅝ x 8¼. 63833-2

THE ABSOLUTE DIFFERENTIAL CALCULUS (CALCULUS OF TENSORS), Tullio Levi-Civita. Great 20th-century mathematician's classic work on material necessary for mathematical grasp of theory of relativity. 452pp. 5⅝ x 8¼. 63401-9

AN INTRODUCTION TO ORDINARY DIFFERENTIAL EQUATIONS, Earl A. Coddington. A thorough and systematic first course in elementary differential equations for undergraduates in mathematics and science, with many exercises and problems (with answers). Index. 304pp. 5⅜ x 8½. 65942-9

FOURIER SERIES AND ORTHOGONAL FUNCTIONS, Harry F. Davis. An incisive text combining theory and practical example to introduce Fourier series, orthogonal functions and applications of the Fourier method to boundary-value problems. 570 exercises. Answers and notes. 416pp. 5⅜ x 8½. 65973-9

COMPUTABILITY AND UNSOLVABILITY, Martin Davis. Classic graduate-level introduction to theory of computability, usually referred to as theory of recurrent functions. New preface and appendix. 288pp. 5⅜ x 8½. 61471-9

ASYMPTOTIC METHODS IN ANALYSIS, N. G. de Bruijn. An inexpensive, comprehensive guide to asymptotic methods–the pioneering work that teaches by explaining worked examples in detail. Index. 224pp. 5⅜ x 8½ 64221-6

APPLIED COMPLEX VARIABLES, John W. Dettman. Step-by-step coverage of fundamentals of analytic function theory–plus lucid exposition of five important applications: Potential Theory; Ordinary Differential Equations; Fourier Transforms; Laplace Transforms; Asymptotic Expansions. 66 figures. Exercises at chapter ends. 512pp. 5⅜ x 8½. 64670-X

INTRODUCTION TO LINEAR ALGEBRA AND DIFFERENTIAL EQUATIONS, John W. Dettman. Excellent text covers complex numbers, determinants, orthonormal bases, Laplace transforms, much more. Exercises with solutions. Undergraduate level. 416pp. 5⅜ x 8½. 65191-6

Math–Decision Theory, Statistics, Probability

ELEMENTARY DECISION THEORY, Herman Chernoff and Lincoln E. Moses. Clear introduction to statistics and statistical theory covers data processing, probability and random variables, testing hypotheses, much more. Exercises. 364pp. 5⅜ x 8½. 65218-1

STATISTICS MANUAL, Edwin L. Crow et al. Comprehensive, practical collection of classical and modern methods prepared by U.S. Naval Ordnance Test Station. Stress on use. Basics of statistics assumed. 288pp. 5⅜ x 8½. 60599-X

SOME THEORY OF SAMPLING, William Edwards Deming. Analysis of the problems, theory, and design of sampling techniques for social scientists, industrial managers, and others who find statistics important at work. 61 tables. 90 figures. xvii +602pp. 5⅜ x 8½. 64684-X

LINEAR PROGRAMMING AND ECONOMIC ANALYSIS, Robert Dorfman, Paul A. Samuelson and Robert M. Solow. First comprehensive treatment of linear programming in standard economic analysis. Game theory, modern welfare economics, Leontief input-output, more. 525pp. 5⅜ x 8½. 65491-5

PROBABILITY: An Introduction, Samuel Goldberg. Excellent basic text covers set theory, probability theory for finite sample spaces, binomial theorem, much more. 360 problems. Bibliographies. 322pp. 5⅜ x 8½. 65252-1

GAMES AND DECISIONS: Introduction and Critical Survey, R. Duncan Luce and Howard Raiffa. Superb nontechnical introduction to game theory, primarily applied to social sciences. Utility theory, zero-sum games, n-person games, decision-making, much more. Bibliography. 509pp. 5⅜ x 8½. 65943-7

INTRODUCTION TO THE THEORY OF GAMES, J. C. C. McKinsey. This comprehensive overview of the mathematical theory of games illustrates applications to situations involving conflicts of interest, including economic, social, political, and military contexts. Appropriate for advanced undergraduate and graduate courses; advanced calculus a prerequisite. 1952 ed. x+372pp. 5⅜ x 8½. 42811-7

FIFTY CHALLENGING PROBLEMS IN PROBABILITY WITH SOLUTIONS, Frederick Mosteller. Remarkable puzzlers, graded in difficulty, illustrate elementary and advanced aspects of probability. Detailed solutions. 88pp. 5⅜ x 8½. 65355-2

PROBABILITY THEORY: A Concise Course, Y. A. Rozanov. Highly readable, self-contained introduction covers combination of events, dependent events, Bernoulli trials, etc. 148pp. 5⅜ x 8¼. 63544-9

STATISTICAL METHOD FROM THE VIEWPOINT OF QUALITY CONTROL, Walter A. Shewhart. Important text explains regulation of variables, uses of statistical control to achieve quality control in industry, agriculture, other areas. 192pp. 5⅜ x 8½. 65232-7

Math–Geometry and Topology

ELEMENTARY CONCEPTS OF TOPOLOGY, Paul Alexandroff. Elegant, intuitive approach to topology from set-theoretic topology to Betti groups; how concepts of topology are useful in math and physics. 25 figures. 57pp. 5⅜ x 8½. 60747-X

COMBINATORIAL TOPOLOGY, P. S. Alexandrov. Clearly written, well-organized, three-part text begins by dealing with certain classic problems without using the formal techniques of homology theory and advances to the central concept, the Betti groups. Numerous detailed examples. 654pp. 5⅜ x 8½. 40179-0

EXPERIMENTS IN TOPOLOGY, Stephen Barr. Classic, lively explanation of one of the byways of mathematics. Klein bottles, Moebius strips, projective planes, map coloring, problem of the Koenigsberg bridges, much more, described with clarity and wit. 43 figures. 210pp. 5⅜ x 8½. 25933-1

CONFORMAL MAPPING ON RIEMANN SURFACES, Harvey Cohn. Lucid, insightful book presents ideal coverage of subject. 334 exercises make book perfect for self-study. 55 figures. 352pp. 5⅝ x 8¼. 64025-6

THE GEOMETRY OF RENÉ DESCARTES, René Descartes. The great work founded analytical geometry. Original French text, Descartes's own diagrams, together with definitive Smith-Latham translation. 244pp. 5⅜ x 8½. 60068-8

PRACTICAL CONIC SECTIONS: The Geometric Properties of Ellipses, Parabolas and Hyperbolas, J. W. Downs. This text shows how to create ellipses, parabolas, and hyperbolas. It also presents historical background on their ancient origins and describes the reflective properties and roles of curves in design applications. 1993 ed. 98 figures. xii+100pp. 6½ x 9¼. 42876-1

THE THIRTEEN BOOKS OF EUCLID'S ELEMENTS, translated with introduction and commentary by Thomas L. Heath. Definitive edition. Textual and linguistic notes, mathematical analysis. 2,500 years of critical commentary. Unabridged. 1,414pp. 5⅜ x 8½.. Three-vol. set. Vol. I: 60088-2 Vol. II: 60089-0 Vol. III: 60090-4

GEOMETRY OF COMPLEX NUMBERS, Hans Schwerdtfeger. Illuminating, widely praised book on analytic geometry of circles, the Moebius transformation, and two-dimensional non-Euclidean geometries. 200pp. 5⅝ x 8¼. 63830-8

DIFFERENTIAL GEOMETRY, Heinrich W. Guggenheimer. Local differential geometry as an application of advanced calculus and linear algebra. Curvature, transformation groups, surfaces, more. Exercises. 62 figures. 378pp. 5⅜ x 8½. 63433-7

CURVATURE AND HOMOLOGY: Enlarged Edition, Samuel I. Goldberg. Revised edition examines topology of differentiable manifolds; curvature, homology of Riemannian manifolds; compact Lie groups; complex manifolds; curvature, homology of Kaehler manifolds. New Preface. Four new appendixes. 416pp. 5⅜ x 8½. 40207-X

History of Math

THE WORKS OF ARCHIMEDES, Archimedes (T. L. Heath, ed.). Topics include the famous problems of the ratio of the areas of a cylinder and an inscribed sphere; the measurement of a circle; the properties of conoids, spheroids, and spirals; and the quadrature of the parabola. Informative introduction. clxxxvi+326pp; supplement, 52pp. 5⅜ x 8½. 42084-1

A SHORT ACCOUNT OF THE HISTORY OF MATHEMATICS, W. W. Rouse Ball. One of clearest, most authoritative surveys from the Egyptians and Phoenicians through 19th-century figures such as Grassman, Galois, Riemann. Fourth edition. 522pp. 5⅜ x 8½. 20630-0

THE HISTORY OF THE CALCULUS AND ITS CONCEPTUAL DEVELOPMENT, Carl B. Boyer. Origins in antiquity, medieval contributions, work of Newton, Leibniz, rigorous formulation. Treatment is verbal. 346pp. 5⅜ x 8½. 60509-4

THE HISTORICAL ROOTS OF ELEMENTARY MATHEMATICS, Lucas N. H. Bunt, Phillip S. Jones, and Jack D. Bedient. Fundamental underpinnings of modern arithmetic, algebra, geometry, and number systems derived from ancient civilizations. 320pp. 5⅜ x 8½. 25563-8

A HISTORY OF MATHEMATICAL NOTATIONS, Florian Cajori. This classic study notes the first appearance of a mathematical symbol and its origin, the competition it encountered, its spread among writers in different countries, its rise to popularity, its eventual decline or ultimate survival. Original 1929 two-volume edition presented here in one volume. xxviii+820pp. 5⅜ x 8½. 67766-4

GAMES, GODS & GAMBLING: A History of Probability and Statistical Ideas, F. N. David. Episodes from the lives of Galileo, Fermat, Pascal, and others illustrate this fascinating account of the roots of mathematics. Features thought-provoking references to classics, archaeology, biography, poetry. 1962 edition. 304pp. 5⅜ x 8½. (Available in U.S. only.) 40023-9

OF MEN AND NUMBERS: The Story of the Great Mathematicians, Jane Muir. Fascinating accounts of the lives and accomplishments of history's greatest mathematical minds–Pythagoras, Descartes, Euler, Pascal, Cantor, many more. Anecdotal, illuminating. 30 diagrams. Bibliography. 256pp. 5⅜ x 8½. 28973-7

HISTORY OF MATHEMATICS, David E. Smith. Nontechnical survey from ancient Greece and Orient to late 19th century; evolution of arithmetic, geometry, trigonometry, calculating devices, algebra, the calculus. 362 illustrations. 1,355pp. 5⅜ x 8½. Two-vol. set. Vol. I: 20429-4 Vol. II: 20430-8

A CONCISE HISTORY OF MATHEMATICS, Dirk J. Struik. The best brief history of mathematics. Stresses origins and covers every major figure from ancient Near East to 19th century. 41 illustrations. 195pp. 5⅜ x 8½. 60255-9

Physics

OPTICAL RESONANCE AND TWO-LEVEL ATOMS, L. Allen and J. H. Eberly. Clear, comprehensive introduction to basic principles behind all quantum optical resonance phenomena. 53 illustrations. Preface. Index. 256pp. 5⅜ x 8½. 65533-4

QUANTUM THEORY, David Bohm. This advanced undergraduate-level text presents the quantum theory in terms of qualitative and imaginative concepts, followed by specific applications worked out in mathematical detail. Preface. Index. 655pp. 5⅜ x 8½. 65969-0

ATOMIC PHYSICS: 8th edition, Max Born. Nobel laureate's lucid treatment of kinetic theory of gases, elementary particles, nuclear atom, wave-corpuscles, atomic structure and spectral lines, much more. Over 40 appendices, bibliography. 495pp. 5⅜ x 8½. 65984-4

A SOPHISTICATE'S PRIMER OF RELATIVITY, P. W. Bridgman. Geared toward readers already acquainted with special relativity, this book transcends the view of theory as a working tool to answer natural questions: What is a frame of reference? What is a "law of nature"? What is the role of the "observer"? Extensive treatment, written in terms accessible to those without a scientific background. 1983 ed. xlviii+172pp. 5⅜ x 8½. 42549-5

AN INTRODUCTION TO HAMILTONIAN OPTICS, H. A. Buchdahl. Detailed account of the Hamiltonian treatment of aberration theory in geometrical optics. Many classes of optical systems defined in terms of the symmetries they possess. Problems with detailed solutions. 1970 edition. xv+360pp. 5⅜ x 8½. 67597-1

PRIMER OF QUANTUM MECHANICS, Marvin Chester. Introductory text examines the classical quantum bead on a track: its state and representations; operator eigenvalues; harmonic oscillator and bound bead in a symmetric force field; and bead in a spherical shell. Other topics include spin, matrices, and the structure of quantum mechanics; the simplest atom; indistinguishable particles; and stationary-state perturbation theory. 1992 ed. xiv+314pp. 6⅛ x 9¼. 42878-8

LECTURES ON QUANTUM MECHANICS, Paul A. M. Dirac. Four concise, brilliant lectures on mathematical methods in quantum mechanics from Nobel Prize–winning quantum pioneer build on idea of visualizing quantum theory through the use of classical mechanics. 96pp. 5⅜ x 8½. 41713-1

THIRTY YEARS THAT SHOOK PHYSICS: The Story of Quantum Theory, George Gamow. Lucid, accessible introduction to influential theory of energy and matter. Careful explanations of Dirac's anti-particles, Bohr's model of the atom, much more. 12 plates. Numerous drawings. 240pp. 5⅜ x 8½. 24895-X

ELECTRONIC STRUCTURE AND THE PROPERTIES OF SOLIDS: The Physics of the Chemical Bond, Walter A. Harrison. Innovative text offers basic understanding of the electronic structure of covalent and ionic solids, simple metals, transition metals and their compounds. Problems. 1980 edition. 582pp. 6⅛ x 9¼. 66021-4